Steels

Steels

Structure, Properties and Design

5th Edition

Harshad K.D.H. Bhadeshia
Professor of Metallurgy
Queen Mary University of London
Emeritus Tata Steel Professor of Metallurgy
University of Cambridge

Robert W.K. Honeycombe[†]
Emeritus Goldsmiths' Professor of Metallurgy
University of Cambridge

ELSEVIER
Butterworth-Heinemann
An imprint of Elsevier

[†] Deceased.

Butterworth-Heinemann is an imprint of Elsevier
125 London Wall, London EC2Y 5AS, United Kingdom
50 Hampshire Street, 5th Floor, Cambridge, MA 02139, United States

ISBN: 978-0-443-18491-8

For information on all Butterworth-Heinemann publications
visit our website at https://www.elsevier.com/books-and-journals

Publisher: Matthew Deans
Acquisitions Editor: Dennis McGonagle
Editorial Project Manager: Tom Mearns
Production Project Manager: Anitha Sivaraj
Cover Designer: Matthew Limbert

Typeset by VTeX

Contents

Biography

Harry Bhadeshia is the Professor of Metallurgy at Queen Mary University of London and Emeritus Tata Steel Professor of Metallurgy at the University of Cambridge. His research is concerned with the theory of solid-state transformations in metals, particularly multicomponent steels, with the goal of creating novel alloys and processes with the minimum use of resources. He is the author or coauthor of more than 700 research papers and 7 books on the subject. He is a Fellow of the Royal Society, Fellow of the Royal Academy of Engineering, the National Academy of Engineering (India) and the American Welding Society, an Honorary Fellow of The Welding Institute and Honorary Member of the Iron and Steel Institute of Japan. In 2015, Professor Bhadeshia was appointed Knights Bachelor in the Queen's Birthday Honours for services to Science and Technology.

Preface

The first edition of this book was published during 1980 by the late Professor Sir Robert Honeycombe, who made extraordinary contributions to the subject and indeed to the well-being of his Department at the University of Cambridge during his long tenure as the Goldsmiths' Professor of Metallurgy.

The book rapidly became a foundational text, with its elegant explanations of the structure and properties of steels. I was delighted when he asked me to coauthor subsequent editions. In this fifth version, I continue to respect Sir Robert's vision, not least because it promotes a proper understanding of what constitutes a successful material. Therefore, the book continues to deal with the structure of iron and its alloys, how that structure depends on a plethora of variables that nurture design, and the consequences of structure on the bank of properties that determine the remarkable success of steels in enhancing the fulfilment of life. This fifth edition has been substantially revised and enhanced with additional chapters to deal with the changes in manufacturing methods together with a different tack on sustainability.

Pure iron is an element that is gifted with an abundance of allotropic and magnetic transitions. It has the right density – not so large that it cannot be used to construct the most powerful of rockets and spaceships, but not so small that its volume becomes a problem. Lest we forget, the majority of civilisation is based on firm ground, with static structures personified by solidity. Iron has universal utility – it is *essential* in the mining, manufacture and processing of all other metals, electronic devices, medicines and even kilns for cement manufacture. No other engineering material matches this versatility that is the hallmark of all things ferrous. Furthermore, it is readily available on Earth and, indeed, everywhere in the Universe where matter has clustered.

Iron has the ability to dissolve quantities of all of the naturally occurring elements of the periodic table. In such mixtures, some solutes snuggle in the space between iron atoms, whereas large atoms substitute for iron. Compounds characterised by narrow composition-ranges may also form. Some consequences of unintentional alloying may be unwelcome – hydrogen and phosphorus do no good, but such is life.

Iron melts at 1811.15 K, which makes it important in machines where thermodynamic efficiency depends on the highest temperature that can be achieved in the practical equivalent of the Carnot cycle. After all, the activation energy for the diffusion, that drives temperature-activated deformation, scales with the melting temperature. Looking to the future, an aluminium or carbon-composite based space vehicle designed for reuse, is unsuitable for entry into the Mars atmosphere at 7.5 km s^{-1} whereas a steel vehicle fitted with a heat shield, would be able to do so.

The utility of iron remains unchallenged, accounting for an indecent 2 billion tons of consumption annually. This enriches the air with unbearable quantities of carbon dioxide, at a time when this is not sensible. An attempt is made in the last chapter of this book to explain how a dramatic reduction in the consumption of iron can be achieved quickly and without complaint.

There is a sickness with science today, the premature publicity that in time causes expectations to be shattered. Take, for example, the statement, made not just in the heat of the moment, that graphene is "the strongest material known to man – more than 200 times stronger than steel". It is not, on both counts. The damage is, however, done – Google's Bard, the conversational service based on "artificial intelligence" repeats the falsity, presumably because it is unable to interpret the misinformation that infects the media on which its algorithms depend.

I hope that this book provides a balanced presentation of the physical metallurgy of iron and its alloys, while recognising impending or neglected problems and suggesting the way forward. As ever, I remain grateful to many people, duly acknowledged in the text, for their prompt assistance in providing images, information and on occasions addressing my queries. This includes the wonderful people I have had the joy of working with over more decades than I care to remember.

Finally, does anything exist that is 10 billion times stronger than steel? And with a breaking strain ten times greater? And can be made large?

Well, yes, do read on.

Harshad K.D.H. Bhadeshia
The year 2023
London and Cambridge

Iron and its interstitial solutions

1

Abstract

Pure iron is remarkable in its complexity, not only because of its many allotropic forms. There are subtle features related to magnetism, that cause, for example, the expansion coefficient of austenite to be much greater than that of more loosely packed ferrite – one consequence of which is that austenitic steels deteriorate when subjected to a combination of stress and thermal fluctuations. We discuss here the choreography of atoms during solid-state phase transformations together with the role and behaviour of interstitial atoms such as carbon, nitrogen and hydrogen. The mobility of substitutional solutes is described to emphasise some counterintuitive observations, such as the fact that heavy atoms like molybdenum actually diffuse faster than the iron in which they are dissolved.

1.1 Introduction

Iron is created in the stars, which at first fuel themselves through the fusion of hydrogen into helium. When the hydrogen is exhausted, the fusion of helium leads to the creation of carbon. This exothermic process of nuclear burning can, in sufficiently massive stars, lead to the formation of even heavier elements, the sequence ending with the creation of iron. Iron is the most stable element in the universe, and further fusion to form even heavier elements does not release energy. As a consequence, the fuel supply becomes exhausted and the star collapses into states that depend on its mass. If the core becomes so heavy that it cannot support its own gravity, a giant explosion (a supernova) occurs that produces in a short time scale, most of the elements beyond iron.

Iron occurs in elemental form on Earth only through human intervention, even though by weight, our Earth is mostly iron that is at the solid–liquid core that formed during the fiery genesis of the planet. The surface, on the other hand, is oxide-rich. However, essentially pure metallic iron is found on the Moon, to the extent that some 150 tons could in principle be magnetically separated from 100,000 tons of lunar soil [1,2]. The crystals of iron are beautifully faceted and typically a few micrometres in size, Fig. 1.1. They are found resting in cavities in lunar rocks, most likely deposited some three billion years ago from hot vapour that is free from oxygen or water. Earth is devoid of elemental iron because of the abundance of oxygen and water. In fact, there was a lot of iron dissolved in the oceans, which consumed the original oxygen produced by stromatolites to form the oxides we mine today. Only then could the oxygen bubble get out to enrich the atmosphere.

As impressive as the cosmological origins and abundance of iron and carbon are, the story truly worth telling can be witnessed in every aspect of terrestrial life. Such is the success of iron that steel forms the "gold-standard" against which emerging materials or supreme acts of endeavour are compared. What is often not realised is that this is a moving standard, with notoriously regular and exciting discoveries being made in the context of iron and its alloys. This is why steel remains the most successful and

Steels. https://doi.org/10.1016/B978-0-44-318491-8.00007-1

Figure 1.1 Scanning electron micrograph of the contents of a fragmented rock from the Apollo 15 mission. The largest iron crystal resting in a cavity within the silicate rock is about three micrometres in size. The perfection of the facets on these crystals indicates slow formation from hot vapour as the rock cooled. The crystals are in the form of deltoidal icositetrahedra truncated by $\{100\}_\alpha$ (square facets), with the other five-sided facets best described as planes of the form $\{229\}_\alpha$ [3]. NASA image S72-55208, catalogued 10th November 1972.

cost-effective of all materials, with close to two billion tons being consumed annually to improve the quality of life. This book attempts to explain why steels continue to take this pre-eminent position, why steel is not challenged in its success by any other material, and examines in detail the phenomena whose exploitation enables the desired properties to be achieved.

One reason for the overwhelming dominance of steels is the endless variety of structures and properties that can be generated by solid-state transformation and processing. And all this at a monetary cost that is readily affordable, although the consequences on the environment are large in spite of the fact that steel is the most recycled metal. When studying steels, it is useful to consider the behaviour of pure iron first, then the iron-carbon alloys and finally the rich varieties possible when further solutes are added.

Pure iron is not easy to produce; entropy opposes extreme purity. It has nevertheless been made with a total impurity content less than 60 parts per million (ppm), of which 10 ppm is accounted for by non-metallic impurities such as carbon, oxygen, sulphur and phosphorus, while the remainder are metallic impurities. Iron of this purity can be extremely weak when reasonably sized samples are tested: the resolved shear stress needed to induce slip in a single crystal at room temperature can be as low as 10 MPa, while the yield stress of a polycrystalline sample at the same temperature can be well below 50 MPa. However, the shear strength of small single crystals has been observed to exceed 19,000 MPa when the size of the sample is reduced to about 2 μm. This is because the chances of finding crystal defects such as dislocations decreases as the size of the crystal is reduced. The ideal tensile strength of a perfect crystal of iron pulled along $\langle 100 \rangle$ is about 14,200 MPa [4].

For comparative purposes, the breaking strength of a very small carbon nanotube or particle of graphene has been estimated to be about 130,000 MPa; this number is so astonishing that it has led to misleading statements about their potential in structural applications. For example, the tubes and graphene are said to be a several hundred times stronger than steel; in fact, there is no carbon tube or graphene particle that can match the strength of iron beyond a scale of 2 mm, because of the inevitable defects, which arise as the tubes or sheets of carbon are grown [5,6]. The beauty of iron is that it is a metal, so atoms can conservatively glide past each other without disrupting the metallic bond. This makes it defect-tolerant, unlike the covalent C–C bonds in graphene, which cannot reform if broken. The lack of plastic deformation makes graphene or nanotubes unsafe for use in structural applications.

The lesson from this is that systems which rely on perfection to achieve strength inevitably fail when scaled to engineering dimensions. Since perfection is thermodynamically impossible to achieve in large samples, steels must in practice be made stronger by other means that are insensitive to size. The mechanisms by which the strength can be increased will be discussed – suffice it to state here that it is possible to produce commercially steel with a strength of 5500 MPa, with sufficient ductility to ensure safe application. Some of the methods by which such impressive combinations of properties are achieved without compromising safety will be discussed, before the wide range of complex structures which determine the properties is dealt with.

1.2 Allotropes of pure iron

At least three allotropes of iron occur naturally in bulk form: body-centred cubic (bcc, α, ferrite), face-centred cubic (fcc, γ, austenite) and hexagonal close-packed (hcp, ε). The phase β in the alphabetical sequence α, β, γ, δ... is missing because the magnetic transition in ferrite was at one time incorrectly thought to be the β allotrope of iron, responsible for hardening when the iron is quenched. Nevertheless, it is true that α-iron is not strictly cubic in its ferromagnetic state below the Curie temperature [7]. This is because the magnetic spins are aligned, say along the z-axis, so that rotations about the x- or y-axes must be combined with time reversal to preserve the directions of the spins. The magnetic point group becomes tetragonal [8]. It follows that the ferrite structure below the Curie temperature is tetragonal, but the extent of tetragonality is very small and is neglected in most experiments on steels containing complex microstructures, where the X-ray diffraction peaks are too broad to detect small differences in the lattice parameters.

In fact, there are magnetic transitions in all of the allotropes of iron [9, p. 3]. The phase diagram for pure iron is illustrated in Fig. 1.2. Each point on any boundary between the phase fields represents an equilibrium state in which two phases can coexist. The *triple point*, where the three boundaries intersect, represents an equilibrium between all three phases which coexist. It is seen that in pure iron, the hcp form is stable only at very high pressures, consistent with its high density. The best comparison of the relative densities of the phases is made at the triple point, where the allotropes are in equilibrium and where the sum of all the volume changes is zero:

$$\left.\begin{array}{l}\Delta V(\text{bcc} \rightarrow \text{hcp}) = -0.34 \\ \Delta V(\text{hcp} \rightarrow \text{fcc}) = +0.13 \\ \Delta V(\text{fcc} \rightarrow \text{bcc}) = +0.21\end{array}\right\} \qquad \text{cm}^3\,\text{mol}^{-1}$$

There may exist a fourth natural allotrope in the core of the Earth, where the pressure is about some three million times that at the surface and where the temperature is estimated to be about 6000°C. The core of the Earth is predominantly iron, and consists of a solid inner core surrounded by a liquid outer core. Knowledge of the core is uncertain, but it has been suggested that the crystal structure of the solid core may be double-hcp, although calculations assuming pure iron indicate that the ε-iron remains the most stable under inner-core conditions. Instead of the $\ldots ABAB \ldots$ stacking of close-packed planes in the hcp structure, the double-hcp has the sequence $\ldots ABACABAC \ldots$. The estimated density of the inner core is 13-14 $\mathrm{g\,cm^{-3}}$ (hcp form) [10,11].

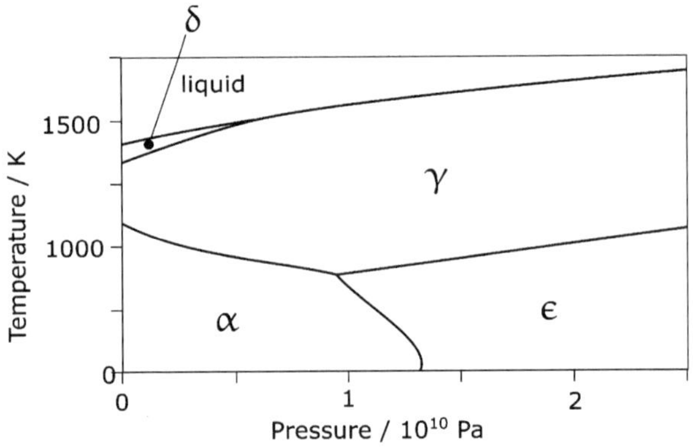

Figure 1.2 Temperature versus pressure equilibrium phase diagram for pure iron based on assessed thermodynamic data. α, γ and ε refer to ferrite, austenite and ε-iron respectively. δ is simply the higher temperature designation of α. However, below the Curie temperature, α-iron is slightly tetragonal with a difference in lattice parameters of just 6×10^{-15} m, determined from magnetostriction experiments [12]. Diagram courtesy of Shaumik Lenka, calculated using the database 'TCFE8', and ThermoCalc version 2015b.

1.2.1 Thin films and isolated particles

There are two further allotropes that can be created in the form of thin films. Face-centred tetragonal iron was prepared by coherently depositing iron as a thin film on a {1 0 0} plane of a substrate such as copper with which the iron has a mismatch. The position of atoms in the first deposited layer in this case replicates that of the substrate. A few monolayers can be forced into coherency in the plane of the substrate with a corresponding distortion normal to the substrate. This gives the deposit a face-centred tetragonal structure. Growing iron on a misfitting {1 1 1} surface of a fcc substrate leads to trigonal iron.

Very thin films of iron retain their ferromagnetic character, but there are special effects due to the small dimensions. The magnetic moment per atom becomes very large: 3.1 Bohr magnetons compared with 2.2 for bulk α-iron. This is due to the lower coordination number of the atoms in a thin film. The second effect is that magnetic anisotropy greatly increases for thin films because the spins tend to align normal to the surface. The Curie temperature is greatly reduced, again because of the change in coordination. For a monolayer of iron, the Curie temperature is just $\simeq$280°C.

Many classical studies on nucleation theory have been conducted on minute (5–1000 nm) particles of iron, where the defects responsible for heterogeneous nucleation can be avoided. Such particles are significant in the manufacture of carbon nanotubes, when they form through the decomposition of ferrocene in chemical mixtures that also contain the ingredients necessary to grow the tubes. Fluids containing particles of iron that are small enough to remain in suspension can exhibit enhanced thermal conductivity, even though the volume fraction of the particles $< 10^{-4}$ [13].

The coarser particles are expected to have the bcc crystal structure of ferrite, but it must be considered that a 5 nm particle has about half its atoms at the surface. Metal surfaces are prone to transform into a variety of two-dimensional structures, complicating the interpretation of the structure of the particle as a whole. The surface also plays another role: it alters the total free energy of the particle, leading to a depression of its melting temperature. It has been estimated that a 5 nm diameter iron particle could melt at a temperature as low as 500°C. Fig. 1.3 illustrates an iron particle inside a carbon nanotube – it has been speculated that its bulbous character is due to melting.

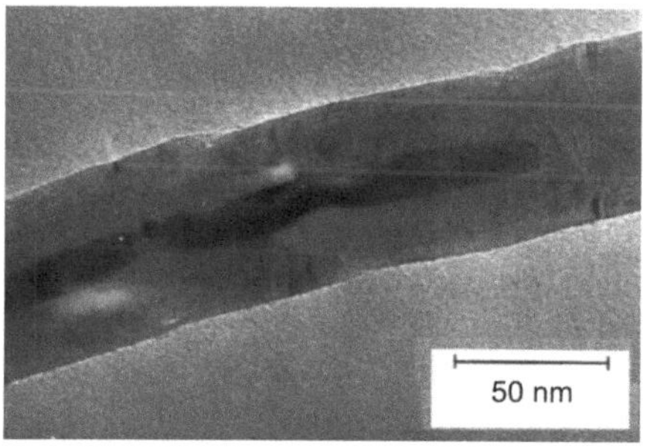

Figure 1.3 A multi-walled carbon nanotube containing a particle of iron (unpublished micrograph courtesy of I. Kinloch).

Small metal particles in the size range 1–5 nm are close to a metal/insulator transition. When observed at the tips of carbon nanotubes using scanning electron microscopy, the iron particles showed a tendency to charge, possibly indicating a loss of metallic behaviour.

1.2.2 Amorphous iron

Simulations indicate that liquid iron would need to be quenched at about $10^{13}\,\mathrm{K\,s^{-1}}$ for pressures up to 20 GPa to cause it to solidify in the glassy state. Pressures greater than 20 GPa would, in the same circumstances, favour the crystalline state, which has a smaller specific volume [14]. Therefore, it does not seem possible to obtain sizeable samples of pure iron, which have an amorphous configuration of atoms.

An amorphous configuration of atoms in a solid alloy was first discovered in a gold-silicon alloy quenched from the liquid state [15]; this work led to an entire field of research on metallic glasses. Whereas iron forms the basis of many metallic glasses, it is not feasible to obtain sizable samples of pure iron in a glassy state. However, small particles about 30 nm in size of almost pure iron can be produced by subjecting an organometallic $Fe(CO)_5$ to intense ultrasound that causes cavities in the liquid to form and collapse, processes associated with momentary high temperatures and pressures. The effective heating and cooling rates in such events exceeds $10^9\,K\,s^{-1}$, resulting in the generation of small, amorphous-iron particles [16]. The particles were found to be ferromagnetic under ambient conditions with a magnetic moment of $1.7\mu_B$ per atom. The amorphous structure crystallises at about 500°C as the iron atoms become mobile; not much is known about the evolution of the crystals within the amorphous matrix, or whether the particle surfaces play a significant role.

There are other chemical methods to produce small, amorphous-iron particles, the crystallisation behaviour of which has been studied using electron diffraction in a transmission electron microscope, Fig. 1.4.

Figure 1.4 Amorphous iron particles obtained from solution by chemical reduction of Fe^{3+}. The diffuse electron-diffraction pattern is consistent with an amorphous state. Image adapted from Falqui et al. [17] under the CC BY 4.0 license, https://creativecommons.org/licenses/by/4.0/.

1.3 Austenite to ferrite transformation

The vast majority of steels rely on just two allotropes, α and γ. Pure iron is a peculiar element in that at ambient pressure, bcc ferrite is stable from all temperatures up to 910°C (the Ae_3 point), when it transforms into the fcc austenite, only to revert to ferrite at 1390°C (the Ae_4 point). This high-temperature ferrite is traditionally labelled δ, although it does not differ in crystal structure from paramagnetic-α. The δ-ferrite remains the stable phase until melting occurs at 1538°C.

Fig. 1.5 shows the phase changes in a plot of the molar volume of iron as a function of temperature. The $\gamma \rightarrow \alpha$ transformation is accompanied by a volume change of approximately 1-3%, the exact amount depending on the temperature because the expansion coefficient of austenite is much greater than that of ferrite.

The detailed geometry of unit cells of α- and γ-iron crystals is particularly relevant to, e.g., the solubility in the two phases of non-metallic elements such as carbon and nitrogen, the diffusivity of alloying elements at elevated temperatures and the general behaviour on plastic deformation. The bcc structure of α-iron is more loosely packed than that of fcc γ-iron (Fig. 1.6). The largest cavities in the bcc structure are the tetrahedral holes existing between two edge and two central atoms in the structure, which together form a tetrahedron. The second largest are the octahedral holes, which occupy the centres of the faces and the ⟨001⟩ edges of the body-centred cube. The surrounding iron atoms are at the corners of a flattened octahedron. It is interesting that the fcc structure, although more densely packed, has larger holes than the bcc structure. These holes are at the centre of the cube edges and are surrounded by six atoms in the form of an octagon, so they are referred to as octahedral holes. There are also smaller tetrahedral interstices. The largest sizes of spheres which enter these interstices are given in Table 1.1.

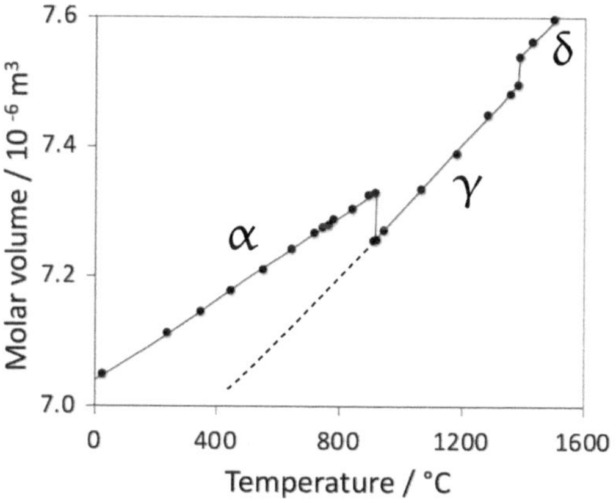

Figure 1.5 Temperature dependence of the volume per mole of atoms in iron crystals (data adapted from [18]). The dashed line illustrates that the thermal expansion coefficient of austenite is larger than that of ferrite, so the magnitude of the volume change accompanying transformation decreases with increasing temperature.

The $\alpha \rightleftharpoons \gamma$ transformation in pure iron occurs very rapidly, so it generally is not possible to retain the high-temperature fcc form at room temperature. Rapid quenching can substantially alter the morphology of the resulting α-iron, but it still retains its bcc structure. It follows that any detailed study of austenite in pure iron must be done at elevated temperatures, e.g., using X-ray or neutron diffraction. The transformation of the austenite on cooling can also be followed using diffraction based on the intense X-rays generated in a synchrotron, or using precision dilatometry. The latter technique relies on the volume change accompanying the transformation from austenite to ferrite.

There are circumstances where it is necessary to study pure austenite at temperatures well below ambient. Pure iron can be retained in its austenitic state to very low temperatures by coherent precipitation in copper. Copper has an fcc crystal structure and hence prevents the coherent particles of austenitic iron from transforming dur-

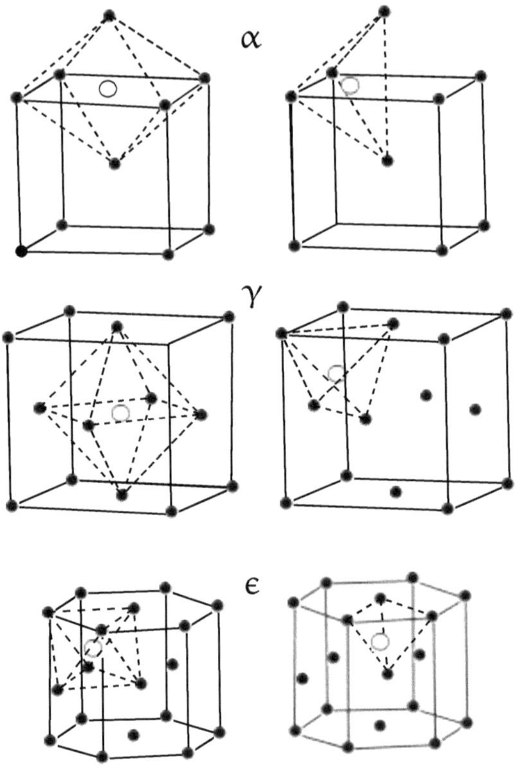

Figure 1.6 Schematic representation of the octahedral and tetrahedral interstices in ferrite (α), austenite (γ) and hexagonal iron (ε). The filled circles represent the positions of iron atoms, whilst the interstices are represented by open circles together with their coordination polyhedra. Adapted from [19].

Table 1.1 Size of largest spheres fitting interstices in bcc and fcc iron. r = atomic radius of iron, N is the number of interstices per iron atom. Typical coordinates of the interstices are expressed as fractions of the cell parameters. The calculations assume hard-sphere models for the crystal structures. The approximate atomic radii of carbon and nitrogen in iron are from Jack [20], to indicate size relative to the space available in the interstices.

	N		**Coordinates**	**Relative radius**	**Radius / Å**
bcc	6	Tetrahedral	$\frac{1}{2},0,\frac{1}{4}$	$0.29r$	0.37
	3	Octahedral	$0,0,\frac{1}{2}$ and $\frac{1}{2},\frac{1}{2},0$	$0.15r$	0.19
fcc	2	Tetrahedral	$\frac{1}{4},\frac{1}{4},\frac{1}{4}$ and $\frac{3}{4},\frac{3}{4},\frac{3}{4}$	$0.23r$	0.28
	1	Octahedral	$\frac{1}{2},\frac{1}{2},\frac{1}{2}$	$0.41r$	0.51
hcp	6	Tetrahedral	$\frac{1}{2},\frac{1}{3},\frac{1}{4}$ and $\frac{1}{2},\frac{1}{3},\frac{3}{4}$	$0.23r$	0.28
	3	Octahedral	$\frac{1}{3},\frac{1}{2},\frac{3}{4}$	$0.41r$	0.51
C					0.66
N					0.68

ing cooling. This technique was used to establish the antiferromagnetic nature of the austenite with a Néel temperature of about −190°C (the austenite can also be ferromagnetic at higher temperatures, with a Curie point of some 1525°C, depending on its density).

1.3.1 Mechanisms of transformation

One of the reasons why there are a large number of microstructures in steels is because the atoms can move in a variety of ways to achieve the same allotropic transition. The transformation can occur either by breaking all the bonds and rearranging the atoms into an alternative pattern (*reconstructive* transformation), or by homogeneously deforming the original pattern into a new crystal structure, i.e., *displacive* or *shear* transformation, Fig. 1.7.

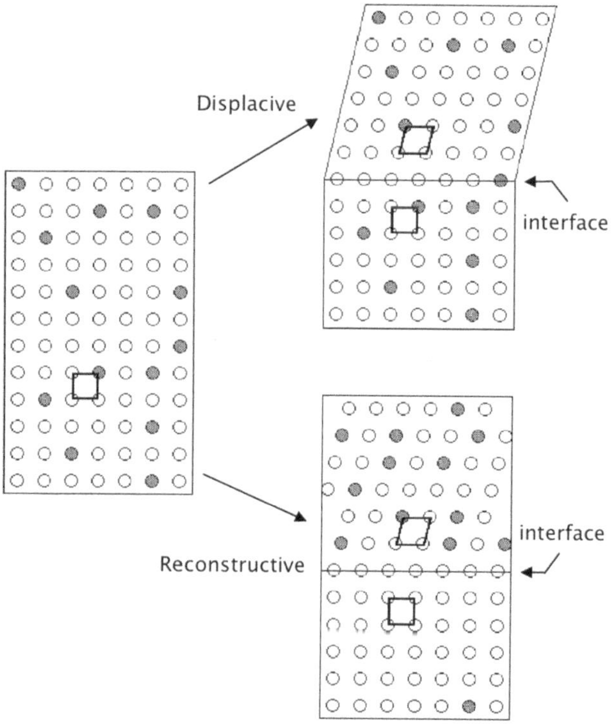

Figure 1.7 Schematic illustration of the mechanisms of transformation. The parent crystal contains two kinds of atoms. The figures on the right represent partially transformed samples with the parent and product unit cells outlined in bold.

In the displacive mechanism, the change in crystal structure also alters the macroscopic shape of the sample when the latter is not constrained. The shape deformation during constrained transformation is accommodated by a combination of elastic and plastic strains in the surrounding matrix. The product phase then grows in the form of

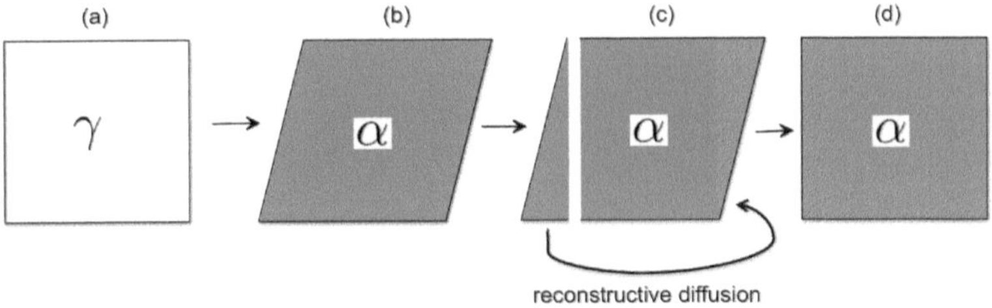

Figure 1.8 Phenomenological interpretation of reconstructive transformation. The virtual operation that eliminates the shape change is the diffusion required during a reconstructive transformation, irrespective of whether it occurs in pure iron or in an alloyed steel [21].

thin plates to minimise the strains. The atoms are displaced into their new positions in a coordinated motion. Displacive transformations can, therefore, occur at temperatures where diffusion is inconceivable within the time scale of the experiment. Displacive transformation will therefore occur in circumstances where the speed of the transformation front exceeds that at which atoms can diffuse ahead of the interface,

$$v > \frac{D}{\lambda} \tag{1.1}$$

where D is the diffusion coefficient and λ is the smallest atomic jump distance normal to the interface, on the order of the interatomic spacing. The term on the right is then the diffusion velocity. The speed v depends on the available driving force and the detailed structure of the interface – after all, the boundary must be able to advance without the requirement for climb of interfacial dislocations. It follows that at high temperatures, where the driving force for the transformation of austenite is small and the diffusion coefficient is large, transformations will tend to occur by a reconstructive mechanism, which in any case, brings the microstructure closer to equilibrium.

Some solutes may be forced into the product phase, a phenomenon known as solute trapping.[1] Both the trapping of atoms and the strains make displacive transformations less favourable from a thermodynamic point of view.

It is the diffusion of atoms that leads to the new crystal structure during a reconstructive transformation. Imagine that the transformation proceeds by the displacive mechanism (Fig. 1.8a,b), but that the resulting shape deformation is eliminated by transporting the segment as in Fig. 1.8c to recover the overall shape shown in (d). This transport is the diffusion necessary for the deformation energy term to be essentially eliminated, as if there were fluid flow in the surrounding matrix. The flow of matter is sufficient to avoid any shear components of the shape deformation, leaving only the effects of volume change. In alloys, the diffusion process may be accompanied by the redistribution of solutes between the phases in a manner consistent with a reduction in the overall free energy.

[1] The formal definition of 'trapping' is that the chemical potential of the solute or solvent atom concerned increases on transfer across the interface into the product phase.

It should be emphasised that the diffusional flow is essential even in pure iron, where the reconstructive ferrite and martensite are distinguished purely by the consequences of the shape deformation in the latter case. Fig. 1.9 shows the shape deformation caused in a single crystal of pure iron when heated sufficiently rapidly to induce the $\alpha \rightarrow \gamma$ transformation by a displacive mechanism [22]. All the phase transformations in steels can be discussed in the context of these two mechanisms (Fig. 1.10). The details are presented in subsequent chapters.

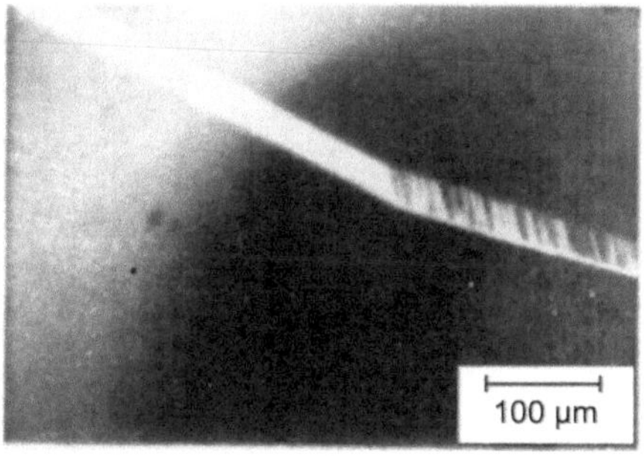

Figure 1.9 A whisker of iron, originally straight, transformed partially to austenite – the image is a frame from a movie taken to show the shape deformation due to the unconstrained $\alpha \rightarrow \gamma$ transformation. The position of the kink represents the interface between the α and the striated γ. After Zerwech and Wayman [22], reproduced with permission of Elsevier.

1.4 Carbon, nitrogen and hydrogen in solution

1.4.1 Solubility in α- and γ-iron

It is said that "ferrite has a solubility for carbon of about 0.02 wt%". Solubility, however, is defined by equilibrium between two or more phases. Therefore, the solubility of carbon in ferrite in equilibrium with austenite is not the same as in equilibrium with cementite, Fig. 1.11. Ferrite, which is in equilibrium with cementite, has a particularly low solubility for carbon at low temperatures. Fig. 1.11c shows the precipitation of cementite occurring in a steel containing only 0.02 wt% of carbon, when the alloy is aged at 240°C. Solubility also depends on external parameters such as pressure, temperature, magnetic fields and stress.

The atomic sizes of carbon and nitrogen (Table 1.2) are sufficiently small relative to that of iron to allow these elements to enter the α- and γ-iron lattices as *interstitial* solute atoms. In contrast, the metallic alloying elements such as manganese, nickel and chromium have much larger atoms, i.e., they are closer in size to those of iron and consequently they enter into *substitutional* solid solution. However, comparison of the atomic sizes of C and N with the sizes of the available interstices makes it clear that some lattice distortion must take place when these atoms enter the iron lattice. Indeed, it is found that C and N in α-iron do not occupy the larger tetrahedral holes but the octahedral interstices, which are more favourably placed for the relief of strain, which occurs by movement of two nearest-neighbour iron atoms.

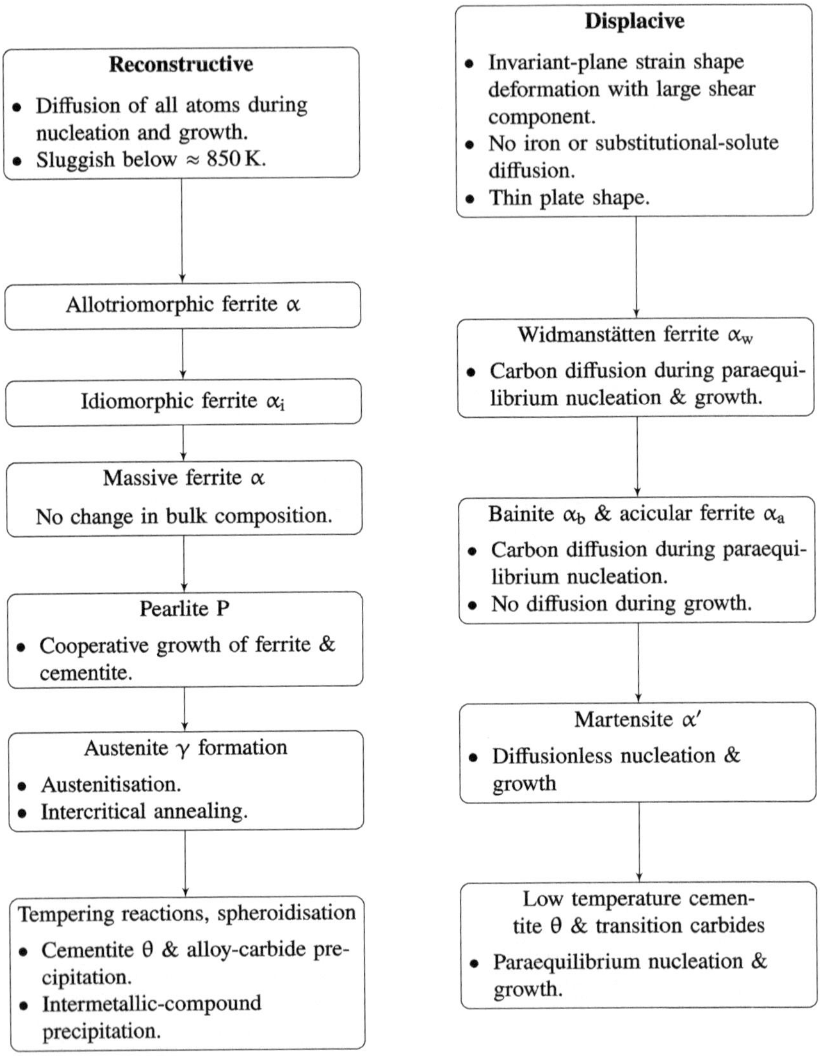

Figure 1.10 Some of the structures generated by the decomposition of austenite. *Paraequilibrium* refers to the case where carbon partitions, but the larger atoms in substitutional sites on the lattice, do not. The substitutional-solute to iron atom ratio is therefore unchanged by transformation. Adapted from [9].

In the case of tetrahedral interstices, four iron atoms are the nearest neighbours and the displacement of all four would require more strain energy. Consequently, these interstices are not preferred sites for carbon and nitrogen atoms. The strain energies for the two locations of carbon in ferrite, when Poisson's effects are neglected, are proportional to [24]:

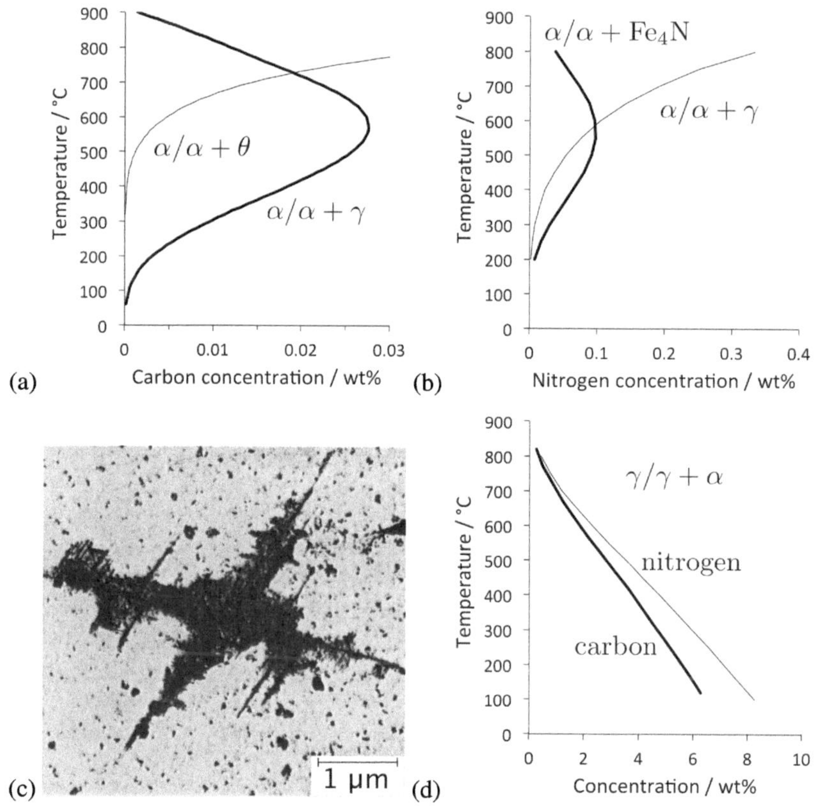

Figure 1.11 (a) The solubility of carbon in ferrite (α) in equilibrium with austenite (γ) or cementite (θ). (b) The solubility of nitrogen in ferrite in equilibrium with austenite or a particular nitride. (c) Cementite precipitation in quench-aged iron containing only 0.02 wt% carbon, 1560 min at 240°C [23]. (d) The solubility of carbon or nitrogen in austenite in equilibrium with ferrite.

$$U_O \propto 2(r_i - r_o)^2 E_{100} \tag{1.2}$$

where r_i is the interstitial-atom radius, r_o is the radius of the undistorted hole and E is the Young's modulus along the crystallographic direction identified in the subscript. For the tetrahedral site,

$$U_{Te} \propto 4(r_i - r_o)^2 E_{210}, \tag{1.3}$$

so for $r = 0.08$ nm, $U_{Te}/U_O = 1.38$. More recent calculations suggest that, in addition, there is a substantial difference in the chemical energy of solution between the octahedral and tetragonal sites, which makes the former interstice the favoured site for carbon atoms [25].

The solubility of both C and N in austenite in equilibrium with ferrite should be greater than in ferrite, because of the larger interstices available, Table 1.1. Fig. 1.11

Table 1.2 Atomic sizes of non-metallic elements in iron.

Element	Atomic radius, r / Å	r/r_{Fe}
α-Fe	1.28	1.00
B	0.94	0.73
C	0.77	0.60
N	0.72	0.57
O	0.60	0.47
H	0.46	0.36

shows that this is so for both elements, with the austenite able to accommodate concentrations in excess of 6 wt%, in contrast to the maximum that can be accommodated in α-iron. These marked differences of the solubilities of the main interstitial solutes in γ and in α are of profound significance in the heat treatment of steels and are fully exploited to increase strength (Chapter 2).

Although the solubility of C and N in ferrite at ambient temperature can be incredibly low, the presence of solute atoms can have profound effects on the mechanical properties through their interaction with defects such as dislocations. Sensitive physical techniques allow the study of very small concentrations of interstitial solute atoms in α-iron. Snoek first showed that internal friction measurements on an iron wire oscillating in a torsional pendulum over a range of temperatures just above ambient give an energy loss peak (Snoek peak) at a particular temperature and frequency. It was shown that the energy loss was associated with the migration of carbon atoms from randomly chosen octahedral interstices to those holes, which were enlarged on application of the stress in one direction, followed by a reverse migration when the stress changed direction and made other interstices larger. This movement of carbon atoms at a critical temperature is an additional form of damping or internal friction: below the critical temperature, the diffusivity is too low for atomic migration, and above it, the migration is too rapid to lead to appreciable damping. The height of the Snoek peak is proportional to the concentration of interstitial atoms, so the technique can be used not only to determine the very low solubilities of interstitial elements in iron, but also to examine the precipitation of excess carbon or nitrogen during an ageing treatment.

1.4.2 Diffusion of solutes in iron

The diffusion coefficient of carbon at 1000°C increases by a factor of five as its concentration changes from about 0.1 to 2 wt%. There are several reasons for this. Diffusion is driven by free energy gradients rather than concentration gradients, so there is a thermodynamic factor related to the way in which the activity of carbon depends on concentration. However, this is not the full explanation because carbon atoms in close proximity repel each other, so in a concentration gradient there is an additional factor promoting diffusion down the gradient. The atoms also cause a concentration-dependent dilatation of the lattice. As a result, the diffusion coefficient is often expressed empirically as [26]:

$$D_{\mathrm{C}}^{\gamma} = 4.53 \times 10^{-7}\left[1 + y_{\mathrm{C}}(1 - y_{\mathrm{C}})\frac{8339.9}{T}\right]$$
$$\times \exp\left\{-\left(\frac{1}{T} - 2.221 \times 10^{-4}\right)(17767 - 26436 y_{\mathrm{C}})\right\} \qquad \mathrm{m^2\,s^{-1}}, \quad (1.4)$$

where y_{C} is the ratio of carbon atoms to all other atoms. There is more rigorous theory available to deal with the diffusion of carbon in austenite, which indicates that the activation energy for diffusion is about $150\,\mathrm{kJ\,mol^{-1}}$ at a low carbon concentration [27,28]. Further diffusion data for which the concentration dependence is minimal are presented in Table 1.3. The activation energy reported for nitrogen diffusion in austenite is similar to that for carbon in that phase. In ferrite, where the concentrations of interstitial solutes are rather limited, internal friction techniques have been used to determine the diffusivities of C and N. The activation energies for diffusion are much smaller than in austenite, Table 1.3.

Table 1.3 Some diffusion data [29–31] conforming to the equation $D = D_0 \exp\{-Q/RT\}$, where R is the gas constant and T is the absolute temperature.

	$D_0/\mathrm{m^2\,s^{-1}}$	$Q/\mathrm{kJ\,mol^{-1}}$
Nitrogen in austenite	7.0×10^{-4}	166
Carbon in ferrite	1.67×10^{-7}	78
Nitrogen in ferrite	1.26×10^{-7}	73
Iron in austenite	7.0×10^{-5}	286
Iron in ferromagnetic ferrite	5.0×10^{-5}	240
Iron in paramagnetic ferrite	1.6×10^{-4}	240

Hydrogen is a pernicious solute in iron in the sense that it leads to dramatic changes in the ability of the metal to absorb energy during fracture, in concentrations so low that it is difficult to avoid the ingress of nascent hydrogen during, for example, corrosion reactions. The dissolution energy of hydrogen in the interstices of iron, calculated using first principles, are [32][2]:

Ferrite	Octahedral	0.34 eV
	Tetrahedral	0.19 eV
Austenite	Octahedral	0.07 eV
	Tetrahedral	0.51 eV.

Unlike carbon, the much smaller hydrogen atom fits best within the larger tetrahedral interstice of ferrite (Table 1.1). For the same reason, the larger octahedral

[2] $1\,\mathrm{eV} = 1.60217662 \times 10^{-19}\,\mathrm{J}$.

interstices in austenite are preferred by hydrogen. The diffusion coefficient of hydrogen in ferrite is much larger than in austenite, Fig. 1.12, whereas the reverse holds for the solubility of hydrogen, which is far greater in austenite than in ferrite. The diffusion of hydrogen through both lattices is complicated by the fact that it interacts strongly with defects such as dislocations and interfaces. The strain fields of these defects are attractors for hydrogen and hence its passage through the defective lattice is retarded. The effect is pronounced in the case of ferrite because of the high intrinsic mobility of hydrogen in body-centred cubic iron, as illustrated in Fig. 1.12.

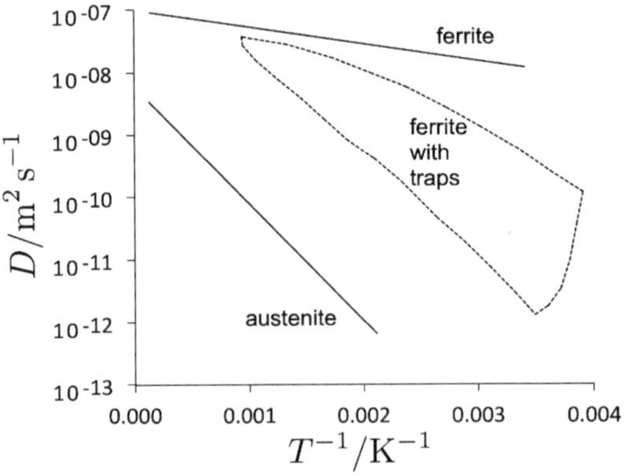

Figure 1.12 Diffusion coefficients for hydrogen in ferrite and austenite. The enclosed region represents diffusion in ferrite containing strong traps. The diffusion of hydrogen in austenite is so slow even in the perfect lattice, that its trapping at defects is less prominent when compared with the corresponding effects in ferrite. Data from [33,34].

The self-diffusion of iron is associated with much larger activation energies, Table 1.3, because unlike the interstitial migration of small atoms, where vacant positions are readily available, self-diffusion must account for both the energy for formation and migration of vacancies. Substitutional and self-diffusion in ferritic iron is complicated by the paramagnetic to ferromagnetic transition at the Curie temperature, changing both the vacancy formation and migration energies in the ferromagnetic state. The activation energy for migration in austenite is larger than in ferrite, presumably because the former is a more densely packed crystal structure. It is worth noting that in Table 1.3, the activation energies for diffusion are all less than $300\,\mathrm{kJ\,mol^{-1}}$. When analysing phenomena such as grain growth or even phase transformations, the activation energy of the process should not be much greater than this. Usually, the very large activation energies sometimes reported are a consequence of the empirical methods employed to analyse the grain growth data or a consequence of overfitting limited experimental data [35]. Substitutional solutes in general have activation energies for diffusion approximately the same as those for the self-diffusion of iron. This makes the homogenisation of steels that contain long-range, uneven distributions of such solutes difficult [36].

There are some remarkable empirical trends in the diffusivities of substitutional solutes in iron, illustrated in Fig. 1.13 [29]. Many of these solutes, including those that have larger atomic mass than iron, diffuse faster than iron itself. Similar data exist for aluminium, silicon, phosphorus and sulphur, all of which diffuse faster than iron. There is no clear explanation for these observations.

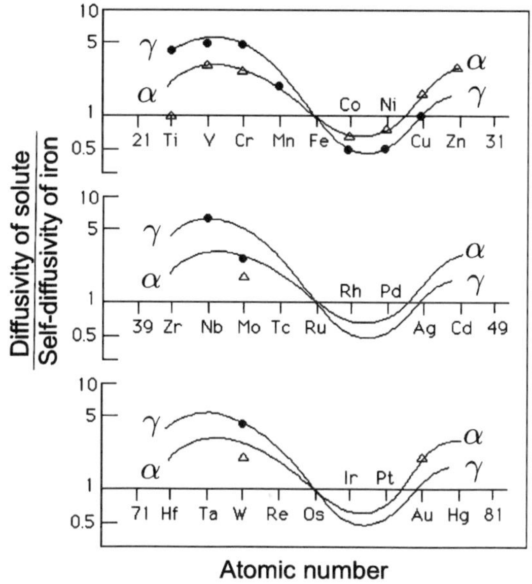

Figure 1.13 Ratio of the diffusivity of solute to that of the self-diffusivity of iron in α or γ as a function of species. Adapted from [29].

1.4.3 Practical consequences of diffusion

Surface treatment

The rapid diffusion of carbon and nitrogen in iron, compared to metallic alloying elements, is exploited in the processes of *carburising* and *nitriding*. Both of these surface modification processes are used extensively in the manufacture of engineering components such as bearings and gears, that require a high hardness at the surface but a tough core. Carburising can be carried out by heating a low-carbon steel in contact with carbonaceous gas, such as carbon monoxide, to the austenitic range, e.g., 1000°C, where the solubility of carbon in austenite that is in equilibrium with gas (g), $c^{\gamma g}$, is substantial. The result is a carbon gradient in the steel, from $c^{\gamma g}$ at the surface in contact with the carbon, to c^{γ} at a depth z. The solution of Fick's second diffusion law for the case where the steel initially contains a carbon concentration $\overline{c}$ is:

$$c^{\gamma} = c^{\gamma g} - (c^{\gamma g} - \overline{c})\mathrm{erf}\left\{\frac{z}{2\sqrt{D_{\mathrm{C}}^{\gamma} t}}\right\}, \tag{1.5}$$

which defines the variation in the carbon concentration as a function of depth from the surface of the steel; t is the carburising time. The diffusion coefficient D_C^γ varies with carbon content (Eq. (1.4)), so the above relationship is not rigorously obeyed. Carburising, whether carried out using carbon, or more efficiently using a carburising gas (*gas carburising*), provides a high-carbon surface on the steel, which, after appropriate heat treatment, is strong and wear resistant. Typical surface hardness values close to 800 HV can be achieved whilst preserving the structural integrity requirements of the core of the component. The heat treatment following carburising involves austenitisation, quenching and tempering at about 200°C, leaving the case in a martensitic state with some austenite retained. The phase transformation also leaves the surface in a state of compression, which is useful in enhancing the resistance to surface-initiated fatigue in components such as shafts and bearings. Fig. 1.14a shows how the optical microstructure of a Fe-0.11C-10.87Cr-0.2Mo-1Cu wt% steel appears due to the carbon enriched region at the top. Some typical data for an aeroengine bearing steel that is case carburised are shown in Fig. 1.14b,c.

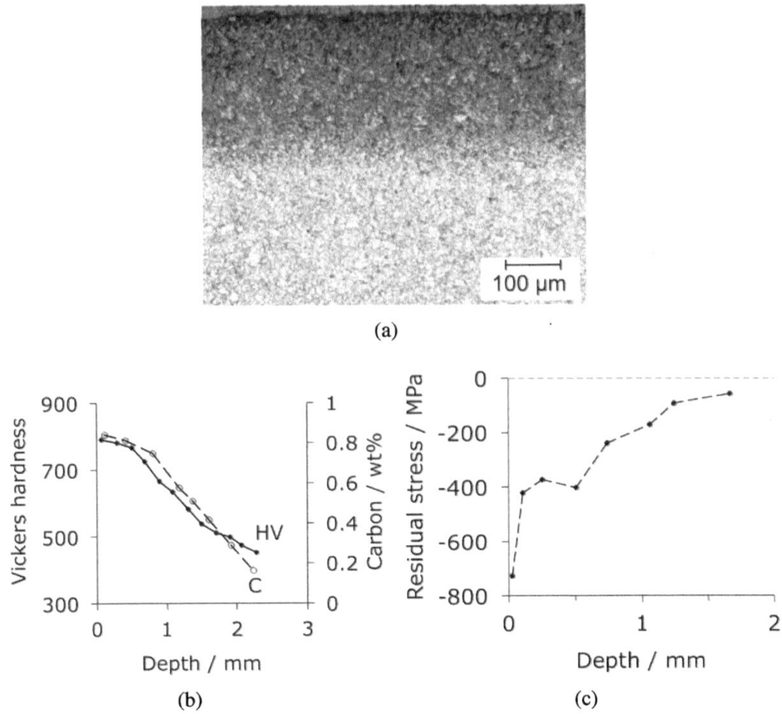

Figure 1.14 (a) The darkened region at the top represents the carburised layer on a power plant steel. The image is from Furukawa and Rouillard [37] with the permission of Elsevier, although it has been truncated here to focus on the etching contrast. (b) Properties of a bearing steel that has been carburised and heat-treated to produce a hard case. Hardness and carbon concentration as a function of depth. (c) Residual stress pattern. Data from [38].

Nitriding is normally carried out in an atmosphere of ammonia, but at a lower temperature (500–550°C) than carburising, consequently the reaction occurs in the ferrite phase, in which nitrogen has a substantially higher solubility than carbon, Fig. 1.11. Nitriding steels usually contain chromium (~0.1 wt%), aluminium (~1 wt%), vanadium or molybdenum (~0.2 wt%), which are nitride-forming elements that contribute to the high hardness, in excess of 1100 HV of the surface layer produced. The hard layer produced is not uniform, as the area closest to the surface becomes a compound, under which there is a mixture of matrix and nitride phases. The compound layer is brittle and hence is removed by grinding the component. In circumstances where the depletion of chromium by the formation of its nitrides can compromise corrosion resistance, plasma nitriding is used because it can be carried out at temperatures lower than 450°C where Fe_4N and $\epsilon-Fe_{2-3}N$ nitrides predominate.

Whereas carburising can produce hardened layers millimetres thick, nitriding penetrates only a few micrometres below the surface. Duplex hardening involves nitriding a component that has already been carburised [39]. This results in a very high surface hardness approaching 1200 HV, which then drops off to carburising levels deeper within the component.

Homogenisation

In cast steels, metallic alloying elements usually are segregated on a microscopic scale, by *coring* of dendrites. Therefore, to obtain a more uniform distribution, *homogenisation* annealing must be carried out, otherwise the chemical segregation persists even after large amounts of mechanical working. The much lower diffusivities of the metallic alloying elements compared to carbon and nitrogen, means that the homogenisation must be carried out at high temperatures (1200–1300°C), approaching the melting point. This may not be economically viable, but hot deformation usually greatly reduces the length scale of the segregation and enhances diffusion. The elongation of solute-rich regions along the hot deformation direction leads to the classical banded microstructures seen in commercial steels.

Assuming that the chemical segregation pattern varies periodically with distance (Fig. 1.15), an approximation that it is sinusoidal leads to the relationship:

$$c = \overline{c} + A \sin \frac{\pi z}{z_0} \exp\left\{-\frac{Dt\pi^2}{z_0^2}\right\} \tag{1.6}$$

where A is the amplitude, $2z_0$ is the wavelength, the decay of the solute wave will occur exponentially with time [40]. The time to reach $1/e$ of the initial amplitude is $z_0^2/\pi^2 D$. The characteristic distance defining the segregation, z_0, features as a square and means that macroscopic segregation, of the kind that occurs in ingots, simply cannot be homogenised on a reasonable time scale [36]. A second corollary is that although the amplitude decreases with time, the length scale of the segregation does not change - this must be altered by deformation.

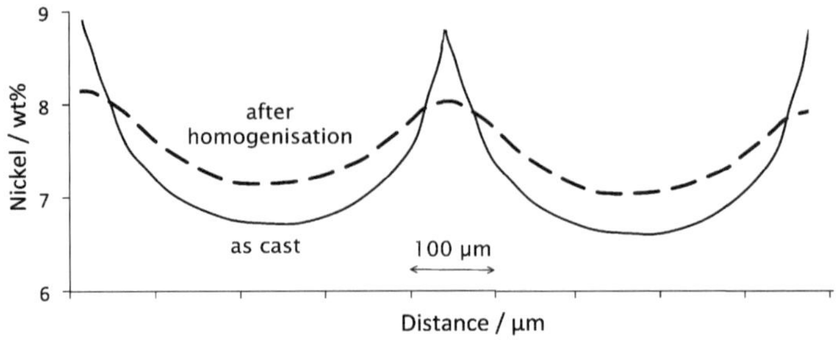

Figure 1.15 Schematic adaptation based on the unpublished work of Jin and Purdy, quoted in [40], but with realistic scales, to show the effect of annealing at 1220°C for 72 h on the chemical segregation pattern.

1.5 Summary

There are few metals that exhibit the peculiarity of allotropic and magnetic transitions that are evident in iron. Plutonium has many more (monoclinic, body-centred monoclinic, face-centred orthorhombic, fcc, bct and bcc). In liquid form, it is even denser than the first solid to form, so there is a sequence of martensitic transformations in structures that lead to densification. Uranium and neptunium each have three allotropic forms. However, these elements are not strictly engineering materials, nor are they available in sufficient quantities – plutonium is essentially an artificially created element on Earth. Nor are they stable compared to iron. Titanium, however, is available in sufficient quantity, although it only has two allotropic forms if the ω-phase [41] is ignored.

And the magnetic transitions in iron are all important because they are responsible for the fact that the most stable form of iron under ambient conditions is not hexagonal close-packed; under ambient conditions, magnetism stabilises the body-centred cubic form. Ruthenium and osmium are iron analogs in the sense that they have the same outer electronic structure (in the same column of the periodic table). However, they do not have the same magnetic characteristics and hence are hcp, a structure that inherently has limited slip systems.

Furthermore, iron undergoes remarkable changes, whether advantageous or detrimental, with the introduction of minute quantities of small atoms that dissolve in the interstices between iron atoms. The manner in which the iron atoms move during allotropic transitions has a great influence on the structures that are developed, and on the influence of interstitial species when they relocate into the new structure. We shall see that alloying and processing introduce a vast variety of possibilities, which will become apparent in the chapters that follow. Nevertheless, it is worth emphasising that iron with a purity in excess of 99.9% is commercially available. It is used, for example, for sacrificial anodes to protect copper-based heat exchangers in desalination equipment; for the yokes of the superconducting magnets that steer particle beams in accelerators designed for high-energy physics, because of the ease with which pure iron can be magnetised by an external magnetic field; it can also be used for magnetic shielding of high-voltage underground power lines.

References

[1] U.S. Clanton, D.S. McKay, R.B. Laughon, G.H. Ladle, Iron crystals in lunar breccias, Geochimica et Cosmochimica Acta, Supplement 4 1 (1973) 925–931.
[2] S. Sastri, Iron and alloys of iron, Space Resources 509 (1992) 281–290.
[3] K.M. Knowles, H.K.D.H. Bhadeshia, Deltoidal icositetrahedron faceting on α-Fe crystals found on the surface of the moon, Journal of Crystal Growth 617 (2023) 127257, https://doi.org/10.1016/j.jcrysgro.2023.127257.
[4] D.M. Clatterbuck, D.C. Chrzan, J.W. Morris Jr, The inherent tensile strength of iron, Philosophical Magazine Letters 82 (2002) 141–147.
[5] H.K.D.H. Bhadeshia, Large chunks of very strong steel, Materials Science and Technology 21 (2005) 1293–1302.
[6] J. Perkins, Doubts cast on graphene strength claims, Materials World (December 2015) 20.
[7] R.M. White, Quantum Theory of Magnetism, Springer, New York, USA, 2007.
[8] K.T. Moore, D.E. Laughlin, P. Söderlind, A.J. Schwartz, Incorporating anisotropic electronic structure in crystallographic determination of complex metals: iron and plutonium, Philosophical Magazine 87 (2007) 2571–2588.
[9] H.K.D.H. Bhadeshia, Theory of Transformations in Steels, CRC Press, Taylor and Francis Group, London, U.K., 2021.
[10] F. Birch, Density and composition of mantle and core, Journal of Geophysical Research 69 (1964) 4377–4388.
[11] O.L. Anderson, Properties of iron at the Earth's core conditions, Geophysical Journal International 84 (1986) 561–579.
[12] Y. Mnyukh, The true cause of magnetostriction, American Journal of Condensed Matter Physics 4 (2014) 57–62.
[13] Y.Y. Song, H.K.D.H. Bhadeshia, D.-W. Suh, Stability of stainless-steel nanoparticle and water mixtures, Powder Technology 272 (2015) 34–44.
[14] J. Mo, H. Liu, Y. Zhang, M. Wang, L. Zhang, B. Liu, W. Yang, Effects of pressure on structure and mechanical property in monatomic metallic glass, Journal of Non-Crystalline Solids 464 (2017) 1–4.
[15] W. Klement Jun., R.H. Willens, P. Duwez, Non-crystalline structure in solidified gold–silicon alloys, Nature 187 (1960) 869–870.
[16] K.S. Suslick, S.-B. Choe, A.A. Cichowlas, M.W. Grinstaff, Sonochemical synthesis of amorphous iron, Nature 353 (1991) 414–416.
[17] A. Falqui, D. Loch, A. Casu, In situ TEM crystallization of amorphous iron particles, Crystals 10 (2020) 10010041.
[18] W. Hume-Rothery, The Structures of Alloys of Iron, Pergamon Press, Oxford, U.K., 1966.
[19] H.K.D.H. Bhadeshia, Carbon in cubic and tetragonal ferrite, Philosophical Magazine 93 (2013) 3714–3715.
[20] K.H. Jack, Binary and ternary interstitial alloys-II. The iron-carbon-nitrogen system, Proceedings of the Royal Society A 195 (1948) 41–55.
[21] H.K.D.H. Bhadeshia, Diffusional formation of ferrite in iron and its alloys, Progress in Materials Science 29 (1985) 321–386.
[22] R.P. Zerwech, C.M. Wayman, On the nature of the $\alpha \rightarrow \gamma$ transformation in iron: a study of whiskers, Acta Metallurgica 13 (1965) 99–107.
[23] F.W. Langer, An investigation of carbide precipitation in iron, Metal Science Journal 2 (1968) 59–66.
[24] D.N. Beshers, An investigation of interstitial sites in the BCC lattice, Journal of Applied Physics 36 (1965) 290–300.

[25] D.H.R. Fors, G. Wahnström, Nature of boron solution and diffusion in α-iron, Physical Review B 77 (2008) 132102.
[26] J. Ågren, A revised expression for the difusivity of carbon in binary Fe-C austenite, Scripta Metallurgica 20 (1986) 1507–1510.
[27] R.H. Siller, R.B. McLellan, Application of first order mixing statistics to the variation of the diffusivity of carbon in austenite, Metallurgical Transactions 1 (1970) 985–988.
[28] H.K.D.H. Bhadeshia, Diffusion of carbon in austenite, Metal Science 15 (1981) 477–479.
[29] J. Fridberg, L.-E. Torndähl, M. Hillert, Diffusion in iron, Jernkontorets Annaler 153 (1969) 263–276.
[30] J.R.G. da Silva, R.B. McLellan, Diffusion of carbon and nitrogen in bcc iron, Materials Science and Engineering 26 (1976) 83–87.
[31] H.J. Grabke, E.M. Petersen, Diffusivity of nitrogen in iron-nickel alloys, Scripta Metallurgica 12 (1978) 1111–1114.
[32] E.J. Song, H.K.D.H. Bhadeshia, D.-W. Suh, Effect of hydrogen on the surface energy of ferrite and austenite, Corrosion Science 77 (2013) 379–384.
[33] F.R. Coe, Welding steels without hydrogen cracking, Technical Report, The Welding Institute, Abingdon, U.K., 1973.
[34] M.L. Hill, E.W. Johnson, The diffusivity of hydrogen in nickel, Acta Metallurgica 3 (1955) 566–671.
[35] H. Pous-Romero, I. Lonardelli, D. Cogswell, H.K.D.H. Bhadeshia, Austenite grain growth in a nuclear pressure vessel steel, Materials Science & Engineering A 567 (2013) 72–79.
[36] E.J. Pickering, Macrosegregation in steel ingots: the applicability of modelling and characterisation techniques, ISIJ International 53 (2013) 935–949.
[37] T. Furukawa, F. Rouillard, Oxidation and carburizing of FBR structural materials in carbon dioxide, Progress in Nuclear Energy 82 (2015) 136–141.
[38] H.J. Böhmer, F.J. Ebert, W. Trojahn, M50NiL bearing material – heat treatment, material properties and performance in comparison with M50 and RBD, Lubrication Engineering 48 (1992) 28–35.
[39] S. Ooi, H.K.D.H. Bhadeshia, Duplex hardening of aerospace bearings, ISIJ International 52 (2012) 1927–1934.
[40] G.R. Purdy, J.S. Kirkaldy, Homogenisation by diffusion, Metallurgical Transactions 2 (1971) 371–378.
[41] H.K.D.H. Bhadeshia, C.M. Wayman, Phase transformations: nondiffusive, in: D.E. Laughlin, K. Hono (Eds.), Physical Metallurgy, Elsevier B. V., North-Holland, 2014, pp. 1021–1072, chap. 9.

Strengthening of iron and its alloys **2**

Abstract

The ability of a metal to sustain plastic flow without permanently disrupting its crystal structure is a testimony to the nature of the metallic bond. The forces necessary to initiate deformation can be controlled by introducing obstacles that interfere with the smooth progress of plastic deformation. These strengthening mechanisms are the tools available to metallurgists in their efforts to design strong steels without compromising other properties. One particular property that is indispensable in safe engineering design and which distinguishes steels from materials such as glass, carbon nanotubes or graphene is the ability to absorb energy by plastic deformation prior to failure. The yearning for strength cannot and should not therefore be considered in isolation, but rather as one component in the holistic design of materials. Some notions about carbon-based materials being 100–300 times stronger than steel have been proven false.

2.1 Introduction

Although pure iron can be weak, steels cover a wide range of the strength spectrum, as illustrated in Fig. 2.1, which shows commercially available alloys that can be manufactured in useful quantities. The vast majority of the 2 billion tonnes of steel produced annually has strength levels far less than 1000 MPa because applications require a combination of properties that cannot usually be achieved in very strong materials. For example, the construction of high-rise buildings requires beams that have sufficient cross-section to be elastically rigid, which is not possible if a thinner section of a strong steel is fully loaded. Similarly, large fabrications are usually put together by welding, which becomes more difficult to implement as steels become stronger. As we will see, strength is achieved by creating particular features in the microstructure that resist plastic flow, and the heat input due to welding can disrupt the carefully engineered microstructure. Research samples can reach strength levels in excess of 10 GPa. This is not difficult to achieve, but the materials usually are prepared as perfect samples, or severely deformed samples, neither of which can be scaled to dimensions or quantities suitable for commercial production, even if a panoply of other design parameters could be achieved. Bearing this in mind, although Fig. 2.1 shows only the strength and elongation of commercially available steels, they possess many other attributes that make them useful.

There are many ways to strengthen steels, which can be exploited in combination. However, there has been limited progress in methods for the mathematical modelling of properties and hence in the deconvolution of overall strength into its components. The basic ways in which iron can be strengthened are discussed first, by reference to simple systems. These results should then be helpful in examining the behaviour of more complex alloys.

Like other metals, iron can be strengthened by several mechanisms, the most important of which are:

Steels. https://doi.org/10.1016/B978-0-44-318491-8.00008-3

- Work hardening.
- Solid solution strengthening by interstitial atoms.
- Solid solution strengthening by substitutional atoms.
- Refinement of grain size.
- Dispersion strengthening, including lamellar and random dispersed structures.
- Order hardening.
- Size and shape effects.

The most distinctive aspect of the strengthening of ferritic iron in particular, is the role of the interstitial solutes carbon and nitrogen. These solutes have powerful interactions with dislocations, but also can combine preferentially with some of the metallic alloying elements used in steels to enhance properties and processing.

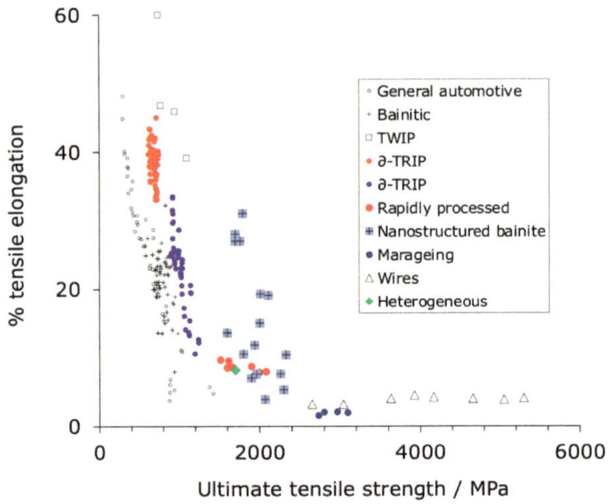

Figure 2.1 A selection of the strength and elongation of steels; each of these has a range of other properties, not illustrated, which make them commercially viable in useful quantities. TRIP and TWIP stand for transformation-induced and twinning-induced plasticity, respectively. The deliberately heterogeneous steel is one where chemical segregation inherited during solidification is exploited in order to engineer hard and soft regions within the microstructure [1]; this steel is not yet commercially available.

2.2 Work hardening

Work hardening is an important process for strengthening steel, particularly for obtaining high strength levels in rod and wire, both in plain carbon and alloy steels. For example, the tensile strength of an 0.05 wt% C steel subjected to 95% reduction in area by wire drawing is raised by no less than 550 MPa, while higher carbon steels are strengthened by up to twice this amount. Indeed, without the addition of special alloying elements, plain carbon steels can reach strength in excess of 1500 MPa simply by the phenomenon of work hardening.

Basic work on the deformation of ferritic iron has tended to focus on the other end of the strength spectrum, namely pure single crystals and polycrystals subjected to small, controlled deformations. This approach has shown that the slip plane in α-iron is not unique. Slip occurs on several planes, $\{101\}$, $\{112\}$ and $\{123\}$, but always in the close packed $\langle 11\bar{1}\rangle$ direction which is common to each of these planes (Fig. 2.2). The diversity of slip planes leads to rather irregular, wavy slip bands in deformed crystals, as the dislocations can readily move from one type of plane to another by cross-slip, provided they share a common slip direction (Fig. 2.2d). The Burgers vector of the slip dislocations would thus be expected to be $\frac{1}{2}\langle 111\rangle a_\alpha$, which has been confirmed by thin-foil electron microscopy.

The yield strength of α-iron single crystals is particularly sensitive to both temperature and strain rate, and a similar dependence has been found for less pure polycrystalline iron. Fig. 2.3 is a plot of flow stress σ_T as a function of the test temperature T, illustrating that both single crystal and polycrystalline iron of different interstitial content give values falling on the same curve. Therefore, the temperature sensitivity cannot be attributed to interstitial impurities. Sensitivity to temperature originates in the Peierls–Nabarro stress, which is needed to move free dislocations between equilibrium positions within the crystal. Crystals are periodic and the Peierls–Nabarro stress is that required for the dislocation to move from one equilibrium position in the lattice to the adjacent equivalent position. It is larger when the slip plane is less densely packed, and is therefore larger in bcc-iron than in fcc-iron, giving rise to the more pronounced temperature sensitivity of flow stress in the former allotrope. Direct observation of screw dislocations in iron in the electron microscope has shown that their mobility decreases strongly with decreasing temperature.

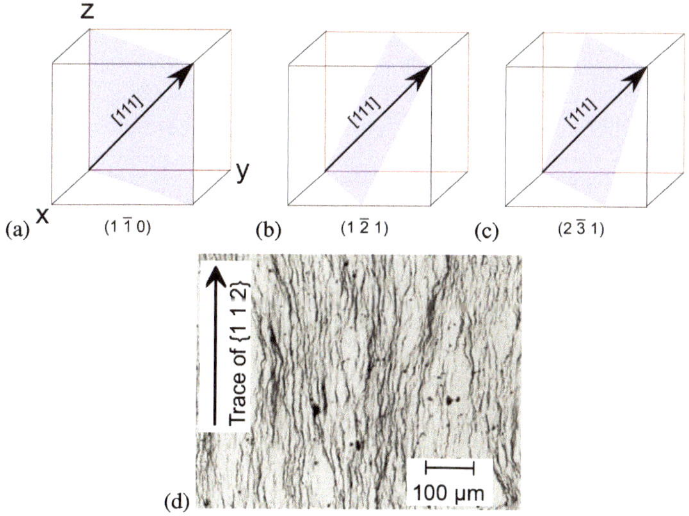

Figure 2.2 (a–c) The slip systems in the bcc structure. (d) Wavy slip traces on a crystal of iron following plastic deformation. Image adapted from Kch [2], with the permission of Taylor & Francis Ltd http://www.tandfonline.com.

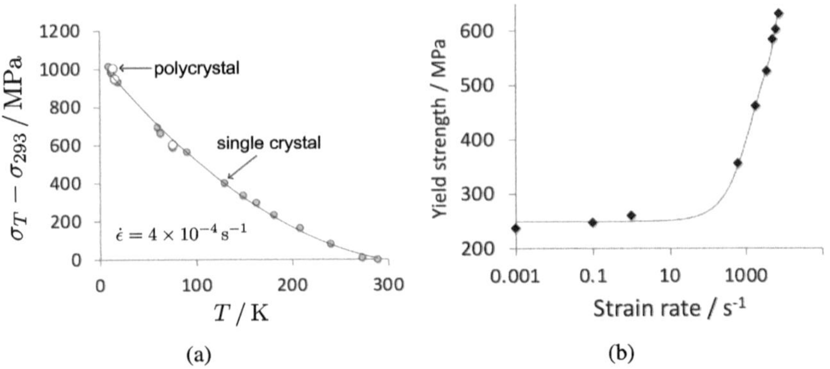

Figure 2.3 (a) Temperature dependence of the flow stress of single crystals and polycrystals of pure bcc-iron. Adapted using selected data from Altshuler and Christian [3]. (b) Strain rate dependence of the strength of iron at ambient temperature. Adapted using selected data from Bao et al. [4]. Strain rates used during routine tensile testing are on the order of $10^{-3}\,\text{s}^{-1}$.

If the shear stress at any point of the stress–strain curve is considered, the measured shear stress τ for further deformation comprises two quantities:

$$-\tau = \tau_o + \tau^*. \tag{2.1}$$

The negative sign in Eq. (2.1) indicates that τ is opposed by τ_o and τ^*. The shear stress, τ_o arises from the interaction of the dislocations with short-range obstacles, e.g., isolated dislocations. This stress is strongly temperature dependent, as thermal activation is helpful in moving dislocations around short-range obstacles [5]. On the other hand, τ^* is the internal stress arising from long-range obstacles such as grain boundaries, cell walls and other complex dislocation arrays [6]. In these circumstances, thermal fluctuations are of no use. The two component stresses are defined as follows:

$$\tau_o = \frac{1}{v^*}\left[H^* + kT \ln \frac{l\dot{\varepsilon}}{\rho_m A_s b f^*}\right], \tag{2.2}$$

$$\tau^* = C_1 \mu b \sqrt{\rho_d}, \tag{2.3}$$

where v^* = activation volume
H^* = activation enthalpy at $\tau = 0$
k = Boltzmann's constant
T = temperature
l = length of dislocation line activated
$\dot{\varepsilon}$ = strain rate
ρ_m = mobile dislocation density
ρ_d = dislocation density
A_s = area of glide plane covered by dislocation

b = magnitude of Burgers vector
f^* = frequency of vibration of dislocation line length
C_1 = constant
μ = shear modulus

The initial flow stress or yield stress with its large temperature dependence arises primarily from τ_o, while the increment in flow stress resulting from work hardening is largely independent of temperature and is caused by the change in τ^* with increasing strain and dislocation density ρ.

To summarise, work hardening in conventional materials is due largely to the creation of crystal defects, primarily dislocations, during plastic deformation. The hardening can reach saturation once the defect creation and annihilation rates balance each other. However, this emphasis on dislocations is not entirely justified because the grain structure of the material changes significantly, not just in adopting a shape compatible with the deformation, but there is an increase in the amount of surface per unit volume (S_V). For grains that initially are equiaxed, in the form of space-filling tetrakaidecahedra, during plane-strain deformation typical of rolling [7,8]:

$$\frac{S_V}{S_{V_0}} = \frac{\eta_{11} + 3(\eta_{11}\sqrt{1+2\eta_{33}^2} + \sqrt{\eta_{11}^2 + 2\eta_{33}^2} + \eta_{33}\sqrt{2(1+\eta_{11}^2)}}{3(2\sqrt{3}+1)} \tag{2.4}$$

where S_{V_0} is the initial grain surface per unit volume, η_{11} and η_{33} are the principal distortions, with the true strains along the principal directions given by $\varepsilon_{ii} = \ln \eta_{ii}$. The grain size is proportional inversely to S_V. This becomes particularly important during processes involving severe deformation, for example, wire drawing, where the change in the space through which dislocations can travel is reduced dramatically [9].

Work hardening has an important consequence on ductility. During tensile testing, the sample inevitably contains features, which cause the stress to concentrate and hence to initiate necking. The reduced cross-sectional area at the neck increases the stress in the necked region. In the absence of work hardening to help resist localised deformation, the neck becomes unstable and the sample fractures with poor overall ductility. To encourage uniform elongation, the work hardening must raise the yield strength at a rate greater than the increase in stress due to the reduced area at the neck.

Given the origin of work hardening, microstructures in which the dislocation density does not substantially increase during deformation should lack ductility. A pertinent example based on extremely fine grains is discussed in Section 2.5. A different mechanism, in which mechanical twins partition the parent grains into ever finer compartments, thus initiating work hardening, is discussed in Section 10.5.

2.3 Interstitial solid-solution hardening or softening

Carbon and nitrogen have a disproportionate influence on the strength of ferritic iron and a relatively minor effect on that of austenitic iron. The following relations between the change in Vickers hardness with interstitial content illustrate this vividly [10–13][1]:

$$\begin{array}{llcllcl} \text{Ferrite} & \Delta HV_\alpha & \approx & 950 \times w_C & \Delta HV_\alpha & \approx & 630 \times w_N \\ \text{Austenite} & \Delta HV_\gamma & \approx & 52 \times w_C & \Delta HV_\gamma & \approx & 240 \times w_N \end{array}$$

where w_i refers to the weight percent of the element identified as the subscript, and the relationships are valid for ambient conditions. The data for ferrite are consistent broadly with the interstitial/dislocation-core interactions energies, which are 0.6 eV for nitrogen and 0.8 eV for carbon, the former having a weaker interaction; similar data do not seem to be available for austenite.

Solid solution strengthening occurs when the strain fields around misfitting solutes interfere with the motion of dislocations. Atoms which substitute for iron cause local expansions or contractions; these strains are isotropic and therefore can interact only with the hydrostatic components of the strain fields of dislocations. In contrast, an interstitial atom located in the irregular octahedral interstice in ferrite causes a tetragonal distortion (Fig. 1.6), which has a powerful interaction with the shear that is the dominant component of a dislocation strain field. This is why interstitial solid solution strengthening is so potent in ferrite. The corresponding interstitial site in austenite is the regular octahedron. An interstitial atom in austenite therefore behaves like a substitutional solute surrounded only by hydrostatic strains. This is why carbon is much less effective in strengthening austenite.

Ferritic iron has such a low solubility for carbon and nitrogen (Fig. 1.11a) that there is an overwhelming tendency for the these solutes to segregate to defects. This leads to another significant effect in α-iron, that carbon and nitrogen can promote heterogeneous deformation by making it difficult to initiate plastic flow. This is the yield point effect, described next.

2.3.1 The yield point

Carbon and nitrogen in iron, even in concentrations as low as 0.005 wt%, lead to a sharp transition between elastic and plastic deformation in a tensile test performed on ferritic iron. Decarburisation of the iron results in the elimination of this sharp transition or *yield point*, which implies that the solute atoms are in some way responsible for this striking behaviour. The load frequently drops dramatically at the *upper* yield point 'a' (Fig. 2.4a) to another value referred to as the *lower* yield point 'b'. Under some experimental conditions, yield drops of about 30% of the upper yield stress can be obtained.

Following the lower yield point, there is frequently a horizontal section of the stress-strain curve 'b→c' during which the plastic deformation propagates at a front, which can move uniformly along the specimen. This front is referred to as a Lüders band [15] (Fig. 2.4b) and the horizontal portion bc of the stress-strain curve as the Lüders extension. The development of Lüders bands can be much less uniform; in

[1] The coefficients given should not be taken too seriously because they depend on a variety of factors other than the interstitial content. However, they serve to show that interstitial strengthening is far more potent in ferrite than in austenite.

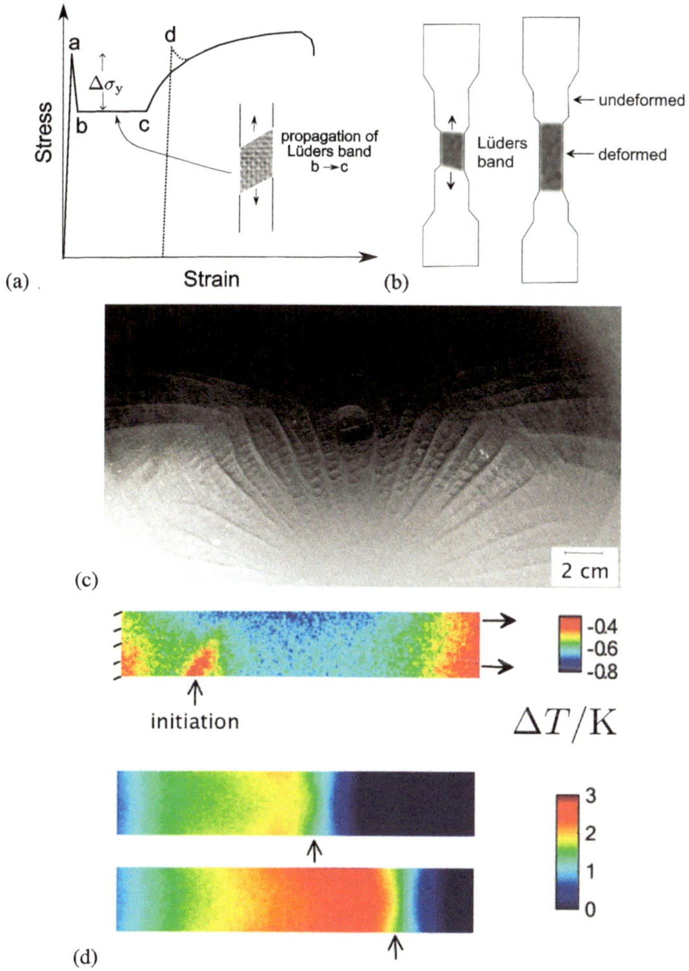

Figure 2.4 (a) Illustration of the yield phenomenon as discussed in the text. $\Delta\sigma_y$ is the difference in the upper and lower yield stresses. (b) Schematic illustration of the propagation of Lüders fronts during tensile testing. (c) Stretcher strains on a biaxially stretched component. (d) Thermal effects (change in temperature) observed on the gauge length of a tensile specimen, due to thermo-elastic cooling followed by adiabatic heating of the sample during Lüders band initiation and propagation along the sample. The arrows indicate the position of the band as deformation progresses. Part (d) adapted with permission of Elsevier Masson SAS from Dumoulin et al. [14].

pressings where the stress is far from uniaxial, complex arrays of bands can be observed (Fig. 2.4c). These are often referred to as *stretcher strains*, but they still are Lüders bands. Stretcher strains clearly undesirable in the manufacture of components that require a smooth surface visible to the user, such as a painted car body.

The concentrated deformation at the Lüders band leads to localised heating that spreads along the sample as the deformed portion increases in size (Fig. 2.4d). When the whole specimen has yielded, general work hardening commences and the stress-strain curve begins to rise in the normal way. However, if this deformation is interrupted and the specimen allowed to rest either at room temperature or for a short time at 100–150°C, on reloading a new yield point is observed (Fig. 2.4d) due to the diffusion of interstitial atoms to dislocations that once again are pinned. This return of the yield point is referred to as *strain ageing*. A rather clever application of this to "bake hardening" automotive steels is described on page 50.

Fabrication processes also are affected by the heterogeneous deformation associated with the yield phenomenon. Fig. 2.5 illustrates a steel pipe (typical composition Fe-0.1C-0.5Mn wt%) that was thinly-coated with a brittle material prior to being subjected to pure bending. The purpose of the coating was to reveal localised deformation. It is evident that the bending deformation leads to nonuniform localised deformation, which intensifies as the bend curvature increases. If continued, this would lead to macroscopic wrinkling on the pipe surface. One conclusion from these experiments was that the larger the magnitude of $\Delta\sigma_y/\Delta\epsilon_{\text{Lüders}}$, the greater is the instability during deformation; $\Delta\sigma_y$ is the drop in stress between the upper yield point and the stress during the Lüders extension $\Delta\epsilon_{\text{Lüders}}$, Fig. 2.4a [16].

Steel pipelines used for the transmission of fossil fuels are hot epoxy-coated, a process associated with a temperature rise of about 200°C consistent with strain ageing; the coating serves to protect against corrosion. During pipe laying, it sometimes is necessary to bend such pipes, but the yield point effect can cause localised plastic deformation, which fractures the coating [17,18], so remedial measures must then be implemented.

2.3.2 Role of interstitial elements in yield phenomena

The sharp upper and lower yield points in iron are eliminated by annealing in wet hydrogen, which greatly reduces the concentration of carbon and nitrogen. However, substantial strain ageing can occur at carbon levels around 0.002 wt%, and as little as 0.001-0.002 wt% N can result in severe strain ageing. Nitrogen is more effective than carbon in this regard, because its residual solubility near room temperature is substantially greater than that of carbon (Fig. 1.11).

Cottrell and Bilby first showed that interstitial atoms such as carbon and nitrogen would interact strongly with the strain fields of dislocations [19]. The misfitting interstitial atoms themselves have strain fields around them, but when such atoms move into the dislocation strain fields, there should be an overall reduction in the total strain energy. This leads to the formation of interstitial concentrations or *atmospheres* in the vicinity of dislocations, which in extreme cases can amount to lines of interstitial atoms decorating the cores of the dislocations (*condensed atmospheres*), e.g., in edge dislocations in the region of the strain field where there is maximum dilation (Figs. 2.6a,b). The binding energy between a dislocation in iron and a carbon atom is typically about $\approx 10^{-19}$ J. Consequently, dislocations can be locked in position by carbon atoms, thus substantially raising the stress, which would be necessary to in-

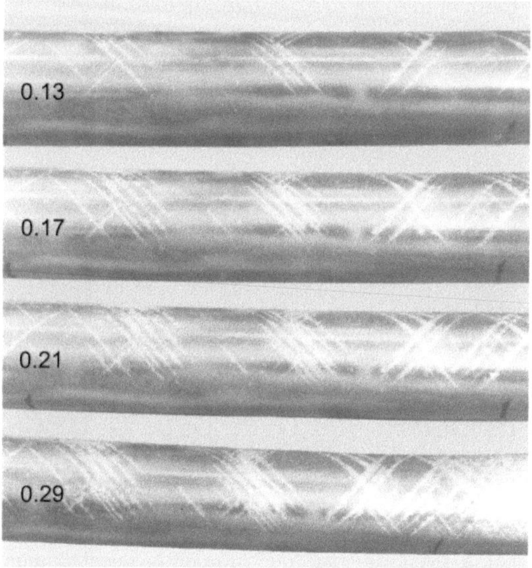

Figure 2.5 The sequence of four photographs reveal the evolution of Lüders deformation as the curvature of the pipe is increased – the numbers on the image are measures of this curvature. After Aguirre et al. [16], reproduced with the permission of Elsevier.

duce dislocation movement. A particular attraction of this theory is that only a very small concentration of interstitial atoms is needed to produce locking along the whole length of all dislocation lines in annealed iron. For a typical dislocation density of $10^{12}\,\mathrm{m}^{-2}$ in annealed iron, a carbon concentration of 10^{-6} wt% would be sufficient to provide one interstitial carbon atom per atomic plane threaded by each dislocation and saturate all the dislocation cores in this manner. Consequently, this theory can explain the observation of yield phenomena at very low carbon and nitrogen concentrations.

The formation of interstitial atmospheres at dislocations requires the diffusion of the solute. As both carbon and nitrogen diffuse rapidly in iron, it is not surprising that strain ageing sets in readily in the range 20–150°C. The fractional occupancy x_i of interstitial sites with binding energy U_i is at equilibrium given by:

$$\frac{x_\mathrm{i}}{1-x_\mathrm{i}} = \frac{\overline{x}}{1-\overline{x}}\exp\left\{\frac{U_\mathrm{i}}{kT}\right\}, \tag{2.5}$$

where $\overline{x}$ is the corresponding average occupancy of interstitial sites in the matrix far from the influence of the dislocation [21]. This approach in general leads to a Maxwellian distribution of solute about the dislocation, but for carbon and nitrogen in steel, the elastic interaction energy U between solute and dislocation is so large that $U \gtrsim kT$. Consequently, the atmosphere condenses to form rows of interstitial atoms along the cores of the dislocations.

The maximum solute-dislocation interaction is when carbon occupies the dislocation core with $U_\mathrm{i} = U_\mathrm{max}$. The temperature T_crit at which half of the core sites are

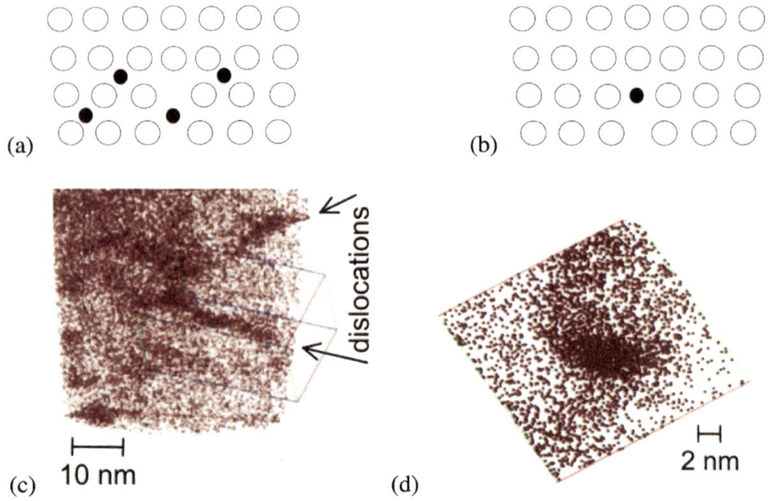

Figure 2.6 Schematic diagram of interstitial atoms (black) in the vicinity of an edge dislocation (a) random distribution, (b) condensed. (c) Atom probe tomographic-image where each dot is a carbon atom, showing dislocations in ferrite, with dislocation-lines parallel to the arrows, decorated with carbon atmospheres. (d) Atom probe image showing an image plane normal to the dislocation line, illustrating the extent of the atmosphere. Figures c,d are adapted from Caballero et al. [20], reproduced with the permission of Elsevier.

occupied is then given by:

$$T_{\text{crit}} = \frac{U_{\max}}{k \ln\{(1-\overline{x})/\overline{x}\}}. \tag{2.6}$$

It is expected therefore that for $\overline{x} = 10^{-4}$, $U_{\max} = 10^{-19}$ J, 10% of the dislocation core will be occupied by carbon, even if the temperature is as high as 1000 K. This does not necessarily mean that the decorated dislocations will be pinned, because the carbon atoms may be able to keep up with the dislocations by diffusion at high temperatures. Fig. 2.7a therefore shows that the zone of yielding is well defined at the lower testing temperatures, becoming less regular as the temperature increases, until it is replaced by fine serrations along the whole stress-strain curve. This phenomenon is referred to as *dynamic strain ageing*, in which the serrations represent the replacement of the primary yield point by numerous localised yield points within the specimen. These arise because the temperature is high enough for interstitial atoms to diffuse during deformation and form atmospheres around dislocations generated throughout the deformation process. Steels tested under these conditions also show low ductilities, due partly to high dislocation density and partly to the nucleation of carbide particles on the dislocations where the carbon concentration is high. The phenomenon is often referred to as *blue brittleness* because blue is the interference colour of the steel surface when oxidised in this temperature range.

Dynamic strain ageing is not limited to ferritic steel. Fig. 2.7b shows how the interaction of carbon with dislocations in an austenitic steel (Fe-0.6C-18Mn wt%) causes serrated flow – a similar effect is found when nitrogen is added to this alloy.

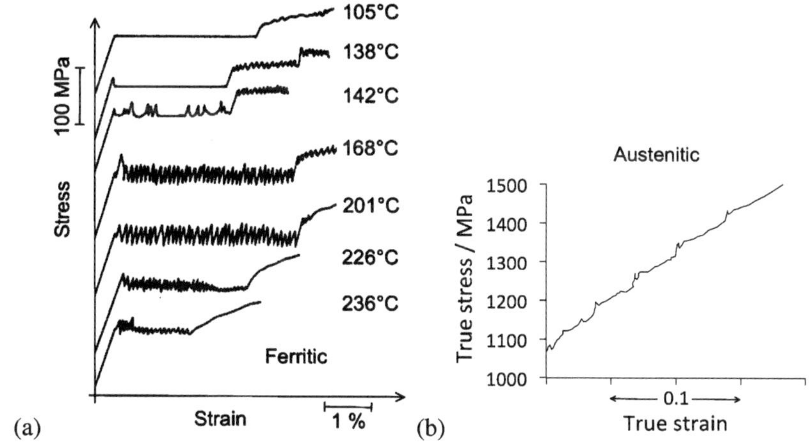

Figure 2.7 (a) Typical stress–strain curves for mild steel at elevated temperatures (adapted from Hall [22]). (b) Dynamic strain ageing in austenitic steel. Adapted from Lee et al. [23], reproduced with the permission of Elsevier.

The break-away of dislocations from their carbon atmospheres as a cause of the sharp yield point became a controversial aspect of the theory, because it was found that the provision of free dislocations, e.g., by scratching the surface of a specimen, did not eliminate the sharp yield point. An alternative theory was developed which assumed that, once condensed carbon atmospheres are formed in iron, the dislocations remain locked, and that the yield phenomena arise from the generation and movement of newly formed dislocations [24]. The velocity of movement v of these fresh dislocations is related to the applied stress as follows:

$$v = \left(\frac{\sigma_y}{\sigma_r}\right)^m, \tag{2.7}$$

where σ_r is the stress corresponding to a unit velocity, σ_y is the yield stress and m is an empirical index characteristic of the material, varying between 1 and 60. The strain rate $\dot{\varepsilon}$ can be defined in terms of the movement of dislocations, as

$$\dot{\varepsilon} = n_A v b, \tag{2.8}$$

where n_A is the number of mobile dislocations per unit area, v is their average velocity and b is the magnitude of the Burgers vector.

Using Eq. (2.8) the strain rates at the upper yield point ($\dot{\varepsilon}_U$) and the lower yield point ($\dot{\varepsilon}_L$) can be defined as follows:

$$\dot{\varepsilon}_U = \rho_U v_U b,$$

where ρ_U is the mobile dislocation density at the upper yield point and

$$\dot{\varepsilon}_L = \rho_L v_L b,$$

where ρ_L is the mobile dislocation density at the lower yield point, so

$$\frac{v_U}{v_L} = \frac{\rho_L}{\rho_U},$$

and using Eq. (2.7),

$$\frac{\sigma_U}{\sigma_L} = \left(\frac{\rho_L}{\rho_U}\right)^{\frac{1}{m}}. \tag{2.9}$$

Thus the ratio σ_U/σ_L will be large, i.e. there will be a large yield drop, for small m and when ρ_L is much greater than ρ_U. Consequently, if at the upper yield stress the density of mobile dislocations is low, e.g., as a result of solute atom locking, a large drop in yield stress occurs when many new dislocations are generated. Observations indicate that the dislocation density just after the lower yield stress is much higher than that observed at the upper yield point.

To summarise, the occurrence of a sharp yield point depends on a sudden increase in the number of mobile dislocations. However, the precise mechanism by which this takes place depends on the effectiveness of the locking of the pre-existing dislocations. If the pinning is weak, then the yield point can arise as a result of unpinning. However, if the dislocations are strongly locked, either by interstitial atmospheres or precipitates, the yield point will result from the rapid generation of new dislocations.

Under conditions of dynamic strain ageing, where atmospheres of carbon atoms form continuously on newly generated dislocations, it would be expected that a higher density of dislocations would be needed to complete the deformation, assuming that most dislocations attracting carbon atmospheres are permanently locked in position. Electron microscopy observations have shown that in steels deformed at 200°C, the dislocation densities are an order of magnitude greater than in specimens similarly deformed at room temperature.

2.3.3 Strengthening at large interstitial concentrations

Austenite, when in equilibrium with cementite (Fig. 3.1), can dissolve about 9 at% of carbon, which can then be retained in solid solution by rapid quenching. However, in these circumstances, the phase transformation takes place not to ferrite but to a body-centred tetragonal structure referred to as *martensite* (Chapter 5). This phase forms as a result of a diffusionless shear transformation leading to characteristic laths or plates, which normally appear acicular in planar metallographic sections. If the quench is sufficiently rapid, the martensite is essentially a supersaturated solid solution of carbon in a tetragonal iron matrix, and as the carbon concentration can be greatly in excess of the equilibrium concentration in ferrite, the strength is raised very substantially. High carbon martensites are normally very hard but brittle, with yield strength reaching as much as 1500 MPa. Much of this increase can be attributed directly to increased interstitial solid solution hardening, but there is also a contribution from the large dislocation density characteristic of martensitic transformations in iron–carbon alloys. Martensite is discussed in more detail in Chapter 5, which shows that by subjecting it to a further heat treatment known as *tempering*, at temperatures below Ae_1, a proportion of the strength is retained, with a substantial gain in the toughness and ductility of the steel. We note that it now is possible to obtain tough, untempered, high-carbon martensite (p. 250).

Very large concentrations of carbon (15 at%) or nitrogen (38 at%) can be introduced into the surfaces of austenitic steels either by diffusion or using techniques such as ion implantation [25,26]. The lattice parameter of the austenite then expands dramatically by 10% in the case of the nitrogen-enriched austenite and 3% for the carbon-enriched version. The resulting surface layers can be extremely hard and corrosion resistant, making the process suitable for components where wear resistance is paramount. Nanohardness values in excess of 1100 HV have been reported. There is sharp transition observed between the lattice parameters of the expanded austenite and that of the unaffected substrate. Fig. 2.8 shows two distinct sets of X-ray diffraction peaks from the two kinds of austenite. The expanded austenite is found to be ferromagnetic, whereas the unaffected austenite is paramagnetic. It is known that the expanded austenite is crystallographically contiguous with the substrate austenite, so a very high dislocation density needed to accommodate the misfit.

2.3.4 Interstitial-solution softening

There is good evidence that in dilute solutions and at temperatures below ambient, the presence of interstitial atoms can lead to overall softening. For example, during tensile testing at 190 K, the proof strength of pure iron single-crystals drops from ≈200 MPa to just ≈100 MPa when 200 ppm of nitrogen is added to the solution [29]. The nitrogen hardens the crystals at even lower temperatures or when tested under ambient conditions.

This phenomenon is generic to many body-centred cubic metals and arises because screw dislocations, in particular, propagate by the movement of kinks, Fig. 2.9. These are steps on the dislocation line where one segment lies in a particular Peierles–Nabarro potential minimum and the remainder of the line lies in a parallel, adjacent minimum. The kinks, which have edge character, are more mobile than the translation of a straight screw dislocation. The edge character means that they can hold a greater concentration of interstitial solutes, which reduces their defect-energy. As long as the interstitials remain mobile, this means that there is a net softening because double-kink formation becomes easier. The return of hardening at even lower temperatures is a consequence of the reduced mobility of the interstitials, which implies that the kinks must be able to pull away from their atmosphere. With large concentrations of solutes, the time required for double-kink to form becomes greater than that for the kinks to migrate to the ends of the dislocation, leading to hardening [30]. The accumulation of solute at dislocations increases if the iron is aged prior to testing, leading again to hardening.

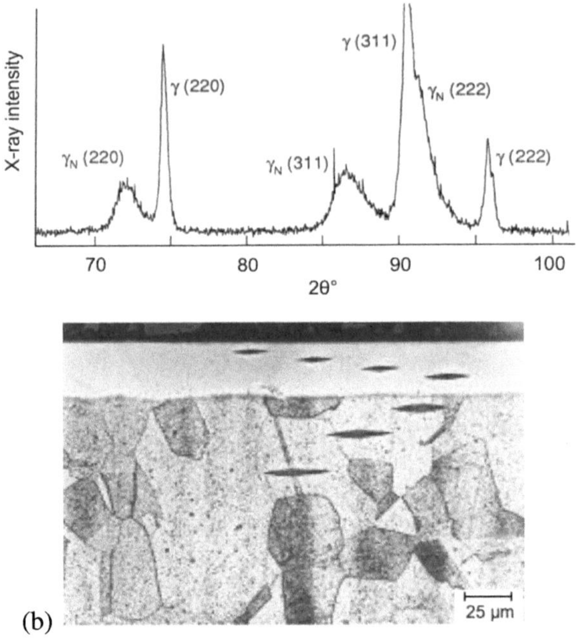

Figure 2.8 (a) X-ray diffraction pattern of a mixture of nitrogen-expanded austenite (γ_N) and the unaffected austenitic stainless steel. After [27], reproduced with the permission of Elsevier. (b) A carbon-rich layer of expanded austenite produced by low temperature plasma carburising, with indents that revealed a Knoop hardness of about 900 at the surface. Image reproduced from [28] with the permission of Maney Publishing.

2.4 Substitutional-solution hardening and softening

Many metallic elements form solid solutions in γ- and α-iron. These are invariably substitutional solid solutions, but for a constant atomic concentration of alloying elements there are large variations in strength. Using single-crystal data for several metals, Fig. 2.10 shows that an element such as vanadium has a weak strengthening effect on α-iron at low concentrations (<2 at%), while silicon and molybdenum are much more effective strengtheners. Other data indicate that phosphorus, manganese, nickel and copper are also effective strengtheners. However, it should be noted that the relative strengthening may alter with the temperature of testing and concentrations of interstitial solutes present in the steels.

The strengthening achieved by substitutional solute atoms depends, for example, on the difference in atomic size of the solute from that of iron, because it is the strain field around the misfitting atoms that determines some of the interaction with dislocations. However, the elastic properties and bonding within the metal are also likely to be influenced and may play a role in the potency of strengthening. Thus, the thermal component of shear stress τ_o (Eq. (2.1)) may increase or decrease due to solute additions. The mechanism by which a solute atom might reduce τ_o is simply that the local distortion helps the dislocation overcome a barrier; nickel, for example, softens

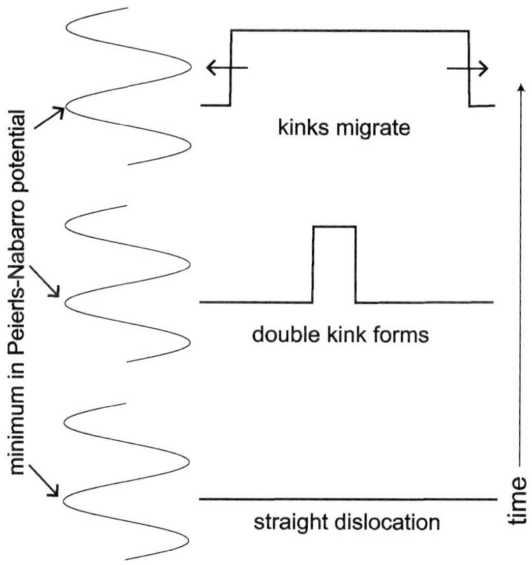

Figure 2.9 Mechanism by which a screw dislocation which lies in a deep Peierls–Nabarro potential well can, under the influence of an appropriate shear stress, advance into the next parallel well by forming a double-kink that can propagate laterally.

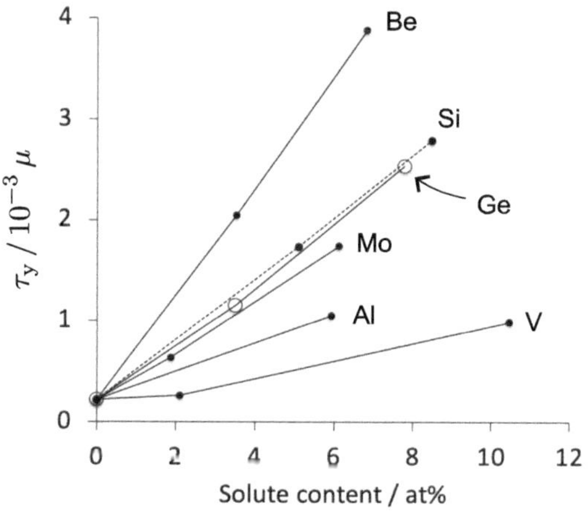

Figure 2.10 Solid solution strengthening of α-iron crystals by substitutional solutes. Ratio of the resolved shear stress τ_y at yield and at 300 K, to shear modulus μ, as a function of atomic concentration. Data from [31].

iron at low temperatures. On the other hand, the athermal component τ^* always increases on alloying because of the contribution of solute atoms to long-range internal stresses [32]. These concepts apply to polycrystalline substances as long as the key deformation mechanism involves the passage of dislocations through significant lengths in individual crystals. They are less relevant when mechanisms such as grain boundary sliding and rotation are dominant. Simulations indicate that solutes that stiffen the grain interiors are most effective in increasing the strength of nanocrystalline alloys that deform by sliding and grain rotation [33].

In practical terms, the contribution to strength from solid solution effects is superimposed on hardening from other sources, e.g., grain size and dispersions. Also, it is a strengthening increment, like that due to grain size, which need not adversely affect ductility. In industrial steels, solid solution strengthening is a far from a negligible factor in the overall strength, where it is achieved by a number of familiar alloying elements, e.g., manganese, silicon, nickel, molybdenum, several of which are frequently present in a particular steel and can be additive in their effect. These alloying elements are usually added for other reasons, e.g., Si to achieve deoxidation, Mn to combine with sulphur or Mo, to promote hardenability. Therefore, the solid solution hardening contribution can be viewed as a useful bonus.

2.5 Grain size

2.5.1 Hall–Petch effect

The refinement of the grain size of ferrite provides one of the most important strengthening routes in the heat treatment of steels.

Yielding of imperfect single crystals depends largely on the orientation of the crystal relative to the applied stress, the operative slip systems and the friction provided by the lattice to the motion of dislocations. In contrast, individual crystals within a polycrystalline aggregate are constrained by their surroundings and there are discontinuities presented to the motion of dislocations at the junctions between crystals, i.e., the grain boundaries. Macroscopic yielding of a polycrystalline aggregate is therefore defined as the point where yield can be stimulated in adjacent grains when a particular crystal is induced to deform so that plasticity can be propagated. The grain size effect on the yield stress can therefore be explained by assuming that a dislocation source operates within a crystal, causing dislocations to move and eventually to pile-up at the grain boundary. The pile-up represents a concentration of stress that extends beyond the grain boundary into the surrounding crystals.

The number of dislocations (n) that can participate in a pile up for an applied stress σ is proportional to the grain size measured as a mean lineal intercept $\overline{L}$, basically scaling with the size of the unobstructed slip planes. It also scales with the stress, which causes the dislocations in the pile up to be closer to each other as the stress is increased. However, the stress of interest is the applied stress minus the friction stress σ_0 that opposes the motion of a dislocation in a lattice. Therefore, $n \propto \overline{L}(\sigma - \sigma_0)$. This pile-up of dislocations causes a concentration of the stress rather like the situation at a crack tip. The stress concentration from the pile up, at some distance in the adjacent grain is proportional to the effective stress and the number of dislocations in the pile-up, i.e., $n \times (\sigma - \sigma_0) \equiv \overline{L}(\sigma - \sigma_0)^2$. When the stress in the adjacent grain reaches a critical value σ_c, it triggers a dislocation source in the adjacent grain and hence causes macroscopic yielding so that the applied stress σ is set to the yield strength σ_y. Therefore, $\sigma_c \propto \overline{L}(\sigma_y - \sigma_0)^2$, which on rearrangement yields $(\sigma_y - \sigma_0) \propto (\overline{L})^{-\frac{1}{2}}$.

This analysis can be used to interpret the first study carried out on pure iron by Hall and Petch, that led to the relationship between the yield stress σ_y and the grain size,

$$\sigma_y = \sigma_0 + k_y \frac{1}{\sqrt{\overline{L}}}, \tag{2.10}$$

where k_y is a constant. This type of relationship holds for a wide variety of irons and steels, as well as for many non-ferrous metals and alloys. A typical set of results for mild steel is given in Fig. 2.11, where the linear relationship between σ_y and $(\overline{L})^{-\frac{1}{2}}$ is seen.

The friction stress σ_0 represents that required to move free dislocations along the slip planes in the bcc crystals, and can be regarded as the yield stress of a very large single crystal. This stress is particularly sensitive to temperature and composition. The k_y term, which is the slope of the σ_y versus $(\overline{L})^{-\frac{1}{2}}$ plot, is not to sensitive to temperature, as shown in Fig. 2.11, nor to the composition and strain rate.

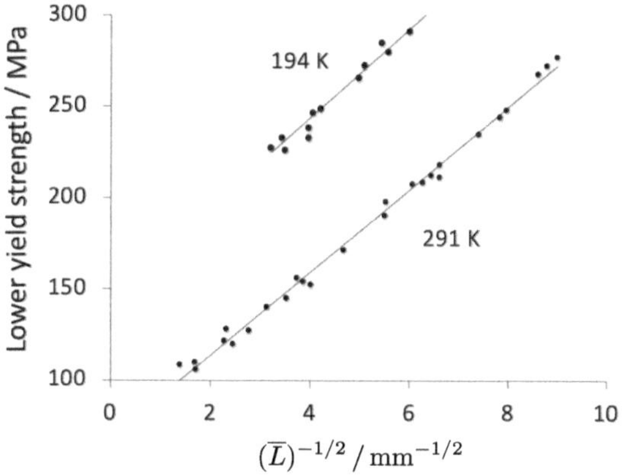

Figure 2.11 Dependence of the lower yield stress of mild steel on grain size. The slopes of the lines are similar in spite of the change in test-temperature, indicating that k_y remains constant over this temperature range. Data from [34].

In line with the Cottrell–Bilby theory of the yield point, involving the breaking away of dislocations from interstitial carbon atmospheres, k_y is referred to as the unpinning parameter. However, the insensitivity of k_y to temperature suggests that unpinning rarely occurs, and emphasises the theory that new dislocations are generated at the yield point. This is consistent with the theories explaining the yield point in terms of the movement of new dislocations, the velocities of which are stress dependent (Section 2.3.2).

In practical terms, the finer the grain size, the higher the yield stress and, as a result, in modern steel working much attention is paid to the final ferrite grain size. While a coarse grain size of $\overline{L} = 0.25$ mm $[(\overline{L})^{-\frac{1}{2}} = 2\,\text{mm}^{-1/2}]$, gives a yield stress in mild steels of around 100 MPa, grain refinement to $\overline{L} = 0.0025$ mm $[(\overline{L})^{-\frac{1}{2}}) = 20\,\text{mm}^{-1/2}]$, raises the yield stress to over 500 MPa, so achieving grain sizes in the range 2-10 μm is extremely worthwhile. Developments in rolling practice and the addition of low concentrations of particular alloying elements to mild steels have resulted in dramatic improvements in the mechanical properties of this widely used engineering material (Chapter 9).

2.5.2 Nanostructured steels

Modern technologies allow steels to be made routinely and in large quantities with grain sizes of about 2-10 μm. Limited manufacturing processes, generally involving severe thermomechanical treatments, have been developed to achieve nanostructured ferrite grains in steel with a size in the range 20–100 nm. To understand the behaviour at such incredibly small crystal sizes, it is pertinent to examine the volume fraction V_V^B of material occupied by the boundaries:

$$V_V^B \simeq \delta \times 2/\overline{L}, \tag{2.11}$$

where $\overline{L}$ is the mean linear intercept defining the grain size, so that $2/\overline{L}$ gives the amount of grain surface per unit volume, and δ is the half-thickness of the boundary layer. Clearly, the fraction of atoms located at the grain surfaces is very large in the nanostructured materials, facilitating diffusional processes such as dislocation absorption and grain sliding (Fig. 2.12a).

The Hall–Petch equation (2.10) appears to fail (for some metals other than iron) when the equiaxed grain size reaches the nanostructure scale less than about 20 nm. Instead, an *inverse Hall–Petch* relation is observed, in which the material softens as the grain size is refined. One explanation for this behaviour is that the probability (p_{dis}) of dislocations are absorbed by grain boundaries increases with decreasing grain size. So a modified form of the Hall–Petch equation is proposed:

$$\sigma_y = \sigma_0 + \underbrace{k_y'(1 - p_{\text{dis}})}_{\text{slope}}(\overline{L})^{-\frac{1}{2}} \tag{2.12}$$

The absorption of the dislocations reduces the number that can participate in pile-ups and hence as p_{dis} increases, the slope of the plot of σ_y versus $(\overline{L})^{-\frac{1}{2}}$ becomes a function of $\overline{L}$, with negative values when the grain size is small enough [36].

It must be emphasised, however, that inverse Hall–Petch effect may not be real, but rather an artifact of making samples with small grain sizes and other experimental and interpretation difficulties highlighted in a recent review [37]. In any event, the inverse Hall–Petch equation is not encountered in practice when considering any commercial steel, but there are other significant consequences of very fine grain sizes. Although the nanostructured steels are strong, they tend to exhibit unstable plasticity after yielding (Fig. 2.12b). The plastic instability occurs in both tension and in compression testing, with shear bands causing failure in the latter case. It is as if the capacity of the material to work harden following yielding diminishes. The consequence is an unacceptable reduction in ductility as the grain size is reduced in the nanometre range. At

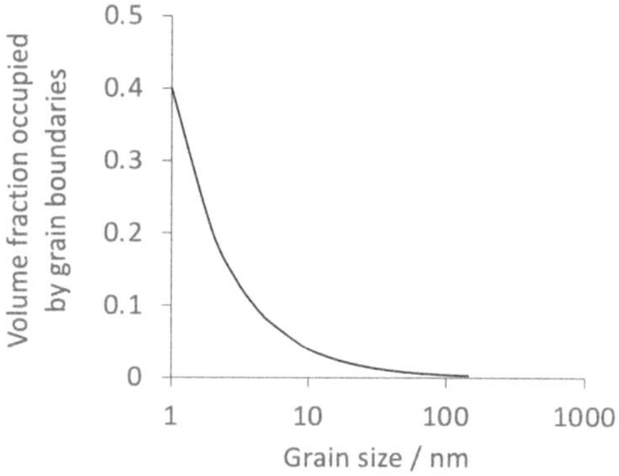

(a)

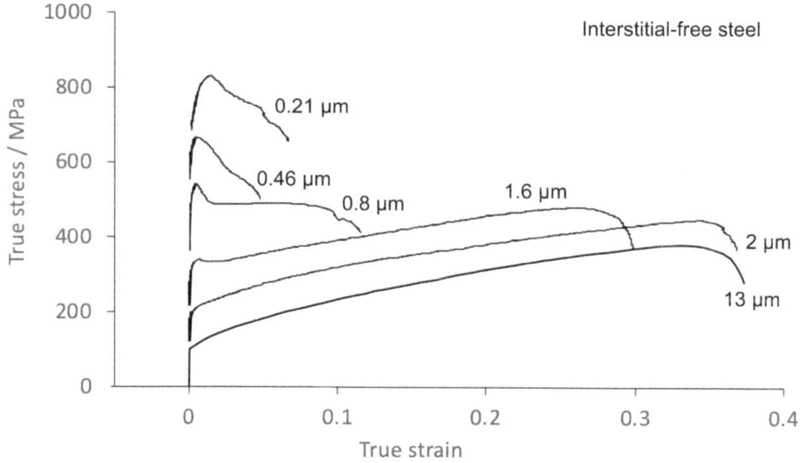

(b)

Figure 2.12 (a) The volume fraction of grain boundary as a function of the grain size ($\overline{L}$), with $\delta = 0.2$ nm, calculated using Eq. (2.11). (b) Loss of ductility with increase in strength due to drastic reduction in grain size in "interstitial-free" iron. Adapted using selected data from Tsuji et al. [35].

very fine grain sizes, the conventional mechanisms of dislocation multiplication fail because of the proximity of the closely space boundaries. It then becomes impossible to accumulate dislocations during deformation. Grain boundaries are also good sinks for defects. This would explain the observed inability of nanostructured materials to work harden. One way to overcome this difficulty is described in Chapter 14.

The difficulty that nanocrystalline grains have in deforming by a dislocation mechanism is highlighted in some interesting experiments [38], where nanocrystals of ferrite were forced to deform in shear. The crystals underwent a shear transformation into austenite.

2.6 Dispersion strengthening

In all steels, there normally is more than one phase present, and indeed it is often the case that several phases can be recognised in the microstructure. The polycrystalline matrix, which usually is ferrite or austenite strengthened by grain boundaries, dislocations and solutes in solution, is further strengthened by controlling the dispersions of the other phases in the microstructure. The most common other phases are carbides, formed as a result of the low solubility of carbon in α-iron (Fig. 2.13). In plain carbon steels, the carbide normally is Fe_3C (cementite), which can occur in a wide range of morphologies, from a coarse apparently-lamellar form in pearlite to fine, plate-like or spheroidal precipitates in tempered martensite. In alloy steels, the same range of structures is encountered, except that in many cases, the iron-rich carbide can be replaced by alloy carbides such as Mo_2C that are thermodynamically more stable. Other dispersed phases include nitrides, intermetallic compounds and in cast irons, graphite, which causes a deterioration in strength [39]. There are special alloys made by mechanical alloying that contain deliberate fine-dispersions of extremely stable oxides that enable the production of unusual grain structures during recrystallisation (e.g., [40]).

Most dispersions lead to strengthening, but they often can have adverse effects on ductility and toughness. In the case of fine dispersions of ideally small spheres randomly dispersed in a matrix, there are well-defined relationships between the yield stress, or initial flow stress, and the parameters of the dispersion. The simplest is that due to Orowan and Ashby, relating the change in the shear yield stress necessary to force a dislocation line between non-deforming spherical particles to be:

$$\Delta\tau \approx \frac{0.81\mu b}{2\pi(1-\nu)^{1/2}(L_p - 2r_p)} \ln\left\{\frac{2r_p}{b}\right\}, \tag{2.13}$$

where L_p is the average centre-to-centre spacing of adjacent particles on a slip plane; this differs from the interparticle-spacing calculated for a three-dimensional random distribution of particles by a factor of 0.81. Since in this model the dislocations do not penetrate and pass through the particles, the effective distance through which the dislocation can bow is $(L_p - 2r_p)$. The dislocation core-radius is taken to equal the magnitude of the Burgers vector b and r_p is the particle radius. When a dislocation bows around a particle, the bulging dislocation lines on either side of the particle form a dipole since they have opposite signs; the term $\ln\{2r_p/b\}$ accounts for the reduction in the stress required due to this dipole effect. The volume fraction of the particles features indirectly because at constant size, it would lead to a reduction in the particle spacing [43–45]. The yield stress increment due to particles therefore varies inversely as the spacing between the particles at constant r_p. It follows that if the dispersion is coarsened by further heat treatment, the strength of the alloy decreases.

We have seen in Fig. 2.13 that the dispersoids are not always spherical and may not be randomly dispersed. They may also prefer to be on certain crystallographic planes where the fit between the precipitate lattice and matrix is optimised. In such cases, the parameters L_p and r_p can be modified to account for specific geometries and arrangements prior to substitution into Eq. (2.13), [46].

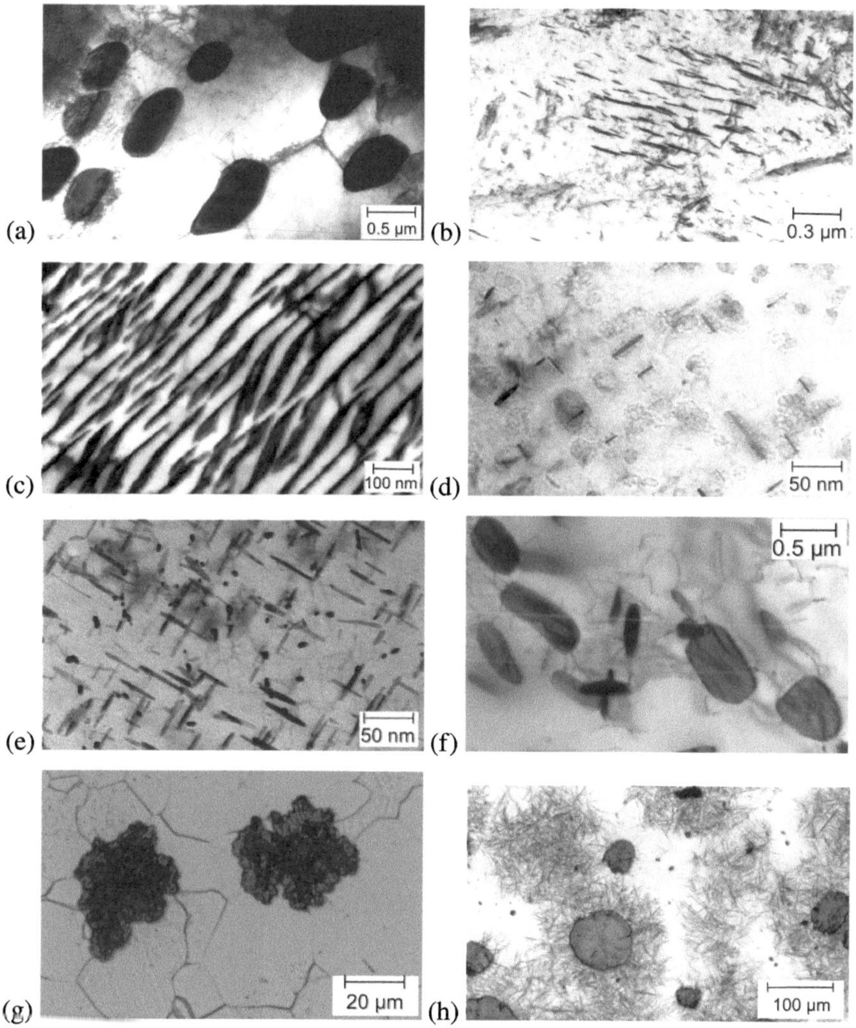

Figure 2.13 A few of the rich variety of dispersoids that help achieve the myriad of properties possible in steels. (a) Spheroidised cementite in a severely tempered high-carbon steel. (b) Fine platelets of cementite in lightly tempered martensite. (c) Lamellae of cementite in fine pearlite (Jaramillo et al. [41], reproduced with the permission of Elsevier). (d) Three orientations of V_4C_3 in secondary hardened steel, one set, the discs, lying in the plane of the image. (e) Needles of molybdenum carbide in secondary hardened steel - the dot-like precipitates are sections of needles normal to the image plane (d,e, courtesy of S. Yamasaki). (f) Fe_2Nb intermetallic precipitates in austenitic stainless steel (Yamamoto et al. [42], with permission from Elsevier). (g) Graphite in annealed steel (courtesy of C. Hulme-Smith). (h) Graphite in austempered ductile cast iron (courtesy of M. A. Yescas-Gonzalez).

Perhaps the most familiar structure in steels is that of the eutectoid pearlite, usually approximated as a lamellar mixture of ferrite and cementite. In the context of dispersions, the cementite is an integral part of the formation of pearlite rather than a phase that is induced to precipitate by heat treatment or other stimuli. It also has dramatically different mechanical properties; being very hard, it can be assumed that the initiation of plasticity during deformation happens first in the ferrite. The cementite lamellae therefore play the role of formidable barriers to the transmission of slip between the lamellae of ferrite and into other colonies of pearlite. The lamellar spacing can be varied over wide limits, and again the strength is sensitive to such changes (Chapter 3). When the coarseness of the pearlite is represented by a mean uninterrupted free ferrite path in the pearlitic ferrite, it has been shown that the flow stress is related to the mean free path by a Hall–Petch type relationship.

There has been much discussion of whether other relationships fit the data equally well or are better justified from a fundamental point of view, but detailed analysis of all the models and the large amount of data available shows the Hall–Petch equation is the most plausible representation of the proof strength of fully pearlitic steels, assuming that deformation begins in the much softer ferrite [47]. The friction stress deduced by the application of the Hall–Petch relation is physically reasonable, whereas that is not the case when other relationships are considered. Some of the problems perceived originally with the Hall–Petch equation in previous work are associated with overfitting the data and/or the use of limited data. An interesting discovery is that ferrite in interstitial-free iron is, at the same length scale, stronger in yield than the ferrite within pearlite, Fig. 2.14; its sensitivity to grain size is also much greater than that of pearlite to interlamellar spacing.

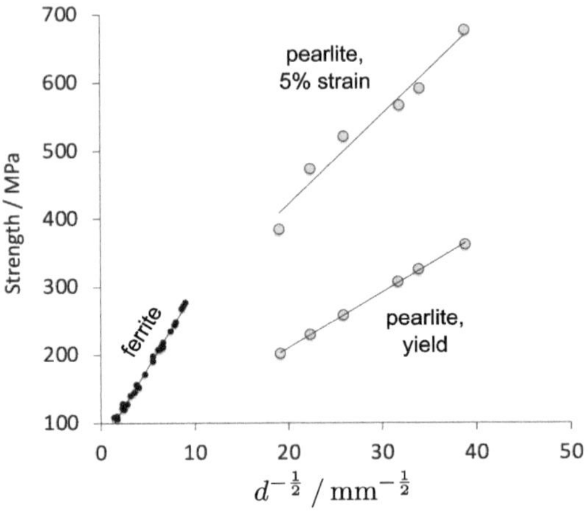

Figure 2.14 Dependence of the flow stress on the mean free path within the ferrite that is a constituent of pearlite (data from [48]). For comparison, data for ferrite grains from Fig. 2.11 are also plotted. The term d therefore represents the ferrite grain size in the latter case, but the mean free path within the ferritic component of pearlite for the remaining data.

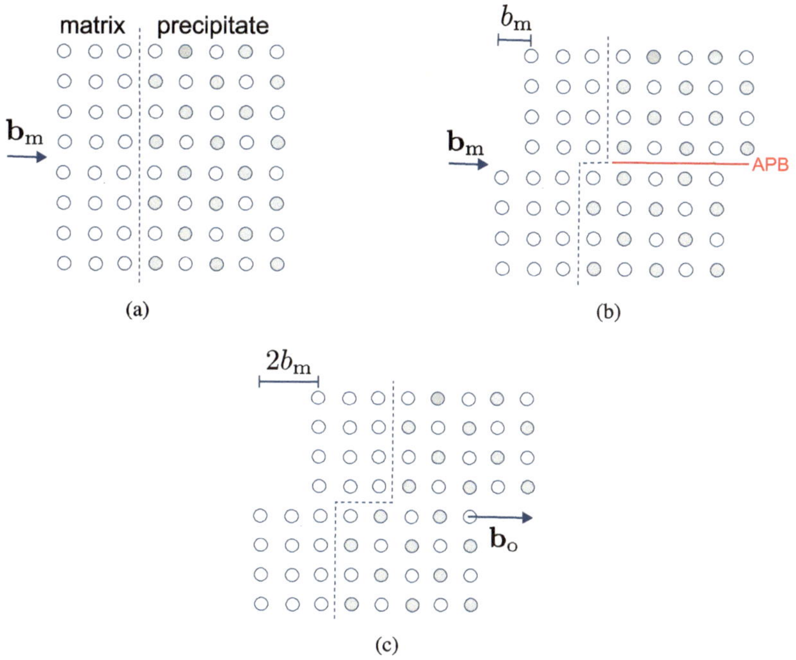

Figure 2.15 Two-dimensional representation of the deformation of an ordered precipitate coherent with the matrix. (a) The dashed line is the coherent interface between the matrix and precipitate prior to deformation. (b) The situation after the passage of a matrix dislocation with Burgers vector $\mathbf{b}_\mathrm{m}$, leading to the creation of an anti-phase domain boundary (APB) in the precipitate. The step due to slip is of magnitude b_m. (c) After the passage of $\mathbf{b}_\mathrm{m} + \mathbf{b}_\mathrm{m} \equiv \mathbf{b}_\mathrm{o}$ dislocations, with order restored in the precipitate. The step due to slip is of magnitude $2b_\mathrm{m}$.

2.7 Order strengthening

A dislocation moving through an ordered crystal ideally causes displacements that do not disrupt the ordered sequences of atoms. This means that its Burgers vector $\mathbf{b}_\mathrm{o}$ must be lattice vector of the ordered structure. However, when ordered precipitates are enclosed in a disordered matrix, deformation may cause a matrix dislocation with Burgers vector $\mathbf{b}_\mathrm{m}$ to penetrate the precipitate, Fig. 2.15. If $\mathbf{b}_\mathrm{m}$ is not a lattice vector of the ordered crystal, then its passage displaces the part of the crystal above its slip plane into a position that disrupts the ordered sequence of atoms. This leaves a two-dimensional trailing fault parallel to the slip plane, known as an *anti-phase domain boundary*. The energy required to create the boundary thus makes it more difficult for matrix dislocations to penetrate the precipitate, leading to the phenomenon of order strengthening. For this process, as illustrated in Fig. 2.15, to be feasible, there clearly must exist a level of crystallographic continuity between the matrix and the precipitate, in terms of the crystal structure, the orientation relationship between the two phases and hence a sufficient level of coherency at the precipitate-matrix interface. There are austenitic steels that satisfy these criteria, with the precipitation of ordered γ'-$Ni_3(Al, Ti)$. The orientation relationship with the austenite is "cube-cube", i.e., $[100]_\gamma \parallel [100]_{\gamma'}$ and $[010]_\gamma \parallel [010]_{\gamma'}$; both phases also have a similar lattice parameter, thus ensuring the required level of coherence to permit the passage of matrix dislocations into the precipitate.

A typical chemical composition for steel strengthened by ordered precipitates, is Fe-0.06C-16Cr-24Ni-0.4Al-2.5Ti wt% [49,50] with the Al/Ti ratio adjusted to give optimal fit between the two phases at the appropriate service temperature. The austenitic matrix has a face-centred cubic lattice with a random distribution of atoms at the lattice points, whereas the γ' has a primitive cubic lattice, Fig. 2.16a,b. A matrix dislocation with $\mathbf{b}_\mathrm{m} = \frac{1}{2}\langle 110\rangle_\gamma$ is, as shown on Fig. 2.16c, held back by the γ' particles because of the work done in creating anti-phase domain boundaries, with the stress required for such motion being proportional directly to the anti-phase domain boundary energy [51], which is $\approx 0.2\,\mathrm{J\,m^{-2}}$ for the alloy considered here [50]. A second dislocation on the same slip plane can glide and eliminate those faults, so matrix dislocations will tend to move in pairs known as *superdislocations* that have a net Burgers vector equal to $\mathbf{b}_\mathrm{o} = \langle 110\rangle_{\gamma'}$, which is a lattice vector of the γ'.

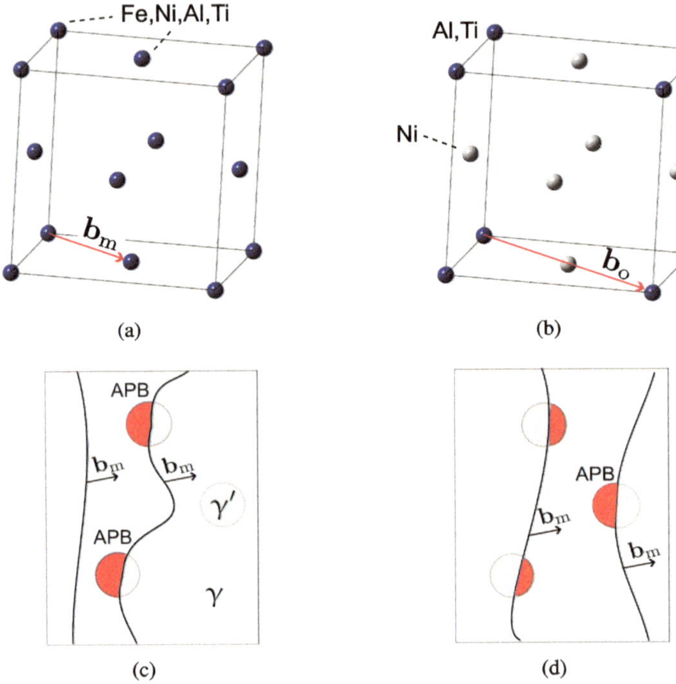

Figure 2.16 (a) Face-centred cubic unit cell of disordered γ with atoms randomly mixed on lattice sites. (b) Ordered γ'-Ni_3(Al, Ti). Other elements are present in negligible concentrations [52]. (c) Pair of dislocations, each with Burgers vector $\mathbf{b}_\mathrm{m}$. Since $\mathbf{b}_\mathrm{m} \neq \mathbf{b}_\mathrm{o}$, the lead dislocation leaves a trail of APBs within the precipitates, which then are eliminated when the second dislocation passes through the cut γ' particles, as in (d).

Order-hardened austenitic stainless steels have applications in superchargers, gas turbines, nuclear reactor-vessel bolts, etc., because of their combination of good corrosion resistance, ability to maintain strength at high temperatures, and ease of fabrication. After quenching from the austenitic state, the alloy usually is heat treated at about 700°C for 16 h so that about 0.1 volume fraction of fine γ' particles form. The 0.2% proof strength is 650 MPa at ambient temperature, decreasing to about 600 MPa at 700°C. The microstructure following this heat treatment and subjecting the sample to 2% plastic deformation is illustrated in Fig. 2.17 – it is obvious that dislocations pair, consistent with the mechanism of order hardening.

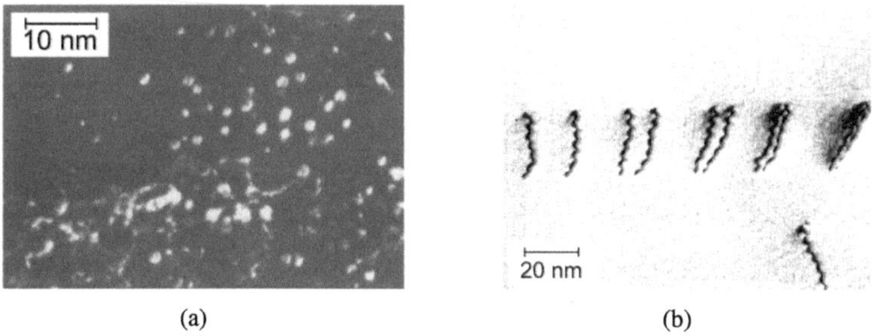

Figure 2.17 (a) Dark-field transmission electron micrograph illustrating the γ' particles. (b) Tendency for dislocations to pair. Images reproduced with the permission of Elsevier from Thompson and Brooks [50].

2.7.1 Aluminides

Alloys of iron and aluminium have the advantage of a reduced density relative to iron together with oxidation resistance due to the formation of an alumina scale. The system forms a variety of ordered phases that have been investigated as candidates for elevated temperature applications, particularly FeAl and Fe_3Al; a part of the iron-aluminium phase diagram is illustrated in Fig. 2.18, which shows in particular that they can have a range of compositions so cannot strictly be defined as compounds, but the perfection of atomic order achieved must depend on the ratio of Fe:Al.

The crystal structures of stoichiometric FeAl and Fe_3Al are illustrated in the form of projections on to their respective (001) planes in Fig. 2.19. If, for example, the actual composition of FeAl is better represented by $Fe_{1+x}Al_{1-x}$, then a fraction $(1+x)^{-1}$ of iron atoms are on the aluminium sites [55]. Properties such as lattice parameters, vacancy concentrations, diffusion coefficients, magnetism and hardness were found to vary with the deviation from stoichiometry.

The ordered intermetallics in the Fe-Al system have not been successful in commercial applications [56]. They are not intrinsically brittle, but have poor ductility due to reaction of the aluminium in the alloys with moisture in the environment, which liberates nascent hydrogen that in turn causes embrittlement [57]. It is not clear how this problem can be resolved in a reliable way. Furthermore, mechanical properties such as yield strength, creep rupture strength and toughness are not particularly outstanding compared to commercially available steels.

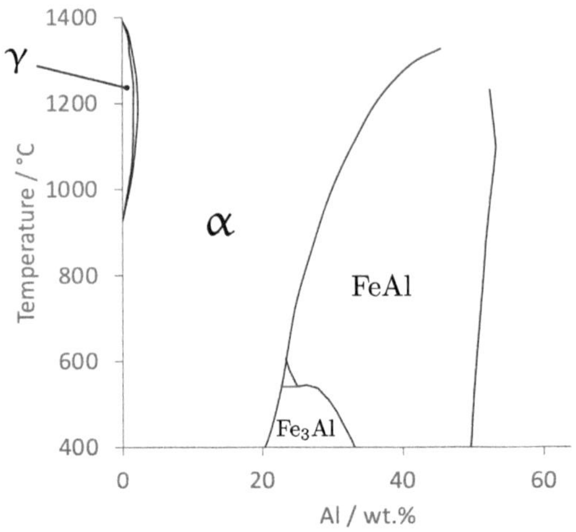

Figure 2.18 Part of the iron-aluminium phase diagram, presented in simplified form to illustrate the wide range of compositions associated with the ordered compounds. Data from Stein and Palin [53,54].

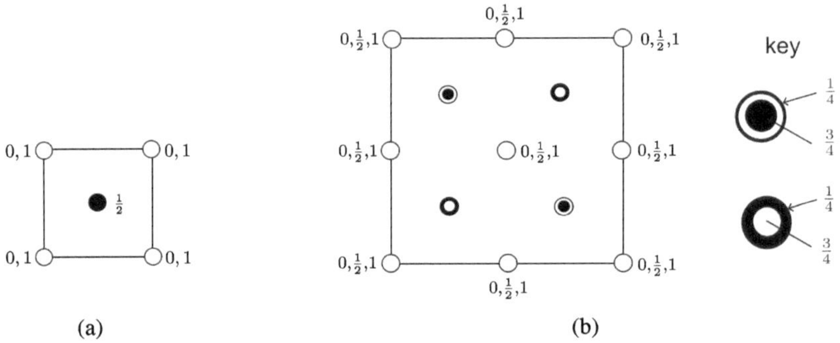

Figure 2.19 Projection along the z-axis of ordered lattices in the Fe-Al system, with iron atoms presented as white and aluminium coloured black. Each number represents the location of an atom as a fraction of the lattice parameter along the z-axis of the cell. (a) FeAl, space group $Pm3m$. (b) Fe_3Al, space group $Fm3m$. The unit cell is about double the size of the one for FeAl.

2.8 Overall strength

Strength in steels arises from several phenomena, which usually contribute collectively to the observed set of mechanical properties. The heat treatment of steels is aimed at adjusting these contributions to achieve the required balance of mechanical properties. Fortunately, the γ/α change allows a great variation in microstructure, so that a wide range of mechanical properties can be obtained even in plain carbon steels. The additional use of metallic alloying elements, primarily as a result of their influence on the transformation, provides an even greater control over microstructure, with consequent benefits in the mechanical properties.

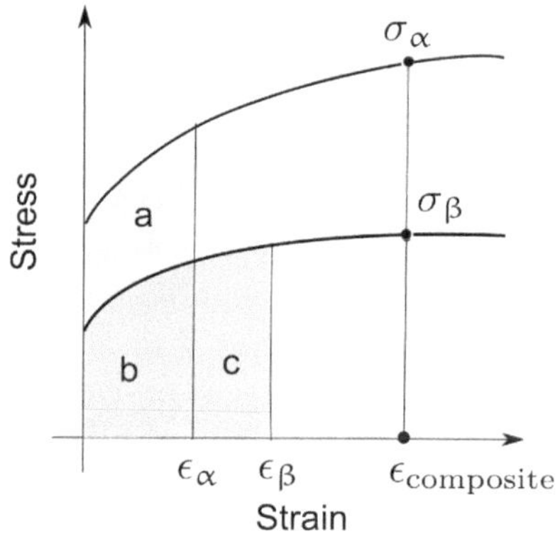

Figure 2.20 An illustration of the equal work and equal plastic strain methods of estimating the deformation behaviour of mixtures of phases.

We have not discussed thus far the strengthening and deformation behaviour of mixed microstructures, such as the dual-phase steels, which consist of ferrite and harder martensite. This can radically alter the stress versus strain behaviour, with the deformation being heterogeneous on a microscopic scale with complex constraint and compatibility issues governing plasticity. Some of these aspects of mixed microstructures are described in Chapter 16 as one of the two case studies. However, there are some general and useful approximations that often capture the essence of the problem without introducing excessive complexity [58]. The first requirement is the knowledge of the constitutive equations of each of the phases, i.e., the stress-strain behaviour of the single phase in isolation. Fig. 2.20 shows such curves for phases α and β with volume fractions V_V^α and V_V^β, respectively. In the equal work method, the strain experienced by each phase is different, determined by the areas $a + b = b + c$ (Fig. 2.20), so the softer phase is deformed to a greater extent. The process must be implemented incrementally so that the work hardening characteristics of each phase are properly accounted for. At any stage of deformation, the strength $\sigma_{composite}$ of the composite microstructure is given by

$$\sigma_{composite} = V_V^\alpha \sigma_\alpha\{\epsilon_\alpha\} + V_V^\beta \sigma_\beta\{\epsilon_\beta\}, \tag{2.14}$$

where $\sigma_\alpha\{\epsilon_\alpha\}$ is the stress in α at a plastic strain ϵ_α. In contrast, the equal strain model has the composite strain $\epsilon_{composite}$ in all phases and Eq. (2.14) applies in the calculation of the overall stress. The models clearly are approximate and their applicability depends on many factors, for example, the extent of the difference in the strengths of the component phases and whether the properties of the pure phases vary with volume fraction within the composite. This is particularly the case when solute partitions between the phases during transformation or when the properties of a phase change due to deformation-induced phase transformation.

Similar difficulties arise when combining the contributions of different strengthening mechanisms to generate the overall strength of a single phase. The extent of dislocation strengthening may depend on the segregation of carbon to dislocations, so that the solution hardening term for carbon becomes uncertain. Assuming that the individual strength components (σ_i) may simply be summed linearly can lead to a large overestimation of strength [59]. A non-linear summation may then be justified, $\sigma^k = \sum \sigma_i{}^k$, where the exponent k has some fundamental meaning but usually is applied as a fitting constant, e.g., [60,61].

2.9 Some practical aspects

The presence of a sharp yield point in a steel can be detrimental to its behaviour, e.g., when used for pressing, where complex patterns of Lüders bands can produce rough surfaces (Fig. 2.4) and lead to poor workability. The severity of the yield point is related directly to the amount of carbon and nitrogen in solution in the ferrite, so that steps taken to reduce these concentrations are helpful. Unfortunately, yield points can be obtained with very low concentrations of carbon and nitrogen, making it impracticable in industrial conditions to obtain steels below these limits. However, any heat treatment that reduces the concentration of the interstitial solute that is in solid solution is beneficial, e.g., slow cooling in annealing treatments. The yield point can be more reliably eliminated prior to working by a small amount of cold rolling (0.5-2%), referred to as *temper rolling*, with the added benefit sometimes of enhancing flatness and reducing surface roughness. The rolling deformation is often referred to as a "skin pass" because the small deformation is localised to the steel surfaces. As both nitrogen and carbon diffuse appreciably in ferrite at ambient temperatures, it is desirable to fabricate steels soon after the temper rolling.

While carbon and nitrogen can both cause strain ageing and consequently a yield point, the higher solubility of nitrogen in ferrite means that it provides the greater problem in steels used for deep drawing and pressing. Steps are taken during steelmaking to keep the nitrogen level down, but to minimise its effects, the easiest solution is to add low concentrations of strong nitride formers such as aluminium, titanium or vanadium, which reduce the nitrogen in solution to very low concentrations.

The occurrence of strain ageing can be beneficial to structural steels used for structural purposes by increasing yield stress and tensile strength. Furthermore, the fatigue properties are improved, both at room temperature and in the range up to 350°C. The existence of a well-defined fatigue limit in steels, i.e., a fatigue stress limit below which failure does not occur, has been linked to the occurrence of strain ageing during the test, but very pure iron also shows this behaviour. It should be emphasised that even in a relatively simple low carbon steel, the strength arises not only from these effects of carbon and nitrogen, but also from the solid solution hardening of elements such as silicon and manganese, and potentially from the refinement of the ferrite grain size by various means.

Bake-hardening steels [62,63] take advantage of the migration of carbon atoms to dislocations. They essentially have a ferritic microstructure with less than 25 ppm of carbon in solid solution. The steels are used in the manufacture of automobile bodies, which after forming into shape are painted. When the paint is baked in the temperature range 150-200°C, the dissolved carbon migrates to pin any free dislocations introduced during the forming operation. The increase in strength illustrated in Fig. 2.21

is significant, given that the tensile strength of the steel is only about 300 MPa. It is not surprising that the strengthening is greatest at the temperatures where carbon migration is possible.

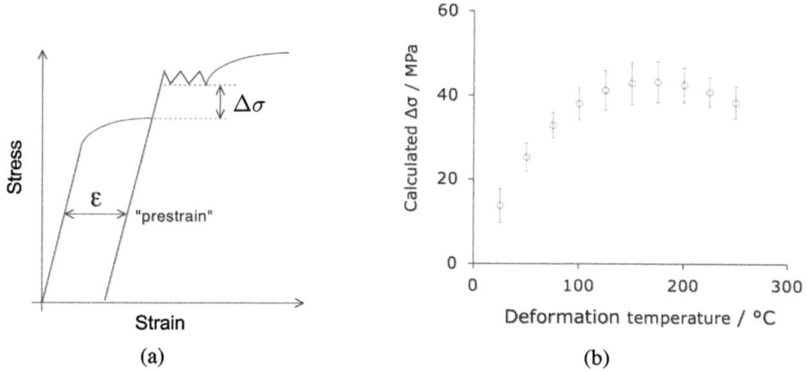

Figure 2.21 (a) Schematic illustration of the increase in strength, $\Delta\sigma$, following plastic deformation and bake-hardening. (b) Dependence of $\Delta\sigma$ on the deformation temperature [63].

2.10 Limits to strength

Strength is not always the most useful entity. It may not in fact be safe to load an engineering structure to the full capability of the material. To illustrate this and some other limits of scale, a comparison is presented here between the potential strength of steel and that of carbon nanotubes and graphene, which are the subject of much contemporary discussion.

Before embarking on a detailed discussion, it is worth noting that there are fundamental differences in the bonding between atoms in a metal and in objects such as the nanotubes or graphene. A metal can slip without any consequence on the bonding – i.e., there are no broken bonds as a result of the plastic deformation implicit in slip. Even the defects created within the metal are tolerated until a point is reached where fracture occurs. It is important to note that the work done during plastic deformation is absorbed by the metal or emitted as heat, the plastic strain is not recovered at fracture. It is this absorption of energy during deformation that makes metals safe as engineering materials.

In contrast, a nanotubes or graphene have bonds that stretch elastically and then break. All the energy is released when the elastic strain goes to zero, no energy is absorbed apart from the minute quantity associated with the permanent fracture of the bonds. The concept of toughness does not therefore apply. The material will simply recoil violently on fracture; these materials are intrinsically brittle and there is nothing that can be done to make them less so, given the nature of the covalent bonding involved. It should not therefore be surprising that there has been no application whatsoever of graphene or carbon nanotubes in structural engineering.

2.10.1 Theoretical strength

The strength of crystals increases sharply as they are made smaller. This is because the chances of avoiding defects become greater as the volume of the sample decreases. In the case of metals, imperfections in the form of dislocations can facilitate shear at much lower stresses than would be the case if whole planes of atoms had to collectively slide across each other. Defects are very difficult, if not impossible, to avoid, but supposing that perfection can somehow be achieved, the strength in the absence of defects is equivalent to that of an ideal crystal.[2]

In an ideal crystal, the tensile strength $\sigma_t \simeq 0.1E$ where E is the Young's modulus. The corresponding ideal shear-strength is $\sigma_s \simeq b\mu/2\pi d_{hkl}$, where μ is the shear modulus, b is a repeat period along the displacement direction and d_{hkl} is the spacing of the slip planes. For ferritic iron, $\mu = 80.65$ GPa and $E \simeq 208.2$ GPa. It follows that the ideal values for tensile and shear strength should be about 21 and 11 GPa, respectively. In fact, tensile strengths approaching the theoretical values were achieved by Brenner as long ago as 1956 (Fig. 2.22a) during the testing of whiskers of iron with diameters less than 2 μm. It is interesting that these stress levels fall out of the regime where Hooke's law applies (Fig. 2.22b).[3]

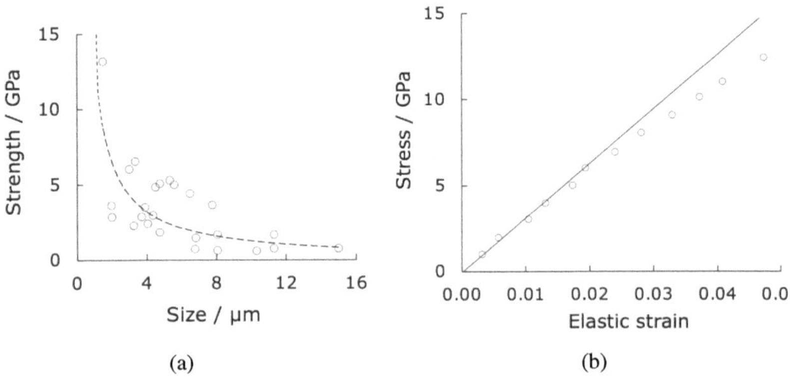

Figure 2.22 (a) The tensile strength of whiskers of iron. (b) Non-linear elasticity at large stresses. Data from Brenner [65,66].

The whisker experiments showed that the strength decreases sharply with increasing sample size (Fig. 2.22a), because the chances of finding defects increases with increasing sample size. It was therefore recognised seven decades ago that it is not wise to rely on perfection as a method of designing strong materials, although it remains the case that incredible strength can be achieved by reducing dimensions, in the case of iron, to a micrometre scale.

[2] In the case of graphene, this is referred to as the intrinsic strength.

[3] In fact, the ideal tensile strength of iron may be about 14.2 GPa along its weakest crystallographic direction of ferrite, because by stretching elastically, the bcc cell of ferrite at that stress transforms into austenite [64]. However, the strength of iron in its fcc form may approach the ideal estimate of 21 GPa.

In this context, we now examine the notion that large-scale engineering structures can be designed using very long carbon nanotubes [67] and address claims regarding the strength of graphene.

2.11 Hundreds of times stronger than steel

Steel is one of many materials available to create new technologies and concepts. It is, in the scientific and popular literature, treated as the "gold standard" to which new developments are compared. The comparisons are, however, either incorrect, misleading or based on flawed science.

Spider thread has a tensile strength of about 700 MPa, but a density of only $1.3\,\mathrm{g\,cm^{-3}}$ giving it a specific strength that is six times greater than steel of identical tensile strength [68]. Spider's thread is typically 10 μm in diameter. It is in fact possible to obtain commercially, steel wire of similar diameter that has a strength of 5.5 GPa, i.e., $\approx 8 \times 700\,\mathrm{MPa}$ and a greater specific strength than the thread [69]. It is claimed that spider thread is tougher than steel on the basis of the area under the stress versus strain curve. Thus, with a strength of 700 MPa and elongation of 30%, the elastic energy per unit volume is $105\,\mathrm{MJ\,m^{-3}}$. The 5.5 GPa wire, at an elastic strain of 0.026, would absorb at least $145\,\mathrm{MJ\,m^{-3}}$ before fracture.

Carbon nanotubes can be imagined to be constructed from sheets of graphene consisting of sp^2 carbon arranged in a two-dimensional hexagonal lattice (Fig. 2.23). The sheets, when rolled up and with the butting edges appropriately bonded, are the nanotubes, which may or may not be capped by fullerene hemispheres. The actual form can be complex, e.g., with occasional pentagonal rings of carbon atoms instead of hexagonal to accommodate changes in shape.

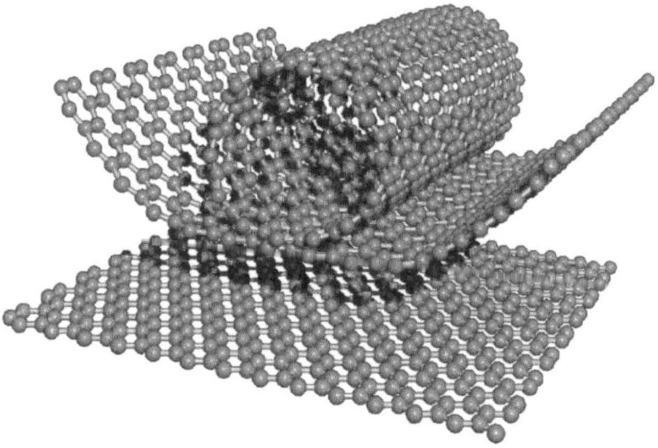

Figure 2.23 Schematic diagram showing how a sheet of graphene might be rolled to form a tube (courtesy of M. Endo).

The breaking strength of such a tube has been estimated to be an extraordinary 130 GPa; this number is astonishing and has led to many exaggerated comparisons against steel. However, this is the strength of an invisibly small nanotube. Larger tubes contain defects, which lead to a gross deterioration of strength, rather like the behaviour of whiskers of iron. Some of these defects are already present at equilibrium and hence are unavoidable. For example, it is known that metals contain an equilibrium concentration of vacancies. The enthalpy change associated with the formation of a vacancy opposes its existence, whereas the change in configurational entropy due to the formation of a vacancy favours its formation.

The total change in free energy on forming n vacancies in a crystal with N atoms is given by [70]:

$$\Delta G = n\Delta g - kT[(N+n)\ln\{N+n\} - N\ln\{N\} - n\ln\{n\}],$$

where k is the Boltzmann constant, T is the absolute temperature, N is the number of atoms, $\Delta g = \Delta h - T\Delta s$, Δh is the enthalpy of formation of one vacancy and Δs is the entropy of formation of a vacancy excluding any contribution from configurational entropy, which is the second term in the equation. The equilibrium mole fraction of vacancies (x) is obtained by writing $\partial\Delta G/\partial n = 0$ giving:

$$x = n/N \simeq \exp\{-\Delta g/kT\}.$$

On this basis, taking the energy of a vacancy in a nanotube as 7 eV, and setting T to be the manufacturing temperature of the tubes (2000-4000 K), it is possible to show that a carbon nanotube strand appropriate for a space elevator, weighing 5000 kg, would contain approximately 10^{10}-10^{20} defects. It is not therefore possible to scale the dimensions of a nanotube by some 18 orders of magnitude and assume that the strength will be retained. The particular structure referred to here is the speculated space elevator to replace rocket launches, which would require carbon tube ropes that are some 120,000 km in length. Fig. 2.24 shows how the strength of a carbon nanotube collapses as a function of its size, so a sound-rope of space-elevator dimensions is impossible.

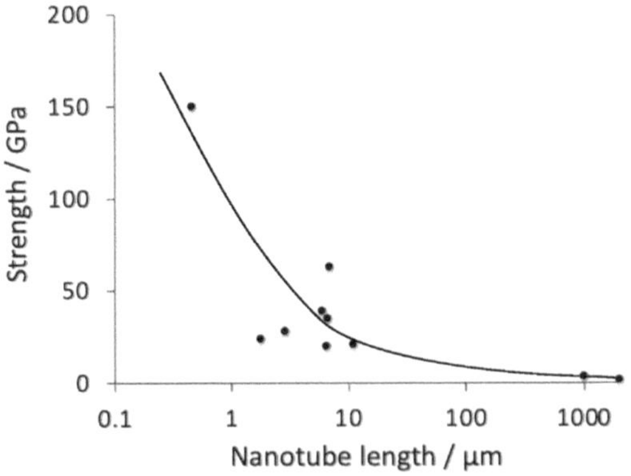

Figure 2.24 Variation in the strength of carbon nanotubes as a function of length. Data collated from [71–73].

This emphasises again that systems which rely on perfection to achieve strength inevitably fail when scaled to engineering dimensions. Indeed, there is no carbon tube that can match the strength of iron beyond a scale of 2 mm. These considerations apply also to graphene sheets.

Graphene has been claimed to be between 100-300 times stronger than steel. Lee and co-workers determined the strength of a monolayer of graphene about 1 μm in diameter by a nanoindentation method. They measured the intrinsic breaking strength of the perfect layer as $42\,N\,m^{-1}$ and converted this into a strength of 130 GPa, the same value as reported for carbon nanotubes [74], which may not be surprising given that the data in both cases refer to carbon–carbon bonds in perfect samples. Suppose that 130 GPa represents the true strength of two-dimensional graphene. Brenner [65] has measured the tensile strength of a 1.6 μm whisker to be 13.4 GPa, so the intrinsic tensile strength of iron, along its weakest crystallographic direction is likely to be 14.2 GPa [64]. If follows that pristine graphene is at best about 9 times stronger than steel. When scaled to sizes greater than the micrometre dimensions, it is likely to suffer the same fate as carbon nanotubes, i.e., lose its integrity.

Fracture

Suppose that gigatubes of carbon could be made capable of supporting a stress of 130 GPa. Would this allow for safe engineering design? One aspect of safe design is that fast fracture should be avoided; most metals absorb energy in the form of plastic deformation before ultimate fracture. Energy absorption in an accident is a key aspect of automobile safety. Carbon nanotubes are not in this sense defect tolerant; their deformation prior to fracture is entirely elastic. The stored energy density in a tube stressed to 130 GPa, given an elastic modulus along its length of $E = 1.2$ TPa is in excess of that associated with dynamite (Table 2.1). Dynamite is explosive because of its high energy density and because this energy is released rapidly, the detonation front propagating at some $6000\,m\,s^{-1}$. The speed of an elastic wave in the carbon is given by $\sqrt{E/\rho}$ where ρ is the density. In the event of fracture, the rate at which the stored energy would be released is much greater than that of dynamite (Table 2.1), meaning that fracture is unlikely to occur in a safe manner.

It follows that structures in tension, which reversibly store energy far in excess of their ability to do work during fracture, must be regarded as unsafe. Strength can only be exploited in a safe manner if the material is capable of absorbing sufficient energy during fracture.

Table 2.1 Comparison of a fully loaded carbon nanotube and dynamite [75].

	Stored energy $J\,g^{-1}$	**Detonation velocity** $m\,s^{-1}$
Dynamite	4650	6000
Carbon nanotube	5420	21,500

2.12 Ten billion times stronger than steel

When the core of a star no longer generates sufficient heat to counter its own gravity, it collapses. The consequences of the collapse depend on the size of the star. Subrahmanyan Chandrasekar in 1931, at the University of Cambridge, predicted that if the core-size of the star exceeded 1.4 times the mass of our Sun, its swan song would involve massive explosions (*supernovae*) that create either a neutron star or, if the original mass was even greater, than a black hole. Supernova explosions, incidentally, also create the heavy elements such as iron.

A neutron star is composed of tightly-packed neutrons with none of the "empty-space" characteristic of the structure of an atom with a nucleus and electrons. The tight packing and gravitational forces make the density of its core ($10^{17}\,\mathrm{kg\,m^{-3}}$) greater than that of the nucleus of an atom, though this is not the case for the outer layer, which is referred to as the crust. It turns out that the breaking strength of the crust of a neutron star is about 10^{10} times greater than that of steel, a consequence of the enormous pressure-induced density of the crust [76]. During simulated breaking, the material fails abruptly in a collective manner, unlike the gradual yielding typical of steel. The breaking strain is estimated to be about 0.1. The strength is so incredible that the crust can support mountains large enough to emit gravitational waves, which in principle are detectable on Earth.

2.13 Summary

The strengthening mechanisms described in this chapter apply to all metals and alloys. However, the yield point effect is less widespread; aluminium containing magnesium shows yielding and serrated flow in appropriate circumstances because magnesium, even though it is a substitutional solute, is quite mobile when dissolved in Al. Crystallography also matters – the yield point effect is much weaker in austenite than in ferrite because of the lower symmetry of the preferred interstice in ferrite.

Although strengthening mechanisms are well understood, it is not simple to combine their contributions to derive an overall strength. There is further complexity in dealing with multiphase alloys because the strains may not then be uniformly distributed.

When looking at combinations of strength and elongation, it simply has not been possible to develop viable steels that are stronger than $\approx$3 GPa and yet have ductility exceeding about 5%. This should not be surprising, because strength is achieved by hindering deformation, whereas sustained elongation requires a work-hardening capacity that delays plastic instabilities. It may not ever be possible to create steels that are very strong and very ductile. Nevertheless, steels stronger than 3 GPa with an elongation less than 2% are in commercial use in the form of ropes. How is it possible to tolerate such a small ductility? By twining thin strands into ropes; the failure of individual strands does not then lead to catastrophes. Such ropes are in effect steel-air composites and this principle could be exploited in developing laminated steels for the mass market. The utility of a material depends not just on its intrinsic properties, but also on how it is deployed. No one, for example, seems to have thought of using wire ropes for the crash-resistant infrastructure of automobiles.

There are advantages in creating polycrystalline steels where individual crystals are fine, as this raises both strength and toughness. But in nanostructured metals, where the crystal size is much smaller than 100 nm, can lack ductility. There are other issues that are reserved for Chapter 14. But it is pertinent here to note that claims about the phenomenal strength of various forms of carbon relative to steel are essentially an attempt to mislead.

References

[1] J.H. Kim, G. Gu, M.H. Kwon, M. Koo, E.Y. Kim, J.K. Kim, J.S. Lee, D.W. Suh, Microstructure and tensile properties of chemically heterogeneous steel consisting of martensite and austenite, Acta Materialia 223 (2022) 117506.

[2] A.S. Keh, Work hardening and deformation sub-structure in iron single crystals deformed in tension at 298 K, Philosophical Magazine 12 (1965) 9–30.

[3] T.L. Altshuler, J.W. Christian, Mechanical properties of pure iron tested in compression over the temperature range 2-293 K, Proceedings of the Royal Society A 261 (1967) 253–287.

[4] W. Bao, Z. Xiong, X. Ren, F. Wang, Effect of strain rate on mechanical properties of pure iron, Advanced Materials Research 705 (2013) 21–25.

[5] K. Conrad, On the mechanism of yielding and flow in iron, Journal of the Iron and Steel Institute 198 (1961) 364–375.

[6] J.T. Michalak, Stress during the plastic deformation of high-purity iron, Acta Metallurgica 13 (1965) 213–222.

[7] S.B. Singh, H.K.D.H. Bhadeshia, Topology of grain deformation, Materials Science and Technology 15 (1998) 832–834.

[8] Q. Zhu, C.M. Sellars, H.K.D.H. Bhadeshia, Quantitative metallography of deformed grains, Materials Science and Technology 23 (2007) 757–766.

[9] J.D. Embury, R.M. Fisher, The structure and properties of drawn pearlite, Acta Metallurgica 14 (1966) 147–159.

[10] R. Blondeau, P. Maynier, J. Dollet, B. Vieillard-Baron, Estimation of hardness, strength and elastic limit of C- and low-alloy steels from their composition and heat treatment, Mémoires Scientifiques de la Revue de Métallurgie 72 (1975) 759–769.

[11] P. Ferguson, Mechanical properties of nitrogen-ferrite, Metallurgical Transactions A 16 (1985) 45–50.

[12] W. Horvarth, W. Pranti, H. Strossnigg, E.A. Werner, Microhardness and microstructure of austenite and ferrite in nitrogen alloyed duplex steels between 20 and 500°C, Materials Science & Engineering A 256 (1998) 227–236.

[13] B.M. Wilson, W.N. Weins, Retained austenite and tooling failure case studies, in: K. Funatani, G.E. Totten (Eds.), Heat Treating, Proceedings of the 20th Conference, ASM International, Ohio, USA, 2000, pp. 566–573.

[14] D. Dumoulin, H. Louch, O.S. Hopperstad, T. Borvik, Heat sources, energy storage and dissipation in high-strength steels: experiments and modelling, European Journal of Mechanics A/Solids 29 (2010) 461–474.

[15] W. Lüders, On the manifestation of elasticity in steel-like iron rods and steel rods, and on a molecular movement observed when such rods are bent (translation from German), Dinglers Polytechnisches Journal 5 (1860) 18–22.

[16] F. Aguirre, S. Kyriakides, H.D. Yun, Bending of steel tubes with Lüders bands, International Journal of Plasticity 20 (2004) 1199–1225.

[17] J. Han, C. Lu, B. Wu, J. Li, Y. Lu, Q. Gao, Innovative analysis of Lüders band behaviour in X80 pipeline steel, Materials Science & Engineering A 683 (2017) 123–128.
[18] J. Han, J. Wang, S.M. Manladan, Y. Cai, Q. Wang, Z. Zhu, L. Zhu, L. Lu, Z. Jiang, Effect of Lüders bands by strain ageing on strain distribution, microstructure and texture evolution of high-strength pipe steel, Acta Metallurgica Sinica 34 (2021) 657–667.
[19] A.H. Cottrell, B.A. Bilby, Dislocation theory of yielding and strain ageing of iron, Proceedings of the Physics Society A 62 (1949) 49–62.
[20] F.G. Caballero, C. García-Mateo, C. Capdevila, C.G. de Andrés, Advanced ultrahigh strength bainitic steels, Materials and Manufacturing Processes 22 (2007) 502–506.
[21] R.G.A. Veiga, M. Perez, C.S. Becquart, C. Domian, Atomistic modeling of carbon Cottrell atmospheres in bcc iron, Journal of Physics. Condensed Matter 25 (2013) 025401.
[22] E.O. Hall, Yield Point Phenomena in Metals and Alloys, Plenum Press, New York, USA, 1970.
[23] S. Lee, J. Kim, S.J. Lee, B.C. de Cooman, Effect of nitrogen on the critical strain for dynamic strain aging in high-manganese twinning-induced plasticity steel, Scripta Materialia 65 (2011) 528–531.
[24] J.J. Gilman, W.G. Johnston, Dislocations in lithium fluoride crystals, Solid State Physics 13 (1962) 147–222.
[25] H. Dong, S-phase surface engineering of Fe-Cr, Co-Cr and Ni-Cr alloys, International Materials Reviews 55 (2010) 65–98.
[26] D. Wu, H. Kahn, J.C. Dalton, G.M. Michal, F. Ernst, A.H. Heuer, Orientation dependence of nitrogen supersaturation in austenitic stainless steel during low-temperature gas-phase nitriding, Acta Materialia 79 (2014) 339–350.
[27] M.P. Fewell, D.R.G. Mitchell, J.M. Priest, K.T. Short, G.A. Collins, The nature of expanded austenite, Surface & Coatings Technology 131 (2000) 300–306.
[28] Y. Sun, Z. Li, T. Bell, Low temperature plasma carburising of austenitic stainless steels for improved wear and corrosion resistance, Surface Engineering 15 (1999) 49–54.
[29] T. Sato, K. Tomomoto, Effects of the structures to impact properties for high carbon steel, Tetsu-to-Hagané 66 (1980) 231–238.
[30] R. Kirchheim, Solid solution softening and hardening by mobile solute atoms with special focus on hydrogen, Scripta Materialia 67 (2012) 767–770.
[31] S. Takeuchi, Solid solution strengthening in single crystals of iron alloys, Journal of the Physical Society of Japan 27 (1969) 929–940.
[32] R.J. Arsenault, Solid solution strengthening and weakening of B.C.C. solid solutions, Acta Metallurgica 17 (1969) 1291–1297.
[33] T.J. Rupert, Solid solution strengthening and softening due to collective nanocrystalline deformation physics, Scripta Materialia 81 (2014) 44–47.
[34] N.J. Petch, The ductile-cleavage transition in α-iron, in: B.K. Averbach, D.K. Felbeck, G.T. Hahn, D.A. Thomas (Eds.), Fracture, John Wiley & Sons, Inc., New Jersey, USA, 1959, pp. 54–64.
[35] N. Tsuji, Y. Ito, Y. Saito, Y. Minamino, Strength and ductility of ultrafine grained aluminium and iron produced by ARB and annealing, Scripta Materialia 47 (2002) 893–899.
[36] C.E. Carlton, P.J. Ferreira, What is behind the inverse Hall–Petch effect in nanocrystalline materials?, Acta Materialia 55 (2007) 3749–3756.
[37] S.N. Naik, S.M. Walley, The Hall-Petch and inverse Hall-Petch relations and the hardness of nanocrystalline metals, Journal of Materials Science 55 (2020) 2661–2681.
[38] Y. Ivanisenko, I. MacLaren, R.Z. Valiev, H.J. Fecht, First observation of a shear-induced bcc to fcc transformation in nanocrystalline ferrite, Advanced Engineering Materials 7 (2005) 1011–1014.

[39] Y. Liu, Y. Li, S. Wang, B. Zheng, D. Tao, W. Li, Effect of graphite morphology on the tensile strength and thermal conductivity of cast iron, Materials Characterization 144 (2018) 155–185.
[40] H.K.D.H. Bhadeshia, Recrystallisation of practical mechanically alloyed iron base and nickel base superalloys, Materials Science & Engineering A A223 (1997) 64–77.
[41] R.A. Jaramillo, S.S. Babu, G.M. Ludtka, R.A. Kisner, J.B. Wilgen, G. Mackiewicz-Ludtka, D.M. Nicholson, S.M. Kelly, M. Murugananth, H.K.D.H. Bhadeshia, Effect of 30 Tesla magnetic field on transformations in a novel bainitic steel, Scripta Materialia 52 (2004) 461–466.
[42] Y. Yamamoto, M. Takeyama, Z.P. Lu, C.T. Liu, N.D. Evans, P.J. Mziasz, M.P. Brady, Alloying effects on creep and oxidation resistance of austenitic stainless steel alloys employing intermetallic precipitates, Intermetallics 16 (2008) 453–462.
[43] E. Orowan, Discussion of internal stresses, in: Symposium on Internal Stresses in Metals and Alloys, Institute of Metals, London, U.K., 1948, pp. 451–453.
[44] M.F. Ashby, The mechanical effects of a dispersion of second phase, in: 2nd International Conference on Strength of Metals and Alloys, vol. 2, ASM, Ohio, USA, 1970, pp. 507–541.
[45] J.W. Martin, Micromechanisms in Particle-Hardened Alloys, Cambridge University Press, Cambridge, U.K., 1980.
[46] J.F. Nie, Effects of precipitate shape and orientation on dispersion strengthening in magnesium alloys, Scripta Materialia 48 (2003) 1009–1015.
[47] H.K.D.H. Bhadeshia, A.R. Chintha, Critical assessment: the strength of undeformed pearlite, Materials Science and Technology 38 (2022) 1291–1299.
[48] T. Takahashi, M. Nagumo, Flow stress and work-hardening of pearlitic steel, Transactions of the JIM 11 (1970) 113–119.
[49] D. Raynor, J.M. Silcock, Strengthening mechanisms in γ' precipitating alloys, Metal Science 4 (1970) 121–130.
[50] A.W. Thompson, J.A. Brooks, The mechanism of precipitation strengthening in an iron-base superalloy, Acta Metalllurgica 30 (1982) 2197–2203.
[51] J.P. Hirth, J. Lothe, Theory of Dislocations, Krieger Publishing Company, New York, USA, 1992.
[52] Y. Yan, Y. Yan, Y. He, J. Li, Y. Su, L. Qiao, The mechanism of precipitation strengthening in Fe–Ni austenitic alloy electron beam weldment, Materials Science & Engineering A 630 (2015) 85–89.
[53] F. Stein, M. Palin, Re-determination of transition temperatures in the Fe–Al system by differential thermal analysis, International Journal of Materials Research 98 (2007) 580–588.
[54] M. Palin, Fe–Al materials for structural applications at high temperatures: current research at MPIE, International Journal of Materials Research 100 (2009) 277–287.
[55] G.K. Wertheim, J.H. Wernick, Mössbauer effect study of bcc structure alloys, FeAl and FeTi, Acta Metalllurgica 15 (1967) 297–302.
[56] M. Palm, F. Stein, G. Dehm, Iron aluminides, Annual Review of Materials Research 49 (2019) 297–326.
[57] C.G. McKamey, J.H. DeVan, P.F. Tortorelli, V.K. Sikka, A review of recent developments in Fe_3Al-based alloys, Journal of Materials Research 6 (1991) 1779–1805.
[58] S. Golling, R. Östlund, M. Oldenburg, A study on homogenization methods for steels with varying content of ferrite, bainite and martensite, Journal of Materials Processing Technology 228 (2016) 88–97.
[59] H.S. Hasan, M.J. Peet, M.-N. Avettand-Fénoël, H.K.D.H. Bhadeshia, Effect of tempering upon the tensile properties of a nanostructured bainitic steel, Materials Science & Engineering A 615 (2014) 340–347.

[60] K.L. Kendig, D.B. Miracle, Strengthening mechanisms of an Al-Mg-Sc-Zr alloy, Acta Materialia 50 (2002) 4165–4175.
[61] S. Schänzer, E. Nembach, The critical resolved shear stress of γ'-strengthened nickel-based superalloys with γ'-volume fractions between 0.07 and 0.47, Acta Metallurgica et Materalia 40 (1992) 803–813.
[62] L.J. Baker, S.R. Daniel, J.D. Parker, Metallurgy and processing of ultralow carbon bake hardening steels, Materials Science and Technology 18 (2002) 355–367.
[63] S. Das, S.B. Singh, O.N. Mohanty, H.K.D.H. Bhadeshia, Understanding the complexities of bake hardening, Materials Science and Technology 24 (2008) 107–111.
[64] D.M. Clatterbuck, D.C. Chrzan, J.W. Morris Jr, The inherent tensile strength of iron, Philosophical Magazine Letters 82 (2002) 141–147.
[65] S.S. Brenner, Tensile strength of whiskers, Journal of Applied Physics 27 (1956) 1484–1491.
[66] S.S. Brenner, The growth of whiskers by the reduction of metal salt, Acta Metallurgica 4 (1956) 62–74.
[67] B.C. Edwards, Design and deployment of a space elevator, Acta Astronautica 47 (2000) 735–744.
[68] M. Euler, Hooke's law and material science projects: exploring energy and entropy springs, Physics Education 43 (2008) 57–61.
[69] H.K.D.H. Bhadeshia, H. Harada, High-strength (5 GPa) steel wire: an atom-probe study, Applied Surface Science 67 (1993) 326–333.
[70] J.W. Christian, Theory of Transformations in Metals and Alloys, Part I, 3rd ed., Pergamon Press, Oxford, U.K., 2003.
[71] M.F. Yu, O. Lourie, M.J. Dyer, K. Moloni, T.F. Kelly, R.S. Rouff, Strength and breaking mechanism of multiwalled carbon nanotubes under tensile load, Science 287 (2000) 637–640.
[72] S. Xie, W. Li, Z. Pan, B. Chang, L. Sun, Mechanical and physical properties on carbon nanotube, Journal of Physics and Chemistry of Solids 61 (2000) 1153–1158.
[73] Z.W. Pan, S.S. Xie, L. Lu, B.H. Chang, L.F. Sun, W.Y. Zhou, G. Wang, D.L. Zhang, Tensile tests of ropes of very long aligned multiwall carbon nanotubes, Applied Physics Letters 74 (1999) 3152–3154.
[74] B.I. Yakobson, R.E. Smalley, Fullerene nanotubes: C1,000,000 and beyond, American Scientist 85 (1997) 324–337.
[75] H.K.D.H. Bhadeshia, Large chunks of very strong steel, Materials Science and Technology 21 (2005) 1293–1302.
[76] C. Horowitz, K. Kadau, Breaking strain of neutron star crust and gravitational waves, Physical Review Letters 102 (2009) 191102.

Iron-carbon equilibrium and plain carbon steels

3

Abstract

The binary Fe-C system at ambient pressure offers at least three crystalline forms, i.e., ferrite, austenite and cementite. Simple permutation would suggest that there are seven possible phase equilibria (six binary and a ternary), but if δ and α are distinguished then the total is nine. This neglects the magnetic transitions that occur in all these phases. The richness of the system increases further when precise atomic mechanisms are invoked, dependent on whether the large and small atoms can move during the evolution of microstructure. We have the allotriomorphic and idiomorphic ferrite, where local equilibrium is maintained at the transformation front, while the pearlite cleverly avoids the accumulation of carbon in the residual austenite. As the deviation from equilibrium increases, Widmanstätten ferrite sets in as a hybrid reaction where the change in crystal structure is achieved by a homogeneous deformation of the austenite, but at a rate controlled by the diffusion of carbon. Just a few remarkable engineering applications of these incredible microstructures are discussed.

3.1 Iron-carbon equilibrium phase diagram

A study of the constitution and structure of all steels and irons can begin with the iron-carbon equilibrium diagram, Fig. 3.1. Many of the basic features of this system influence the behaviour of even the most complex alloy steels. For example, the phases found in the simple binary Fe-C system persist in complex steels, but it is necessary to examine the effects of alloying elements on the formation and properties of these phases. The iron-carbon diagram provides a valuable foundation on which to build knowledge of both plain carbon and alloy steels in their immense variety.

The normal equilibrium diagram really represents a metastable equilibrium between iron and iron carbide (Fe_3C, cementite). Cementite is metastable relative to the equilibrium between iron and graphite. Although graphite occurs extensively in some cast irons (2-4 wt% C, Fig. 2.13h), it usually is difficult, though not impossible, to achieve this equilibrium in steels (0.03-1.5 wt% C) without annealing for a long time. Therefore, the metastable equilibrium between iron and cementite should be considered as it is relevant to the behaviour of most steels in practice.

The crystallography of cementite is interesting – the unit cell is primitive orthorhombic, space group *Pnma*, with lattice parameters $a = 0.50837$ nm, $b = 0.67475$ nm and $c = 0.45165$ nm; the cell includes twelve iron atoms and four carbon atoms. It can be visualised as in Fig. 3.2a, where pleated layers of iron atoms are stacked in ...ABAB... formation, rather as in the hcp crystal structure, with carbon atoms in the interstices between the layers [3]. Given that the unit cell is primitive with large lattice parameters, the Burgers vectors of slip dislocations in cementite are also expected to be large, meaning that dislocations are difficult to move, rendering the carbide hard and relatively brittle. The common slip system appears to be (010)[001],

Steels. https://doi.org/10.1016/B978-0-44-318491-8.00009-5

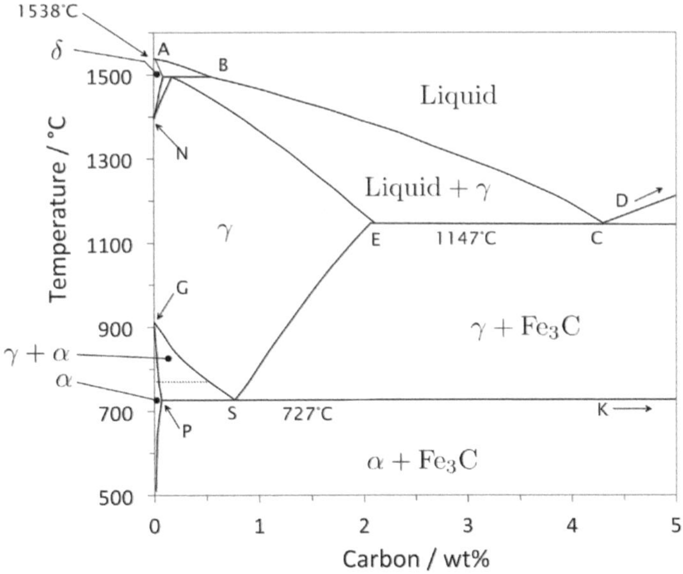

Figure 3.1 The iron-carbon phase diagram showing the equilibrium between the liquid, ferrite, austenite and cementite phases, adapted from [1]; the dotted line marks the Curie temperature of ferrite. The eutectoid composition is 0.76C wt%. Cementite has the composition ≈6.67 wt% carbon corresponding to 25 at%. However, the composition of cementite becomes iron-rich at high temperatures. As a consequence, ferrite precipitates within the cementite produced by the high-temperature eutectic reaction when it cools to lower temperatures [2].

though other systems also operate. The hardness under ambient conditions is in excess of 1000 HV and the fracture toughness less than 3 MPa $m^{1/2}$. Nevertheless, when enclosed within the microstructure of many steels as a minor phase, it exhibits considerable plasticity dependent on the mode of deformation. The crystal structure of cementite makes it far more anisotropic in its mechanical properties compared to the allotropic structures of iron.

The structure illustrated in Fig. 3.2 is based on a stoichiometric Fe_3C composition, whereas its carbon concentration can vary when it is in equilibrium with ferrite or austenite, in the range ≈ 24 to 25 at.%, with the largest deviation at high temperatures. Indeed, when large precipitates of cementite are cooled, ferrite precipitates inside the cementite, which approaches the stoichiometric composition at lower temperatures, Fig. 3.2c.

The larger phase field of γ-iron compared with that of α-iron reflects the much greater solubility of carbon in austenite, with a maximum value of just over 2 wt% at 1147°C when it is in equilibrium with liquid and cementite (E, Fig. 3.1). This relatively high solubility of carbon in γ-iron is of extreme importance in heat treatment, when solution treatment in the γ-region followed by rapid quenching to room temperature allows a supersaturated solid solution of carbon in iron to be formed.

The α-iron phase field is severely restricted, with a maximum carbon solubility of 0.02 wt% at 727°C when the ferrite is in equilibrium simultaneously with austen-

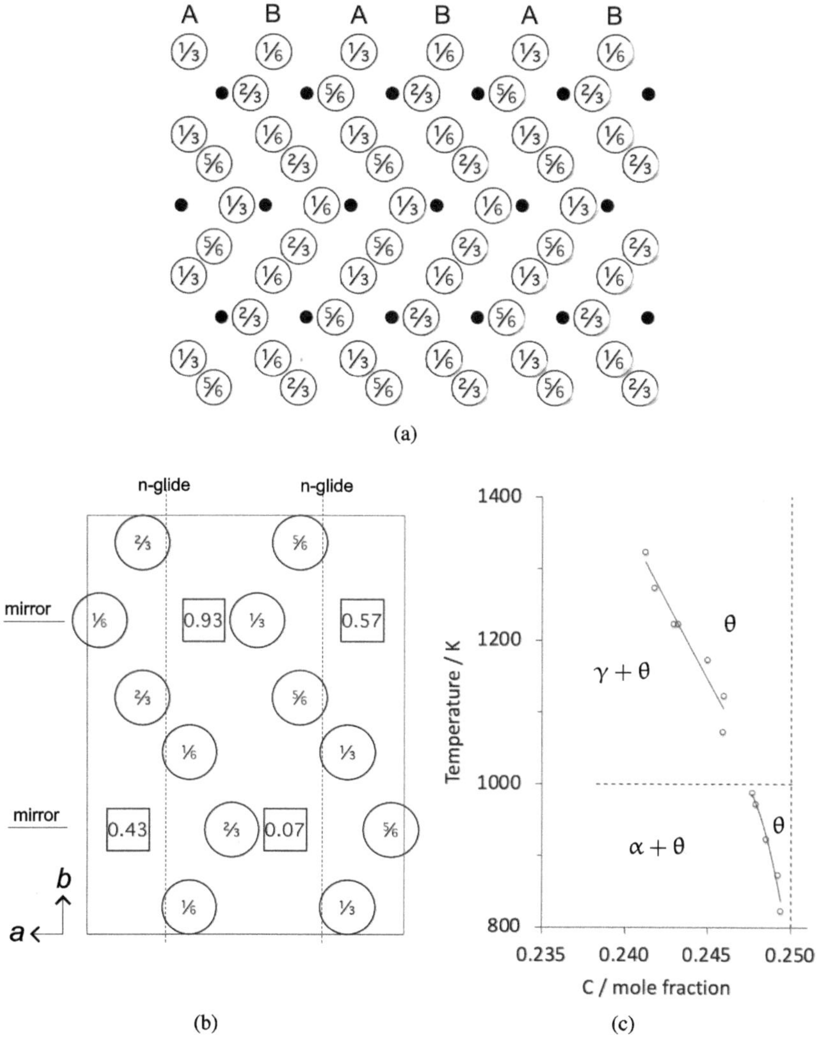

Figure 3.2 (a) A projection of the crystal structure of cementite with the z-axis normal to the plane of the diagram. The larger atoms represent iron, although the relative sizes are not to scale; fractional z-coordinates included only for iron atoms to illustrate that the puckered layers are stacked in . . .ABAB. . . sequence, noting that the atoms in the B layer have different z-coordinates than those in the A layer. (b) The unit cell of cementite with Fe atoms as circles and carbon as squares to present the fractional z-coordinates with clarity. Eight of the iron atoms are in symmetry related positions and the other four iron atoms, together with the carbon atoms, are on mirror planes parallel to $(010)_\theta$. (c) The composition of cementite in equilibrium with austenite or with ferrite in an Fe-C alloy. The data are from Leineweber et al. [4].

ite and cementite ('P', Fig. 3.1); if the steel contains more than this level of carbon, the ferrite exists in equilibrium either with cementite ($T \leq 727°C$) or with austenite ($712 < T < 912°C$). Similarly, the δ-phase field exhibits only limited solubility for carbon when in equilibrium with liquid or with austenite at temperatures in the range 1390 and 1538°C.

There are several temperatures or critical points in Fig. 3.1 that are important, both from the basic and from the practical point of view. Firstly, there is the Ae_1 temperature at which the eutectoid reaction occurs (P-S-K), which is 727°C in the binary diagram. Secondly, there is the Ae_3 temperature when α-iron transforms to γ-iron. For pure iron, this occurs at 912°C, but the transformation temperature is progressively lowered along the line GS by the addition of carbon. The Ae_1 temperature has the more generic meaning that it is the lowest temperature at which the α-ferrite and austenite can exist in equilibrium. Similarly, the Ae_3 temperature can be identified with the temperature at which a hypereutectoid steel becomes fully austenitic. The third point is Ae_4, at which γ-iron transforms to δ-iron, 1390°C in pure iron, but this value increases when carbon is added.

Iron atoms contain electrons that are not spin-paired, making then magnetic ions with their moments strongly coupled when the lattice is body-centred cubic. When ferrite exists at low temperatures, the ions order spontaneously with the magnetic dipoles aligned parallel to $\langle 100 \rangle_\alpha$. The transition from disordered dipoles to ordered ones occurs when the temperature is reduced to below the Curie temperature (Ae_2, 769°C), when the ferritic iron changes from the paramagnetic to the ferromagnetic state. This causes a very slight change in crystal structure, because ferromagnetic ferrite is in fact body-centred tetragonal, but the difference between the c and a lattice parameters is so small that it can be neglected in most circumstances. The tetragonality arises because the magnetic dipoles are aligned along only one of the $\langle 100 \rangle_\alpha$ directions, say [001]. This in turn means that two-fold rotation symmetry about [100] or [010] direction, implicit in the tetragonal lattice, is lost because the rotation would switch the direction of the dipoles by 180°, i.e., it would not restore the lattice to its original condition. If a spin cannot be represented correctly, but simply as a current flowing in a loop around an axis representing the direction of the magnetic dipole, Fig. 3.3, then the orientation of the dipole can be flipped by reversing the current, i.e., what is often confusingly called "turning time backwards". Thus, if the two-fold rotation about the [100] or [010] direction is accompanied by time reversal, the tetragonal symmetry is restored.

The Ae_1, Ae_3 and Ae_4 points are easily detected by thermal analysis or dilatometry during cooling or heating experiments conducted at rates slow enough to achieve equilibrium, because of the significant enthalpy and density changes accompanying transformation. When equilibrium experiments are not conducted, two further values are associated with each equilibrium temperature, i.e., *Ac* for heating *(chauffage)* and *Ar* for cooling *(refroidissement)*. It is emphasised that the *Ac* and *Ar* values will be sensitive to the rates of heating and cooling, as well as to the presence of alloying elements. Therefore, these temperatures should always be quoted in association with rate information.

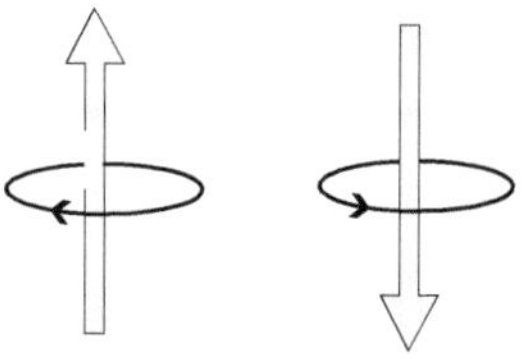

Figure 3.3 The magnetic moment inverts if the direction of the current is reversed, achieved by turning time backwards. The space group of paramagnetic ferrite is $Im\overline{3}m$, whereas that of the ferromagnetic state is $I4/mm'm'$, where the symbol m' represents the operation of a mirror plane with time inversion.

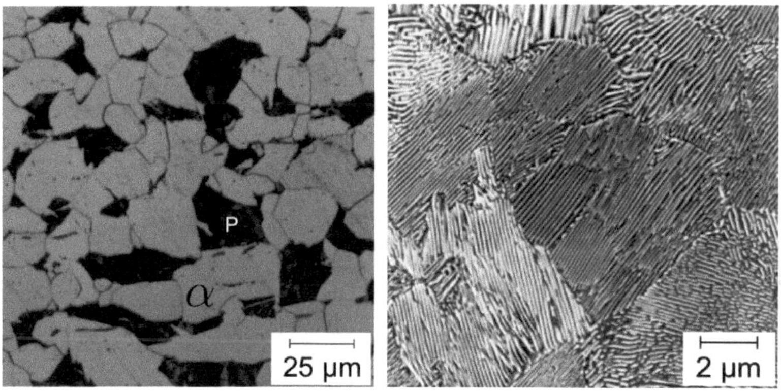

(a) (b)

Figure 3.4 (a) 0.2 wt% C steel containing a mixture of ferrite and pearlite (courtesy R. C. Cochrane). The pearlite (P) in this case is not resolved, but etches relatively dark because of the high density of α/θ interfaces within each colony. (b) Fully pearlitic 0.76 wt% C steel (courtesy of S. S. Babu). The structure often is interpreted as consisting of alternating lamellae of cementite and ferrite, but each colony is in fact a bi-crystal of the two interpenetrating phases in three dimensions [5].

Given that α-iron has a much smaller solubility for carbon than the γ phase, the precipitation of α in hypo-eutectoid steel leads to an increase in the carbon concentration of the residual austenite. If this process continues during cooling to the eutectoid temperature, then any remaining austenite can decompose simultaneously to a mixture ferrite and cementite, usually as pearlite. The microstructure of a hypo-eutectoid steel therefore consists of a mixture of pro-eutectoid ferrite and pearlite as the prime constituents (Fig. 3.4a), the relative fractions of which are determined by the lever rule. Steels that have a eutectoid carbon concentration transform completely into pearlite.

For austenite with 0.76-2.06 wt% carbon, slow cooling in the temperature interval from 1147 to 727°C, leads first to cementite precipitation, which progressively depletes the austenite in carbon, until at 727°C, the austenite is left with the eutectoid concentration of carbon and transforms completely into pearlite, Fig. 3.4b.

The three phases ferrite, cementite and pearlite are thus the principal constituents of the microstructure of plain carbon steels, provided they have been subjected to sufficiently slow cooling to avoid the formation of other phases that dominate when transformation occurs in circumstances where the atomic mobility is reduced.

3.2 Austenite-ferrite transformation

Pro-eutectoid ferrite forms in iron-carbon alloys containing less than the eutectoid concentration of carbon. It is possible for it to precipitate at < 912°C in pure iron, but takes place between 912°C and 727°C in iron-carbon alloys. However, by supercooling the austenitic state to temperatures below the eutectoid temperature, ferrite can be formed in reasonable time periods down to temperatures as low as 600°C. In fact, such ferrite can grow at even lower temperatures, but at incredibly slow rates. In an Fe-0.3C-4.08Cr wt% steel it took 43 days at 478°C for the ferrite to grow by 0.1 μm [6].

There are pronounced morphological changes as the transformation temperature is lowered, of a type that applies to both hypo- and hyper-eutectoid phases, although in each case there are variations due to the precise crystallography of the phases involved. Thus, the same principles apply to the precipitation of cementite from austenite, but it is not difficult to distinguish ferrite from cementite morphologically.

As a result of a survey of the behaviour of plain carbon steels, Dubé proposed a useful classification of morphologies of ferrite, which occur as the γ/α transformation temperature is lowered. Dubé recognised four well-defined morphologies based on optical microscopy, later extended by Aaronson[1]:

1. *Grain boundary allotriomorphs*: an allotriomorph has a shape that does not reflect its internal crystalline symmetry. This is because it tends to nucleate at the austenite grain surfaces, forming layers that follow the grain boundary contours (Fig. 3.5a). Furthermore, allotriomorphic ferrite is able to grow into all the austenite grains with which it has contact; its growth is not hindered by austenite grain boundaries. Each layer may be constituted of several different crystals; the morphology arises because the austenite grain surfaces are good nucleation sites with relatively easy diffusion along the boundaries compared to diffusion through the volume of the grains. An allotriomorph is in contact with at least two of the austenite grains and often has a random orientation with one of them, but an orientation which is more coherent with the other [7–9]. It may, therefore, be crystallographically faceted on one side but with a curved boundary on the other side (Fig. 3.5b). If austenite grains are crystallographically textured, then the probability of ferrite developing a good-fit orientation relationship with all the austenite grains with which it is in contact at the nucleation stage increases [10]. It also is possible to obtain cementite allotriomorphs in hyper-eutectoid steels, as shown in Fig. 3.5c. This kind of ferrite is the predominant phase in the vast majority of steels in existence today.

[1] Some of the micrographs presented in this chapter are not necessarily from plain carbon steels because without other austenite stabilising solutes, they transform so rapidly that it becomes difficult to preserve the morphology of individual precipitates in the partially transformed state.

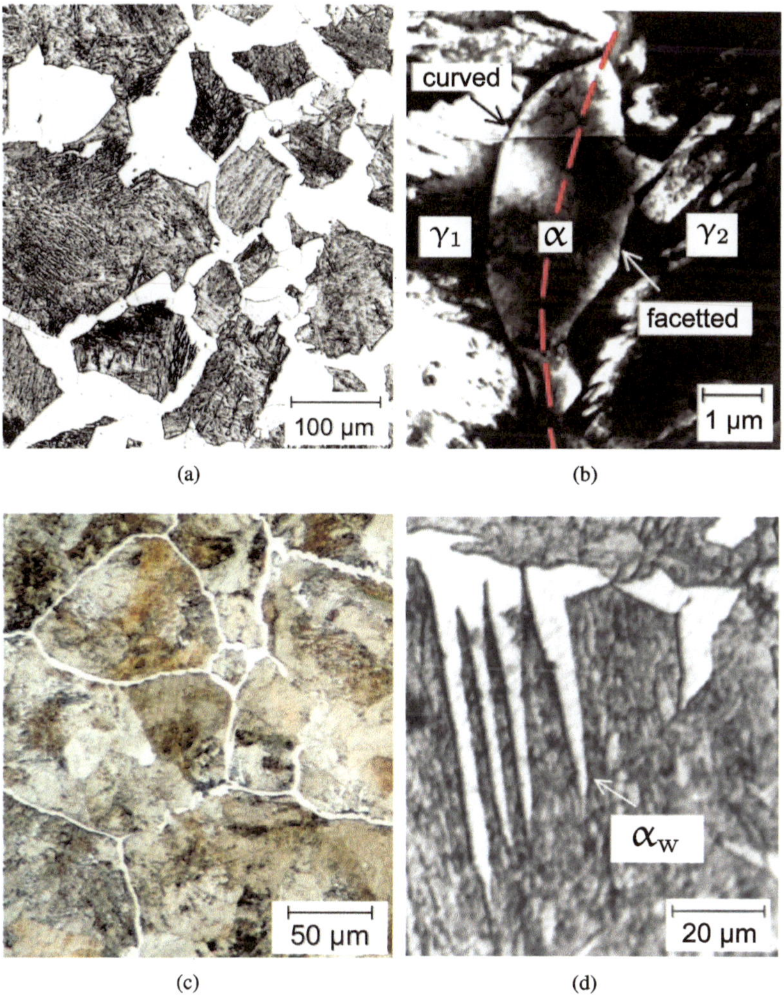

Figure 3.5 (a) Allotriomorphs of ferrite decorating prior austenite grain boundaries. (b) Transmission electron micrograph of ferrite allotriomorph at an austenite grain boundary between γ_1 and γ_2. There is a crystallographically good match between α and γ_2. The red line marks the prior austenite grain boundary, emphasising the fact that the ferrite grows into both of the γ-grains with which it has contact. (c) Allotriomorphs of cementite in a hypereutectoid steel (courtesy of R. C. Cochrane). (d) Widmanstätten ferrite plates.

2. *Intragranular idiomorphs*: These are equiaxed crystals, which nucleate inside the austenite grains (Fig. 3.6), usually on non-metallic inclusions present in the steel. An idiomorph forms without contact with the austenite grain surfaces and has a shape, which sometimes shows crystallographic facets.
3. *Widmanstätten ferrite plates*: these plates develop along well-defined planes of the austenite and are unable to grow across the austenite grain boundaries. Pri-

Figure 3.6 Intragranularly nucleated idiomorphic ferrite and Widmanstätten ferrite plates. Allotriomorphic ferrite layers decorate the austenite grain surfaces. Notice the very large austenite grains necessary for intragranular nucleation to occur unhindered by transformation products originating at the austenite grain surfaces. Courtesy of Ashraf Ali.

mary Widmanstätten ferrite initiates directly from the austenite grain surfaces, whereas secondary Widmanstätten ferrite develops from allotriomorphs of ferrite already present in the microstructure (Fig. 3.5d).

4. *Intragranular plates*: These plates are similar to those growing from the grain boundaries, but they nucleate entirely within the austenite grains, usually heterogeneously at inclusions (Fig. 3.6).

Grain boundary allotriomorphs are the first morphology to appear over the whole range of composition and temperature. However, at the highest temperatures (above 800°C) they predominate, growing along the boundaries and also into the grains, giving a well-defined grain structure. As the transformation proceeds, it is inevitable that grains nucleated at different locations will impinge. As a result, the structure evolves at large fractions of transformation to grains that are generally referred to as equiaxed ferrite because they have similar dimensions in all directions. The allotriomorphs nucleate having a reproducible orientation relationship that approximates as the Kurdjumov–Sachs orientation with one austenite grain (γ_2, Fig. 3.5b):

$$\{111\}_{\gamma_2} \parallel \{110\}_{\alpha},$$
$$\langle 1\bar{1}0\rangle_{\gamma_2} \parallel \langle 1\bar{1}1\rangle_{\alpha}.$$

But they also grow into the adjacent austenite grain (γ_1, Fig. 3.5b), to which they normally should have a random orientation relationship. The disordered boundary responsible for this growth should migrate more readily at high temperatures because such interfaces contain a greater free volume relative to more coherent interfaces.

At lower transformation temperatures, the mobility of curved or random γ/α boundaries decreases, while the coherent interfaces become more dominant for two reasons: first, displacive transformations are kinetically favoured at low temperatures and second, such transformations require that interfaces be relatively coherent to move without diffusion (Chapter 5). For example, plates of ferrite grow from protuberances on the allotriomorphic ferrite on the side of the coherent boundary. Thus, the plates develop into austenite with which they have the Kurdjumov–Sachs type relationship. They can also grow from clean austenite grain boundaries, the net result being a structure which normally is referred to as primary Widmanstätten ferrite. This structure is promoted by large austenite grain sizes, which limit interference with allotriomorphic ferrite, to allow Widmanstätten ferrite room to grow, Fig. 3.7.

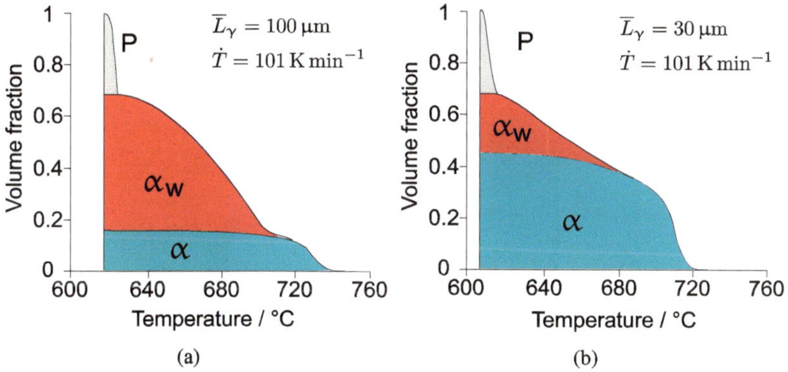

Figure 3.7 The evolution of microstructure in Fe-0.18C-1.15Mn wt% steel during cooling at the constant rate indicated [11]. The symbol 'P' represents pearlite. (a) Coarse austenite grain size. (b) Fine austenite grain size.

An important structural feature found in Widmanstätten ferrite is that the formation of plates is accompanied by surface relief effects in the form of invariant-plane strains with a large shear component, reflecting a disciplined motion of atoms during transformation (Section 3.5).

3.3 Austenite-cementite transformation

The Dubé classification applies to some of the morphologies of cementite formed at progressively lower transformation temperatures. The initial development of grain boundary allotriomorphs is very similar to that of ferrite, and the growth of side plates or Widmanstätten cementite follows the same pattern (Figs. 3.5, 3.8). Idiomorphs of cementite are not observed [12], possibly reflecting the difficulty of heterogeneous nucleation on inclusions.

Unlike ferrite, the austenite does not have a unique orientation relationship with the proeutectoid cementite, even when experimental errors or scatter are taken into account; the following is indexed using the *Pnma* space group setting of cementite. The orientation relationship with austenite tends to be one in which the close-packed planes in austenite, $(1\overline{1}\overline{1})_\gamma$, are parallel to the most densely packed planes in the cementite, $(0\overline{3}1)_\theta$ or $\{220\}_\theta$, and with directions in the latter which fit well with the close-packed direction in austenite being parallel, as follows [13]:

Pitsch	$[100]_\theta \parallel [110]_\gamma$	$(0\overline{3}1)_\theta \parallel (1\overline{1}\overline{1})_\gamma$
Zhang and Kelly	$[001]_\theta \parallel [121]_\gamma$	$(220)_\theta \parallel (\overline{1}1\overline{1})_\gamma$

In the Pitsch orientation, the $[100]_\theta$ is not only parallel to $[110]_\gamma$, but its length is almost identical.

As in the case of ferrite, most of the Widmanstätten cementite plates originate from grain boundary allotriomorphs, but some also nucleate at annealing-twin boundaries in the austenite.

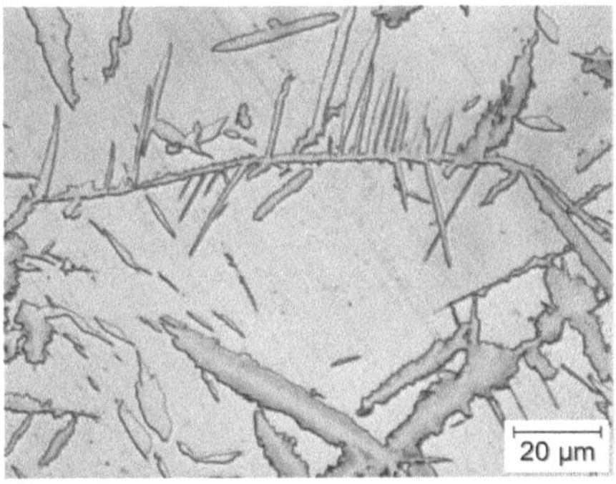

Figure 3.8 Cementite allotriomorphs and Widmanstätten cementite plates in austenite. After Zhang and Kelly, [13], reproduced with the permission of Elsevier.

3.4 Kinetics of the $\gamma \to \alpha$ transformation

The transformation of austenite in steels can be studied during continuous cooling using various physical measurements, e.g., dilatometry, thermal analysis, electrical resistivity, hot-stage microscopy, etc., however, the results obtained are sensitive to the cooling rate used. Davenport and Bain first introduced the isothermal transformation approach in combination with good microscopy to show that by studying the reaction isothermally at a series of temperatures, a characteristic time-temperature-transformation or TTT curve can be obtained for each particular steel. In their simplest form, these transformation curves have a well-defined 'C' shape (Fig. 3.9), where there is a temperature at which the reaction proceeds most rapidly, slowing down at both higher and lower temperatures. This temperature at which the reaction rate is fastest is often referred to as the nose of the C-curve. This can be explained in general terms as follows. For a eutectoid steel transformed close to the eutectoid temperature, the degree of undercooling, ΔT, is low so the driving force for the transformation is small. However, as ΔT increases, so does the driving force making the reaction faster until the maximum rate at the nose of the curve. Below this temperature, the driving force for the reaction continues to increase, but the reaction is now impeded by the slow diffusivity of the rate-controlling element, which in plain carbon steels may be carbon or iron, depending on the transformation conditions.

One of the simplest examples of a TTT curve is that for a steel of composition approximating the eutectoid carbon content. In Fig. 3.9, the beginning and end of transformation over a wide temperature range is plotted to produce two curves making up the diagram. When the carbon content of the steel is lowered, the ferrite reaction also takes place. This is represented by another curve, which frequently is imposed on the same diagram, and which normally precedes the pearlite reaction. Similarly, the cementite reaction can be recorded in hyper-eutectoid steels. The TTT curve strictly applies to the nucleation and growth of one phase in austenite, but at the lower temperatures other constituents can appear, e.g., the displacive transformation products Widmanstätten ferrite, bainite, martensite. These have quite different characteristics to ferrite and pearlite that form by a reconstructive mechanism that entails the diffusion of all atoms, the minimisation of strain energy and the partitioning of solutes so that their chemical potentials become equal in all phases. Martensite and bainite will be dealt with separately (Chapters 5 and 6).

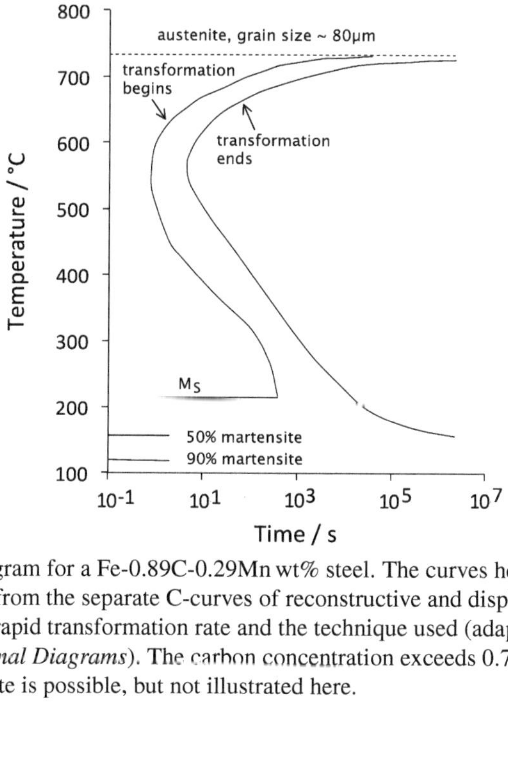

Figure 3.9 TTT diagram for a Fe-0.89C-0.29Mn wt% steel. The curves here represent overlapping information from the separate C-curves of reconstructive and displacive transformations because of the rapid transformation rate and the technique used (adapted from US Steel Co., *Atlas of Isothermal Diagrams*). The carbon concentration exceeds 0.76 wt% so some proeutectoid cementite is possible, but not illustrated here.

3.4.1 Growth kinetics of ferrite

Both the lengthening and thickening of grain boundary allotriomorphs have been studied. The latter process is considered first, represented as the one-dimensional thickening of allotriomorphs into the γ-grains in a direction normal to the austenite grain boundaries, controlled by the diffusion of carbon in the matrix ahead of the interface.

Ferrite has a lower solubility $c^{\alpha\gamma}$ for carbon when in equilibrium with austenite of composition $c^{\gamma\alpha}$, so if the average concentration $\overline{c} > c^{\alpha\gamma}$, carbon is partitioned into the austenite during transformation. As the ferrite grows, so does the extent of its diffusion field in the austenite. This retards growth because the solute then has to diffuse over ever larger distances. As will be proven, the thickness of the ferrite increases with the square root of time, i.e., the growth rate slows down as time increases. Following Zener, it is assumed in the derivation that the concentration gradient in the matrix is constant, and that the far-field concentration $\overline{c}$ never changes (i.e., the matrix is semi-infinite normal to the advancing interface). This is to simplify the mathematics without loosing any of the insight into the problem.

For isothermal transformation in a plain carbon steel, the concentrations at the interface are given by a tie-line of the phase diagram, as shown in Fig. 3.10a. This is because it is assumed that local equilibrium exists at the transformation front, so that the maximum concentration that can be tolerated in the austenite that is in contact with ferrite, is $c^{\gamma\alpha}$, Fig. 3.10b. Therefore, for the interface to be able to advance, any partitioned solute must diffuse away so that $c^{\gamma\alpha}$ is never exceeded. The diffusion flux of solute from the interface must equal the rate at which solute is partitioned:

$$\underbrace{(c^{\gamma\alpha} - c^{\alpha\gamma})\frac{\partial z^*}{\partial t}}_{\text{Rate solute partitioned}} = \underbrace{-D_{\mathrm{C}}^{\gamma}\frac{\partial c}{\partial z}}_{\text{Diffusion flux from interface}} \simeq D_{\mathrm{C}}^{\gamma}\frac{c^{\gamma\alpha} - \overline{c}}{\Delta z}, \tag{3.1}$$

where z is a coordinate normal to the interface with a value z^* at the position of the interface.

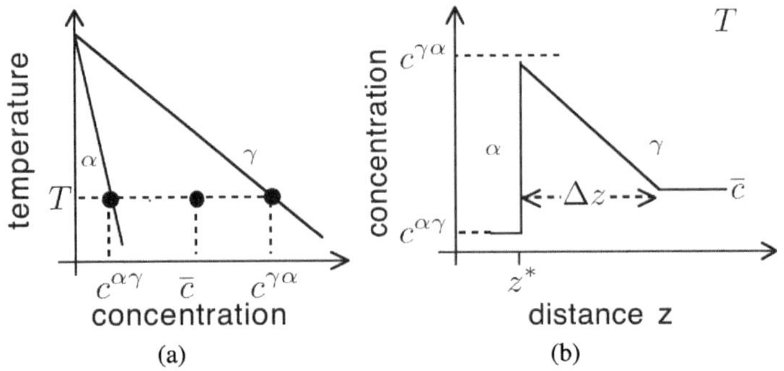

Figure 3.10 Phase diagram and its relationship to the concentration profile at the α/γ interface during diffusion-controlled growth.

A second equation can be derived by considering the overall conservation of mass:

$$(\overline{c} - c^{\alpha\gamma})z^* = \frac{1}{2}(c^{\gamma\alpha} - \overline{c})\Delta z.$$

By combining these expressions to eliminate Δz, we get:

$$\frac{\partial z^*}{\partial t} = \frac{D_{\mathrm{C}}^{\gamma}(c^{\gamma\alpha} - \overline{c})^2}{2z^*(c^{\gamma\alpha} - c^{\alpha\gamma})(\overline{c} - c^{\alpha\gamma})}. \tag{3.2}$$

It follows that:

$$z^* = \underbrace{\sqrt{D_{\mathrm{C}}^{\gamma}}\frac{(c^{\gamma\alpha} - \overline{c})}{[2(c^{\gamma\alpha} - c^{\alpha\gamma})(\overline{c} - c^{\alpha\gamma})]^{\frac{1}{2}}}}_{\text{parabolic rate constant } \alpha_1} \times \sqrt{t}. \tag{3.3}$$

Since the thickness of an allotriomorph increases parabolically with time (Fig. 3.11a), the growth rate decreases as the ferrite thickens. This is because as the ferrite thickens, more and more carbon is trapped in the austenite. Given that $c^{\gamma\alpha}$ is fixed, there is a reduced flux of carbon away from the transformation front and an increase in the diffusion distance Δz (Fig. 3.11b).

Eq. (3.3) contains the term $c^{\gamma\alpha} - \overline{c}$ that in effect represents the driving force for transformation, since growth ceases when the composition of the austenite is uniformly $c^{\gamma\alpha}$. Therefore, the growth rate in the Fe-0.07C wt% is slower than in the Fe-0.05C wt% steel.

Fig. 3.12 illustrates the fact that the growth rate is maximum at an intermediate temperature because the driving force increases with undercooling whereas the diffusivity decreases as the transformation temperature is reduced.

The approximations involved in this simple analysis designed to illustrate the essence of the problem are as follows:

- It is assumed that the far-field concentration $\overline{c}$ is unaffected by the growth of ferrite. This may well be true during the early stages of transformation, but eventually the accumulation of solute in the austenite will lead to an increase in $\overline{c}$ towards $c^{\gamma\alpha}$. The rate of growth will then be lower than predicted by Eq. (3.2). Growth ceases when $\overline{c} = c^{\gamma\alpha}$.
- The concentration gradient in the austenite is not in fact constant, but decreases with distance ahead of the transformation front, mathematically according to an error function [14].The concentration gradient in Eq. (3.1) must then be evaluated at the position of the interface ($z = z^*$).
- The diffusion coefficient of carbon in austenite is not constant, Fig. 3.12b, but increases dramatically with concentration [15,16].
- The shape of the ferrite may not strictly be in the form of a layer, in which case, growth might involve diffusion in two or three dimensions [17].
- There is an implicit assumption that the growth of ferrite is diffusion controlled. Other factors can be rate limiting, for example, the transfer of atoms across the α/γ interface, which becomes of overwhelming importance at low transformation temperatures, typically below about 600°C.

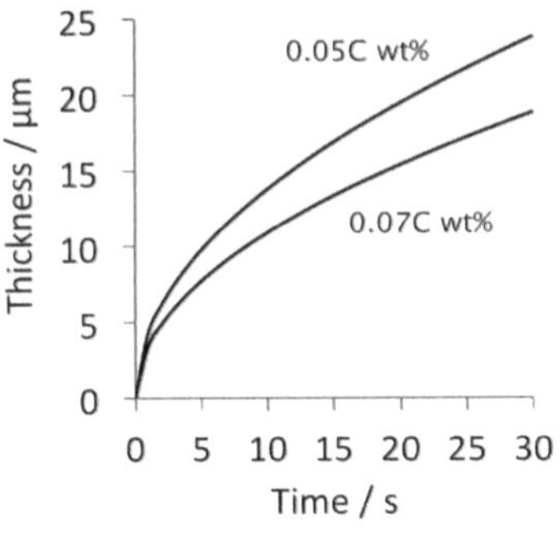

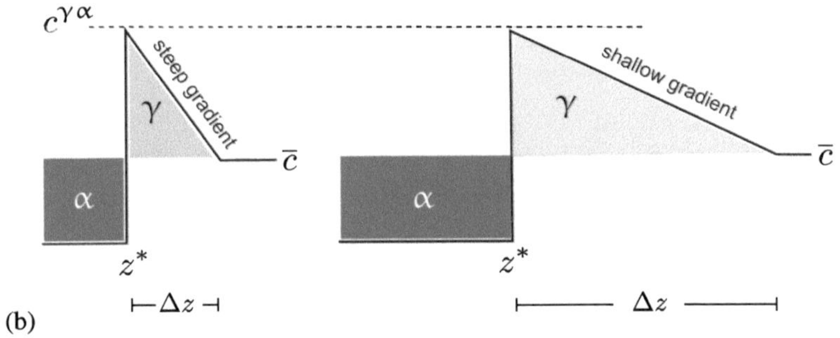

Figure 3.11 Kinetics of allotriomorphic ferrite growth (thickening) in Fe-C alloys. (a) The thickness increases parabolically with time. The growth rate is greater for the lower carbon steel. (b) The dark-shaded region is the deficit of carbon in ferrite relative to the average concentration $\bar{c}$, and the light-shaded region represents the corresponding accumulation of carbon partitioned into the austenite. As the ferrite thickens, i.e., as Δz increases, the gradient of concentration in the austenite becomes more shallow, thus reducing the flux of carbon away from the interface. If the far-field concentration $\bar{c}$ increases, growth would slow down even more because the gradient of concentration would become shallower and eventually, zero as the concentration of carbon in austenite becomes uniformly $c^{\gamma\alpha}$.

- Fig. 3.13 shows that when the allotriomorphic ferrite has a good-fit orientation relationship with the parent austenite, the transformation proceeds by the translation of steps on an otherwise stationary interface. The parabolic growth rate model described here does not then apply. Although there is considerable theory for the diffusion-controlled motion of individual steps or trains of steps in binary systems [17], there is no method that predicts the height or frequency of steps. In other words, it is impossible at the moment to predict the kinetics of allotriomorphic ferrite growth by a ledge mechanism.

With the exception of the consequences of the step mechanism of growth, there are ways to deal with all the approximations listed above [14,17].

The theory described thus far deals with the thickening of allotriomorphs, with thickness varying with $t^{\frac{1}{2}}$ because of the progressive accumulation of solute ahead of the transformation front necessitating ever larger diffusion distances (Δz). The length-

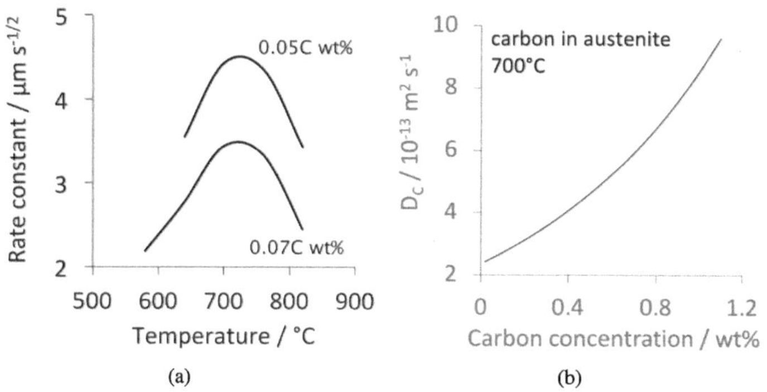

Figure 3.12 (a) Kinetics of allotriomorphic ferrite thickening in Fe-C alloys. The parabolic rate constant as a function of transformation temperature. (b) Illustration of the concentration dependence of the diffusivity of carbon in austenite.

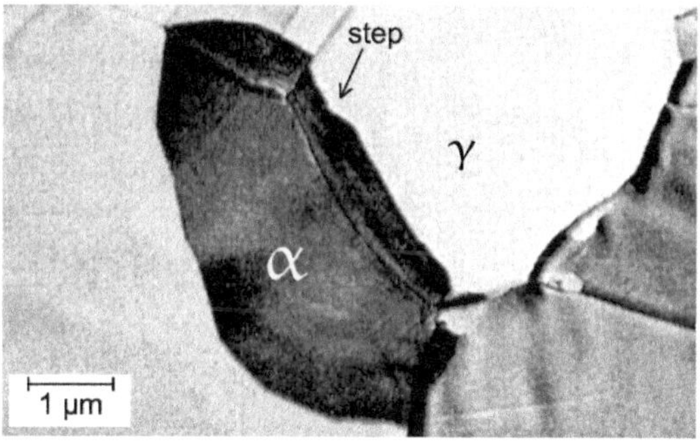

Figure 3.13 Allotriomorphic ferrite forming in austenite with planar and stepped interfaces in a low carbon alloy steel (Edmonds and Honeycombe). Photoemission electron micrograph.

ening of an allotriomorph along an austenite grain boundary was thought initially to vary linearly with time because the carbon that is partitioned is left behind the advancing tip of the ferrite. However, measurements [18] show that the lengthening also occurs parabolically with time, because the solute is in fact partitioned in two dimensions, both in the thickening and lengthening direction, leading to its accumulation on all transformation fronts. The aspect ratio, i.e., the thickness to length of an allotriomorph is about a third so the lengthening of these rather fat allotriomorphs cannot be treated in the same way as the lengthening of a thin plate.

3.5 Widmanstätten ferrite

3.5.1 Morphology

Primary Widmanstätten ferrite initiates and grows directly from the austenite grain surfaces, whereas the secondary form evolves from allotriomorphic ferrite that may be present in the microstructure (Fig. 3.14). Widmanstätten ferrite can form at temperatures close to the Ae_3 temperature and hence can occur at small driving forces; the undercooling needed amounts to a free energy change of only $\approx 50\,\mathrm{J\,mol^{-1}}$ [19]. This is much less than required to sustain diffusionless transformation.

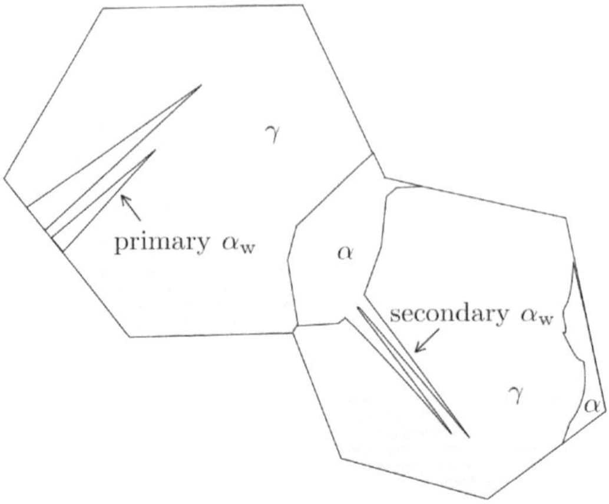

Figure 3.14 Morphology of primary and secondary Widmanstätten ferrite. See also Fig. 3.5d.

3.5.2 Shape change

The growth of a single plate of martensite is accompanied by an invariant-plane strain of the type illustrated in Fig. 3.15a. However, at the high temperatures (low undercoolings) at which Widmanstätten ferrite grows, the driving force available is insufficient to support the strain energy associated with a single plate. Widmanstätten ferrite formation therefore involves the simultaneous and adjacent cooperative-growth of two plates, which are crystallographic variants, such that their shape deformations mutually accommodate (Fig. 3.15b). This cancels out much of the strain energy [19].

It follows that what is seen as a single plate in an optical microscope is actually a combination of two variants, usually separated by a low-misorientation boundary (Fig. 3.15d,e). Widmanstätten ferrite has an irrational habit plane that is close to $\{5\,5\,8\}_\gamma$ [20]. Hence, the two plates α_{w1} and α_{w2}, which have different variants of this habit with the austenite, together form the thin-wedge-shaped plate characteristic of Widmanstätten ferrite. A self-consistent set [21] describing the crystallography [20] of a plate of Widmanstätten ferrite has the habit plane normal:

$$\mathbf{p} = (0.5057\ 0.4523\ 0.7346)_\gamma,$$

the orientation relationship being irrational but close to Kurdjumov–Sachs:

$$(1\ 0\ 1)_{\alpha_w} \parallel (0.5916\ 0.5772\ 0.5628)_\gamma$$
$$[1\ 1\ \overline{1}]_{\alpha_w} \parallel [0.6984\ \overline{0.7157}\ 0.0001]_\gamma$$

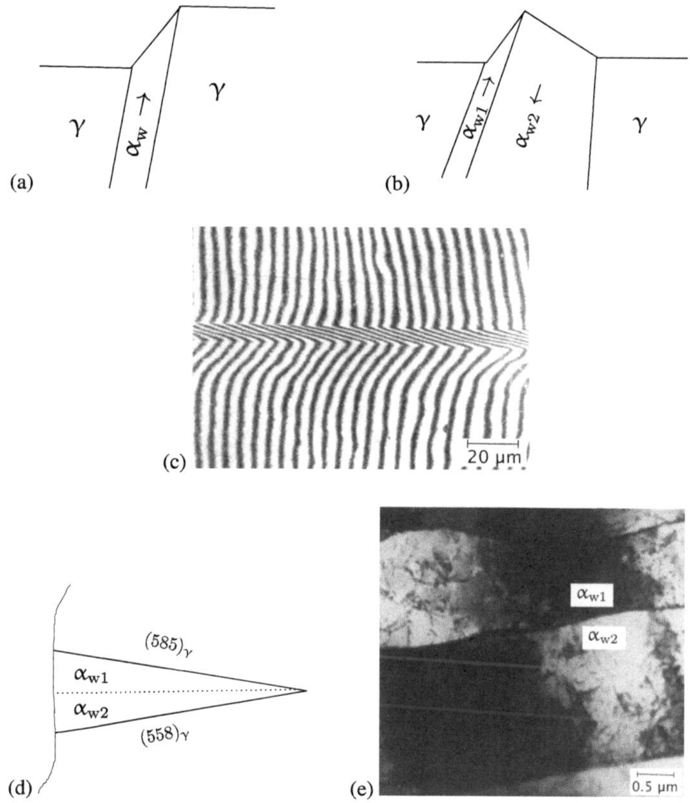

Figure 3.15 (a) A single invariant-plane strain shape deformation. (b) The combined effect of two mutually accommodating, back-to-back IPS deformations. (c) Tolansky interference micrograph showing the surface relief as illustrated schematically in (b). After Watson and McDougall [20], reproduced with permission of Elsevier. (d) The morphology of two plates, with different habit plane variants, growing together in a mutually accommodating manner. (e) Transmission electron micrograph showing two plates that appear in an optical microscope to be just one wedge-shaped plate. After Bhadeshia [19].

and the average magnitude of the shape deformation and direction:

$$m = 0.36$$
$$\mathbf{d} = [\overline{0.8670}\; 0.4143\; 0.2770]_\gamma.$$

The displacement vector **d** does not lie precisely within the habit plane **p**, because it describes both the shear and dilatational strains, the latter being directed normal to the habit plane (see Fig. 5.3).

Because Widmanstätten ferrite forms at low undercoolings, it is required thermodynamically that the carbon be redistributed during growth. α_w therefore always has an equilibrium carbon content and grows at a rate which is controlled by the diffusion of carbon in the austenite ahead of the plate-tip.[2] For plates, diffusion-controlled growth can occur at a constant rate because solute is partitioned to the sides of the thin plate, whereas the growing tip can advance into fresh austenite. Since the transformation is nevertheless, displacive, substitutional atoms do not partition and maintain an atomic correspondence between the parent and product lattices.

3.5.3 Growth kinetics of Widmanstätten ferrite

For isothermal growth without the partitioning of substitutional solutes, the growth rate is governed by the rate at which carbon diffuses ahead of the Widmanstätten ferrite plate tip. In the first approximation, the concentrations at the interface are given by a tie-lie of the phase diagram after allowing for the strain energy due to the shape change, as in Eq. (3.1).

The plate can be described as a parabolic cylinder in three dimensions (Fig. 3.16), a shape, which is preserved as the plate lengthens. Assuming that the diffusion distance Δz is equal to the plate tip radius r (Fig. 3.16), then from Eq. (3.1) it follows that the lengthening rate $v_\ell = \partial z^* / \partial t$ is given by

$$v_\ell \approx \frac{D_C^\gamma}{r} \frac{c^{\gamma\alpha} - \overline{c}}{c^{\gamma\alpha} - c^{\alpha\gamma}}. \tag{3.4}$$

This leads to the obvious difficulty that $v_\ell \rightarrow \infty$ as $r \rightarrow 0$, caused by the fact that the creation of additional interfacial area as the plate grows is neglected in the derivation.

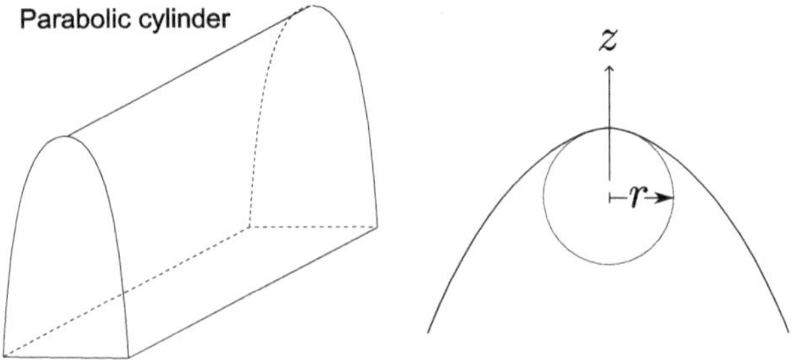

Figure 3.16 Widmanstätten ferrite plate, represented as a parabolic cylinder, with tip radius r.

Given that the change in surface area per atom added to the plate is v_a/r, the corresponding increase in the free energy per atom due to the creation of additional interface is $\sigma v_a/r$, where σ is the interfacial energy per unit area and v_a is the volume per atom.[3] Therefore, the net free energy change per atom, Δg_r, as the plate grows is

$$\Delta g_r = \Delta g_\infty - \frac{\sigma v_a}{r}, \tag{3.5}$$

[2] The concept of a T_0 temperature, above which transformation without a composition change is not possible, will be introduced in Chapter 5.

[3] Zener treated the tip of a plate to be a cylinder of radius r and length l. Since the change in area $\mathrm{d}A = 2\pi l\,\mathrm{d}r$ and the corresponding change in volume is $\mathrm{d}V = 2\pi l r\,\mathrm{d}r$, $\mathrm{d}A/\mathrm{d}V = 1/r$.

where Δg_∞ represents the free energy change per atom, driving the transformation in the absence of interface creation. At a critical radius r_c, $\Delta g_r = 0$ so that $\Delta g_\infty = \sigma v_a / r_c$ and $v_\ell = 0$. Eq. (3.5) can therefore be written as

$$\Delta g_r = \frac{\sigma v_a}{r_c} - \frac{\sigma v_a}{r} \qquad \text{or} \qquad \frac{\Delta g_r}{\Delta g_\infty} = 1 - \frac{r_c}{r}. \tag{3.6}$$

The velocity scales with the driving force when the latter is small, so Eq. (3.4) can be rewritten to account for the interface creation as follows:

$$v_\ell \approx \frac{D_C^\gamma}{r}\left(\frac{c^{\gamma\alpha} - \bar{c}}{c^{\gamma\alpha} - c^{\alpha\gamma}}\right) \times \left(1 - \frac{r_c}{r}\right). \tag{3.7}$$

The accounting for interfacial energy in this manner is known as the *capillarity effect* [22], which governs the equilibrium between a curved particle and the matrix. Fig. 3.17 shows how the lengthening rate now goes through a maximum, and it is often assumed that the plate picks a radius consistent with the maximum growth rate. This is approximately consistent with experimental measurements as shown in Fig. 3.17b [23].

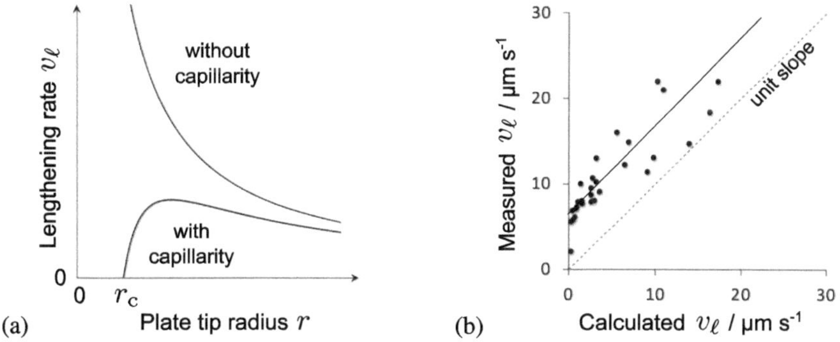

Figure 3.17 (a) Influence of interface curvature on plate lengthening rate. (b) Comparison of measured versus calculated lengthening rates.

3.5.4 Summary

The essential features of Widmanstätten ferrite all relate to the fact that it is a displacive transformation, but controlled by the rate at which carbon diffuses, because while the substitutional lattice is deformed into the final structure, the interstitial carbon must partition in order for there to be a sufficient free energy change to permit the ferrite to grow. An atomic correspondence is therefore maintained for substitutional atoms during the growth of Widmanstätten ferrite, but it is a true example of paraequilibrium transformation.

To cope with the limited free energy available to drive the transformation at small undercoolings, the growth occurs by the simultaneous and cooperative formation of a pair of adjacent, self-accommodating plates of Widmanstätten ferrite. This is how the strain energy due to the shape change is mitigated, but the drawback is that two appropriate crystallographic variants must nucleate simultaneously, thus reducing the nucleation rate and leading to a relatively coarse microstructure that is not conducive to good toughness.

3.6 Austenite-pearlite reaction

Pearlite is probably the most familiar microstructural feature in the whole science of metallography (Fig. 3.18). It was discovered and characterised by Sorby in 1886, who called it a *pearly compound* and correctly assumed it to be a lamellar mixture of "iron free from carbon" and "iron combined with a larger amount of carbon" [24]. Pearlite is an extremely common constituent of a wide variety of steels, where it provides a substantial contribution to strength, so it is not surprising that this phase has received intensive study. Lamellar eutectoid structures of this type are widespread in metallurgy, so pearlite often is used as a generic term to describe them. These structures have much in common with the cellular precipitation reactions where a supersaturated parent Γ' decomposes into an equilibrium mixture of $\Gamma + \Phi$ at a common transformation front, with Γ' and Γ having the same crystal structure. Both types of reaction occur by nucleation and growth (Fig. 3.18), and, more importantly, rely on the cooperative growth of phases during diffusional growth. Pearlite nuclei occur on austenite grain boundaries, but it is clear that they can also be associated with both pro-eutectoid ferrite and cementite.

In some steels, pearlite nodules can nucleate on non-metallic inclusions such as MnS, Fig. 3.19. The exact mechanism by which the MnS acts as a heterogeneous nucleation site is not known, especially since there are two phases involved in the product. The experimental observations rely on steels that are slow-transforming and richly-alloyed. It is not clear that the intragranular nucleation of pearlite in common commercial steels plays a significant role in the development of microstructure, given that the austenite grain sizes are usually kept small prior to transformation. There is a greater discussion of the functioning of inclusions and compounds as nucleation sites in Chapter 7, which deals with the intragranular formation of ferrite plates.

3.6.1 The morphology of pearlite

The idealised view of pearlite is a hemispherical nodule nucleated at an austenite grain boundary and growing gradually into the austenite (Fig. 3.20). Apart from examining possible sites for nucleation, the following information is needed to understand the development of pearlite:

a) how the lamellae increase in number,
b) the crystallographic relationships between the phases,
c) the nature of the pearlite/austenite interface,
d) the rate-controlling process.

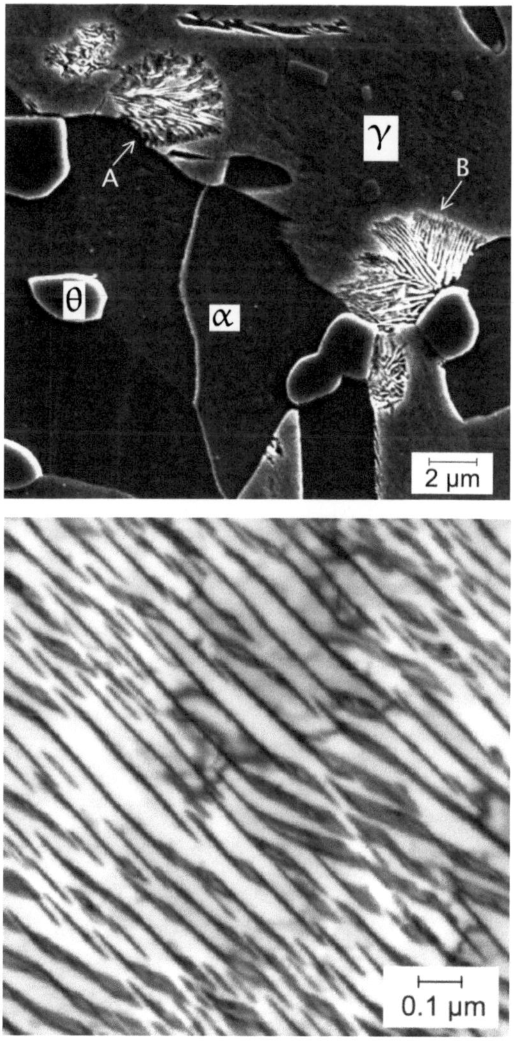

Figure 3.18 (a) The initiation of pearlite colonies during isothermal transformation from a mixture of austenite, ferrite and proeutectoid cementite (θ) (courtesy H. S. Hasan). The colony marked A nucleates on ferrite whereas that marked B on cementite. (b) Transmission electron micrograph of a colony of pearlite showing the intimate mixture of cementite and ferrite (courtesy of S. S. Babu).

Not all these questions can yet be fully answered, but the essentials have been established. Following the classical work of Mehl and colleagues, Hillert and co-workers were able to show that pearlite could be nucleated either by ferrite or by cementite, depending on whether the steel was hypo- or hyper-eutectoid in composition. They came to this conclusion after observing lattice continuity between the ferrite in pearlite and

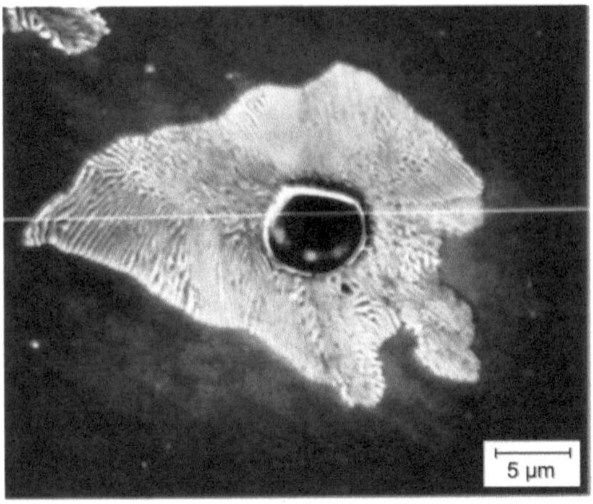

Figure 3.19 Multiple pearlite colonies nucleating and growing from a manganese sulphide inclusion in a Fe-12Mn-0.8C wt% steel that was warm-rolled and transformed isothermally at 600°C for 1 h. Reproduced from Furuhara et al. [25], with permission of © WILEY-VCH Verlag GmbH & Co. KGaA, Weinheim.

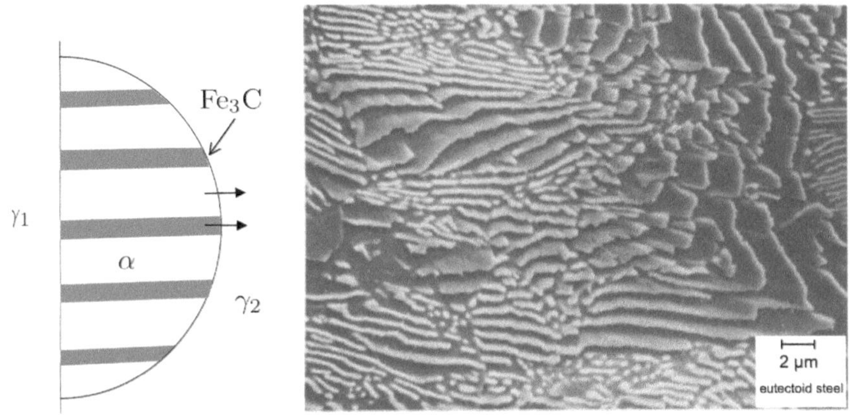

(a) (b)

Figure 3.20 (a) Idealised pearlite nodule at austenite grain boundary. In this scenario by Mehl and co-workers, the pearlite grows by sideways nucleation and edge-ways growth. Each layer has to nucleate separately in this model. (b) Deep-etched sample of pearlite, showing the complexity of the real structure in three dimensions. The cementite lamellae are not continuous or flat plates, they may contain holes allowing the ferrite to penetrate through. Image reproduced from Tian and Kraft [26], with the permission of Springer Nature, © 1987.

pro-eutectoid ferrite, as well as between cementite in pearlite and hyper-eutectoid cementite.

Mehl and co-workers took the view that pearlite nodules formed by sideways nucleation and edge-ways growth, as illustrated in Fig. 3.20, which also shows that the real three-dimensional morphology of pearlite is quite complex.[4] In this way, the rapid increase in the number of lamellae in a nodule which occurred during growth could be explained, but Modin indicated that this could equally well result from the branching of lamellae during growth. Thin-foil electron microscopy work by Dippenaar and Honeycombe on Fe-13Mn-0.8C wt% steel allowed the examination of small nodules at an early stage of growth in an austenitic matrix rendered stable by addition of manganese. This steel is hyper-eutectoid, i.e., grain boundary cementite forms prior to nucleation of pearlite, which frequently takes place on the cementite. This work showed conclusively the continuity of grain boundary and pearlitic cementite (Fig. 3.21), and also indicated that both the cementite and ferrite possessed unique orientations within a particular nodule. Fig. 3.21 also shows the beginning of branching of the Fe_3C lamella. However, in other nodules, sideways nucleation of laths of cementite and ferrite was observed. Nucleation of pearlite also took place on clean austenite boundaries. Hillert has shown that nucleation also occurs on ferrite, so all three types of sites are effective, and the predominant sites will be determined primarily by the composition. Alternatively, as illustrated Fig. 3.18a, when the reaction begins from a mixed microstructure containing ferrite, austenite, and proeutectoid cementite, the pearlite can initiate in contact with just ferrite (region A) or cementite (region B).

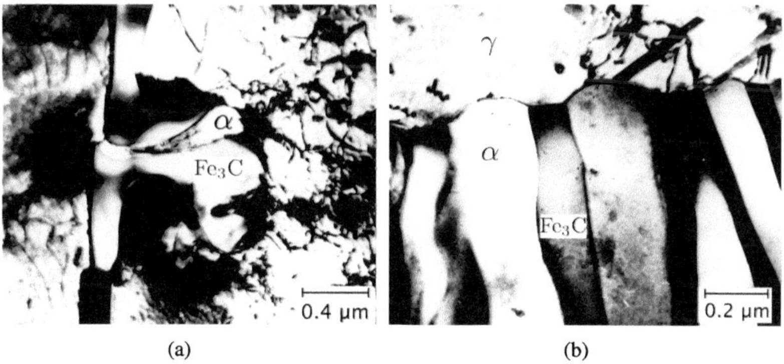

Figure 3.21 Fe-13Mn-0.8C wt% partly transformed at 600°C. Austenite is retained in conjunction with ferrite and cementite: (a) nucleation of a pearlite nodule on grain boundary cementite, (b) interface of nodule with austenite. Thin foil electron micrographs (courtesy of Dippenaar).

C. S. Smith first pointed out that the moving pearlite interface in contact with austenite is an incoherent, high-energy interface growing into a grain with which the pearlitic ferrite and cementite had no reproducible orientation relationship, presumably because this leads to more mobile interfaces in the context of a reconstructive transformation. Electron microscopy observations have confirmed that the pearlite interface to austenite is not coherent. The level of incoherency depends also on the orientation of the normal to the interface plane; if there is significant orientation

[4] An excellent review of the evolution of pearlite theory is available [27].

dependence of interfacial energy, then the transformation front will translate by the movement of ledges (step mechanism), even if the motion periodically traverses different phases (i.e., α and θ) [28]. Fig. 3.21b shows a typical interface on a 13Mn-0.8C wt% steel, where the untransformed austenite has been retained at room temperature.

3.6.2 The crystallography of pearlite

In a typical pearlite nodule, there are two interpenetrating single crystals of ferrite and cementite, neither of which may be orientation related to the austenite grain in which they are growing. The level of reproducibility that characterises displacive transformations is not essential when the mechanism is reconstructive, especially when the activation energy for nucleation is similar for a variety of relative orientations. Many of the reported data or orientation relations within pearlite have not been determined accurately. Focusing on the more precise data that permit distinctions to be made, several reproducible orientation relationships have been observed between cementite ($Pnma$) and ferrite that form pearlite [29]:

	"New-2"			"New-3"	
$(0\,3\,\bar{1})_\theta$	$\parallel$	$(\bar{1}\,0\,1)_\alpha$	$(0\,3\,\bar{1})_\theta$	$\parallel$	$(\bar{1}\,0\,1)_\alpha$
$[1\,0\,0]_\theta$	8.5° from	$[1\,3\,1]_\alpha$	$[\bar{2}\,2\,0]_\theta$	2.4° from	$[1\,\bar{3}\,1]_\alpha$
$[\bar{1}\,1\,3]_\theta$	$\parallel$	$[1\,\bar{1}\,1]_\alpha$	$[1\,1\,3]_\theta$	$\parallel$	$[1\,\bar{1}\,1]_\alpha$

	"New-4"			"New-5"	
$(1\,0\,2)_\theta$	$\parallel$	$(\bar{1}\,0\,1)_\alpha$	$(0\,3\,\bar{1})_\theta$	$\parallel$	$(\bar{1}\,0\,1)_\alpha$
$[1\,1\,3]_\theta$	5.95° from	$[1\,0\,1]_\alpha$	$[1\,1\,3]_\theta$	8.5° from	$[1\,\bar{1}\,1]_\alpha$
$[2\,1\,\bar{1}]_\theta$	$\parallel$	$[1\,3\,1]_\alpha$	$[1\,0\,0]_\theta$	$\parallel$	$[1\,3\,1]_\alpha$

The New-2, New-3 and New-5 relations are close, but significantly different, from that attributed to Pitsch and Petch,

$(0\,1\,0)_\theta$	$\parallel$	$(5\,\bar{2}\,\bar{1})_\alpha$
$[0\,0\,1]_\theta$	2.6° from	$[1\,3\,\bar{1}]_\alpha$
$[1\,0\,0]_\theta$	2.6° from	$[1\,1\,3]_\alpha$

but as Zhang and Kelly [29] pointed out, neither the Pitsch–Petch nor the Bagaryatskii orientations were found, and suggested that they may not actually exist if accurate measurements are made; only the Isaichev relation was observed in addition to the new orientations relationships listed above. The θ/α orientation relationship found by Isaichev [30], using rational indices, can be expressed as follows:

$$
\begin{array}{lll}
\langle 1\,0\,0\rangle_\theta & \parallel & \langle 1\,\bar{1}\,\bar{1}\rangle_\alpha \\
\{0\,3\,1\}_\theta & \approx\parallel & \{1\,0\,1\}_\alpha.
\end{array}
$$

The approximation sign is significant, because these planes are not exactly parallel, as pointed out in the original paper by Isaichev [30]. There is an angle of some

1.5-2° between $\{0\ 3\ 1\}_\theta$ and $\{1\ 0\ 1\}_\alpha$. The Isaichev orientation relationship is close to that of Bagaryatskii making them difficult to distinguish using conventional electron diffraction. It deviates from Bagaryatskii by a rotation of 3.8° about the a-axis of the cementite [31]. Accurate measurements on tempered martensite have repeatedly identified the Isaichev orientation relationship and this has led to the suggestion that the Bagaryatskii orientation does not exist [32,33]. There is a good reason why the Isaichev orientation persists, because it leads to greater coherence at the interface [34].

The mechanisms required to explain these observations depend on the carbon concentration of the steel [29]. In hypoeutectoid steel, the ferrite precedes the cementite and therefore influences its nucleation, leading to the Isaichev orientation relationship. In hypereutectoid steels, the cementite nucleates first while in contact with austenite, leading to the New-5 orientation relationship with the ferrite; in this case, the orientation of the ferrite within the pearlite is determined by its relationship with the austenite. The mechanisms determining the New-2, New-3 and New-4 relationships are not as clear, but apparently occur in eutectoid steel and are dominated by the relationship of the closest-packed planes with the cementite with the closest-packed planes in austenite and ferrite.

3.6.3 Kinetics of pearlite growth

A colony of pearlite, viewed in three dimensions, consists of an interpenetrating bicrystal of ferrite and cementite [5]. In planar sections, the phases appear as lamellae which grow at a common front with the austenite. Cementite is rich in carbon, whereas ferrite accommodates very little when it is in equilibrium with either cementite or austenite. It therefore is necessary for carbon to be redistributed at the transformation front. This can happen by diffusion in the austenite in a direction parallel to the transformation front (Fig. 3.22a). The growth must occur at a constant rate because in plain carbon steel, the average chemical composition of the pearlite is identical to that of the parent austenite. Some measurements are illustrated in Fig. 3.22b. The rate of growth is often measured by reacting a series of samples for increasing times at a particular temperature. As a result of measurements on polished and etched sections, the radius of the largest pearlite area, assumed to be a projection of the first nodule to nucleate, can be plotted against time. Normally a straight line is obtained, the slope of which is the growth rate v_P (Fig. 3.22b).

It is notable in Fig. 3.22b that the growth speed of pearlite is far greater than the thickening rate of allotriomorphic ferrite. The reason for this is simple, that the diffusion distance involved in the growth or pearlite is less than its interlamellar spacing, whereas that for the ferrite increases with time because (unlike pearlite) there then is an accumulation of carbon ahead of the transformation front. The free energy reduction accompanying the formation of pearlite will also be larger than that for ferrite precipitation in the same steel, because cementite also precipitates in the former case.

It has been found that v_P is structure insensitive, i.e., structural changes such as γ-grain size, presence or absence of carbide particles have little effect. However, the rate is markedly dependent on temperature, specifically the degree of cooling ΔT below the eutectoid temperature T_E, and increases with increasing degree of undercooling until the nose of the TTT curve is reached. The growth rate is also strongly influenced by the concentration of alloying elements present – the detailed role of substitutional solutes will be discussed in Chapter 4.

Fig. 3.23 illustrates the Hultgren extrapolation [35]; if the average carbon content, $\bar{c}$, of the steel lies between $c^{\gamma\theta}$ and $c^{\gamma\alpha}$ then the steel is able to transform into a fully pearlitic microstructure even though $\bar{c}$ is not equal to the eutectoid concentration (0.76 wt% in Fe-C). The fraction of cementite within the pearlite is lower than normal when $\bar{c} < 0.76$ wt%, and vice versa. Within the Hultgren extrapolation, the austenite is supersaturated with respect to both cementite and ferrite; this is an essential condition for the development of cooperative growth with a common transformation front with the austenite. When $\bar{c}$ does not fall into the region defined by the Hultgren extrapolation, the pearlite obtained in hypoeutectoid is degenerate, i.e., the cementite lamellae are not continuous [36]. Such a degenerate structure does not strictly represent pearlite in the sense that cooperative growth does not occur. Indeed, the mechanical properties of "degenerate pearlite" reflect the lack of continuity of the cementite.

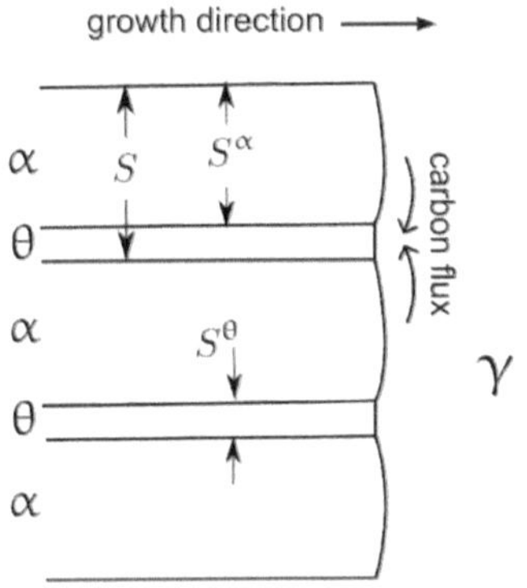

(a)

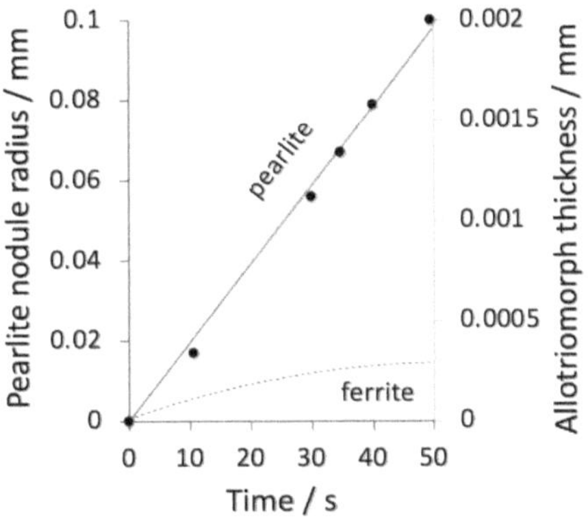

(b)

Figure 3.22 (a) Diffusion flux in the austenite parallel to the advancing interface. S is the interlamellar spacing. The thickness S^θ of the cementite relative to that of the ferrite (S^α) is determined by the mean carbon concentration of the steel. (b) The flux illustrated in (a) carries carbon that is partitioned into the austenite as the ferrite grows, to be absorbed by the cementite that grows simultaneously. There is therefore no change in the average composition of the austenite during pearlite growth in binary Fe-C alloys. As a consequence, the pearlite grows at a constant rate. The data are for a steel of approximately eutectoid composition transformed isothermally at 680°C in order to measure the maximum pearlite nodule size (adapted from Mehl and Hagel [37]). Also included is a calculated plot of the much lower growth rate of allotriomorphic ferrite in a eutectoid steel at the same temperature.

We shall assume that growth is controlled by the diffusion of carbon in the austenite ahead of the interface. The diffusion distance parallel to the interface can be approximated as aS where $a < 1$ to allow for the fact that the distance is a fraction of the

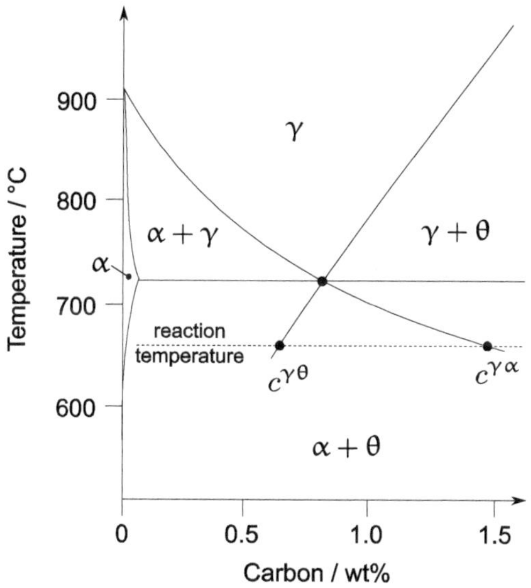

Figure 3.23 Phase diagram with extrapolated phase boundaries to identify the concentrations in the austenite in equilibrium with cementite or ferrite. The construction below the eutectoid temperature is known as the *Hultgren extrapolation* [35], the region where supercooled austenite can be transformed into a fully pearlitic structure, with the ferrite and cementite maintaining a common transformation front with the austenite.

interlamellar spacing S, since the diffusion of carbon occurs in both directions parallel to the transformation front (Fig. 3.22a). By analogy with Eq. (3.1), it follows that the rate at which solute is absorbed by the cementite must equal the amount arriving there by diffusion, so that

$$v_{\mathrm{P}}(c^{\theta\gamma} - c^{\gamma\theta}) = D_{\mathrm{C}}^{\gamma} \frac{(c^{\gamma\alpha} - c^{\gamma\theta})}{aS}, \tag{3.8}$$

where v_{P} is the speed of the growth front, D_{C}^{γ} is the diffusivity of carbon in austenite and the concentration terms are self-explanatory.

However, there is an additional process that consumes energy, the creation of cementite/ferrite interfaces within the pearlite colony. The minimum value of interlamellar spacing possible is a critical spacing $S_{\mathrm{C}} = 2\sigma^{\alpha\theta}/\Delta G$ where $\sigma^{\alpha\theta}$ is the interfacial energy per unit area and ΔG is the magnitude of the driving force for transformation in Joules per unit volume.[5] Growth ceases when $S = S_{\mathrm{C}}$ since all of the available driving force is consumed to form α/θ interfaces. To allow for the energy consumed in the process of interface creation, and following the procedure outlined in Section 3.5.3,

[5] If a cube with sides of length $2S$ contains three parallel lamellae (two of them forming the opposite faces of the cube) then the amount of α/θ interface is $4 \times 4S^2$ in a volume $8S^3$. Therefore, the interfacial area per unit volume is $S_{\mathrm{V}} = 2/S$ so the energy due to interfaces is $2\sigma^{\alpha\theta}/S$. This equals ΔG when $S = S_{\mathrm{C}}$, so that $S_{\mathrm{C}} = 2\sigma^{\alpha\theta}/\Delta G$.

Eq. (3.8) is modified by a term $(1 - [S_C/S])$ as follows:

$$v_P = \frac{D_C^\gamma}{aS}\frac{(c^{\gamma\alpha} - c^{\gamma\theta})}{(c^{\theta\gamma} - c^{\gamma\theta})}\left(1 - \frac{S_C}{S}\right). \tag{3.9}$$

With this modification, the velocity reaches a maximum as a function of the interlamellar spacing, Fig. 3.24. We now need to specify the value that S adopts during growth, and one assumption is that distance is that which coincides with the maximum growth rate, i.e., when $S = 2S_C$.

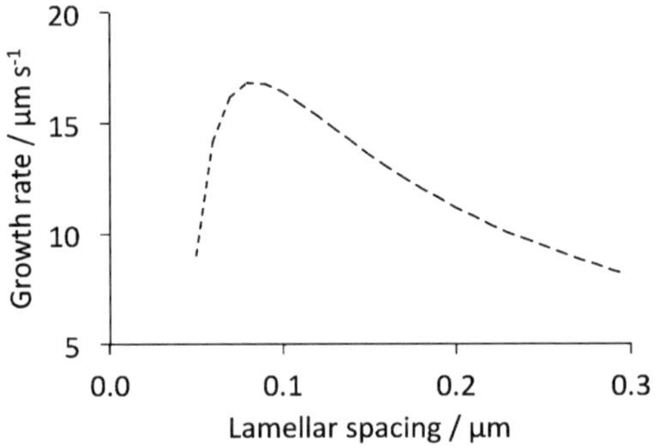

Figure 3.24 The growth rate of pearlite as a function of the interlamellar spacing. The calculations assume that $D_C^\gamma = 3.44 \times 10^{-11}\ \mathrm{m^2\,s^{-1}}$, the concentration term in Eq. (3.9) is 0.00824, $S_C = 0.042\ \mu\mathrm{m}$, $a = 1$.

The kinetic theory predicts that the interlamellar spacing within pearlite should decrease with the transformation temperature, i.e., as the driving force ΔG for transformation increases, and this indeed is the case, Fig. 3.25. At the eutectoid temperature T_E, $\Delta G = 0$, so that ΔH, the enthalpy change is equal to $T_E\Delta S$, where ΔS is the entropy of transformation, assumed to be constant. Therefore, as an approximation, $\Delta G = T_E\Delta S - T\Delta S$, i.e., ΔG is proportional to the undercooling $T_E - T$.

Suppose now that the diffusion of carbon during the growth of pearlite occurs by a combination of flux through the volume of the austenite and through the interface itself, as illustrated in Fig. 3.26a. To account for both of these fluxes, Eq. (3.8) is modified as follows [40]:

$$v_P(c^{\theta\gamma} - c^{\gamma\theta}) = D_{C,V}^\gamma\frac{(c^{\gamma\alpha} - c^{\gamma\theta})}{aS} + D_{C,B}\delta\frac{(c^{\gamma\alpha} - c^{\gamma\theta})}{a'S}, \tag{3.10}$$

where $D_{C,V}^\gamma$ and $D_{C,B}$ are the volume and boundary diffusion coefficients for carbon, δ is the thickness of the boundary (Fig. 3.26a) and a' is a constant. With certain assumptions about the constants, a solution of this then leads to the velocity equation [39,40]:

$$v_P = \left(\frac{c^{\gamma\alpha} - c^{\gamma\theta}}{c^{\theta\gamma} - c^{\alpha\gamma}}\right)\left(2D_{C,V}^\gamma + \frac{12D_{C,B}\,\delta}{S}\right)\frac{S}{S^\alpha S^\theta}\left(1 - \frac{S_C}{S}\right). \tag{3.11}$$

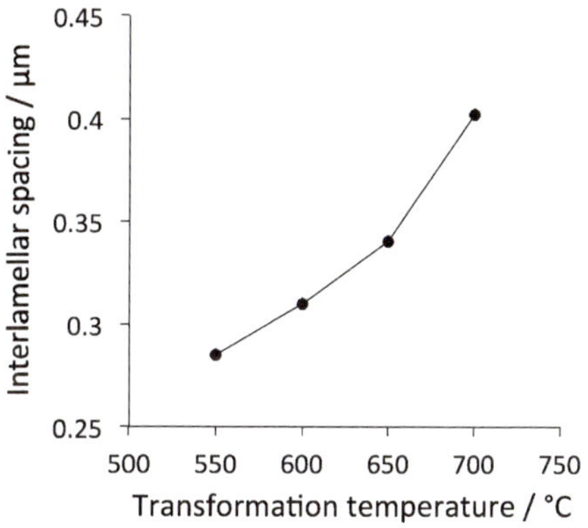

Figure 3.25 Interlamellar spacing in pearlite as a function of the undercooling below the eutectoid temperature. The steel composition is Fe-0.85C-0.67Mn-0.18Si wt%. Data from Ray and Mondal [38].

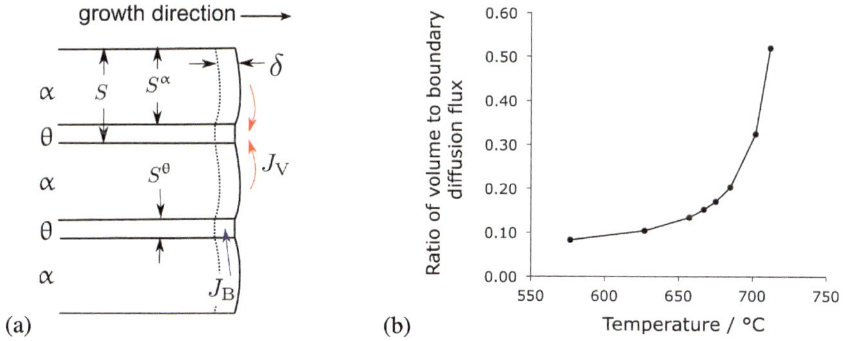

Figure 3.26 (a) Geometry of pearlite colony. The dashed arrows indicate the volume and interface diffusion processes. The thickness of the boundary is written δ and there is an additional flux through it, labelled J_B. (b) Relative contributions of volume and boundary diffusion fluxes during the formation of pearlite in Fe-0.8C wt% steel. Adapted from Pandit and Bhadeshia, [39].

Fig. 3.26b illustrates how the flux through the interface dominates that through the volume of the austenite as the transformation temperature is reduced, because although the area through which the boundary flux occurs is much smaller than the volume of austenite available, $D_{C,B} \gg D^{\gamma}_{C,V}$ at low temperatures.

3.6.4 Divorced pearlite

The formation of lamellar pearlite, in which the α and θ grow cooperatively at a common front, can be suppressed entirely by ensuring that there is an appropriate distribution of proeutectoid cementite particles present in the austenite before it reaches the eutectoid temperature. During the cooling of such a mixture, a divorced eutectoid occurs in which the proeutectoid cementite particles absorb the carbon that is partitioned by ferrite at the advancing γ/α interface [41–43]. The cementite particles coarsen in the process of absorbing the partitioned carbon, as illustrated in Fig. 3.27. A fine particle spacing and low undercooling favour the divorced form over the lamellar structure.

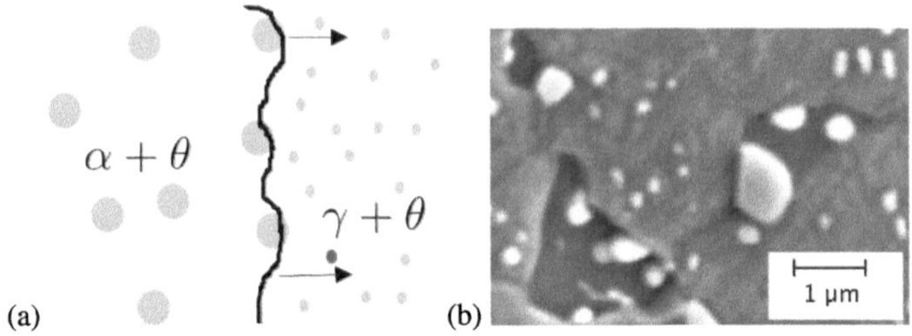

Figure 3.27 (a) The mechanism of the divorced eutectoid transformation of a mixture of austenite and fine cementite [41]. (b) The dark regions are ferrite containing coarse cementite (i.e., the divorced pearlite), surrounded by the austenite and fine proeutectoid cementite particles. The steel has a carbon concentration of about 1 wt% and yet does not transform into hard lamellar pearlite on cooling through the eutectoid temperature. Image courtesy of Danyi Luo.

This kind of a heat treatment is particularly useful when strong steels need to be soft during the fabrication process; a divorced eutectoid is the simplest way of achieving a soft steel that, after shaping into the appropriate form, is hardened to a level consistent with its application. The process is routine in the manufacture of bearings containing 1 wt% of carbon, that must be extremely hard when they enter service but are required to be softer than about 250 HV during the fabrication stage [44].

3.6.5 Overall kinetics of pearlite formation

The formation of pearlite is a good example of a nucleation and growth process. The pearlite nucleates at preferred sites in the austenite and the nuclei then grow until the nodules impinge on each other. The process is both time and temperature dependent, as it is controlled by the diffusivity of the relevant atoms. Johnson and Mehl first applied a detailed analysis of nucleation and growth to the pearlite reaction, which assumed that the fraction of austenite transformed (X) could be expressed in terms of a rate of nucleation $\dot{N}$ defined as the number of nuclei per unit volume of untransformed austenite formed per second, and a rate of growth of these nuclei v_P, expressed as radial growth in cm s^{-1}. They made certain simplifying assumptions, the most significant were:

a) Nucleation was regarded as a random event.
b) The rate of nucleation per unit volume, I_V, was assumed to be constant with time.
c) The rate of growth v_P was, as is the case for pearlite (Fig. 3.22b), assumed to be constant with time.
d) The nuclei were regarded as spherical and during growth they would eventually impinge on neighbouring spheres.

To model the development of volume fraction, as opposed to individual growth rates or nucleation rates, requires impingement between particles to be taken into account.

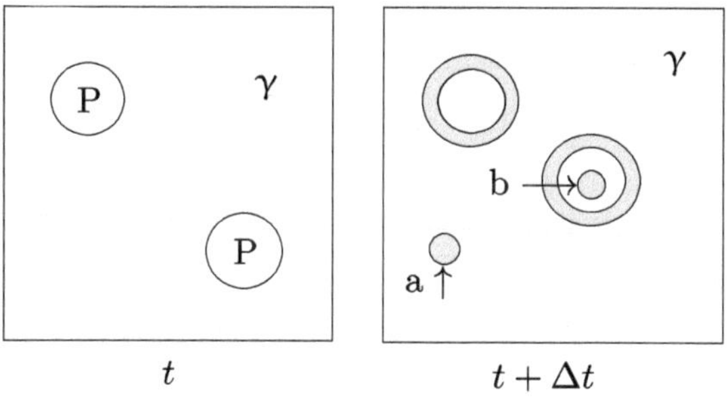

Figure 3.28 An illustration of the concept of extended volume. Two precipitate particles have nucleated together and grown to a finite size in the time t. New regions (grey) are formed as the original particles grow, but "a" and "b" are new particles, of which "b" formed in a region that is already transformed. "P" represents a pearlite nodule.

This is done using the extended volume concept of Kolmogorov, Johnson, Mehl and Avrami [45–49]. Referring to Fig. 3.28, suppose that two particles exist at time t; a small interval Δt later, new regions marked in grey are formed, assuming that they are able to grow unrestricted in extended space whether or not the region into which they grow is already transformed. However, only those components which lie in previously untransformed matrix can contribute to a change in the real volume of the product phase pearlite. Particles such as "b" cannot in fact nucleate in regions that already have transformed. To correct for this, the change in real volume is related to an extended volume change as follows:

$$\mathrm{d}V^{\mathrm{P}} = \left(1 - \frac{V^{\mathrm{P}}}{\mathrm{V}}\right)\mathrm{d}V_{\mathrm{e}}^{\mathrm{P}}, \tag{3.12}$$

where it is assumed that the microstructure develops at random. The subscript e refers to extended volume, V^{P} is the volume of pearlite and V is the total volume. Multiplying the change in extended volume by the probability of finding untransformed regions has the effect of excluding regions such as "b", which clearly cannot contribute to the real change in volume of the product. For a random distribution of precipitated particles, this equation can easily be integrated to obtain the real volume fraction,

$$\frac{V^{\mathrm{P}}}{\mathrm{V}} = 1 - \exp\left\{-\frac{V_{\mathrm{e}}^{\mathrm{P}}}{\mathrm{V}}\right\}. \tag{3.13}$$

The extended volume $V_{\mathrm{e}}^{\mathrm{P}}$ is straightforward to calculate using nucleation and growth models and neglecting completely any impingement effects. Assuming that pearlite grows isotropically at a constant rate v_{P} with a nucleation rate per unit vol-

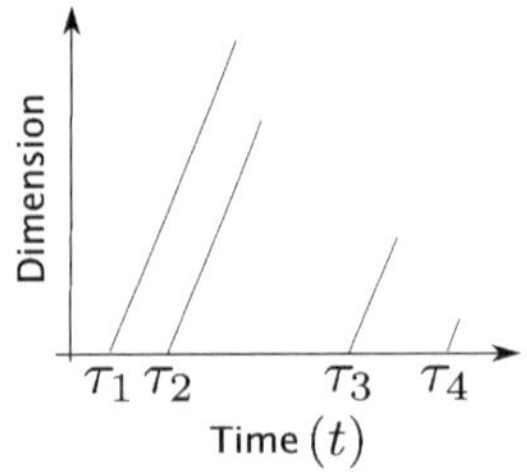

Figure 3.29 An illustration of the incubation time τ for each particle. For example, particle 1 does not exist before $t = \tau_1$. It also is assumed that each particle grows without impingement with any other.

ume, I_V, the volume of a nodule nucleated at time $t = \tau$ (Fig. 3.29) is given by

$$v_\tau = \frac{4}{3}\pi v_P^3 (t-\tau)^3. \tag{3.14}$$

The change in extended volume over the interval τ and $\tau + d\tau$ is

$$dV_e^P = \frac{4}{3}\pi v_P^3 (t-\tau)^3 \times I_V \times V \times d\tau. \tag{3.15}$$

By substituting into Eq. (3.13) and writing $\xi = V^P/V$, we get

$$dV^P = \left(1 - \frac{V^P}{V}\right)\frac{4}{3}\pi v_P^3 (t-\tau)^3 I_V V \; d\tau,$$

so that $$-\ln\{1-\xi\} = \frac{4}{3}\pi\, v_P^3\, I_V \int_0^t (t-\tau)^3\, d\tau$$

and $$\xi = 1 - \exp\{-\pi\, v_P^3\, I_V\, t^4/3\}, \tag{3.16}$$

which is the equation derived by Johnson and Mehl for the evolution of the volume fraction of pearlite during isothermal transformation. Here ξ is the volume fraction of the growing phase can consume all of the austenite, but in general the volume fraction normalised by the equilibrium volume fraction.

This relationship gives a sigmoidal type of curve, when ξ is plotted against t for chosen values of I_V and v_P. A typical curve is shown in Fig. 3.30a for particular values of I_V and v_P. If ξ is plotted against the logarithm of $\sqrt[4]{(I_V v_P^3)}t$, a sigmoidal master curve is obtained which expresses the basic kinetic behaviour expected of a nucleation and growth process in a given alloy (Fig. 3.30b).

The assumptions made in reaching Eq. (3.16) are not in fact necessary, for example, that the nucleation sites are randomly distributed. Rate laws have been derived for grain surface, edge and corner nucleation, while accounting for site saturation, i.e., the complete exhaustion of particular kinds of initiation sites [50]. A thorough review of overall transformation theory is available in [14]. There are further complications if the steel contains substitutional solutes, which will be dealt with in Chapter 4.

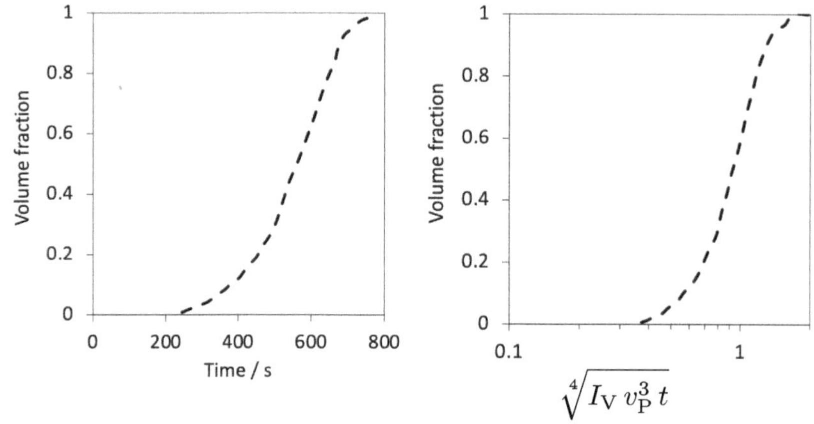

(a) (b)

Figure 3.30 Kinetics of pearlite reaction: (a) calculated curve for specific $I_V = 1000\,\text{cm}^{-3}\,\text{s}^{-1}$ and $v_P = 0.003\,\text{cm}\,\text{s}^{-1}$, (b) master reaction curve for general nucleation (after [37]).

3.6.6 The strength of pearlite

Pearlite on its own can be incredibly strong with commercial wires available routinely with strength in the range 3-4 GPa. The longest single-span suspension bridge in the world, the Akashi–Kaikyo Bridge, utilises exceptionally strong pearlitic ropes to suspend the deck. The bridge connects Kobe with Awaji Island and has a span of 1.9 km between the towers. There is enough steel wire used in the bridge to circle the earth seven times, with the bridge being designed to withstand an earthquake of Richter 8.5 magnitude. The bridge represents a magnificent triumph of engineering and steel in the form of pearlite (Fig. 3.31a). The strong wires do not use pearlite in its as-transformed condition, but in a cold-deformed state so that dislocation cell structures are created that reduce the spacing between structural barriers even more. The deformation is carried out by wire drawing in which case the cementite also experiences substantial plastic deformation.

The strength of pearlite would be expected to increase as the interlamellar spacing S is reduced. This is indeed the case, as illustrated in Fig. 3.31b,where the yield stress is plotted as a function of $S^{-\frac{1}{2}}$ to respect the Hall–Petch relationship described in Section 2.5. This would imply that macroscopic yielding propagates by the stimulation of dislocation sources in adjacent ferrite lamellae in the first instance, with the cementite only undergoing plastic deformation as the ferritic component work hardens and the strength incompatibilities between the different phases diminish in magnitude.

The situation is rather different for lower carbon steels, i.e., below 0.3 wt%, where pearlite occupies a substantially smaller volume of the microstructure. In these steels, the yield stress is not markedly affected as the proportion of pearlite is increased, provided other factors, e.g., ferrite grain size are kept constant. However, the tensile strength is quite sensitive to the pearlite content, which is explained by the fact

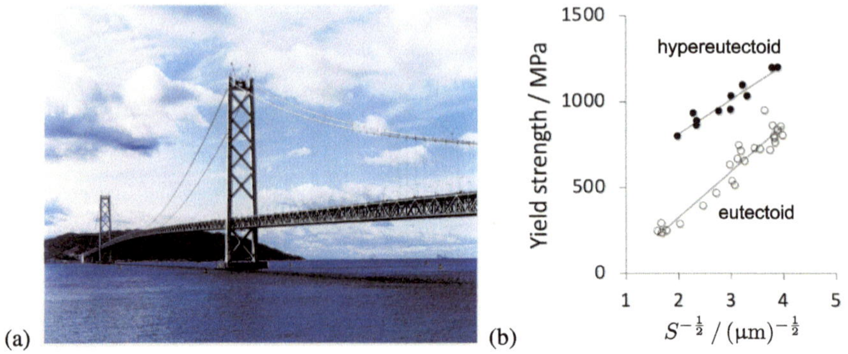

Figure 3.31 (a) The Akashi–Kaikyo Bridge in Japan, the longest single-span suspension bridge, which relies on huge cables made from pearlitic steel. Photograph courtesy of Professor Nobutaka Yurioka. (b) Strength of eutectoid and hypereutectoid pearlite as a function of the interlamellar spacing. Adapted from [51].

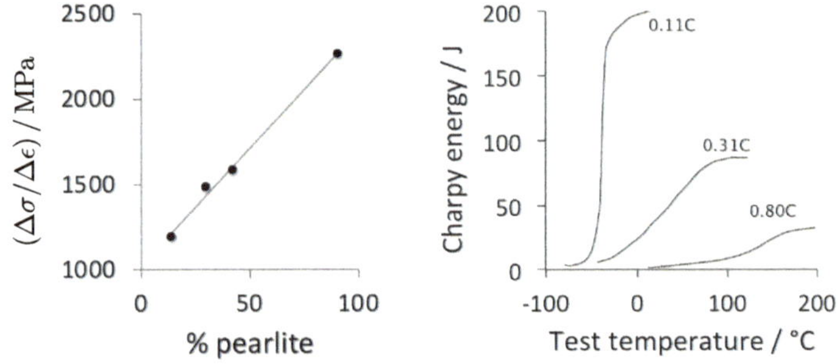

(a) (b)

Figure 3.32 (a) Effect of pearlite content on work hardening rate, expressed as the increase in strength for a unit increase in true strain. After [52]. (b) Effect of pearlite on toughness measured by Charpy impact transition temperature. After [52].

that there is a linear relationship between work hardening and the pearlite content (Fig. 3.32a), which arises because pearlite has a larger work hardening coefficient than the ferrite – it is generally the case in two-phase deformation theory that the composite work hardening coefficient is somewhere between that of the individual phases because of strain partitioning.

Pearlite has, however, an adverse effect on ductility and toughness of plain carbon steels. For example, the impact transition temperature (Chapter 11) is raised substantially as the carbon content is increased (Fig. 3.32b), and quantitative studies have shown that 1% by volume of pearlite raises the transition temperature by about 2°C. The presence of pearlite in the microstructure provides sites of easy nucleation of cracks, particularly at the ferrite-cementite interfaces. However, as a crack can only propagate in ferrite a short distance before encountering another cementite lamella, energy is absorbed during propagation. The result is that there is a wide transition temperature range (Fig. 3.32b). In contrast, the low energy absorbed overall in impact tests on pearlitic structures arises from the fact that many crack nuclei can occur at the pearlitic interfaces, which, together with the high work hardening rate, restricts plastic deformation in the vicinity of the crack. These naturally are mechanistic explanations, but a simple interpretation is that the strength increases with the volume fraction of pearlite, so the toughness is expected to decrease unless another mechanism is invoked whereby the entire structure is refined – how this can be achieved will be discussed in Chapter 10, suffice it to note here the ferrite-pearlite steels are incredibly successful in a vast number of applications that improve the quality of life.

3.7 Ferrite-pearlite steels

A high proportion of the steels used in industry have a ferrite-pearlite structure. These include a wide range of plain carbon steels where alloying additions are primarily made for steel-making purposes, although they do have a strengthening role as well. For example, manganese is added to combine with sulphur, but it is also a strengthener, while manganese and silicon are deoxidisers and aluminium is used as a deoxidiser and as a grain refiner, and therefore a strengthener. Many low and medium alloy steels, e.g., those with nickel, give ferrite-pearlite structures, but here only essentially plain carbon steels will be dealt with.

Most plain carbon steels are not subject to heat treatment in the sense of quenching followed by tempering, but they are cooled at different rates to obtain a range of structures. Two important treatments are normalising and annealing, which have special but not very precise meanings when applied to steels.

Normalising

In the process of normalising, the steel is reheated about 100°C above the Ac_3 temperature to form austenite, followed by air cooling through the phase transformation. This avoids the excessive growth of γ-grains while promoting a reasonably fine ferrite and pearlite microstructure. It is often used after hot rolling, where a high finishing temperature can lead to a coarse microstructure.

The rate of cooling during normalising is dependent on the dimensions of the steel, but some control can be exerted by using forced air cooling.

Slow cooling and annealing

It is often the case that following austenitisation at a fairly high temperature, the steel is cooled slowly, for example in a furnace. This results in transformation at the higher temperatures in the pearlite range, giving a coarse pearlite which is conducive to good machinability. After all, during the fabrication of components, it is useful to deal with a soft material that can later be heat treated to generate a finer pearlite.

There are other types of heat treatments that are commonly practiced, e.g., cooling from the austenitic state to just below the transformation temperature and holding the temperature there to permit isothermal transformation into a mixture of ferrite and coarse pearlite. Spheroidising annealing is applied to higher carbon pearlitic steels to improve their machinability. The steel is then held at a temperature just below Ae_1 for sufficient time to permit the cementite lamellae of the pearlite to spheroidise. This is a relatively slow process because the driving force is simply the reduction in the total α/θ interfacial energy per unit volume.

Ferrite-pearlite steels are essentially those which depend for their properties on the presence of carbon and manganese. The carbon content can be varied from 0.05-1.0 wt% while the manganese content is from 0.25 wt% up to about 1.7 wt%. Fig. 3.33 shows the effect on the tensile strength of varying the concentration of these two elements. It has also been possible by regression analysis to determine the relative contributions to the strength of the three important mechanisms: solid solution hardening, grain size and dispersion strengthening from lamellar pearlite. The results plotted are from steels in the normalised condition, which ensures that the austenite grain sizes are roughly comparable. Variation of the carbon at constant manganese level causes a substantial increase in strength, which is almost entirely due to an increasing proportion of pearlite in the structure. The situation is rather more complex when manganese is varied at constant carbon content, as all three strengthening mechanisms are influenced. Manganese causes the eutectoid composition to occur at lower carbon contents, and so increases the proportion of pearlite in the microstructure. Manganese is also an effective solid solution strengthener and has a grain refining influence.

It is clear that carbon provides a very cheap way of strengthening normalised steels, but the extent to which this approach can be used depends on whether the steel is to be welded or not. Welding of higher carbon steels makes them susceptible to cracking within the weld heat-affected zone, where brittle, untempered martensite may be generated, so it usually is necessary to limit the carbon content to not greater than 0.2 wt%. In these circumstances, additional strength can then be obtained by solid solution hardening by raising the manganese content to between 1 and 1.5 wt%.

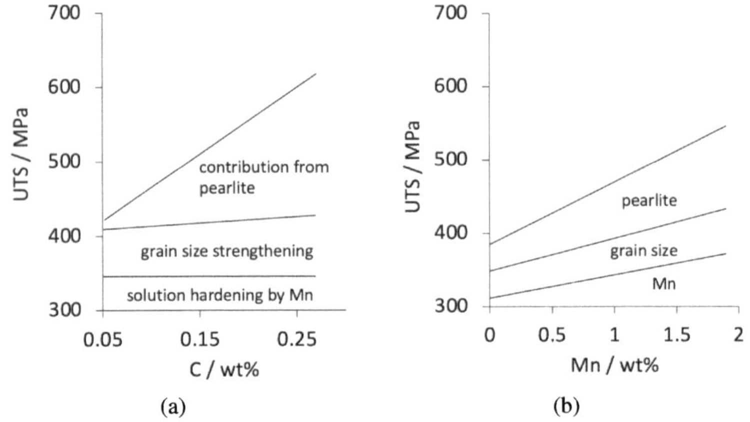

Figure 3.33 Factors contributing to the strength of C-Mn steels. Data from Irvine [53].

Alternatively, refinement of the grain size can be achieved by minor alloying additions such as aluminium, vanadium, titanium and niobium, in concentrations not normally exceeding 0.1 wt% (Chapter 9). Aluminium forms a stable dispersion of AlN particles, some of which remain in the austenite grain boundaries at high temperatures and by pinning these boundaries prevent excessive grain growth. On transformation to ferrite and pearlite, grain sizes around 5-6 μm can be achieved with as little as 0.03 wt% AlN in the steel. Vanadium, titanium and niobium form very stable carbides, which also lock austenite grain boundaries, and thus allow much finer ferrite grain sizes to be achieved when the austenite (or deformed austenite) transforms (Chapter 10).

Plain-carbon steels are used mostly in the hot-finished condition, i.e., straight from hot rolling without subsequent cold rolling or heat treatment. This represents the cheapest form of steel, containing less than about 0.25 wt% carbon to avoid the formation of martensite in the heat-affected zone during welding. The most important group of hot-finished plain carbon steels contains less than 0.25 wt% carbon and is used in structural shapes such as plates, I-beams, angles, etc., in buildings, bridges, ships, pressures vessels and storage tanks. Hot-rolled low carbon steel sheet is an important product and used extensively for fabrication where surface finish is not of prime importance. Cold rolling is used for finishing where better finish is required, and the additional strength from cold working is needed. However, for high-quality sheet to be used in intricate pressing operations, it is necessary to anneal the cold-worked steel to cause the ferrite to recrystallise. This is done below the Ae_1 temperature (subcritical annealing).

Carbon steels are also used extensively for closed die or drop forgings, usually containing 0.2-0.5 wt% carbon, and covering a wide range of applications, e.g., shafts and gears. The other important field of application of plain carbon steels is as castings. Low-carbon cast steels containing up to 0.25 wt% C are used for miscellaneous cast-components as reasonable strength and ductility levels are readily obtained. Yield strengths of 240 MPa and elongations of 30% are fairly typical for this type of steel.

3.8 Summary

The carbon atom is small and hence it resides in the spaces between the iron atoms. The allotropic forms of iron are sufficiently different in their crystal structures to have vastly different capacities to dissolve carbon. This has major consequences when transformations occur, because carbon must then be partitioned into the phase where it is more soluble. The transformation front can then only move at a rate consistent with the diffusion of the excess carbon into the far-field, such that equilibrium is maintained at the boundary between the parent and product crystals. This basic principle leads to the kinetic theory for diffusion-controlled growth, in which the shape of the transformation product also matters. Thus, allotriomorphic ferrite grows at a diminishing rate, Widmanstätten ferrite at a constant rate because of its plate shape, and pearlite at a constant rate because there in no net partitioning of carbon. When nucleation and growth theory can be combined with a model for the impingement of particles that originated from different sites, it becomes possible to calculate time-temperature-transformation diagrams that depict the isothermal evolution of structure as a function of time.

There are many approximations involved in all of the theory presented in this chapter, but the essence of the principles involved is clear and more elaborate theory is available [14,54].

Returning now to the iron-carbon equilibrium diagram, it is intriguing that the concentration scale in most publications does not extend beyond cementite, which contains 25 at.% or 6.67 wt% carbon. The domain beyond cementite consists of carbides that are even richer in carbon concentration than cementite. Fig. 3.34a shows one such carbide, known as Hägg carbide, which has the formula Fe_5C_2 ($\approx$ 30 at.% carbon) forms when the steel is exposed to a sufficient carbon activity during carburisation [55]. The Hägg carbide forms after cementite grows on the iron surface. Given the micrograph in Fig. 3.34a where all three phases (α, θ and Hägg) are present, it is possible suggest that the Fe-C equilibrium diagram beyond the concentration corresponding to 100% cementite might appear as in Fig. 3.34b. In such a phase diagram, other carbides may also be present that have a concentration greater than cementite, for example, ε-carbide, which has a hexagonal crystal structure, so that the actual diagram might be more complex.

Hägg carbide in isolation is known to decompose into a mixture of cementite and carbon during appropriate high-temperature heat treatment [56]. It is assumed here that any pure carbon, which is precipitated, has the form of graphite.

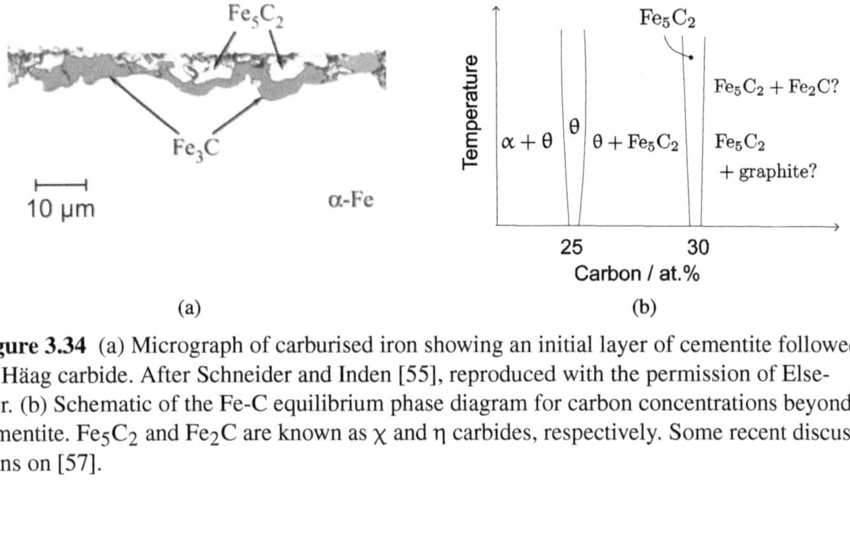

Figure 3.34 (a) Micrograph of carburised iron showing an initial layer of cementite followed by Häag carbide. After Schneider and Inden [55], reproduced with the permission of Elsevier. (b) Schematic of the Fe-C equilibrium phase diagram for carbon concentrations beyond cementite. Fe_5C_2 and Fe_2C are known as χ and η carbides, respectively. Some recent discussions on [57].

References

[1] M. Hansen, K. Anderko, Constitution of Binary Alloys, 2nd ed., McGraw Hill, New York, USA, 1958.

[2] T. Okamoto, H. Matsumoto, Precipitation of ferrite from cementite, Metal Science 9 (1975) 8–12.

[3] H.L. Yakel, Crystal structure of stable and metastable iron-containing carbides, International Metals Reviews 30 (1985) 17–40.

[4] A. Leineweber, S.L. Shang, Z.K. Liu, C-vacancy concentration in cementite, Fe_3C_{1-z}, in equilibrium with α-Fe[C] and γ-Fe[C], Acta Materialia 86 (2015) 374–384.

[5] M. Hillert, The formation of pearlite, in: V.F. Zackay, H.I. Aaronson (Eds.), Decomposition of Austenite by Diffusional Processes, Interscience, New York, USA, 1962, pp. 197–237.
[6] H.K.D.H. Bhadeshia, Bainite: the incomplete-reaction phenomenon and the approach to equilibrium, in: M.S.H.I. Aaronson, D.E. Laughlin, C.M. Wayman (Eds.), Solid-Solid Phase Transformations, TMS-AIME, Materials Park, Ohio, USA, 1981, pp. 1041–1048.
[7] A.D. King, T. Bell, Crystallography of grain boundary proeutectoid ferrite, Metallurgical Transactions A 6 (1975) 1419–1429.
[8] A.D. King, T. Bell, Morphology and crystallography of proeutectoid ferrite, Metallography 9 (1976) 397–413.
[9] D.W. Kim, R.S. Qin, H.K.D.H. Bhadeshia, Transformation texture of allotriomorphic ferrite in steel, Materials Science and Technology 25 (2009) 892–895.
[10] D.W. Kim, D.W. Suh, R.S. Qin, H.K.D.H. Bhadeshia, Dual orientation and variant selection during diffusional transformation of austenite to allotriomorphic ferrite, Journal of Materials Science 45 (2010) 4126–4132.
[11] S. Jones, H.K.D.H. Bhadeshia, Kinetics of the simultaneous decomposition of austenite into several transformation products, Acta Materialia 45 (1997) 2911–2920.
[12] M.V. Kral, G. Spanos, Three-dimensional analysis of proeutectoid cementite precipitates, Acta Materialia 47 (1999) 711–724.
[13] M.X. Zhang, P.M. Kelly, Crystallography and morphology of Widmanstätten cementite in austenite, Acta Materialia 46 (1998) 4617–4628.
[14] J.W. Christian, Theory of Transformations in Metals and Alloys, Part I, 3rd ed., Pergamon Press, Oxford, U.K., 2003.
[15] C.S. Smith, Microstructure, Transactions of ASM 45 (1953) 533–575.
[16] H.K.D.H. Bhadeshia, Diffusion of carbon in austenite, Metal Science 15 (1981) 477–479.
[17] H.K.D.H. Bhadeshia, Diffusional formation of ferrite in iron and its alloys, Progress in Materials Science 29 (1985) 321–386.
[18] J.R. Bradley, J.M. Rigsbee, H.I. Aaronson, Growth kinetics of grain boundary ferrite allot. in Fe-C alloys, Metallurgical Transactions A 8 (1977) 323–333.
[19] H.K.D.H. Bhadeshia, Rationalisation of shear transformations in steels, Acta Metallurgica 29 (1981) 1117–1130.
[20] J.D. Watson, P.G. McDougall, The crystallography of Widmanstätten ferrite, Acta Metallurgica 21 (1973) 961–973.
[21] H.K.D.H. Bhadeshia, Comments on "The mechanisms of the fcc-bcc martensitic transformation revealed by pole figures", Scripta Materialia 64 (2011) 101–102.
[22] J.W Christian, Theory of Transformations in Metals and Alloys, Part I, 2nd ed., Pergamon Press, Oxford, U.K., 1975.
[23] H.K.D.H. Bhadeshia, Critical assessment: diffusion controlled growth of ferrite plates in plain carbon steels, Materials Science and Technology 1 (1985) 497–504.
[24] H.C. Sorby, On the application of very high powers to the study of the microscopical structure of steel, Journal of the Iron and Steel Institute 1 (1886) 140–147.
[25] T. Furuhara, N. Kimura, T. Maki, Enhancement of pearlite nucleation by deformation of austenite matrix in an Fe–12Mn–0.8C alloy, Steel Research International 83 (2012) 358–362.
[26] Y.L. Tian, R.W. Kraft, Mechanisms of pearlite spheroidization, Metallurgical Transactions A 18 (1998) 1403–1414.
[27] P.R. Howell, The pearlite reaction in steels: mechanisms and crystallography, Materials Characterization 40 (1998) 227–260.
[28] S.A. Hackney, G.J. Shiflet, The pearlite-austenite growth interface in an Fe-0.8C-12Mn alloy, Acta Materialia 35 (1987) 1007–1017.

[29] M.-X. Zhang, P.M. Kelly, Accurate orientation relationships between ferrite and cementite in pearlite, Scripta Materialia 37 (1997) 2009–2015.
[30] I.V. Isaichev, Orientation of cementite in tempered carbon steel (Orientatsiya tsementita v otpushchennoi uglerodistoi stali), Zhurnal Tekhncheskoi Fiziki 17 (1947) 835–838.
[31] F.G. Wei, K. Tsuzaki, Crystallography of [011]/54.7° lath boundary and cementite in tempered 0.2C steel, Acta Materialia 53 (2005) 2419–2429.
[32] M.-X. Zhang, P.M. Kelly, Determination of carbon content in bainitic ferrite and carbon distribution in austenite by using CBKLDP, Materials Characterization 40 (1998) 159–168.
[33] M.-X. Zhang, P.M. Kelly, Crystallography of spheroidite and tempered martensite, Acta Materialia 46 (1998) 4081–4091.
[34] H.K.D.H. Bhadeshia, Solution to the Bagaryatskii and Isaichev ferrite-cementite orientation relationship problem, Materials Science and Technology 34 (2018) 1666–1668.
[35] A. Hultgren, Isothermal transformation of austenite, Transactions of American Society for Metals 39 (1947) 915–1005.
[36] J.P. Houin, A. Simon, G. Beck, Relationship between structure and mechanical properties of pearlite between 0.2% and 0.8%C, Transactions of the Iron and Steel Institute of Japan 21 (1981) 726–731.
[37] R.F. Mehl, W.C. Hagel, The austenite: pearlite reaction, Progress in Metal Physics 6 (1956) 74–134.
[38] K.K. Ray, D. Mondal, Effect of interlamellar spacing on strength of pearlite in annealed eutectoid and hypoeutectoid plain carbon steels, Acta Metallurgica Et Materialia 39 (1991) 2201–2208.
[39] A.S. Pandit, H.K.D.H. Bhadeshia, Mixed diffusion-controlled growth of pearlite in binary steel, Proceedings of the Royal Society A 467 (2011) 508–521.
[40] M. Hillert, The role of interfaces in phase transformations, in: Mechanism of Phase Transformations in Crystalline Solids, in: Monograph and Report Series, vol. 33, Institute of Metals, London, U.K., 1970, pp. 231–247.
[41] J.D. Verhoeven, E.D. Gibson, The divorced eutectoid transformation in steel, Metallurgical & Materials Transactions A 29 (1998) 1181–1189.
[42] A.S. Pandit, H.K.D.H. Bhadeshia, Divorced pearlite in steels, Proceedings of the Royal Society A 468 (2012) 2767–2778.
[43] D. Luo, M.J. Peet, S.W. Ooi, P. Yan, Z. Yin, H.K.D.H. Bhadeshia, Spheroidisation of hypereutectoid state of nanostructured bainitic steel, Materials Science and Technology 30 (2014) 1282–1286.
[44] H.K.D.H. Bhadeshia, Steels for bearings, Progress in Materials Science 57 (2012) 268–435.
[45] A.N. Kolmogorov, On statistical theory of metal crystallisation, Izvestiya Akad. Nauk SSSR (Izvestia Academy of Science, USSR) Ser. Math. 3 (1937) 335–360.
[46] M. Avrami, Kinetics of phase change 1, Journal of Chemical Physics 7 (1939) 1103–1112.
[47] W.A. Johnson, R.F. Mehl, Reaction kinetics in processes of nucleation and growth, TMS-AIMME 135 (1939) 416–458.
[48] M. Avrami, Kinetics of phase change 2, Journal of Chemical Physics 8 (1940) 212–224.
[49] M. Avrami, Kinetics of phase change 3, Journal of Chemical Physics 9 (1941) 177–184.
[50] J.W. Cahn, The kinetics of grain boundary nucleated reactions, Acta Metallurgica 4 (1956) 449–459.
[51] E.M. Taleff, J.J. Lewandowski, B. Pouladian, Microstructure-property relationships in pearlitic eutectoid and hypereutectoid carbon steels, Journal of Metals 54 (2002) 25–30.
[52] K.W. Burns, F.B. Pickering, Deformation and fracture of ferrite-pearlite structures, Journal of the Iron and Steel Institute 202 (1964) 889–906.

[53] K.J. Irvine, The development of high strength steel, Journal of the Iron and Steel Institute 200 (1962) 820–828.
[54] J.W. Christian, Theory of Transformations in Metals and Alloys, Part II, 3rd ed., Pergamon Press, Oxford, U.K., 2003.
[55] A. Schneider, G. Inden, Carbon diffusion in cementite (Fe_3C) and Hägg carbide (Fe_5C_2), Computer Coupling of Phase Diagrams and Thermochemistry 31 (2007) 141–147.
[56] E.M. Cohn, L.J.E. Hofer, Some thermal reactions of higher iron carbides, Journal of Chemical Physics 21 (1953) 354–359.
[57] H.K.D.H. Bhadeshia, A.R. Chintha, S. Lenka, Are χ (Hägg), η and ε carbides transition-phases relative to cementite in steels?, Materials Science and Technology 35 (2019) 1301–1305.

Solutes that are substitutes for iron 4

Abstract

If the wealth of structures available in the binary Fe–C system is impressive, the addition of substitutional solutes creates a breathtaking variety of phases and structures, a seemingly endless palette that makes it feasible to custom-design alloys of iron. The substitutional solutes influence the thermodynamics of all transformations, but have the most profound effect when they are required to diffuse. While all of this is very useful, the complexity of the theory necessary to deal with multicomponent steels is also increasing, ameliorated by the availability of modern computer programs and databases. The understanding necessary to deal with such complexity is introduced so that the reader can make intelligent use of the mathematical models implemented in the software.

4.1 General principles

The term *alloying elements* in the context of steels often is used to denote substitutional solutes, which can and do dramatically influence the structure and properties of steels. Indeed, they are responsible for the incredible versatility and utility of steels. The general effects of substitutional solutes are summarised in Fig. 4.1 and discussed in context in this chapter. The solutes affect the relative free energies of the relevant phases, a thermodynamic effect that applies to all of the phase transformations, irrespective of the detailed atomic mechanisms of transformation. Unexpected phases frequently appear, for example, ϵ-iron, intermetallic compounds such as Laves phase, carbides, nitrides and oxides that are rich in substitutional solutes and an array of transition phases such as glassy precipitates. Careful studies often show unexpected structures, for example, an amorphous interface between the crystalline phases CrTaN and ferrite [1]. Some of these phases may be confined to the surface of the steel.

The thermodynamic calculation of phase diagrams has been an incredible asset in alloy design, but the method is empirical in the sense that it relies on measured experimental data. It cannot therefore provide estimates for phases that have not yet been characterised or which may not yet have been discovered. An example is the Z-phase (CrNbN), which came to prominence during long-term creep rupture studies of ferritic steel containing Cr, Mo, V and Nb [2]. This stimulated the development of the relevant thermodynamic database for phase diagram calculations, revealing that all modern, ferritic creep-resistant alloys are susceptible to detrimental Z-phase precipitation, even in alloys free from Nb [3]. However, the kinetics of precipitation could be controlled by adjusting the composition to delay precipitation beyond the service life of the steel.

This leads nicely to the second effect (Fig. 4.1), related to the kinetics of transformations, where the atomic mechanisms become important. Solutes partitioning during reconstructive transformation limit the rate of change, an effect absent during diffusionless transformation. The consequence is illustrated vividly in the TTT diagram of Fig. 4.2, where the change in kinetics due to the addition of manganese is much larger for reconstructive transformations due to the diffusion required, whereas the evidently

Steels. https://doi.org/10.1016/B978-0-44-318491-8.00010-1

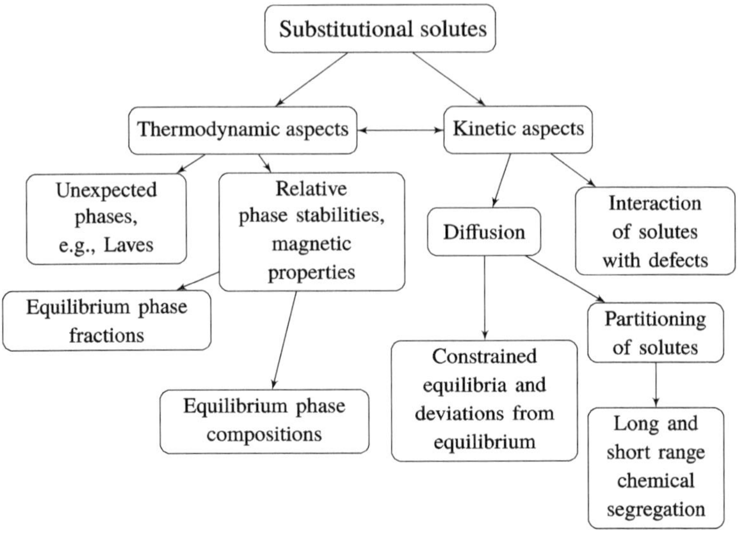

Figure 4.1 The variety of ways in which substitutional solutes can affect phase transformations and phase stabilities in steels. That thermodynamics and kinetics are not entirely interdependent consequences of solutes is highlighted by the double-headed arrow joining the blocks. For example, the free energy of a single-phased system can depend on extent of the homogeneity of composition, which in turn is determined by kinetics.

smaller thermodynamic effect on Mn on the relative stabilities of the γ and α affects both mechanisms of transformation.

4.2 Alloying elements: γ and α phase fields

While it would not be reasonable to include a comprehensive discussion of the thermodynamic effects of substitutional solutes, there are some useful generic concepts that are helpful in classifying their roles, beginning with the consequences on the binary iron-carbon equilibrium diagram.

Commercial alloys of iron are almost never simple binary or ternary mixtures, because specific additions are used to achieve the combination of properties required for specific applications; a typical steel contains ten or more deliberate additions of solutes in controlled concentrations ranging from parts per million to a few weight percent. It is necessary therefore to deal with multicomponent systems, and fortunately it is possible now to calculate with some confidence, multicomponent phase diagrams on a routine basis. Such diagrams may not be as easily visualised as their binary or ternary counterparts, but multicomponent phase diagrams carry the same numerical information, i.e., phase fractions and phase compositions as a function of chemical composition, temperature and pressure. Such data can easily be tabulated and selectively plotted instead of struggling with multidimensional plots or their projections into three dimensions. Consider, for example, a steel of the following chemical

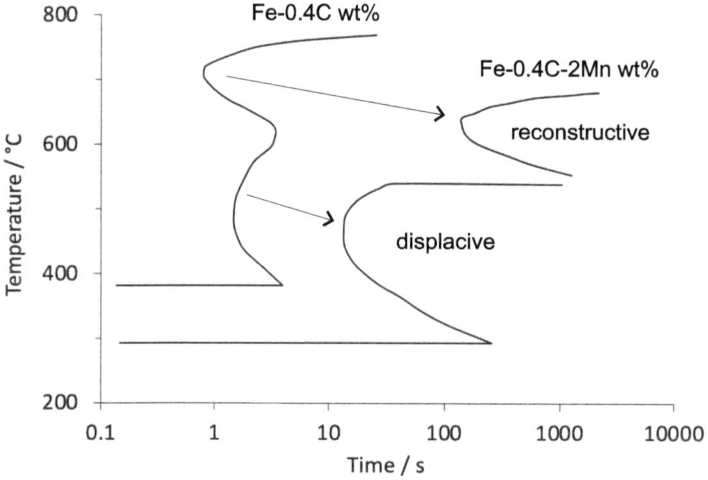

Figure 4.2 TTT diagrams for two steels, one of which contains manganese. Notice that the time scale is logarithmic, which emphasises that the effect of the Mn on the displacive transformation is much smaller than on the reconstructive reactions where solutes partition between the parent and product phases by diffusion.

composition:

$$\text{Fe-1.04C-0.25Si-0.35Mn-0.125Ni-0.05Mo-1.45Cr wt\%.}$$

A multicomponent, multiphase diagram calculation yields the data listed in Table 4.1 and Fig. 4.3 is an example of the selective plot where the amount of the carbide $M_{23}C_6$ that is in equilibrium with ferrite and cementite is plotted as a function of temperature.[1] The plot may need to be supplemented by others that show the chemical composition of $M_{23}C_6$ as a function of temperature, and there may be several other similar plots for the variety of phases present.

Table 4.1 Calculated equilibrium phase data for a multicomponent steel at 400°C. 'M' in $M_{23}C_6$ stands for metal atoms. All the phases listed were permitted to exist in the calculation [4].

Phase	moles of phase per 100 kg	atomic percent					
		C	Si	Mn	Ni	Mo	Cr
α	1517	0.0006	0.59	0.02	0.01	0.00	0.03
Cementite	342	25.00	0.00	1.77	0.04	0.00	7.83
$M_{23}C_6$	5.46	20.69	0.00	0.03	0.52	9.51	0.13
M_6C	0						
γ	0						

[1] M in $M_{23}C_6$ stands for metal atoms.

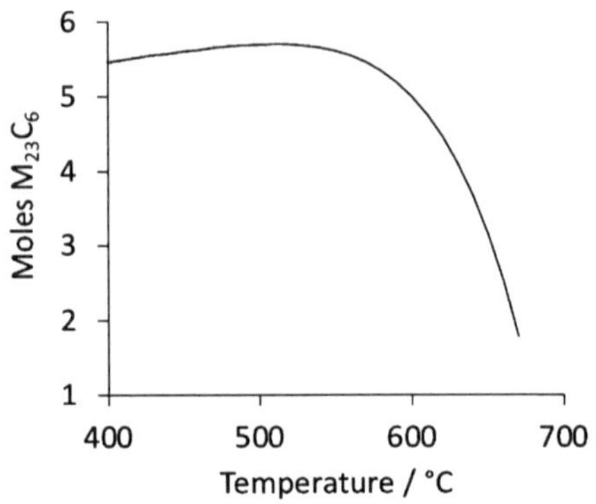

Figure 4.3 Calculated change in the number of moles in 100 kg of steel, of $M_{23}C_6$ that is in equilibrium with α and cementite as a function of temperature. In practice, equilibrium would be hard to achieve at temperatures below about 500°C because of the limited mobility of substitutional solutes. After [4].

Having described the complexity of multicomponent steels, it nevertheless is useful to look at simplifications of the role of alloying elements on the relative stabilities of austenite and ferrite, which after all, are the dominant phases in most steels. Wever pointed out that iron/substitutional-solute binary equilibrium systems fall into four main categories (Fig. 4.4): open and closed γ-field systems, and expanded and contracted γ-field systems [5]. This approach indicates that alloying elements can influence the equilibrium phase-diagram in two ways:

a) By expanding the γ-field and encouraging the formation of austenite over wider compositional limits. These elements are called γ-stabilisers.

b) By contracting the γ-field and encouraging the formation of ferrite over wider compositional limits. These elements are called α-stabilisers.

The form of the diagram depends to some degree on the electronic structure of the alloying elements, which is reflected in their relative position in the periodic classification.

Class 1: Open γ-field To this group belongs the important steel alloying elements manganese and nickel, as well as cobalt and the exotic metals ruthenium, rhodium, palladium, osmium, iridium and platinum. Both nickel and manganese, if added in sufficiently large concentrations, completely eliminate the bcc α-iron phase and replace it, down to room temperature, with the γ-phase or even the ϵ form of iron. So, nickel and manganese depress the phase transformation from γ to α to lower temperatures (Fig. 4.4a), i.e., both Ae_1 and Ae_3 are lowered. It also becomes easier to obtain metastable austenite by quenching from the γ-region to room temperature. Nickel and manganese are therefore useful elements in the formulation of austenitic steels (Chapter 11). Concentrations of nickel or manganese in austenitic steels can be as large as 20 wt%. Commercial manganese-rich alloys of this kind have the peculiar characteristic that there is intense mechanical twinning during plastic deformation at ambient temperature, leading to much increased uniform elongation; these particular alloys, known as the TWIP steels, are discussed in Chapter 10. It is emphasised, however, that both the manganese-rich and nickel-rich alloys have other alloying elements that also play a role in determining the phase stability.

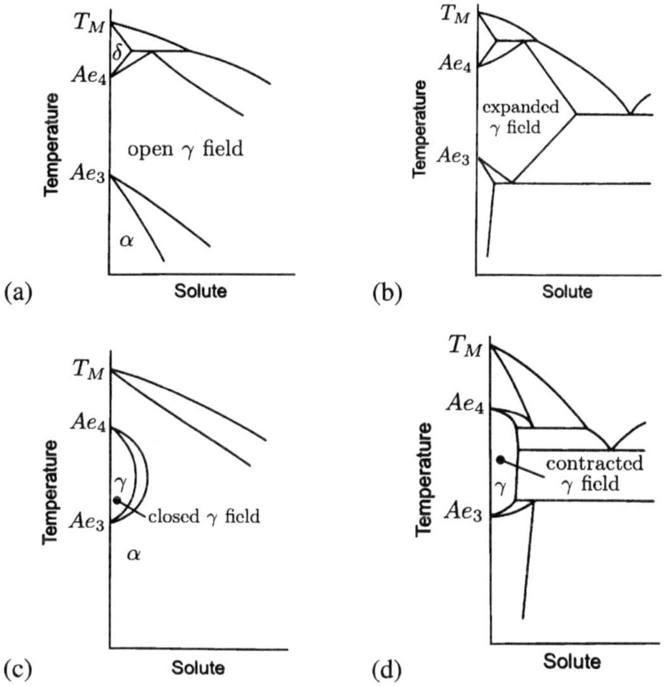

Figure 4.4 Classification of iron alloy phase diagrams: (a) open γ-field; (b) expanded γ-field; (c) closed γ-field; (d) contracted γ-field. After Wever [5].

Class 2: Expanded γ-field Carbon and nitrogen are the most important elements in this group. The γ-phase field is expanded, but its range of existence can be cut short by compound formation (Fig. 4.4b) – copper and gold have a similar influence on the γ-phase field. The expansion of the γ-field by carbon and nitrogen underlies the whole of the heat treatment of steels by allowing the formation of a homogeneous solid solution (austenite) containing up to 2.0 wt% of carbon or 2.8 wt% of nitrogen.

Class 3: Closed γ-field Many elements restrict the formation of γ-iron, causing the γ-region of the diagram to contract to a small area referred to as the gamma loop (Fig. 4.4c). This means that the relevant elements are encouraging the formation of bcc iron (ferrite), and one result is that the δ- and α-phase fields become continuous. Alloys in which this has taken place are, therefore, not amenable to the normal heat treatments involving cooling through the γ/α-phase transformation. Silicon, aluminium, beryllium and phosphorus fall into this category, together with the strong carbide-forming elements titanium, vanadium, molybdenum and chromium. Sometimes it is useful to avoid austenite altogether. A coarse ferrite grain structure is exploited in steels, which have to be magnetically soft for applications in electrical transformers. The δ-ferrite grains form at temperatures close to melting and hence are coarse. By adding 4 wt% Si, austenite is avoided enabling the grains to be retained at room temperature. Ferritic alloys of iron and silicon are the main soft-ferromagnetic materials (easy to magnetise and demagnetise) used in the manufacture of magnetic cores in devices such as transformers, motors, electromagnets and electrical machines in general.

Class 4: Contracted γ-field Boron is the most significant element of this group, together with the carbide-forming elements tantalum, niobium and zirconium. The γ-loop is strongly contracted, but is accompanied by compound formation (Fig. 4.4d).

The overall behaviour can be approximated in thermodynamic terms along the lines developed by Zener [6].

When α and γ are in equilibrium, the chemical potential μ_i of each element is identical in both phases[2]:

$$\mu_i^{\alpha} = \mu_i^{\gamma} \qquad \text{for all } i \tag{4.1}$$

The chemical potential is expanded as $\mu_i = \mu_i^{\circ} + RT\ln\{a_i\}$, where i represents a particular element, μ_i° represents the free energy per mole of the pure component i in the structure of the phase concerned and a_i is the activity of i and may be approximated by the concentration c_i for dilute solutions. Therefore,

$$\mu_i^{\circ,\alpha} + RT\ln\{c_i^{\alpha}\} = \mu_i^{\circ,\gamma} + RT\ln\{c_i^{\gamma}\}$$

$$\frac{c_i^{\alpha}}{c_i^{\gamma}} = \exp\left\{\frac{\mu_i^{\circ,\gamma} - \mu_i^{\circ,\alpha}}{RT}\right\}$$

$$\text{for ferrite formers,} \quad \mu_i^{\circ,\alpha} < \mu_i^{\circ,\gamma} \qquad \Delta\mu_i^{\circ} + ve$$

$$\text{for austenite formers,} \quad \mu_i^{\circ,\alpha} > \mu_i^{\circ,\gamma} \qquad \Delta\mu_i^{\circ} - ve$$

In the simple treatment, two fundamentally different types of equilibrium diagrams are obtained, where the phase boundaries are represented by similar thermodynamic equations, but depending on whether $\Delta\mu_i^{\circ}$ is positive or negative, are mirror images of each other (Fig. 4.5). In the $\Delta\mu_i^{\circ}$ negative case the γ-field is unlimited, while in the $\Delta\mu_i^{\circ}$ positive case, the γ-loop is introduced. $\Delta\mu_i^{\circ}$ will vary widely from element to element. In Fig. 4.6 illustrates the relative strengths of alloying elements in terms of ΔH_i, which is the heat of solution of i in γ minus that in α; at zero Kelvin, it relates to $\Delta\mu_i^{\circ}$. The ferrite formers are listed in (a) and the austenitic formers in (b). Apart from these general considerations, any solute or combination of solutes that increases the free energy of austenite relative to that of ferrite will stabilise the latter phase and accelerate the $\gamma \rightarrow \alpha$ transformation, as illustrated in Fig. 4.7. Elements such as silicon (Co, Al) therefore accelerate transformation, whereas Mn, Ni and Cr are seen to retard transformation.

[2] See [7] for an introduction to equilibria between phases.

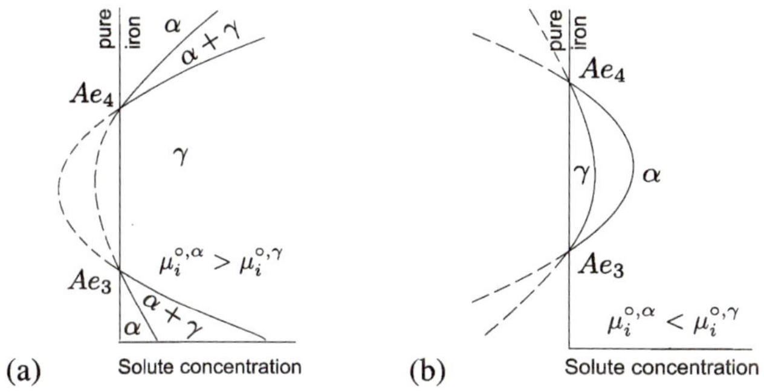

Figure 4.5 Two basic phase diagrams: (a) γ favoured; (b) α favoured (after Zener [6]).

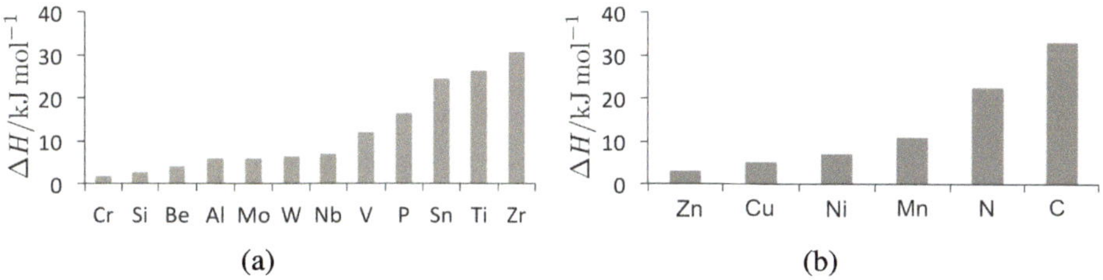

Figure 4.6 Relative potency of alloying elements that are in solid solution, as: (a) ferrite formers; (b) austenite formers (after Andrews [8]).

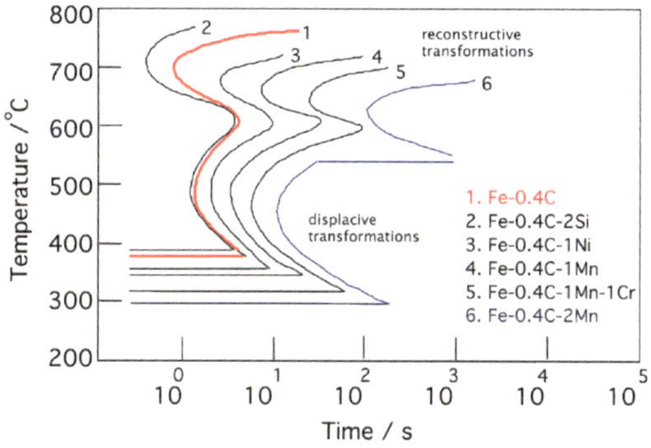

Figure 4.7 Time–temperature transformation diagrams for the initiation of transformation in a variety of steels [9]. Each diagram consists of two C-curves, the one for high temperatures representing reconstructive and that at lower temperatures, displacive transformations. The curves are experimentally difficult to separate if the reaction rate is rapid.

4.3 Distribution of alloying elements in steels

The discussion in Section 4.2 focuses on how substitutional alloying influences the relative stability of austenite and ferrite in binary mixtures. The general trends are in many cases similar when carbon is added to form ternary steels. For a fixed carbon content, as the alloying element is added the γ-field is either expanded or contracted depending on the particular solute. With an element such as silicon, the γ-field is restricted and there is a corresponding enlargement of the α-field. When vanadium is added, the γ-field contracts and vanadium carbide is in equilibrium with ferrite over much of the ferrite field. Nickel does not form a carbide in steel, although it can be manufactured independently as Ni_3C; its important contribution is to expand the γ-field. Normally, elements with opposing tendencies cancel each other out at the appropriate combinations, but in some cases anomalies occur. For example, chromium in addition to nickel in a steel in concentrations around 18 wt% helps to stabilise the γ-phase, as shown by 18Cr-8Ni wt% austenitic steels (Chapter 12).

Many of the solutes used in steels are carbide formers, which at low concentrations can dissolve in cementite or in ferrite. At higher concentrations, they may form alloy carbides which are thermodynamically more stable than cementite. Typical examples are manganese, chromium, molybdenum, vanadium, titanium, tungsten and niobium. The stability of the alloy carbides frequently found in steels relative to cementite is illustrated in Fig. 4.8, where a more negative free energy of formation, ΔG, indicates a greater thermodynamic stability. These plots represent the reaction indicated on the ordinate of Fig. 4.8 when the carbides form from elemental metal and carbon; they do not represent the stability while in equilibrium with other phases within steel, but the trend illustrated is still correct.

Manganese carbide is not found in steels. Instead, manganese enters readily into solid solution in Fe_3C. The carbide-forming elements are usually present greatly in excess of the amounts needed in the carbide phase, which are determined primarily by the carbon content of the steel. The remainder enter into solid solution in the ferrite with the non-carbide-forming elements nickel and silicon. Some of these elements, notably titanium, tungsten and molybdenum, produce substantial solid solution hardening of ferrite.

There are a few elements added to steels which enter the carbide phase. Precipitates such as Fe_3C and $Cr_{23}C_6$ are never pure, but contain mixtures of metal atoms in multicomponent steels, so it is appropriate to express their chemical formulae using M for metal atoms, M_3C and $M_{23}C_6$. Table 4.2 illustrates the compositions expected in these carbides as a function of temperature for a particular steel containing chromium and molybdenum. The cementite is in fact not the equilibrium phase, so its composition is estimated by disallowing other carbides to exist during the computation [11]. Nevertheless, cementite is the first to precipitate, but it dissolves over time, giving way to more stable precipitates Thus, it is relevant to consider it in any scheme where an attempt is made to discover how the steel approaches equilibrium.

Nitrogen in particular forms carbo-nitrides with iron and many alloying elements. However, in the presence of certain strong nitride-forming elements, e.g., titanium and aluminium, separate alloy nitride phases can occur. ϵ-carbide in steels can absorb rogue-hydrogen that may have entered the steel, thus rendering it immobile and hence reducing the hydrogen embrittlement of iron [12]. Fig. 4.9 illustrates direct evidence showing the association of hydrogen with the ϵ-carbide.

It is appreciated that for complex steels an equilibrium diagram can only give a rough guide to the structures likely to be encountered in practice. The diagrams nevertheless provide insight into the interpretation of experimental data. For example, Cr-V steels are used widely in the heat-treated condition, and both chromium and vanadium

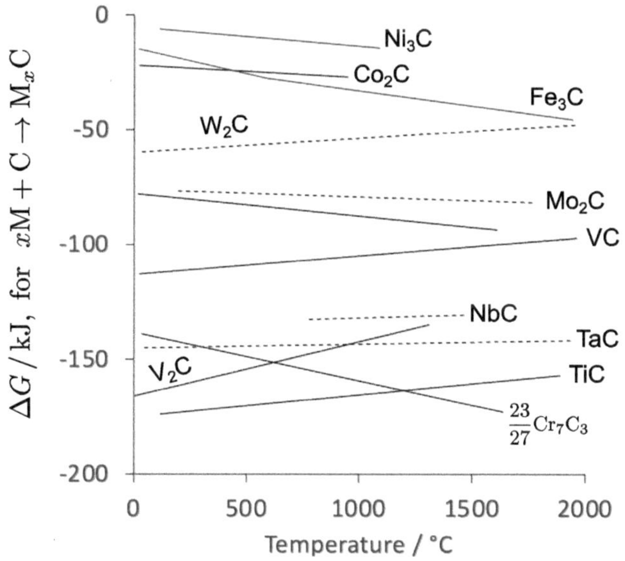

Figure 4.8 The thermodynamic stabilities of carbides presented as an Ellingham diagram based on a compilation by Shatynski [10].

Table 4.2 Estimated equilibrium chemical compositions (wt%) of carbides as a function of temperature for $2\frac{1}{4}$Cr-1Mo steel. Selected data from [11].

	M_3C			**$M_{23}C_6$**		
	650°C	**700°C**	**750°C**	**650°C**	**700°C**	**750°C**
C	6.86	6.83	6.81	5.05	5.07	5.10
Cr	40.70	35.52	30.97	37.87	34.31	30.69
Mo	1.04	1.22	1.44	18.44	16.94	14.69
Fe	48.38	54.01	58.94	38.14	43.12	48.91
Mn	2.98	2.36	1.78	0.33	0.38	0.43

are carbide formers. If a particular carbon level, e.g., 0.2 wt% and a temperature at which equilibrium can be readily reached, e.g., 700°C, is chosen, it is possible to examine a wide range of compositions to identify the carbide phases in equilibrium with ferrite at that temperature. The phase fields can then be plotted on a diagram as a function of chromium and vanadium, as shown in Fig. 4.10. Cementite is only stable up to about 1.5 wt% chromium or 0.6 wt% vanadium and for much of the diagram several alloy carbides replace cementite.

4.4 Alloying & kinetics of γ/α transformation

Since alloying elements have different tendencies relative to austenite, to dissolve in ferrite and carbides, the rate at which the decomposition of austenite occurs should be sensitive to the concentration of solutes. The growth rates of ferrite or pearlite are

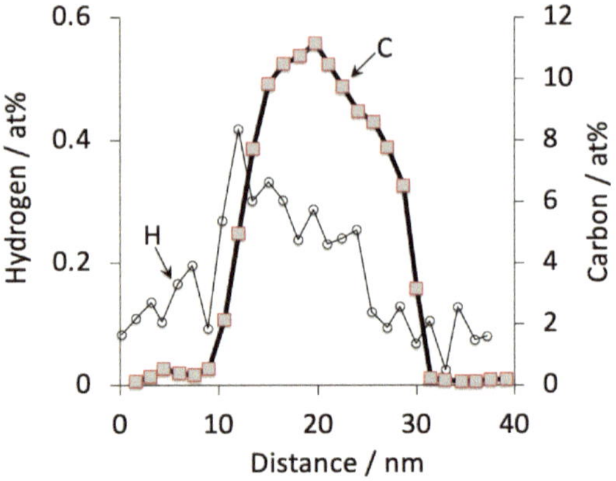

Figure 4.9 Atom probe data taken across and ϵ-carbide particle in a hydrogen-containing steel. Adapted from Zhu et al. [12].

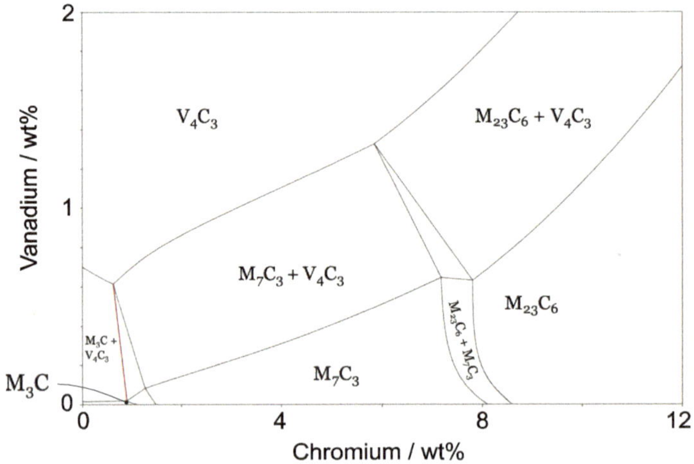

Figure 4.10 Carbide constitution in 0.2 wt%C steels at 700°C as a function of vanadium and chromium content. Calculated using the TCFE11 thermodynamic database, courtesy of Apparao Chintha. Graphite was suspended in the calculation.

significantly affected. Most familiar alloying elements displace the time-temperature-transformation curve for a plain carbon steel towards longer transformation times. However, a small group of elements, e.g., Co and Al, accelerate transformation kinetics.

4.4.1 Effect of alloying elements on the ferrite reaction

Consider now a ternary steel, say Fe-Mn-C, with the growth of ferrite occurring in a manner that maintains local equilibrium at the α/γ interface. It would be necessary to satisfy two equations in the form of Eq. (3.1), simultaneously, one per solute:

$$\left.\begin{array}{l}(c_{\mathrm{C}}^{\gamma\alpha}-c_{\mathrm{C}}^{\alpha\gamma})v=-D_{\mathrm{C}}^{\gamma}\nabla c_{\mathrm{C}}\\(c_{\mathrm{Mn}}^{\gamma\alpha}-c_{\mathrm{Mn}}^{\alpha\gamma})v=-D_{\mathrm{Mn}}^{\gamma}\nabla c_{\mathrm{Mn}}\end{array}\right\}. \tag{4.2}$$

Because $D_{\mathrm{C}}^{\gamma} \gg D_{\mathrm{Mn}}^{\gamma}$, these equations cannot in general be satisfied simultaneously for the tie-line passing through the alloy composition $\overline{c}_{\mathrm{C}}, \overline{c}_{\mathrm{Mn}}$. It is, however, possible to choose other tie lines which satisfy Eq. (4.2). If the tie-line is such that $c_{\mathrm{C}}^{\gamma\alpha} = \overline{c}_{\mathrm{C}}$ (e.g., line *cd* for alloy A of Fig. 4.11a), then ∇c_{C} becomes very small, the driving force for carbon diffusion in effect being reduced, so that the flux of carbon atoms is forced to slow down to a rate consistent with the partitioning of manganese. Ferrite forming by this mechanism is said to grow by a *partitioning, local equilibrium* (or PLE) mechanism, in recognition of the fact that $c_{\mathrm{Mn}}^{\alpha\gamma}$ can differ significantly from $\overline{c}_{\mathrm{Mn}}$, with the long-range diffusion of manganese into the austenite.

An alternative choice of tie-line could allow $c_{\mathrm{Mn}}^{\alpha\gamma} \rightarrow \overline{c}_{\mathrm{Mn}}$ (e.g., line *cd* for alloy B of Fig. 4.11b), so that ∇c_{Mn} is drastically increased, since only very small amounts of Mn are partitioned into the austenite. The flux of manganese atoms at the interface correspondingly increases and manganese diffusion can then keep pace with that of carbon, satisfying Eqs. (4.2). The growth of ferrite in this manner is said to occur by a *negligible partitioning, local equilibrium* (or NPLE) mechanism, in recognition of the fact that the manganese content of the ferrite approximately equals $\overline{c}_{\mathrm{Mn}}$, so that little if any manganese partitions into austenite.

What circumstances determine whether growth follows the PLE or NPLE mode? Fig. 4.12a shows the Fe-Mn-C phase diagram, now divided into domains where either PLE or NPLE is possible, but not both. The domains are obtained by drawing right-handed triangles on each tie-line in the $\alpha+\gamma$ phase field and joining up all the vertices.[3] For example, if an attempt is made to define NPLE conditions in the PLE domain, then the tie-line determining interface compositions will incorrectly show that both austenite and ferrite contain less carbon than $\overline{c}_{\mathrm{C}}$, a circumstance that is physically impossible.

Problem with NP-LE

There is a difficulty with the widely applied negligible-partitioning local-equilibrium concept, as described above, because it loses meaning as the velocity of the interface increases, i.e., at large supersaturations. This is because the extent of the diffusion field for the substitutional solute decreases so much, that it becomes a mathematical formality [13–15]; indeed, there are no experimental data that confirm the existence of the sharp concentration spikes predicted theoretically in domains where the transformation is supposed to occur by the NP-LE mechanism. In steep concentration gradients, it becomes important to account for the dependence of the diffusion coefficient on the concentration gradient itself [16]. This can be achieved using the treatment by Hilliard [17].

[3] Strictly not right-handed, but accounting for isoactivity, as illustrated in Fig. 4.11.

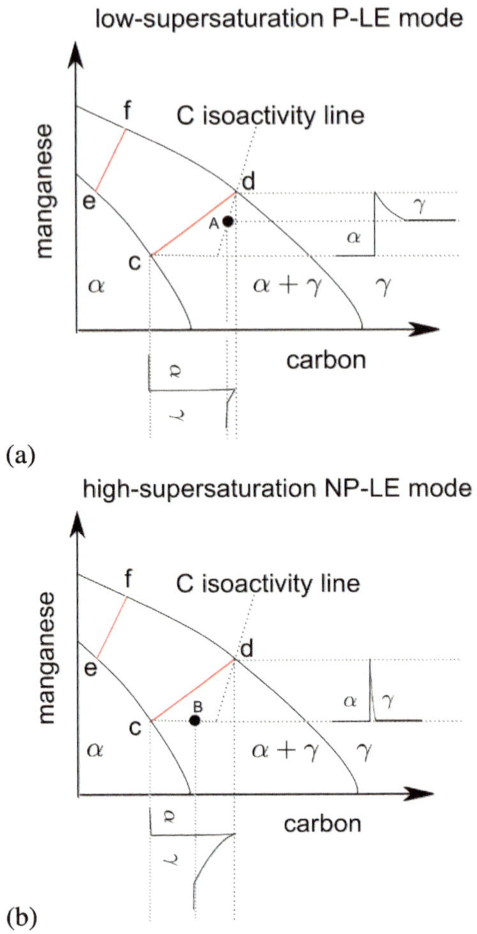

Figure 4.11 Schematic isothermal sections of the Fe-Mn-C system, illustrating ferrite growth occurring with local equilibrium at the α/γ interface. (a) Growth at low supersaturations (PLE) with bulk redistribution of manganese, (b) Growth at high supersaturations (NPLE) with negligible partitioning of manganese during transformation. The bulk alloy compositions $(\bar{c}_{\mathrm{Mn}}, \bar{c}_{\mathrm{C}})$ are designated by the symbol • in each case.

The free energy of a heterogeneous solution can be expressed by a multivariable Taylor expansion (e.g., [18]):

$$g\{a,b,\ldots\} = g\{\bar{c}\} + a\frac{\partial g}{\partial a} + b\frac{\partial g}{\partial b} + \ldots + \frac{1}{2}\left[a^2\frac{\partial^2 g}{\partial a^2} + b^2\frac{\partial^2 g}{\partial b^2} + 2ab\frac{\partial^2 g}{\partial a\,\partial b} + \ldots\right] + \ldots \tag{4.3}$$

in which the variables, $a, b, \ldots$ are the spatial composition derivatives ($\mathrm{d}c/\mathrm{d}z$, $\mathrm{d}^2c/\mathrm{d}z^2$, etc.). For the free energy of a small volume element containing a one-

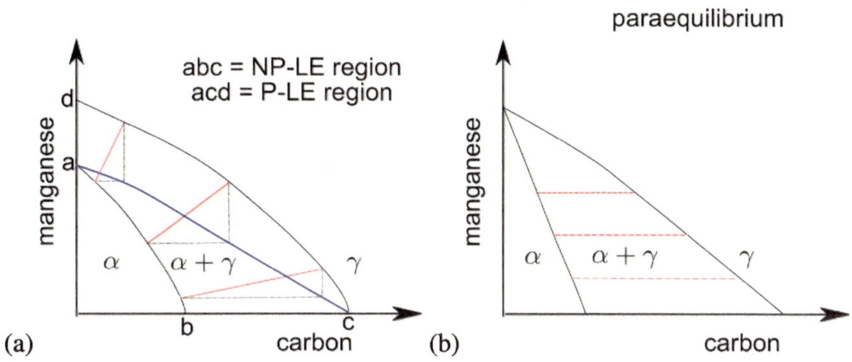

Figure 4.12 (a) Regions of the two-phase field where either PLE or NPLE modes of transformation are possible. The figure is an isothermal section of a ternary phase diagram and the tie-lines are in red. (b) An isothermal section of a para-equilibrium phase diagram. The Mn/Fe ratio remains identical in α and γ, but carbon redistributes subject to that constraint. Since the carbon concentration is usually low, the tie-lines are almost parallel to the carbon axis.

dimensional composition variation (and neglecting third and high-order terms), this gives

$$g = g\{\overline{c}\} + \kappa_1 \frac{\mathrm{d}c}{\mathrm{d}z} + \kappa_2 \frac{\mathrm{d}^2 c}{\mathrm{d}z^2} + \kappa_3 \left(\frac{\mathrm{d}c}{\mathrm{d}z}\right)^2, \tag{4.4}$$

where $\overline{c}$ is the average composition and

$$\kappa_1 = \frac{\partial g}{\partial(\mathrm{d}c/\mathrm{d}z)} \qquad \kappa_2 = \frac{\partial g}{\partial(\mathrm{d}^2 c/\mathrm{d}z^2)} \qquad \kappa_3 = \frac{1}{2}\frac{\partial^2 g}{\partial(\mathrm{d}c/\mathrm{d}z)^2}. \tag{4.5}$$

In this, κ_1 is zero for a centrosymmetric crystal, since the free energy must be invariant to a change in the sign of the coordinate z.

The total free energy per atom, g_{ih} for the inhomogeneous solution, is obtained by integrating over the volume V:

$$g_{\mathrm{ih}} = \int_{\mathrm{V}} \left[g\{\overline{c}\} + \kappa_2 \frac{\mathrm{d}^2 c}{\mathrm{d}z^2} + \kappa_3 \left(\frac{\mathrm{d}c}{\mathrm{d}z}\right)^2 \right]. \tag{4.6}$$

When integrating the third term in this equation by parts:

$$\int \kappa_2 \frac{\mathrm{d}^2 c}{\mathrm{d}z^2} = \kappa_2 \frac{\mathrm{d}c}{\mathrm{d}z} - \int \frac{\mathrm{d}\kappa_2}{\mathrm{d}c}\left(\frac{\mathrm{d}c}{\mathrm{d}z}\right)^2 \mathrm{d}z. \tag{4.7}$$

As before, the first term on the right is zero, so that an equation of the form below is obtained for the free energy per atom of a heterogeneous system:

$$g_{\mathrm{ih}} = \int \left[g\{\overline{c}\} + v_{\mathrm{a}}^3 \kappa (\nabla c)^2 \right] \mathrm{d}V, \tag{4.8}$$

where v_a is the volume per atom and κ is known as the gradient energy coefficient. The term $g\{\overline{c}\}$ is the free energy of a homogeneous solution with the average concentration $\overline{c}$. The interpretation of this equation is that gradients of concentration lead to an increase in the free energy, so the formation of steep gradients will in general be opposed. Fig. 4.13 shows the increase in free energy of the solution as a consequence of the substitutional solute concentration spike in the austenite at the γ/α interface during negligible partitioning local equilibrium growth. The diffusion distance is intended to represent the width of the spike. The calculations assume a gradient energy coefficient of $3.85\times10^{-10}\,\mathrm{J\,m^{-1}}$ based on the Fe-Cr system [19,20] and that $(x-\overline{x})=0.03$.

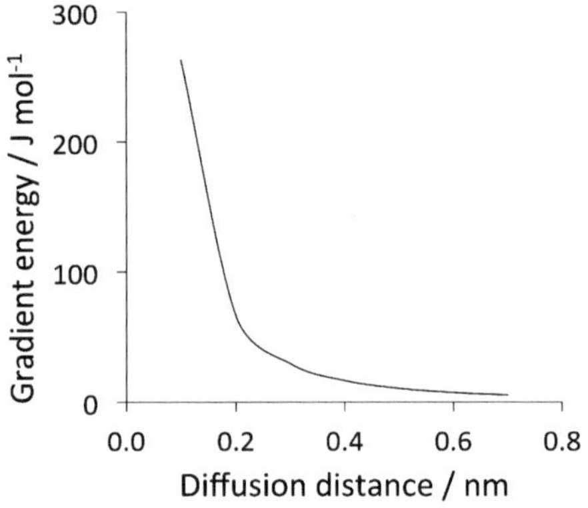

Figure 4.13 Estimate of the penalty on free energy due to the gradient of concentration in the austenite ahead of the α/γ interface [16].

NP-LE calculations, as currently implemented, are therefore wrong because they do not account for the gradient energy term. If such a term is incorporated, the concentration spike would be moderated to larger widths, making the extent of partitioning greater and the growth rate slower.

Para-equilibrium

Para-equilibrium is a constrained equilibrium. It occurs at temperatures where the diffusion of substitutional solutes is not possible within the time scale of the experiment. Nevertheless, interstitials may remain highly mobile. Thus, in a steel, manganese does not partition between the ferrite and austenite, but subject to that constraint, the carbon redistributes until it has the same chemical potential in both phases. Therefore, the tie lines in the phase diagram (Fig. 4.12b) are all virtually parallel to the carbon axis, since the Mn to Fe ratio remains the same everywhere during ferrite growth.

In an isothermal section of the ternary phase diagram, the para-equilibrium phase boundaries must lie within the equilibrium phase boundaries, as illustrated in Fig. 4.14a. This is because there is no difference between the equilibrium and para-equilibrium cases in the absence of manganese, so the phase boundaries meet on the horizontal axis. For para-equilibrium, α and γ have identical manganese content in the absence of carbon, so the $\alpha/\alpha+\gamma$ and $\gamma/\gamma+\alpha$ phase boundaries meet at a point $\overline{c}_{\mathrm{Mn}}$ on the vertical axis. In contrast, the manganese concentrations of these two phases are different at equilibrium even in the absence of carbon.

Fig. 4.14b shows how the thickness of allotriomorphic ferrite changes parabolically with time during para-equilibrium growth. The carbon concentration changes by a regular 0.02 wt% between the different curves, but the change in the rate of transformation becomes very large when the average concentration approaches the solubility of carbon in ferrite that is in equilibrium with austenite ($\overline{c}_{\mathrm{C}} \rightarrow c_{\mathrm{C}}^{\alpha\gamma}$). The rate of diffusion-controlled growth would be infinite when $\overline{c}_{\mathrm{C}} = c_{\mathrm{C}}^{\alpha\gamma}$, but obviously other rate-limiting factors such as the barrier to the transfer of atoms across the α/γ interface then become rate limiting.

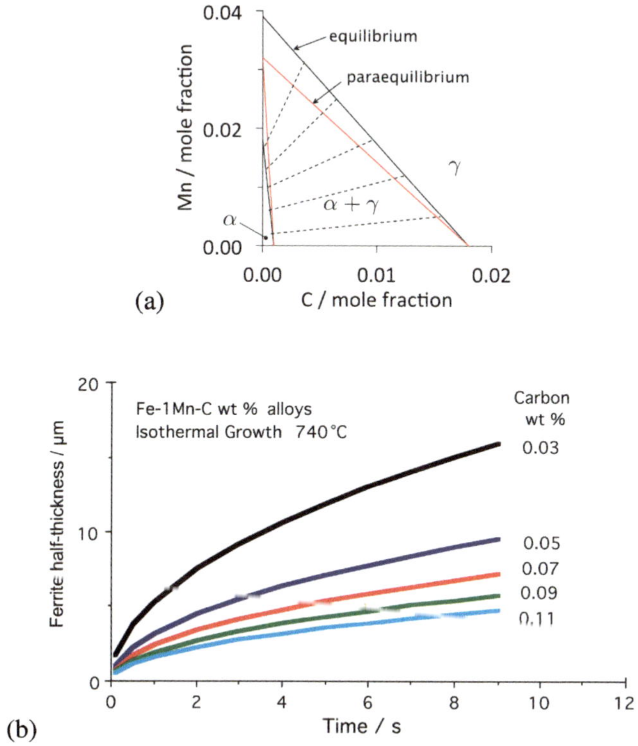

Figure 4.14 (a) Superimposed isothermal sections (1053 K) of the equilibrium and para-equilibrium Fe-Mn-C phase diagrams. The equilibrium tie-lines are marked dashed. (b) An illustration of the thickening of allotriomorphs of ferrite, for steels containing a variety of carbon concentrations. The calculations assume para-equilibrium transformation, so that the manganese influences only the thermodynamics of the phase change.

4.4.2 Effect of alloying elements on the pearlite reaction

The pearlite reaction is a typical nucleation and growth reaction, and under the appropriate experimental conditions rates of nucleation I_V and rates of growth v_P can be determined (Eq. (3.11)). The work of Mehl and coworkers showed that many alloying elements reduce both I_V and v_P, although in rare cases, e.g., cobalt, can accelerate the growth rate (Fig. 4.15). Pearlite is a reconstructive transformation and there are no reported instances where the substitutional solute does not partition between the phases. Therefore, the idea of para-equilibrium pearlite does not apply. The growth rate is therefore influenced by the requirement to diffuse all solutes (including those that are substitutional), and there naturally must also be a thermodynamic effect whereby the driving force for the $\gamma \rightarrow \alpha + \theta$ transformation is changed by the addition of the solute.

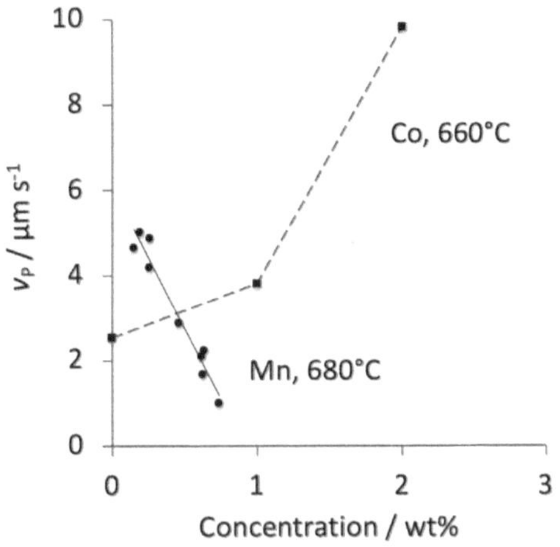

Figure 4.15 The growth rate v_P of pearlite as a function of manganese, which retards the reaction, and cobalt, which accelerates it. Data from Mehl and Hagel [21].

The situation therefore is more complex for Fe-C-X steels containing a substitutional solute (X) in addition to interstitial carbon. Local equilibrium requires the compositions at the interface to be maintained at levels that are consistent with a tie-line of the Fe-C-X phase diagram. At a constant temperature, this is generally not possible to achieve for the tie line passing through $\overline{c}_{Mn}, \overline{c}_C$, because the rate at which each solute is partitioned must then equal to that at which it is carried away from the interface by diffusion. In analogy to Eqs. (4.2), it is necessary therefore that

$$\text{at } \alpha/\gamma \text{ interface:} \begin{cases} v_P(c_C^{\gamma\alpha} - c_C^{\alpha\gamma}) = -D_{C,V}^{\gamma} \nabla c_C \\ v_P(c_{Mn}^{\gamma\alpha} - c_{Mn}^{\alpha\gamma}) = -D_{Mn,V}^{\gamma} \nabla c_{Mn} \end{cases} \tag{4.9}$$

$$\text{at } \theta/\gamma \text{ interface:} \begin{cases} v_P(c_C^{\gamma\theta} - c_C^{\theta\gamma}) = -D_{C,V}^{\gamma} \nabla c_C \\ v_P(c_{Mn}^{\gamma\theta} - c_{Mn}^{\theta\gamma}) = -D_{Mn,V}^{\gamma} \nabla c_{Mn} \end{cases} \tag{4.10}$$

where the subscripts identify the solute. Given that $D^{\gamma}_{\mathrm{Mn,V}} \ll D^{\gamma}_{\mathrm{C,V}}$, it becomes impossible to simultaneously satisfy either Eq. (4.9) or (4.10) if the tie-line passing through $\overline{c}_{\mathrm{Mn}}, \overline{c}_{\mathrm{C}}$ is selected. However, a ternary alloy can, at a constant temperature offer a myriad of other tie lines that would link the $c^{\gamma\alpha} \leftrightarrow c^{\alpha\gamma}$ and separately link $c^{\gamma\theta} \leftrightarrow c^{\theta\gamma}$, the condition for local equilibrium at the α/γ and γ/α interfaces can be satisfied in principle by selecting a suitable set that does not pass through $\overline{c}_{\mathrm{Mn}}, \overline{c}_{\mathrm{C}}$.

Such a set of tie lines can be found by considering the following equations that are analogous to Eq. (3.11), but the number of equations is now two in order to allow for both the Mn and C [22,23]:

$$v^{\mathrm{C}}_{\mathrm{P}} = \left(2D^{\gamma}_{\mathrm{C,V}} + \frac{12sD_{\mathrm{C,B}}\delta}{S}\right)\frac{S}{S_{\alpha}S_{\theta}}\left(\frac{c^{\gamma\alpha}_{\mathrm{C}} - c^{\gamma\theta}_{\mathrm{C}}}{c^{\theta\gamma}_{\mathrm{C}} - c^{\alpha\gamma}_{\mathrm{C}}}\right)\left(1 - \frac{S_{\mathrm{C}}}{S}\right),$$

$$v^{\mathrm{Mn}}_{\mathrm{P}} = \left(2D^{\gamma}_{\mathrm{Mn,V}} + \frac{12sD_{\mathrm{Mn,B}}\delta}{S}\right)\frac{S}{S_{\alpha}S_{\theta}}\left(\frac{c^{\gamma\alpha}_{\mathrm{Mn}} - c^{\gamma\theta}_{\mathrm{Mn}}}{c^{\theta\gamma}_{\mathrm{Mn}} - c^{\alpha\gamma}_{\mathrm{Mn}}}\right)\left(1 - \frac{S_{\mathrm{C}}}{S}\right), \tag{4.11}$$

where the velocities $v^{\mathrm{C}}_{\mathrm{P}}$ and $v^{\mathrm{Mn}}_{\mathrm{P}}$ are calculated on the basis of the diffusion of only carbon or only manganese, respectively. The term s is a length normal to the growth direction, essentially a fitting parameter, so it is common to lump the product sD_{B} together, so, for example,

$$sD_{\mathrm{B,Mn}} = 2.81 \times 10^{-3} \exp\left\{-\frac{164\,434\,\mathrm{J\,mol^{-1}}}{RT}\right\}\mathrm{m^2\,s^{-1}}.$$

Clearly, since there is only one transformation front, the equations must be solved such that $v^{\mathrm{C}}_{\mathrm{P}} = v^{\mathrm{Mn}}_{\mathrm{P}}$. Bearing in mind that the interlamellar spacing is also identical in these equations, a further condition arises that:

$$\frac{D_{\mathrm{C}}}{D_{\mathrm{Mn}}} = \frac{R_{\mathrm{Mn}}}{R_{\mathrm{C}}} \quad \text{with} \quad \left\{\begin{array}{l} D_i \equiv D_{i,\mathrm{V}} + \frac{6sD_{i,\mathrm{B}}\delta}{S} \\ R_i \equiv \frac{c_i^{\gamma\alpha} - c_i^{\gamma\theta}}{c_i^{\theta\gamma} - c_i^{\alpha\gamma}} \end{array}\right. \tag{4.12}$$

The R_i condition ensures that the weighted average of the ferrite and cementite yields the mean composition of the steel. With these two constraints and in addition the local equilibrium condition, it is possible to find unique compositions at the growth front by coupling the conditions and the velocity equations. Some comparisons of the calculations with experimental data for ternary steels are shown in Fig. 4.16; there is a reasonable agreement given the uncertainties that exist in the boundary diffusion coefficients that are necessary for the calculations. The method outlined above can, in principle, be extended to steels containing more than one substitutional solute.

Other effects

Fully pearlitic steel is vital in the manufacture of the strongest ropes available for engineering applications, as discussed in section 3.6.6. Much research has therefore been done to increase the strength of the ropes. One approach is to increase the amount

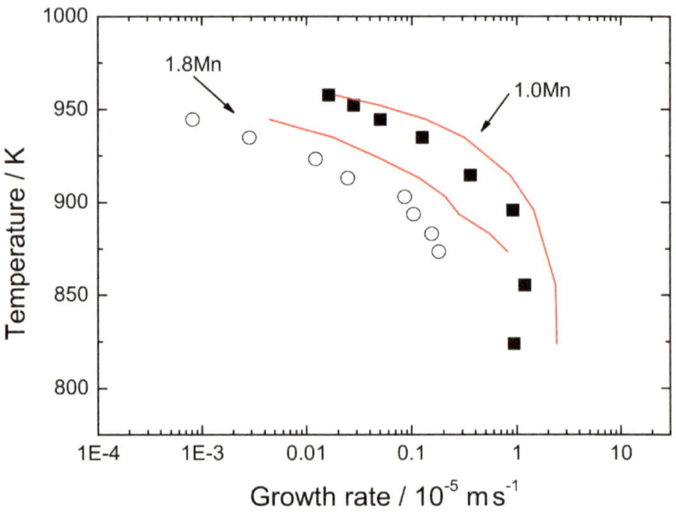

Figure 4.16 Mixed diffusion-controlled model applied to 1.0 and 1.8 wt% Mn eutectoid steels and experimental data for comparison (after Seo et al. [23]).

of carbon without inducing the formation of any proeutectoid cementite. Proeutectoid cementite precipitates as allotriomorphs on the austenite grain boundaries and as Widmanstätten plates (Fig. 3.8); both of these cause a deterioration in the ductility of the steel and therefore should be avoided.

Substitutional solutes can affect the eutectoid composition, which in a plain carbon steel is just below 0.8 wt%; Fig. 4.17a shows that silicon, which does not dissolve in cementite, reduces the eutectoid concentration. Cobalt, on the other hand, has the opposite effect, so with carbon concentrations in excess of 0.8 wt%, it can be made fully pearlitic, Fig. 4.17b. The concept of a eutectoid composition differs, of course, between Fe-C and Fe-C-X because in the latter case the eutectoid reaction has greater degrees of freedom according to the phase rule and hence can occur over a range of temperatures.

It is possible in a ternary Fe-C-X system for austenite, cementite and ferrite to co-exist in equilibrium. The pearlite that grows then does not have the average composition of the steel, so the austenite composition changes as the pearlite grows. This causes the driving force for transformation to decrease as the pearlite grows, to a point where it reaches zero when equilibrium is achieved. The reduction in driving force as the pearlite grows leads to a progressive increase in its interlamellar spacing, so that the microstructure is known as *divergent pearlite* [26], Fig. 4.18.

4.4.3 Alloy pearlite

Once the alloying element concentration reaches a critical level, the cementite within pearlite is replaced by another carbide phase [28]. For example, in a chromium, tungsten or molybdenum steel, the cubic $M_{23}C_6$ carbide can form, where M can include

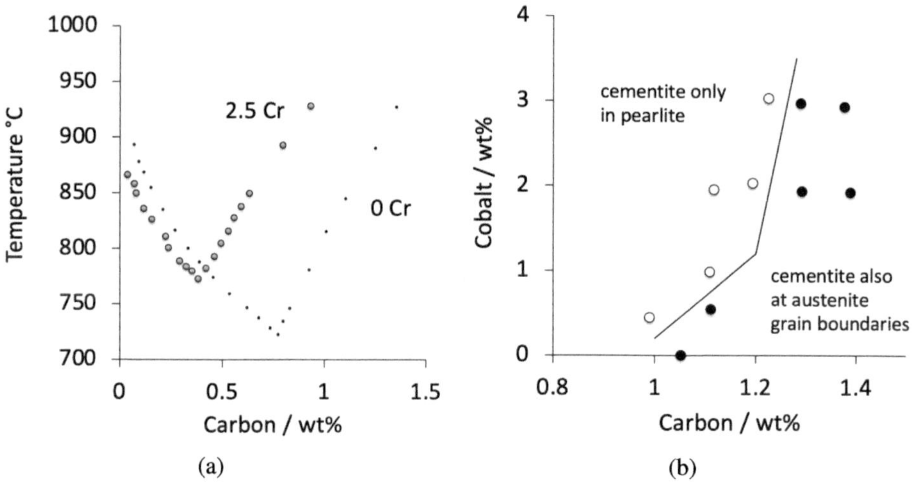

Figure 4.17 (a) Change in the eutectoid composition as the chromium concentration is increased. After [24]. (b) Fully pearlitic state (open circles) obtained in steels containing more than 0.8 wt% carbon, with the addition of cobalt. The filled circles represent cases where allotriomorphic and Widmanstätten cementite was present in addition to pearlite. In all cases, the steels were transformed isothermally at 550°C (adapted using data from Kanetsuki et al. [25]).

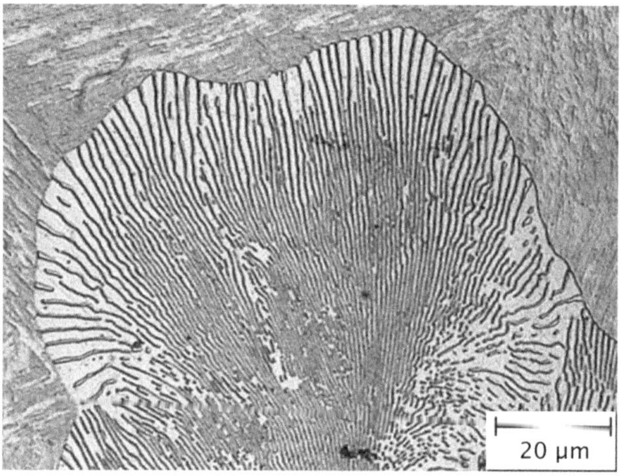

Figure 4.18 Fe-2.5C-5.4Mn at% steel transformed in the $\gamma + \alpha + \theta$ phase field, where the three phases can coexist in equilibrium, showing divergent pearlite. The interlamellar spacing increases as the colony grows because the change in composition of the austenite reduces the available driving force progressively (after Hutchinson et al. [27], reproduced with the permission of Elsevier).

iron, chromium, molybdenum or tungsten (Figs. 4.19 and 4.20). This change in the carbide phase does not necessarily alter the basic pearlitic morphology and consequently *alloy pearlites* are obtained in which an alloy carbide is associated with ferrite (Fig. 4.20). These pearlites occur only in medium and highly alloyed steels, usually at the highest transformation temperatures. At lower transformation temperatures in the same steel, cementitic pearlite may still form because of the inadequate diffusion of the alloying element.

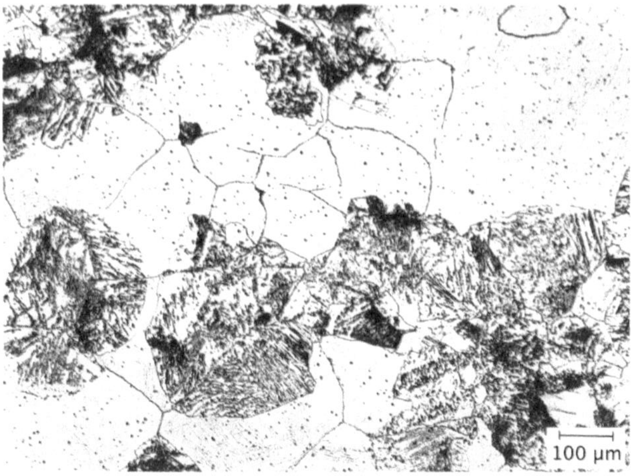

Figure 4.19 Fe-12Cr-0.2C wt% transformed 30 min at 775°C. Pearlite-type reaction involving $M_{23}C_6$ instead of cementite as the carbide phase (courtesy of Campbell).

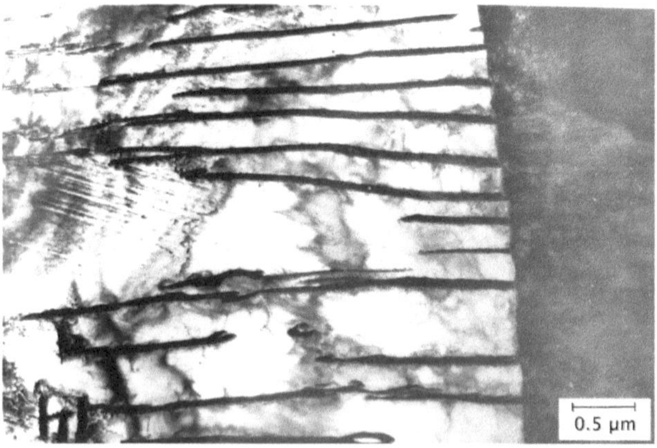

Figure 4.20 Fe-12Cr-0.2C wt% transformed 15 min at 750°C. Alloy pearlite. $M_{23}C_6$/ferrite (courtesy of Campbell).

4.5 Structural changes due to alloying additions

The addition of elements such as nickel, silicon and manganese to iron-carbon alloys that do not form carbides in competition with cementite does not basically alter the microstructures formed after transformation. However, in the case of strong carbide-forming elements such as molybdenum, chromium and tungsten, cementite is replaced by the appropriate alloy carbides, often at relatively low alloying element concentrations. Still stronger carbide-forming elements such as niobium, titanium and vanadium are capable of forming alloy carbides, preferentially at concentrations less than 0.1 wt%. It would therefore be expected that the microstructures of steels containing these elements would be radically altered.

The tendency to form carbides and nitrides can be expressed in terms of bonding. Cottrell was able to explain many of the observed trends in the stability, crystal structure and stoichiometry of the carbides of transition metals in terms of chemical bonds [29]. He points out that Ti, Zr and Hf, which in the periodic table are elements near the beginning of the long periods, form very stable MC carbides, but the affinity for carbon diminishes further along the rows of the periodic table (Fig. 4.21). A part of the reason for this is that more electrons have to be accommodated for elements further along the rows, so antibonding states are progressively filled thereby reducing the bond order.[4] This does not completely explain the trend, because the maximum bond order occurs with Cr, Mo and W and we know that carbides of these elements are less stable.

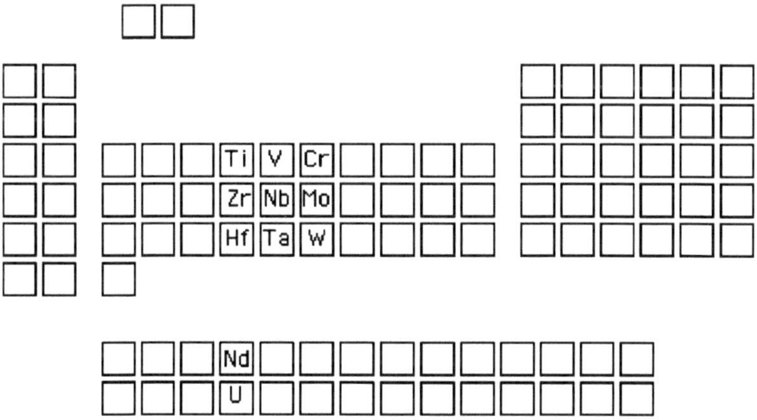

Figure 4.21 The periodic table showing the positions of strong carbide-forming elements.

With MC carbides (where M stands for metal atoms), the metal has to sacrifice four electrons to form the bonds with carbon. Titanium has exactly the right number so that its *d*-orbitals are left empty on forming TiC. This is not the case with V_4C_3, since vanadium has an additional *d*-electron which forms a V-V bond. The electrons in the two kinds of bonds, V-C and V-V mutually repel, leading to a reduction in the

[4] When two hydrogen atoms, each with a single electron, are brought together, they no longer have separate atomic orbitals. Instead, they have a pair of communal orbitals (bonding and antibonding), each of which can hold two electrons. It follows that for H_2, both electrons are in the bonding orbitals, giving a bond order of 2 and a strong molecule. For He_2, on the other hand, the four electrons fill up both the bonding and the antibonding orbitals, so the bond order is zero and the molecule is not formed.

stability of V_4C_3 when compared with TiC. This problem becomes larger along the row of the periodic table until MC carbide formation becomes impossible or unlikely.

Although Cottrell did not consider the carbides in the lanthanide or actinide series of elements, it is possible that the same principles should apply there. Both NdC and UC exist. Remarkably, neodymium nitride has already been incorporated into a ferritic creep-resistant steel by Igarashi and Sawaragi [30] with rather good results. The concentration of neodymium used was only 0.04 wt% but gave an increase in the creep rupture life by a factor of about two during tests at 650°C. They also tried hafnium, but did not recommend it due to a tendency to form coarse particles.

Alloy carbides and nitrides are much less soluble in austenite than is cementite. In Fig. 4.22, the solubility products in austenite of vanadium, titanium and niobium carbides and nitrides are plotted as a function of $1/T$. The equation for the solubility product comes from a consideration of the reaction

$$\mathrm{M}+\mathrm{X} \rightleftharpoons \mathrm{MX} \qquad \Delta G = \Delta G^\circ + RT\ln\left\{\frac{a_{\mathrm{MC}}}{a_{\mathrm{M}}a_{\mathrm{X}}}\right\},$$

with ΔG being the free energy change accompanying the reaction, ΔG° the standard free energy change for the reaction and a stands for the activity. To a good approximation, $\Delta G^\circ = -C_1 + C_2T$ where C_i are positive constants, the activity of the MX is taken as unity, and for dilute solutions it might be assumed that activities are equal to concentrations (indicated by square enclosing brackets). At equilibrium $\Delta G = 0$, so it follows that

$$\Delta G^\circ = -C_1 + C_2T \approx RT\ln\{[\mathrm{M}][\mathrm{X}]\}$$

$$-\frac{C_1}{RT} + C_2 = \ln\{[\mathrm{M}][\mathrm{X}]\},$$

which has the same form as the equation at the top of Fig. 4.22. If the constants on the left hand side are known, and given the [X] value in the austenite, it becomes possible to calculate the concentration [M] of the metal atoms that remain in solution at the chosen temperature.

Solubility products like these are useful in estimating quickly the temperature at which carbides should dissolve in austenite, but a full assessment for a multicomponent steel can be done using phase diagram calculations based on thermodynamic databases. For example, the addition of niobium to a rail steel that contains 0.88 wt% of carbon limits the concentration of niobium to only 0.01 wt% when it should dissolve into austenite at 1200°C [31].

4.5.1 Ferrite/alloy carbide aggregates

The carbide-forming elements substantially restrict the γ-loop (Fig. 4.4), thus depressing the eutectoid composition to lower carbon levels and to higher transformation temperatures. One result is that pearlite can completely disappear from the transformed microstructures, which now exhibit very different ferrite/carbide aggregates, usually

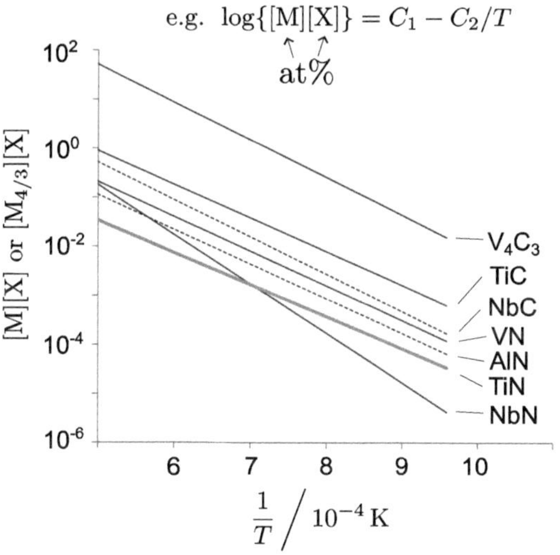

Figure 4.22 Solubility products for carbides and nitrides in austenite as a function of temperature (after Aronson [32]). M and X stand for metal and C or N atoms, respectively. Note that for multicomponent steels, the estimation of carbide solubility with respect to another phase is best done using phase diagram calculation methods [33,34].

on a much finer scale than pearlite. Apart from the alloy carbide based pearlites, particularly found in high chromium steels, there are three morphologies of alloy carbides, which are intimately associated with ferrite in the transformation temperature range in which plain carbon steels form ferrite/pearlite structures.

4.5.2 Alloy carbide fibres and laths

The clearest analogy to pearlite is found when the alloy carbide in lath morphology forms nodules in association with ferrite, Fig. 4.23. These pearlitic nodules often are encountered at temperatures just below Ae_1 in steels which transform relatively slowly. For example, these structures are obtained in chromium steels with between 4 and 12 wt% chromium (Fig. 4.19), and the morphology is analogous to that of cementitic pearlite. It is, however, different in detail because of the different crystal structures of the possible carbides, e.g., M_7C_3 is either orthorhombic or hexagonal depending on the relative magnitudes of its lattice parameters [35, p. 472]; $M_{23}C_6$ is cubic. The structures observed are relatively coarse, but finer than pearlite formed under equivalent conditions, because of the need for the partition of the substitutional solutes, e.g., chromium between the carbide and the ferrite. To achieve this, the interlamellar spacing must be substantially finer than in the equivalent iron-carbon case.

At lower temperatures, the lath morphology is largely replaced by much finer fibrous aggregates, e.g., in high Cr steels, coarse laths of $M_{23}C_6$ can be replaced by fine fibres of the same carbide, usually 500 Å in diameter. Their length, which is determined by the size of the ferrite colony, can be up to 10 μm, with little or no branching.

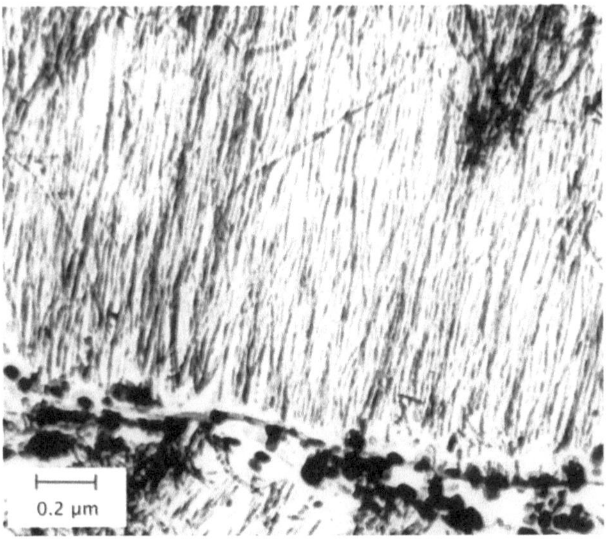

Figure 4.23 Fe-4Mo-0.2C wt% transformed 4 days at 600°C. Fibrous Mo_2C growth from γ boundary (courtesy of Berry). The prior austenite grain boundary is at the bottom of the image, where the coarse, spheroidal Mo_2C particles are present.

Similar morphologies occur, but are much less dominant, in steels containing W, Ti, V and Nb.

4.5.3 Interphase precipitation

Interphase precipitation is associated with the periodic nucleation of carbides at the γ/α interface during the transformation. The precipitate particles form in bands parallel to the interface and adapt to the general direction of the interface, even when it abruptly changes direction. The phenomenon is found in a wide variety of alloys and precipitate types [36]; Fig. 4.24 is a nice example in the context of a vanadium-containing steel.

A further characteristic is the frequent development of only one of the possible Widmanstätten variants, e.g., V_4C_3 platelets in a particular region are all only of one variant of the habit, i.e., that in which the plates are most nearly parallel to the interface. The bands often are associated with planar, low-energy interfaces and the spacing between bands is determined by the height of steps, which move along the interface. The nucleation of the carbide particles occurs normally on the low-energy planar interfaces rather than at the rapidly moving high-energy steps. Fig. 4.25 illustrates rather coarser interphase-precipitation in a chromium-containing steel, where the mechanism involving the translation of steps at the α/γ interface is vividly illustrated.

The need for step movement on the γ/α interface is in contrast to the growth of fibrous carbides behind an interface on which no steps are observed. Indeed if, in these

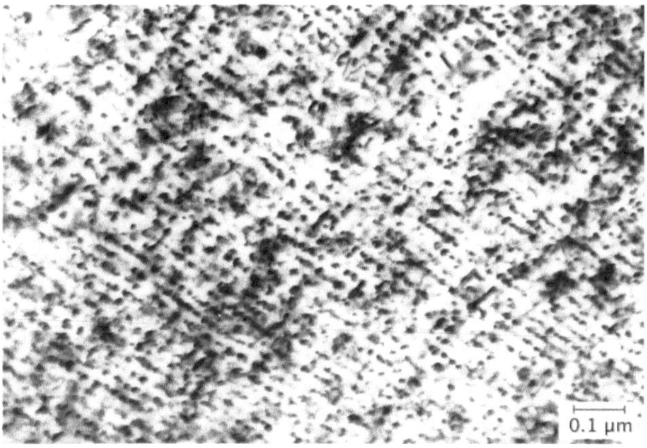

Figure 4.24 Fe-1V-0.2C-0.02Nb wt% alloy transformed 5 min at 775°C. Interphase precipitation of V_4C_3 in ferrite (courtesy of Batte).

(a)
(b)
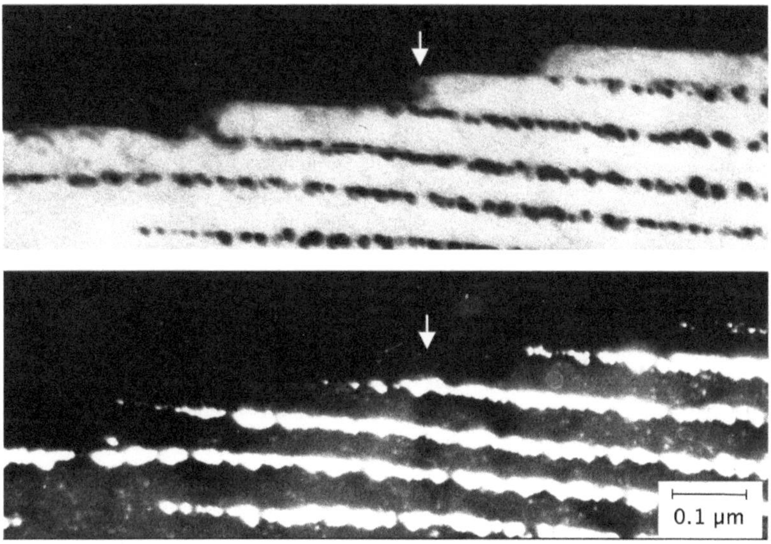

Figure 4.25 Fe-12Cr-0.2C wt% transformed 30 min at 650°C. Precipitation of $M_{23}C_6$ at stepped γ/α interface: (a) bright field; (b) precipitate spot dark field (courtesy of Campbell). The arrows show that the precipitates are not forming on the moving risers of the steps, but rather on the immobile components of the stepped-interface.

circumstances, a step does move along the interface, the fibrous growth is stopped and replaced by interphase precipitation. The step height and therefore the band spacing of the precipitation depends on the temperature of transformation and composition. As the temperature of transformation is lowered, the band spacing is reduced, e.g., in a 1V-0.2C wt% steel, the spacing varies from 25 nm at 825°C to 7.5 nm at 725°C

(Fig. 4.26) and at lower temperatures has been observed to be less than 5 nm. The extremely fine scale of this phenomenon in vanadium steels, which also occurs in Ti and Nb steels, is due to the rapid rate at which the γ/α transformation takes place. At the higher transformation temperatures, the slower rate of reaction leads to coarser structures. Similarly, if the reaction is slowed down by addition of further alloying elements, e.g., Ni and Mn, the precipitate dispersion coarsens. The scale of the dispersion also varies from steel to steel. It is coarsest in chromium, tungsten and molybdenum steels, where the reaction is relatively slow, and much finer in steels in which vanadium, niobium and titanium are the dominant alloying elements and the transformation is rapid.

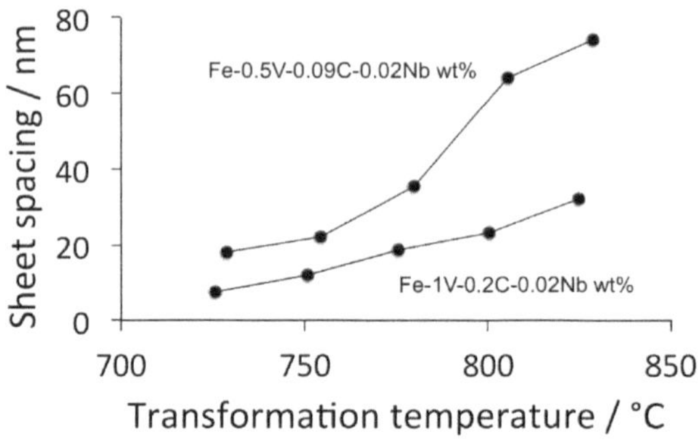

Figure 4.26 Interphase precipitation of V_4C_3 in vanadium steels. Precipitate sheet spacing as a function of transformation temperature (courtesy of Ballinger). The higher vanadium steel has about twice as much of a volume fraction of vanadium carbide.

One of the recent successes in the field of high-strength low-alloy steels has the trade name *Nanohiten* [37] of composition Fe-0.043C-0.2Si-1.6Mn-0.0035N wt%, with titanium in the range 0.018-0.186 wt% and molybdenum from 0–0.37 wt%. The steel contains (Ti,Mo)C in the form of interphase precipitates in the ferrite, which adds about 300 MPa to the total of 780 MPa ultimate tensile strength. Furthermore, the alloy microstructure can be obtained in the hot-rolled state and has an excellent formability, which makes it suitable for large-scale automotive applications. Formability is often compromised when the steel contains fractions of hard and soft phases, such as the dual-phase steels that are a mixture of martensite and ferrite; with the *Nanohiten* steel, the mechanical properties are uniform on the scale of the microstructure.

An example is illustrated in Fig. 4.27. First-principles calculations indicate that the dissolution of molybdenum in TiC is not favoured from a thermodynamic point of view, but that it decreases the misfit between the carbide and the ferrite, thus reducing the activation energy for nucleation [38]. The same influence of molybdenum on the interfacial misfit, which is related to the interfacial energy per unit area, also reduces the coarsening rate of the precipitate. This is important because steels produced on a large scale by hot rolling are coiled while the steel is hot, so that the slow cooling process to room temperature can induce coarsening of the microstructure.

Surprisingly, interphase precipitation can also be found within the ferritic component of pearlite. Fig. 4.28 [40] shows the precipitation of vanadium carbides within

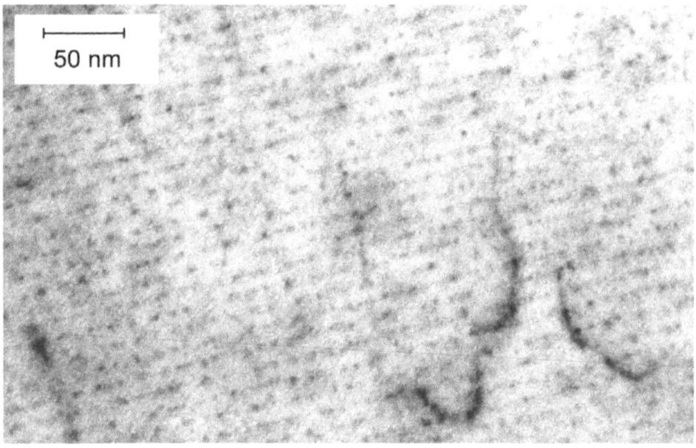

Figure 4.27 Interphase precipitation of (Ti,Mo)C in ferrite within a high-strength low-alloy steel. The chemical composition is Fe-0.06C-1.5Mn-0.1Si-0.1Ti-0.2Mo wt%. The precipitates are about 4.5 nm in length and the precipitate sheet spacing is about 20 nm (after Yen et al. [39], reproduced with the permission of Elsevier).

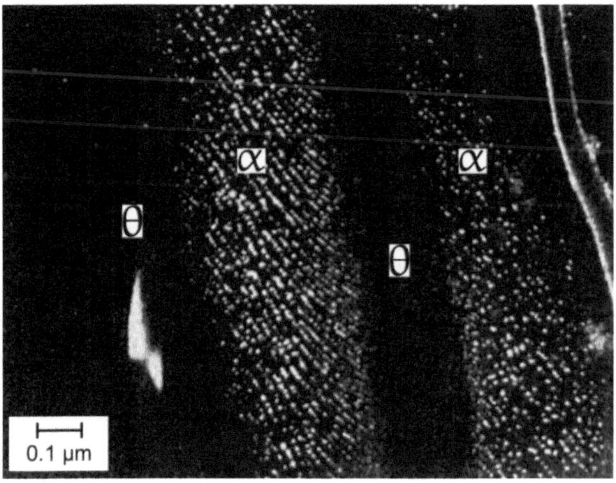

Figure 4.28 A dark-field transmission electron micrograph showing rows of vanadium carbide precipitation within the pearlitic ferrite of a steel (0.5C-0.3Si-0.82Mn-0.3V wt%) cooled to produce a mixture of allotriomorphic ferrite and pearlite. Reproduced with the permission of Maney Publishing from [40].

the pearlitic-ferrite in row formations that are consistent with interphase precipitation, presumably by the movement of steps at the pearlite-austenite interface (section 3.6.1). The precipitation is of technological importance because it offers an additional method of further strengthening pearlite.

4.5.4 Precipitation in supersaturated ferrite

In microalloyed steels, where strong carbide-forming elements are present in concentrations less than 0.1 wt%, it often is possible to obtain the ferrite in a supersaturated condition with little or no carbide precipitation taking place during the γ/α transformation. Instead, while the steel is held at the transformation temperature, carbide precipitates form within the newly formed ferrite grains, usually on dislocations (Fig. 4.29).

Figure 4.29 Fe-0.25V-0.05C transformed and held at 2.5 h at 740°C. V_4C_3 precipitation on dislocations (courtesy of Ballinger).

By careful choice of alloy and experimental conditions, it is possible to obtain each of the above microstructures separately, but in practice they are often all present in transformed alloy steels, provided the steel contains a strong carbide-forming element. Consequently, the microstructures of transformable alloy steels can be very complex. The full extent of these complexities only being revealed when high-resolution electron microscopy is used to study them.

In general, the fibrous morphology represents a closer approach to an equilibrium structure and is therefore more predominant in steels, which have been transformed slowly. In contrast, the interphase precipitation and dislocation nucleated structures occur more readily in rapidly transforming steels, where there is a large driving force, e.g., in microalloyed steels (Chapter 10).

4.6 Transformation diagrams for alloy steels

The transformation of austenite below the eutectoid temperature is best represented in an isothermal transformation diagram (Chapter 3), in which the beginning and end of transformation are plotted as a function of temperature and time. These time–temperature transformation, or TTT, curves are an important source of quantitative information on the heat treatment of steels. In the simple case of an eutectoid plain carbon steel, the curve is roughly C-shaped, with the pearlite reaction occurring down to the nose of the curve and a little beyond. At lower-temperatures bainite and martensite form (Chapters 5 and 6). The diagrams become more complex for hypo- and hyper-eutectoid alloys, as the ferrite or cementite reactions must also be represented by additional lines.

Alloying elements, on the whole, retard both the pro-eutectoid reactions and the pearlite reaction, so that TTT curves for alloy steels are moved increasingly to longer times as the alloy content is increased. Additionally, those elements which expand the γ-field depress the eutectoid temperature, with the result that they also depress the position of the TTT curves relative to the temperature axes. This behaviour is shown by steels containing manganese or nickel. For example, in a 13Mn-0.8C wt% steel, pearlite can form at temperatures as low as 400°C. In contrast, elements which favour the ferrite phase raise the eutectoid temperature and the TTT curves move correspondingly to higher temperatures. The slowing down of the ferrite and pearlite reactions by alloying elements enables these reactions to be more readily avoided during heat treatment, so that the much stronger low-temperature phases such as bainite and martensite can be obtained in the microstructure. The hard martensitic structure is only obtained in plain carbon steels by water quenching from the austenitic condition, while the addition of alloying elements requires a lower critical cooling rate to achieve this condition. Consequently, alloy steels allow hardening during oil quenching or even air cooling if the TTT curve has been sufficiently displaced to longer times.

4.7 Light steels

The density of iron can be reduced by alloying with light elements such as silicon and aluminium and there may be smaller contribution from any expansion of the lattice parameter due to alloying. Early work on the subject was published in the Soviet Union [41,42], where alloys with a density 13-15% lower than that of iron were obtained by adding aluminium to a steel that is otherwise made austenitic because of its large manganese ($\approx$ 30 wt%) and carbon ($\approx$ 0.9 wt%) concentrations. When aged the alloy precipitates a carbide $(Fe,Mn)_3AlC_x$ that leads to strengthening. This is important because it is the density normalised strength (often referred to as specific strength) that features in some engineering designs.

Low-density steels are the subject of intense development, but there are three essential approaches. The first has already been mentioned, i.e., precipitation hardened austenite (e.g., [43]), multiphase steels containing ferrite, austenite and carbides [44], and finally, steels in which ferrite is the major phase (e.g., [45,46]).

Fig. 4.30a shows the microstructure of an Fe-10Al-15Mn-5Ni-0.5C wt% steel which has an austenitic matrix precipitation hardened with an Fe-Al intermetallic compound that has a primitive cubic structure with a motif of one Fe atom at 0,0,0, and an Al atom at $\frac{1}{2}, \frac{1}{2}, \frac{1}{2}$ [43]. This intermetallic compound is a formidable barrier to the penetration of dislocations present in the austenite leading to a large work hardening rate that is conducive to ductility and high strength. The density is just 6.82 g cm^{-3}, so the steel has a specific strength and ductility combination that is greater than that of Ti-6Al-4V, at a cost that is claimed to be a tenth that of titanium alloy.

The banded structure illustrated in Fig. 4.30a is found to exhibit anisotropic ductility and unsatisfactory formability. By adjusting the alloy composition to reduce the fraction of intermetallic phase, it has been possible to obtain a much finer and homogeneous distribution of precipitates (Fig. 4.30b) without any reduction in the yield strength [47]. The specific strength data in Fig. 4.30c are impressive, particularly because the cost of steel is lower than that of aluminium, magnesium or titanium.

The physical metallurgy and mechanical properties of these lightweight steel concepts have been demonstrated, but there are difficulties in producing large quantities using continuous casting. Alumina formation can cause the casting nozzles to be blocked and manganese has a large vapour pressure. Cost can also be an issue. But these problems are not insurmountable, so the future of low-density steels looks promising. High-manganese low-density steels are discussed in context, on page 300.

Alongside the work to create steels with good density-corrected properties, simple engineering solutions can lead to a dramatic reduction in the amount of steel needed in design. Steel frames for buildings have conventionally been constructed using hot-rolled plates and sections that tend to be thick, but "light steel frames" based on stronger cold-formed channel sections that typically are 1–2.5 mm in thickness with shapes that provide stiffness [50]. Some examples of sections and combinations of sections of cold-formed steel are illustrated in Fig. 4.31.

4.8 Summary

Elements that substitute for iron influence the relative stability of austenite and ferrite and at large concentrations can secure the stability of either phase at all temperatures in the solid state. More generally, a thermodynamic effect such as this can affect the rate of transformation through free energy terms that feature explicitly or implicitly (in determining equilibrium compositions) in theories of nucleation and growth. In fact, equilibrium may not be maintained when circumstances arise under which only carbon is sufficiently mobile. Thus, during para-equilibrium, substitutional atoms are trapped at the transformation front because they are unable to move, whereas subject to this constraint, the carbon achieves a uniform chemical potential throughout. Although the theory described in this chapter refers to equilibrium or para-equilibrium, it is quite possible that the conditions at the interface do not satisfy either mechanism.

Beyond thermodynamics, there are profound effects of the introduction of substitutional solutes on reconstructive transformations in particular, which occur at temperature and time combinations that permit atoms to move over substantial distances comparable to the scale of the microstructure. Since the mobility of substitutional solutes is far smaller than that of interstitials, the rate of reconstructive transformation is dramatically retarded. But in addition, alloy-rich carbides may form, giving rise to a variety of microstructures, including pearlite without cementite, interphase precipitation of sheets of carbides in ferrite, fibrous precipitation, heterogeneous precipitation on defects, etc.

In a ternary system, it is possible for equilibrium to be established between α, γ and θ, so that the chemical compositions of all three phases change as pearlite forms, leading to an ever diminishing driving force for transformation as equilibrium is achieved.

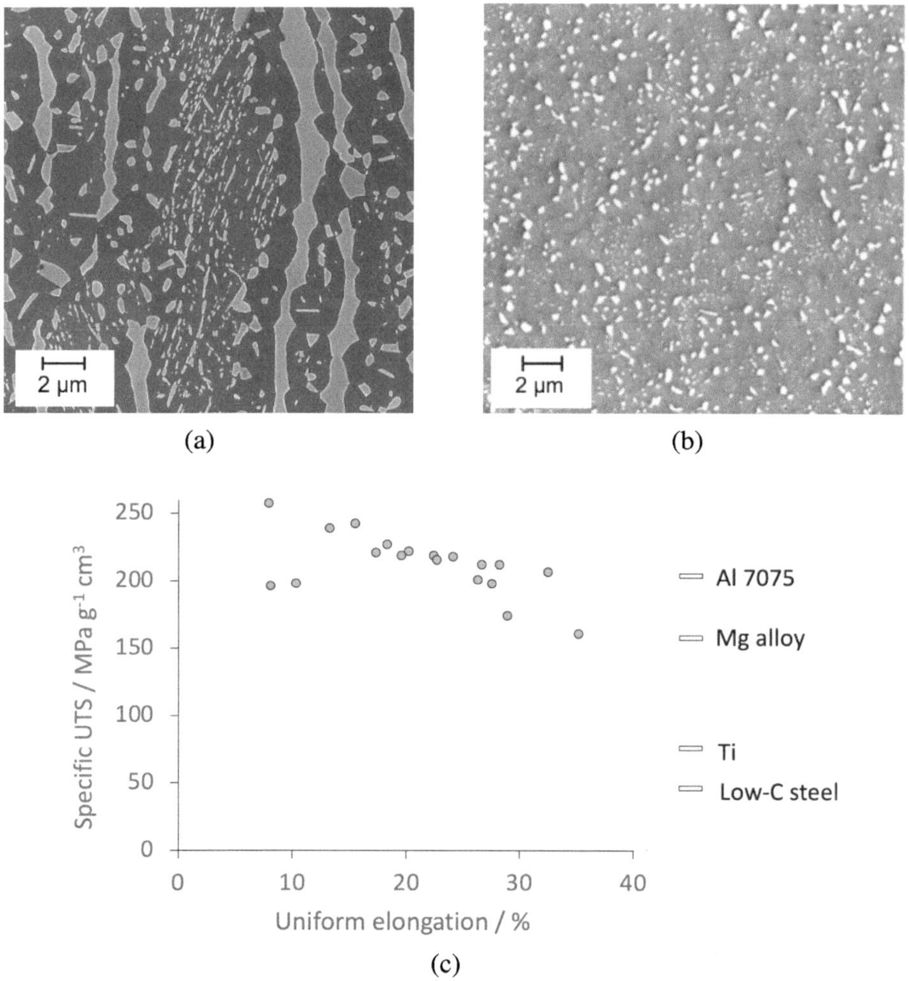

Figure 4.30 (a) The microstructure of a low-density steel (Fe-10Al-15Mn-5Ni-0.5C wt%) containing FeAl intermetallic compound precipitate in the darker austenitic matrix. Image courtesy of Sang-Heon Kim, Hansoo Kim and Nack J. Kim. (b) Fe-8Al-15Mn-5Ni-1C wt% showing the elimination of the banded structure, leaving a finer and isotropic dispersion of the intermetallic compound. Micrograph reproduced from Park et al. [47] with permission of Elsevier. (c) Specific strength of steels containing FeAl intermetallic compound against uniform elongation (selected data from Zagaran et al. [48]). Also included is a comparison against other metals that can be made in sheet form, although the uniform elongation data are not available [49].

The interlamellar spacing therefore increases with the volume fraction of pearlite, giving rise to the so-called divergent pearlite.

Finally, it is this luxuriance of the microstructures that can occur in alloyed steels that makes it possible to tune steels for specific requirements. There are few readily

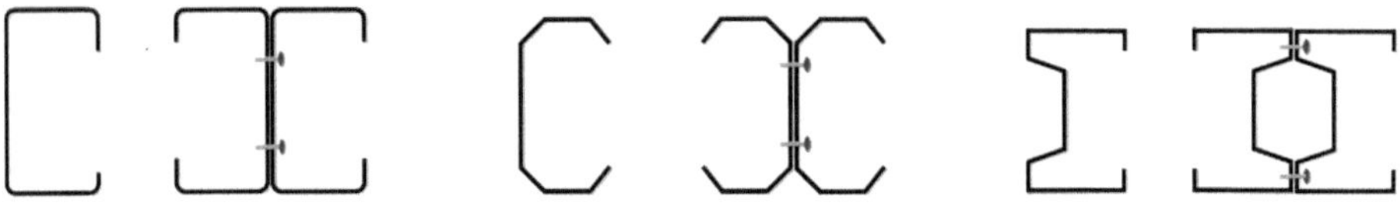

Figure 4.31 Examples of sections and combinations of sections of cold-formed steel used in construction. (a) Typical form. (b) Folded flange. (c) Sigma. Images reproduced from Kanthasamy et al. [51] under the CC-BY license.

available elements that do not dissolve in iron; and those that do not, for example lead, still are utilised, for example, to make free-machining steels.

References

[1] H.K. Danielsen, S. Ladkhodazadeh, F.B. Grumsen, M.A.J. Somers, New amorphous interface for precipitate nitrides in steel, Philosophical Magazine 94 (2014) 2339–2349.
[2] A. Strang, V. Vodarek, Z-phase formation in a martensitic 12CrMoVNb steel, Materials Science and Technology 12 (1996) 552–556.
[3] H.K. Danielsen, J. Hald, Z-phase in 9-12% Cr steels, in: D.G.R. Viswanathan, K. Coleman (Eds.), Advances in Materials Technology for Fossil Power Plants, ASM International, Materials Park, Ohio, USA, 2005, pp. 999–1012.
[4] H.K.D.H. Bhadeshia, Mulitcomponent phase diagram calculation, https://www.phase-trans.msm.cam.ac.uk/mtdataarchive/2009_11_B.pdf, 2023.
[5] F. Wever, On the influence of the elements on the polymorphism of iron, Archiv für das Eisenhüttenwesen 2 (1928–1929) 739–748.
[6] C. Zener, Equilibrium relations in medium alloy steels, Transactions of the American Institute of Mining and Metallurgical Engineers 167 (1946) 513–534.
[7] H.K.D.H. Bhadeshia, Introduction to materials modelling, in: Z. Barber (Ed.), Thermodynamics, London, Maney, London, 2006, pp. 45–72, chap. 4.
[8] K.W. Andrews, The calculation of transformation temperatures and austenite-ferrite equilibrium in steels, Journal of the Iron and Steel Institute 184 (1956) 414–427.
[9] H.K.D.H. Bhadeshia, A thermodynamic analysis of isothermal transformation diagrams, Metal Science 16 (1982) 159–165.
[10] S.R. Shatynski, The thermochemistry of transition metal carbides, Oxidation of Metals 13 (1979) 105–118.
[11] M.A.M. Alshahrani, S.W. Ooi, M.H. Colliander, G.M.A.M. El-Fallah, H.K.D.H. Bhadeshia, High-temperature hydrogen attack on 2.25Cr-1Mo steel: the roles of residual carbon, initial microstructure and carbide stability, Metallurgical & Materials Transactions A 58 (2022) 4221–4232.
[12] X. Zhu, W. Li, T.Y. Hsu, S. Zhou, L. Wang, X. Jin, Improved resistance to hydrogen embrittlement in a high-strength steel by quenching–partitioning–tempering treatment, Scripta Materialia 97 (2015) 21–24.
[13] M. Hillert, Paraequilibrium, Technical Report, Swedish Institute for Metals Research, Stockholm, Sweden, 1953.
[14] M. Hillert, The role of interfaces in phase transformations, in: Mechanism of Phase Transformations in Crystalline Solids, in: Monograph and Report Series, vol. 33, Institute of Metals, London, U.K., 1970, pp. 231–247.

[15] D.E. Coates, Diffusion controlled precipitate growth in ternary systems I, Metallurgical Transactions 3 (1972) 1203–1212.
[16] H.K.D.H. Bhadeshia, Some difficulties in the theory of diffusion-controlled growth in substitutionally alloyed steels, Current Opinion in Solid State & Materials Science 20 (2016) 396–400.
[17] J.E. Hilliard, Spinodal decomposition, in: V.F. Zackay, H.I. Aaronson (Eds.), Phase Transformations, ASM International, Metals Park, Ohio, USA, 1970, pp. 497–560.
[18] J.W. Cahn, Spinodal decomposition, Transactions of the Metallurgical Society of AIME 242 (1968) 166–179.
[19] T. Ujihara, K. Osamura, Kinetic analysis of spinodal decomposition process in Fe–Cr alloys by small angle neutron scattering, Acta Materialia 48 (2000) 1629–1637.
[20] T. Ujihara, K. Osamura, Assessing composition gradient energy effects due to spin interaction on the spinodal decomposition of Fe–Cr, Materials Science & Engineering A 312 (2001) 128–135.
[21] R.F. Mehl, W.C. Hagel, The austenite: pearlite reaction, Progress in Metal Physics 6 (1956) 74–134.
[22] A.S. Pandit, H.K.D.H. Bhadeshia, The growth of pearlite in ternary steels, Proceedings of the Royal Society A 467 (2011) 2948–2961.
[23] S.W. Seo, H.K.D.H. Bhadeshia, D.W. Suh, Pearlite growth rate in Fe-C and Fe-Mn-C steels, Materials Science and Technology 31 (2015) 487–493.
[24] J.S. Kirkaldy, B.A. Thomson, E. Baganis, Prediction of multicomponent equilibrium and transformation diagrams for low alloy steels, in: D.V. Doane, J.S. Kirkaldy (Eds.), Hardenability Concepts with Applications to Steels, TMS-AIME, Materials Park, Ohio, USA, 1978, pp. 82–125.
[25] Y. Kanetsuki, N. Ibaraki, S. Ashida, Effect of cobalt addition on transformation behavior and drawability of hypereutectoid steel wire, ISIJ International 31 (1991) 304–311.
[26] J.W. Cahn, W.C. Hagel, Divergent pearlite in a manganese eutectoid steel, Acta Metallurgica 11 (1963) 561–574.
[27] C.R. Hutchinson, R.E. Hackenberg, G.J. Shiflet, The growth of partitioned pearlite in Fe-C-Mn steels, Acta Materialia 52 (2004) 3565–3585.
[28] K. Campbell, R.W.K. Honeycombe, The isothermal decomposition of austenite in simple chromium steels, Metal Science 8 (1974) 197–203.
[29] A.H. Cottrell, Chemical Bonding in Transition Metal Carbides, Institute of Materials, London, U.K., 1996.
[30] M. Igarashi, Y. Sawaragi, Development of 0.1C-11Cr-3W-3Co-V-Nb-Ta-Nd-N ferritic steel for USC boilers, in: International Conference on Power Engineering 97, Japan Society of Mechanical Engineers, Tokyo, Japan, 1997, pp. 107–112.
[31] A. Ray, H. Bhadeshia, Niobium in microalloyed rail steels, in: HSLA Steels 2015, Microalloying 2015 & Offshore Engineering Steels 2015, Springer International Publishing, Germany, 2015, pp. 33–39.
[32] B. Aronsson, Structure and mechanical properties of some martensitic steels with special consideration of the influence of niobium and molybdenum (translation from German), in: Steel Strengthening Mechanisms, Climax Molybdenum Co., Michigan, USA, 1969, pp. 77–87.
[33] T.G. Chart, J.F. Counsell, G.P. Jones, W. Slough, P.J. Spencer, Provision and use of thermodynamic data for the solution of high-temperature practical problems, International Metals Reviews 20 (1975) 57–82.
[34] J. Ågren, Computer simulations of diffusion-controlled precipitation and dissolution of carbide particles in steel, Materials Science and Engineering 55 (1982) 135–141.

[35] H.K.D.H. Bhadeshia, Theory of Transformations in Steels, CRC Press, Taylor and Francis Group, London, U.K., 2021.
[36] R.W.K. Honeycombe, Ferrite, Metal Science 14 (1980) 201–214.
[37] Y. Funakawa, T. Shiozaki, K. Tomita, Y. Yamamoto, E. Maeda, Development of high strength hot-rolled sheet steel consisting of ferrite and nanometer-sized carbides, ISIJ International 44 (2004) 1945–1951.
[38] J.H. Jang, C.H. Lee, Y.U. Heo, D.W. Suh, Stability of (Ti, M) C (M= Nb, V, Mo and W) carbide in steels using first-principles calculations, Acta Materialia 60 (2012) 208–217.
[39] H.-W. Yen, P.-Y. Chen, C.-Y. Huang, J.-R. Yang, Interphase precipitation of nanometer-sized carbides in a titanium–molybdenum-bearing low-carbon steel, Acta Materialia 59 (2011) 6264–6274.
[40] S.A. Parsons, D.V. Edmonds, Microstructure and mechanical properties of medium carbon ferrite-pearlite steel microalloyed with vanadium, Materials Science and Technology 3 (1987) 894–904.
[41] G.L. Kayak, Fe-Mn-Al precipitation hardening austenitic alloys, Metallovedenie i Termicheskaya Obrabotka Metallov 2 (1969) 13–16.
[42] M.F. Alekseenko, G.S. Krivonogov, L.G. Kozyreva, I.M. Kachanova, L.V. Arapova, Phase composition, structure and properties of low-density steel 9G28Yu9MVB, Metallovedenie i Termicheskaya Obrabotka Metallov (3) (1972) 2–4.
[43] S.-H. Kim, H.S. Kim, N.J. Kim, Brittle intermetallic compound makes ultrastrong low-density steel with large ductility, Nature 518 (2015) 77–79.
[44] G. Frommeyer, U. Brüx, Microstructures and mechanical properties of high-strength Fe-Mn-Al-C light-weight TRIPLEX steels, Steel Research International 77 (2006) 627–633.
[45] S. Chatterjee, M. Murugananth, H.K.D.H. Bhadeshia, δ-TRIP steel, Materials Science and Technology 23 (2007) 819–827.
[46] Y.-U. Heo, Y.-Y. Song, S.-J. Park, H.K.D.H. Bhadeshia, D.-W. Suh, Influence of silicon in low density Fe-C-Mn-Al steel, Metallurgical & Materials Transactions A 43 (2012) 1731–1735.
[47] G. Park, C.H. Nam, A. Zargaran, N.J. Kim, Effect of B2 morphology on the mechanical properties of B2-strengthened lightweight steels, Scripta Materialia 165 (2019) 68–72.
[48] A. Zagaran, T.T. Trang, G. Park, N.J. Kim, κ-carbide assisted nucleation of B2: a novel pathway to develop high specific strength steels, Acta Materialia 220 (2021) 117349.
[49] Anonymous, Specific strength, https://en.wikipedia.org/wiki/Specific_strength, 2022.
[50] B. Shollock, D. Thakur, G. Couchman, Why steel in construction?, MRS Bulletin 41 (2016) 700–706.
[51] E. Kanthasamy, J. Hussain, K. Thirunavukkarasu, K. Poologanathan, K. Roy, G.B.G. Ananthi, T. Suntharalingam, Flexural behaviour of built-up beams made of optimised sections, Buildings 12 (2022) 1868.

Formation of martensite

5

Abstract

A spaceship reenters the Earth atmosphere at about 8000 m s^{-1}. Martensitic transformation in steel has been recorded to occur at 1000 m s^{-1}, i.e., an incredible eighth of the re-entry speed. How can this happen, and sometimes at temperatures where helium liquefies? Clearly, the atoms must move in a choreographed manner that enables an orderly motion of atoms because stochastic motion takes time to achieve the same change in structure. The complete theory of martensite is described in this chapter, with consequences on properties explained in context.

5.1 Introduction

The quenching to room temperature of austenite in a steel can lead to the formation of martensite, a very hard phase in which the carbon, formerly in solid solution in the austenite, remains in solution in the new phase. The velocity of the interface is greater than the ability of carbon to diffuse away into the austenite:

$$v_{\alpha'} > \frac{D_C^\gamma}{\lambda}, \tag{5.1}$$

where λ is the distance between the interstices in which carbon resides, so the term on the right is the maximum speed with which the carbon can diffuse (*cf.* Eq. (1.1)). The carbon is said to be trapped within the martensite, meaning that *its* chemical potential increases on transfer across the interface, even though there is a net reduction in free energy when all potentials are accounted for.

Unlike ferrite or pearlite, martensite forms by a deformation of the austenite lattice without any diffusion of atoms. The deformation causes a striking change in the shape of the transformed region, consisting of a large shear and a unidirectional volume-expansion. Martensite is therefore often referred to as a diffusionless, shear transformation, which has a strong crystallographic character, because it is generated by a specific deformation of the austenite. When the formation of martensite is constrained by its surroundings, it forms as thin plates or laths in order to minimise the strain energy due to the deformation.

5.2 General characteristics

The martensite reaction in steels normally is said to occur athermally, i.e., the fraction transformed depends on the undercooling below a martensite-start temperature, M_S, defined as the highest temperature at which a minimum detectable amount of transformation is achieved; thus, it depends on the sensitivity of the equipment used to monitor the transformation. For example, M_S measured using acoustic emission is known to be greater than that using dilatometry. The proper procedure is, therefore, to state the M_S temperature in conjunction with the sensitivity of the measurement [1].

Steels. https://doi.org/10.1016/B978-0-44-318491-8.00011-3

In this way, the measurements become reproducible and the level of accuracy needed for a particular application, can be assessed.

The extent of transformation does not seem to depend on time, as expressed in the Koistinen and Marburger equation [2], which describes the progress of transformation below M_S:

$$1 - V_V^{\alpha'} = \exp\{\beta(M_S - T_Q)\}, \quad \text{where } \beta \simeq -0.011. \tag{5.2}$$

$V_V^{\alpha'}$ is the volume fraction of martensite and T_Q the temperature below M_S to which the sample is cooled. This apparent athermal character is a pragmatic description – the absence of time in Eq. (5.2) is a consequence of very rapid transformation on time scales much shorter than observable by ordinary perception, combined with the fact that in most steels the maximum extent of transformation is limited by the undercooling below M_S. The reason why the undercooling below M_S must be increased in order to achieve more transformation is because the easiest nucleation sites are triggered first at small undercoolings, followed progressively by the remaining less potent nucleation sites as $M_S - T_Q$ increases.

Martensite, like all other phases that evolve from austenite, is a first-order transformation with nucleation and growth. Both processes require thermal activation, but the activation energy is small because there is no diffusion required during martensitic transformation. Furthermore, the interfacial structure is conducive to easy glide, i.e., with sufficient interfacial coherence to enable the conservative motion of atoms that does not necessitate diffusion.

Nevertheless, the fact that thermal activation is required to overcome the small barriers means that there is a time dependence of transformation, which can be detected readily if the transformation occurs at cryogenic temperatures. Using a richly-alloyed steel, Kurdjumov and Maximova first demonstrated the isothermal formation of martensite, following a classical C-curve behaviour, Fig. 5.1 [3]. On the same reasoning, the transformation to martensite can also be suppressed completely by cooling sufficiently rapidly to very low temperatures. An Fe-33Ni wt% alloy cooled to 4 K remained fully austenitic, but then began to transform into martensite on warming to about 42 K [4].

Nevertheless, for most steels where the M_S temperature is well above ambient, the kinetics of transformation appear independent of time and follow Eq. (5.2), from which it is evident that some austenite remains untransformed when T_Q is set to room temperature. This is referred to as *retained austenite*. It also is clear that there is in principle no martensite-finish temperature, M_F, but for convenience the latter frequently is defined at the point where 95% of the martensitic transformation is completed.

Martensite is not restricted to steels, although its technological importance in steels is unsurpassed. Table 5.1 lists a variety of materials which exhibit martensitic transformation, together with M_S temperatures and hardness values where available.

To obtain martensite, the steel must usually be cooled from the austenite phase field at a rate, which is sufficiently fast to avoid all other solid-state transformations such as ferrite and pearlite. This cooling rate can be very high for plain carbon steels, but quite

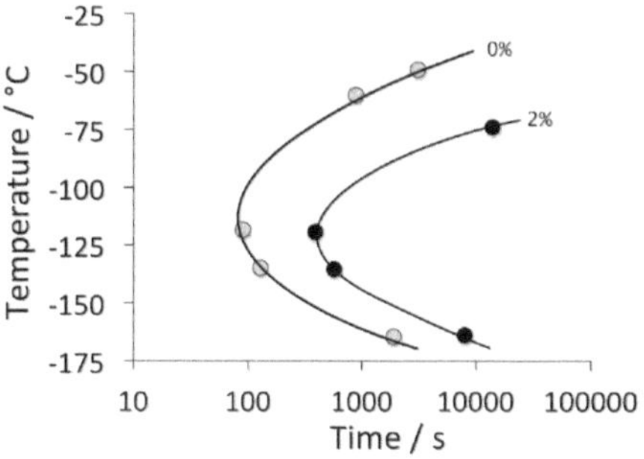

Figure 5.1 A time–temperature transformation diagram for the isothermal formation of martensite in Fe-0.05C-23.6Ni-3.3Mn-0.2Cu wt%. These are the first experiments to reveal the role of thermal activation in martensitic transformation. Selected data from Kurdjumov and Maximova [3].

Table 5.1 The temperature M_S at which martensite first forms on cooling, and the approximate Vickers hardness of the resulting martensite for a number of materials.

Composition	M_S / K	Hardness / HV
ZrO_2	1200	1000
Fe-31Ni-0.23C wt%	83	300
Fe-34Ni-0.22C wt%	<4	250
Fe-3Mn-2Si-0.4C wt%	493	600
Cu-15Al at%	253	200
Ar-40N_2 at%	30	
In-25Tl at%	125	
In-29Tl at%	193	
Gd-35Ce at%	763	

slow for a richly alloyed steel containing large concentrations of austenite stabilising solutes.

Martensite can form at very low temperatures, where diffusion, even of interstitial atoms, is not conceivable over the time period of the experiment. Table 5.1 gives typical values of the martensite-start temperature for a variety of materials. Martensite plates can grow at speeds which are a substantial fraction of the speed of sound in the steel, some $1100\,\mathrm{m\,s^{-1}}$ [5]. Such a high rate of growth is inconsistent with diffusion during transformation. Transformations, which involve diffusion are much slower – the fastest recorded solidification rate is about $80\,\mathrm{m\,s^{-1}}$ in pure nickel [6], illustrating the difference between coordinated and uncoordinated atomic motions. The difference is even greater when the random diffusional jumps happen in the solid state. Thus, the speed of a massive transformation, which is reconstructive but essentially inherits the chemical composition of the austenite has been reported to be just $0.03\,\mathrm{m\,s^{-1}}$ [7].

The chemical composition of martensite can be measured and has been shown to be identical to that of the parent austenite. These observations demonstrate convincingly that martensitic transformations are diffusionless.

5.2.1 Habit plane

The *habit* refers to the interface plane between austenite and martensite as measured on a macroscopic scale (Fig. 5.2). For unconstrained transformations, this interface plane is flat, but the need to minimise strains introduces some curvature when the transformation is constrained by its surroundings. Nevertheless, the macroscopic habit-plane is identical in both cases, as illustrated schematically in Fig. 5.2.

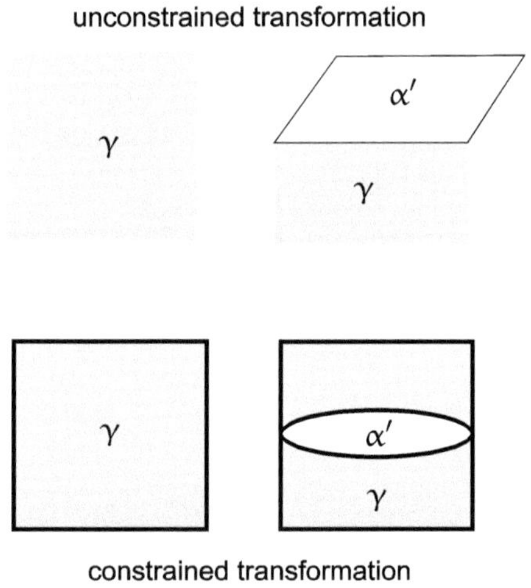

Figure 5.2 Definition of the *habit plane*, the broad interface between the martensite and austenite. The interface is flat during unconstrained transformation, but curved when the displacements due to martensitic transformation are constrained by the surrounding material. The average plane of the martensite is then the habit plane, which is parallel to the flat plane of the unconstrained transformation.

Steels of vastly different chemical composition can have martensite with the same habit plane (Table 5.2) and indeed have other identical crystallographic characteristics.

5.2.2 Orientation relationships

The formation of martensite involves the coordinated movement of atoms. It follows that the austenite and martensite lattices will always be intimately related. All martensitic transformations therefore lead to a reproducible orientation relationship between the parent and product lattices. It frequently is the case that a pair of corresponding closely-packed planes in the ferrite and austenite are parallel or nearly parallel, and it usually is the case that corresponding close-packed directions within these planes are roughly parallel:

Table 5.2 Habit plane indices for martensite. With the exception of ε-martensite, the quoted indices are approximate because the habit planes in general have irrational indices (e.g., $\sqrt{2}$ is not a rational number).

Composition / wt%	Approximate habit plane indices
Low-alloy steels, Fe-28Ni	$\{1\ 1\ 1\}_\gamma$
Plate martensite in Fe-1.8C	$\{2\ 9\ 5\}_\gamma$
Fe-30Ni-0.3C	$\{3\ 15\ 10\}_\gamma$
Fe-8Cr-1C	$\{2\ 5\ 2\}_\gamma$
ε-martensite in 18/8 stainless steel	$\{1\ 1\ 1\}_\gamma$

Kurdjumov–Sachs orientation relationship:

$$\{1\ 1\ 1\}_\gamma \parallel \{0\ 1\ 1\}_\alpha,$$

$$\langle 1\ 0\ \bar{1}\rangle_\gamma \parallel \langle 1\ 1\ \bar{1}\rangle_\alpha,$$

Nishiyama–Wasserman orientation relationship:

$$\{1\ 1\ 1\}_\gamma \parallel \{0\ 1\ 1\}_\alpha,$$

$$\langle 1\ 0\ \bar{1}\rangle_\gamma \text{ about } 5.3^\circ \text{ from } \langle 1\ 1\ \bar{1}\rangle_\alpha \text{ towards } \langle \bar{1}\ 1\ \bar{1}\rangle_\alpha,$$

Greninger–Troiano orientation relationship:

$$\{1\ 1\ 1\}_\gamma \text{ about } 0.2^\circ \text{ from } \{0\ 1\ 1\}_\alpha,$$

$$\langle 1\ 0\ \bar{1}\rangle_\gamma \text{ about } 2.7^\circ \text{ from } \langle 1\ 1\ \bar{1}\rangle_\alpha \text{ towards } \langle \bar{1}\ 1\ \bar{1}\rangle_\alpha.$$

Note that these have been stated approximately: the true relations are irrational, meaning that the indices of the parallel planes and directions cannot be expressed using rational numbers. One consequence is that there always will be 24 crystallographic variants of α' martensite possible in a single grain of austenite. The exact Nishiyama–Wasserman orientation relationship would lead only to 12 variants, but the exact relationship does not exist.

5.2.3 Structure of the interface

Any process that contributes to the formation of martensite cannot rely on much assistance from thermal activation. There must, therefore, exist a high level of continuity across the interface between martensite and austenite. The interface must be coherent or semicoherent. In the latter case, the dislocations in the interface must be able to glide as the martensite grows. It turns out that a stress-free coherent interface cannot be found between austenite and α'-martensite in steel, the best that can be achieved is semicoherency, with one direction within the interface remaining fully coherent. This direction is known as an *invariant-line* since it is unrotated and undistorted.

5.2.4 The shape deformation

The passage of a slip dislocation through a crystal causes the formation of a step where the glide plane intersects the free surface (Fig. 5.3a). The passage of many such dislocations on parallel slip planes causes macroscopic shear (Fig. 5.3b). Slip causes a change in shape, but not a change in the crystal structure, because the Burgers vectors of the dislocations are lattice vectors.

During martensitic transformation, the pattern in which the atoms in the parent crystal are arranged is deformed into that appropriate for martensite, so there must be a corresponding change in the macroscopic shape of the crystal undergoing transformation. The dislocations responsible for the deformation are in the α'/γ interface, with Burgers vectors such that in addition to deformation they also cause the change in crystal structure. The deformation is such that an initially flat surface, on a macroscopic scale, becomes uniformly tilted about the line formed by the intersection of the interface plane with the free surface. Any scratch traversing the transformed region is similarly deflected, though the scratch remains connected at the α'/γ interface. These observations, and others, confirm that the measured shape deformation belongs to a class of deformations known as invariant plane strains (Figs. 5.3c–e), with martensite being the most general case in this class with a combined shear (s) and dilatational strain (ζ), the latter being directed normal to the habit plane.

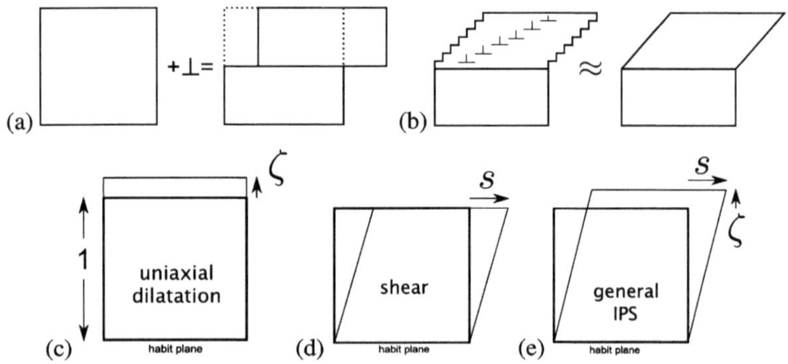

Figure 5.3 (a) Step caused by the passage of a slip dislocation. (b) Many slip dislocations, causing a macroscopic shear. (c) An invariant-plane strain with a uniaxial dilatation. (d) An invariant-plane strain which is a simple shear. (e) An invariant-plane strain, which is the combined effect of a uniaxial dilatation and a simple shear.

Evidence that the transformation involves large shears can easily be obtained by polishing a surface, or better two surfaces, at right angles prior to transformation. After martensite plates form, the surface reveals relief including large shear displacements and any scratches present prior to transformation are themselves displaced (Fig. 5.4). The characteristic surface relief can also be analysed by using two-beam interferometry or atomic force microscopy, and quantitative data obtained from the displacement of the fringe patterns or surface contours, respectively. Experiments like these reveal that the shear strain is of the order of $s \simeq 0.26$ and the dilatational strain normal to the habit plane is typically $\zeta \simeq 0.03$.

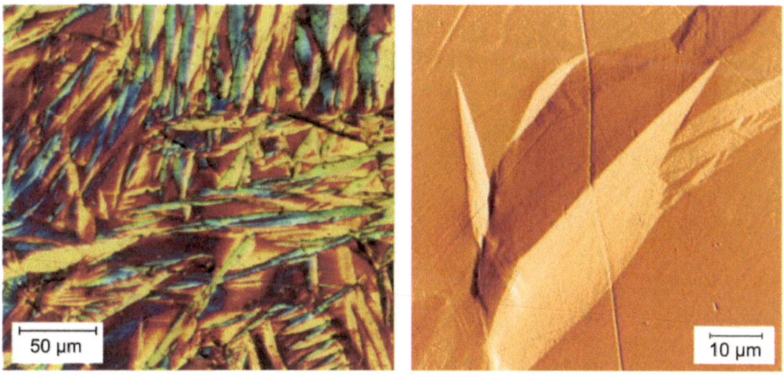

(a) (b)

Figure 5.4 Samples of steel polished flat followed by transformation into martensite. (a) Nomarsky interference contrast due to a large fraction of martensitic transformation in a low alloy steel. (b) Atomic force microscope image showing the upheavals caused by transformation together with the shear displacement of scratches. Notice that the scratches are continuous across the habit plane, showing that there is no expansion in that plane. The regions surrounding the martensite are untransformed austenite. Image courtesy of Stefan Forsik.

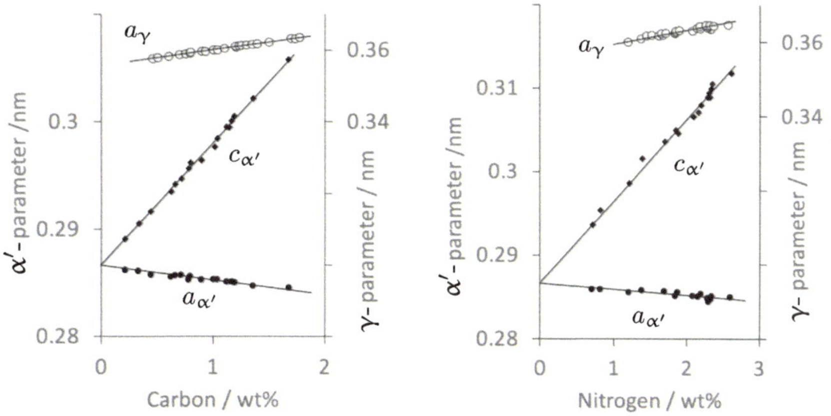

(a) (b)

Figure 5.5 Lattice parameters of austenite and martensite as a function of the interstitial concentration in iron. Adapted using compilation of data by Cheng et al. [8]. (a) Carbon. (b) Nitrogen.

5.3 Crystal structure of martensite

Martensite in steels is a supersaturated solid solution of carbon in ferritic iron. The deformation (the Bain strain) that changes the lattice of austenite, which has just one octahedral interstice per iron atom, into one of the three octahedral interstices per iron atom in the lattice of martensite, necessarily leaves it in a tetragonal state. Given thermal activation, the carbon atoms may disorder amongst the three sets of octahedral interstices if the carbon content is low and the transformation temperature high [9]. For alloys which have a low martensite-start temperature or a high carbon concentration, the carbon atoms remain ordered so that the crystal structure becomes body-centred tetragonal (bct). The tetragonality of the ordered martensite, measured by the ratio between the axes, $c_{\alpha'}/a_{\alpha'}$, increases with interstitial content:

$$\frac{c_{\alpha'}}{a_{\alpha'}} = 1 + 0.0445\, w_{\mathrm{C}} \tag{5.3}$$

$$\frac{c_{\alpha'}}{a_{\alpha'}} = 1 + 0.0367\, w_{\mathrm{N}} \tag{5.4}$$

so at zero interstitial content, the structure would be body-centred cubic. The effect of carbon and nitrogen on the lattice parameter of austenite, and on the $c_{\alpha'}$ and $a_{\alpha'}$ parameters of martensite is shown in Fig. 5.5, the straight lines represented as follows:

$$a_{\alpha'}/\,\mathrm{nm} = 0.28664 - 0.00136 w_{\mathrm{C}} \qquad c_{\alpha'}/\,\mathrm{nm} = 0.28664 + 0.01134 w_{\mathrm{C}}$$
$$a_{\gamma}/\,\mathrm{nm} = 0.3556 \;\; + 0.00442 w_{\mathrm{C}}$$

$$a_{\alpha'}/\,\mathrm{nm} = 0.28664 - 0.00072 w_{\mathrm{N}} \qquad c_{\alpha'}/\,\mathrm{nm} = 0.28664 + 0.00978 w_{\mathrm{N}}$$
$$a_{\gamma}/\,\mathrm{nm} = 0.3556 \;\; + 0.00376 w_{\mathrm{N}}$$

It is interesting to observe that carbon and nitrogen in interstitial solid solution expand the fcc iron lattice uniformly, but with bcc iron the expansion is non-symmetrical giving rise to a tetragonal distortion. To understand this important difference in behaviour, it is necessary to compare the interstitial sites for carbon in the two lattices. In each case, carbon atoms occupy octahedral sites, indicated for martensite in black in Fig. 5.6, and have six near-neighbour iron atoms. In the fcc lattice, the six iron atoms around each interstitial carbon atoms form a regular octahedron, whereas in the bcc case the corresponding octahedra are not regular but are shortened along the z-axis (Fig. 1.6). These compressed octahedra only have four-fold symmetry along the shortened axis in each case, in contrast to the fcc structure in which the regular octahedra have three four-fold axes of symmetry.

Analysis of the distortion produced by carbon or nitrogen atoms in the several types of site available in the fcc and bcc lattices, has shown that in the fcc structure the distortion is completely symmetrical, whereas in the bcc one, interstitial atoms in z positions give rise to much greater expansion of iron–iron atomic distances along z than in the x and y positions.

Assuming that the fcc $\rightarrow$ bcc transformation occurs in a diffusionless way, there is no opportunity for carbon atoms to move. The ferrite contains three octahedral interstices per iron atom, whereas the austenite has only one octahedral interstice per iron atom. Each of the three sets of octahedral interstices (the three sublattices) in the ferrite is associated with one of the cube edges of the ferrite unit cell. Upon diffusionless transformation, all the carbon atoms in the austenite end up on just one of the octahedral sublattices of ferrite, causing a tetragonal distortion of the bcc lattice into a bct lattice. Thus, in Fig. 5.6, since only the z sites are common to both the

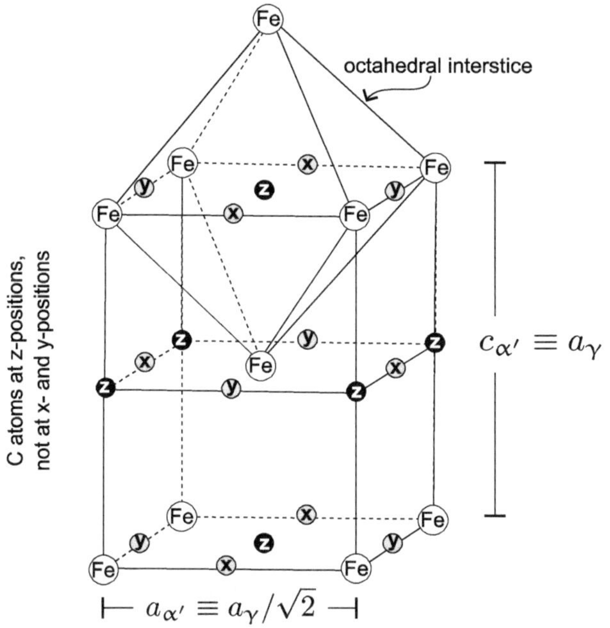

Figure 5.6 The relationship between austenite and martensite prior to the martensitic transformation (relationship also illustrated in Fig. 5.7). The diagram illustrates the fact that one octahedral interstice per γ-iron atom (z) changes into three such interstices per α'-iron atom (x, y, z). As a result, the carbon atoms located in austenite are become located on just one set of octahedral interstices in martensite, leading to a body-centred tetragonal α'-lattice. The z-set is here fully occupied by carbon atoms (adapted from Cohen [10]).

fcc and bcc lattices, on transformation the z-axis becomes the c-axis of the tetragonal form.

Therefore, the tetragonality of martensite arises as a direct result of interstitial solution of carbon atoms in the bcc lattice, together with the preference for a particular type of octahedral site imposed by the diffusionless character of the reaction. Tetragonal martensite can occur in interstitial-free iron alloys if substitutional solutes have long-range order in the austenite or if the austenite contains ordered precipitates [11].

5.4 Crystallography of martensitic transformations

Martensitic transformation is diffusionless, so the change in crystal structure is achieved by a homogeneous deformation of austenite. The strain needed to transform the fcc lattice of γ into the bcc or bct lattice of α' was first proposed by Bain in 1924 and hence is known as the *Bain strain* (Fig. 5.7). There is a compression of about 17% along the $[0\ 0\ 1]_\gamma$, corresponding to the c-axis of the martensite cell, and a uniform expansion of about 12% in the $(0\ 0\ 1)_\gamma$ plane.

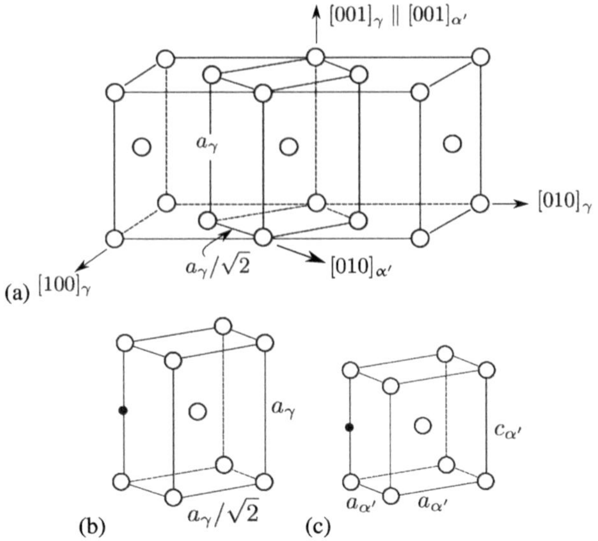

Figure 5.7 The Bain strain. (a) The lattice correspondence for formation of martensite from austenite, showing a single carbon atom in an octahedral interstice on the $[001]_\gamma$ axis. (b) Tetragonal unit cell outlined in austenite. (c) Lattice deformation (compression along *c*-axis) to form martensite with an appropriate *c/a* ratio.

The Bain strain implies the following orientation relationship between the parent and the product lattices:

$$[0\ 0\ 1]_\gamma \parallel [0\ 0\ 1]_{\alpha'} \quad [1\ \bar{1}\ 0]_\gamma \parallel [1\ 0\ 0]_{\alpha'} \quad [1\ 1\ 0]_\gamma \parallel [0\ 1\ 0]_{\alpha'},$$

but this is inconsistent with the observed orientation relationship, which is irrational, and has corresponding closest-packed planes and close-packed directions approximately parallel. The reason is that the Bain strain on its own is not the complete deformation, because it is necessary to ensure a high degree of coherency in the interface. It is a requirement that the deformation, which changes austenite into martensite leaves at least one line fully coherent, i.e., it must be unrotated and undistorted, an invariant line. Such a deformation is said to be an invariant-line strain (ILS).

In Figs. 5.8a,b, the austenite is represented as a sphere which, as a result of the Bain strain **B**, is deformed into an ellipsoid of revolution which represents the martensite. There are no lines which are left undistorted or unrotated by **B**. The lines wx and yz are undistorted, but are rotated to the new positions $w'x'$ and $y'z'$. Such rotated lines are not invariant. However, the combined effect of the Bain strain **B** and the rigid body rotation **R** is indeed an ILS, because it brings yz and $y'z'$ into coincidence (Fig. 5.8c). This is the reason why the observed irrational orientation relationship differs from that implied by the Bain strain. Indeed, the rotation required to convert **B** into an ILS precisely corrects the Bain orientation into that which is observed experimentally.

There remains a further discrepancy. As can be seen from Fig. 5.7c, there is no rotation, which can make **B** into an invariant-plane strain, since this would require

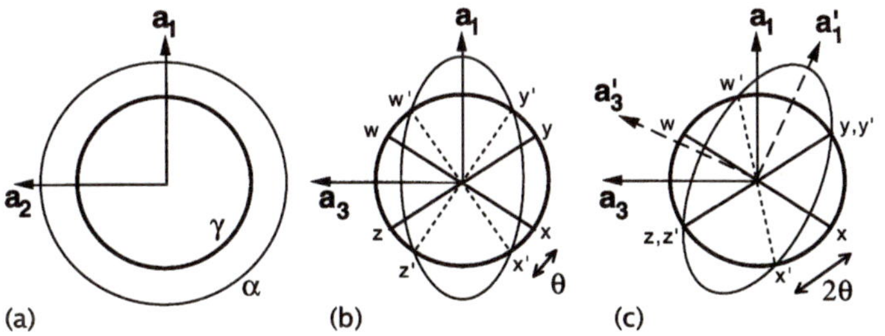

Figure 5.8 (a) and (b) show the effect of the Bain strain on austenite, which when undeformed is represented as a sphere of diameter $wx = yz$ in three-dimensions. The strain transforms it into an ellipsoid of revolution. (c) Shows the ILS obtained by combining the Bain strain with a rigid body rotation through an angle θ. $\mathbf{a}_1$, $\mathbf{a}_2$ and $\mathbf{a}_3$ refer to $[100]_\gamma$, $[010]_\gamma$ and $[001]_\gamma$ axes, respectively.

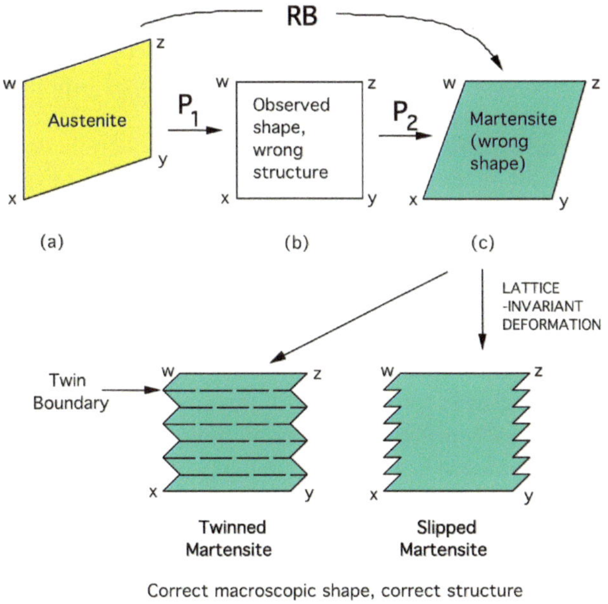

Figure 5.9 The phenomenological theory of martensite crystallography.

two non-parallel invariant lines. It follows that austenite cannot be transformed into martensite by a homogeneous strain, which leaves a plane invariant. And yet, the observed shape deformation leaves the habit plane apparently undistorted and unrotated, i.e., it is an invariant-plane strain.

The phenomenological theory of martensite crystallography elegantly solves this remaining problem (Fig. 5.9). The Bain strain converts the structure of the parent phase into that of the product phase. When combined with an appropriate rigid-body rotation, the net homogeneous lattice deformation **RB** is an ILS (step *a* to *c* in Fig. 5.9). However, the observed shape deformation is an invariant-plane strain $\mathbf{P}_1$

(step *a* to *b* in Fig. 5.9), but this gives the wrong crystal structure. If a second homogeneous shear $\mathbf{P_2}$ is combined with $\mathbf{P_1}$ (step *b* to *c*), then the correct structure is obtained, but the wrong shape since:

$$\mathbf{P_1 P_2} = \mathbf{RB}. \tag{5.5}$$

These discrepancies are all resolved if the shape-changing effect of $\mathbf{P_2}$ is cancelled macroscopically by an inhomogeneous lattice-invariant deformation, which may be slip or twinning, as illustrated in Fig. 5.9.

The theory explains all the observed features of the martensite crystallography. The orientation relationship is predicted by deducing the rotation needed to change the Bain strain into an ILS. The habit plane has no rational indices because the amount of lattice-invariant deformation needed to recover the correct macroscopic shape is not usually rational. The theory predicts a substructure in plates of martensite (either twins or slip steps) as is observed experimentally. The transformation goes to all the trouble of ensuring that the shape deformation is macroscopically an invariant-plane strain, because this reduces the strain energy when compared with the case where the shape deformation might be an ILS.

Fig. 5.10 shows schematically the two types of lattice-invariant deformation occurring within a martensite plate. It should be noted that the block of martensite formed has produced a surface tilt and that the observed habit is preserved by the accommodation provided by either slip (Fig. 5.10a) or twinning (Fig. 5.10b). The result is a macroscopically planar interface which would clearly have irregularities on a very fine scale.

The above theoretical approach had considerable success in predicting the observed habit planes, the orientation relationships between the matrix and the martensite, as well as the shape deformation for a number of martensitic transformations, including ferrous martensites. It is, however, necessary to have accurate data from plates that are isolated in the matrix, free from the elastic fields of other plates. The habit planes of individual martensite plates can be directly associated with a specific orientation relationship of the plate to the adjacent matrix. For example, Greninger and Troiano used an Fe-22Ni-0.8C wt% alloy in which austenite was retained in association with martensite to predict successfully the correct habit plane, which in this alloy is an irrational plane near $\{3\ 10\ 15\}_\gamma$ and also the shape change and the orientation relationship between martensite and austenite.

An important point to note from Eq. (5.5) is that the shape deformation, habit plane and the orientation relationship are uniquely mathematically-connected. In other words, it is not in general possible for two plates to have the same habit plane but different shape deformations, etc.

5.5 Morphology of ferrous martensites

The two-shear theory of martensite formation was first confirmed by crystallographic measurements of the two phases, but the existence of the inhomogeneous lattice invariant deformation could only be directly established by microscopic examination.

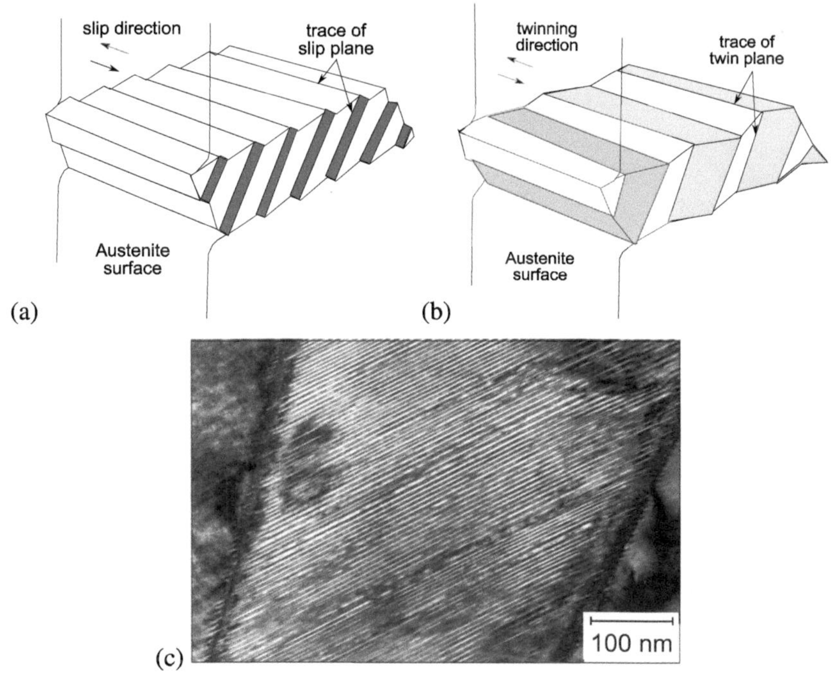

Figure 5.10 Formation of martensite plate, illustrating schematically, two types of lattice deformation. The broad faces in these schematics represent the habit planes: (a) slip and (b) twinning (adapted from Christian [12]). (c) Transmission electron micrograph showing a finely twinned plate of martensite in Fe-22Ni-20Co-0.6C wt% steel. Micrograph courtesy of A. Shibata and T. Maki.

Examination of a number of non-ferrous martensite transformations in the optical microscope revealed that the martensite lamellae contained numerous very fine twins in uniform arrays. For example, the martensite reaction in the indium-thallium system has some similar characteristics to ferrous martensites in so far as the transformation is from fcc to a face-centred tetragonal lattice. The martensitic lamellae are very uniform and contain fine twins on a single variant in each lamella.

Martensitic plates in steel frequently are not parallel sided, but are lenticular. This is the result of constraints in the matrix, which oppose the shape change resulting from the transformation. This is one of the reasons why it is difficult to identify precisely the habit planes in ferrous martensite.

Another feature of higher carbon martensites is the burst phenomenon, in which one martensite plate nucleates a sequence of plates, presumably as a result of stress concentrations set up when the first plate reaches an obstruction such as a grain boundary or another martensite plate (Fig. 5.11).

Perhaps the most striking advances in the structure of ferrous martensites occurred when thin-foil electron microscopy was first used on this problem. The two modes of plastic deformation needed for the inhomogeneous deformation part of the trans-

Figure 5.11 Lenticular martensite illustrating the burst phenomenon in a nickel-rich steel.

formation, i.e., slip and twinning, were both observed by Kelly and Nutting. Most ferrous martensites show high dislocation densities on the order of 10^{11}-10^{12} cm^{-2}, Fig. 5.12b, which are similar to those of heavily cold-worked alloys. Thus, it is usually impossible to analyse systematically the planes on which the dislocations occur or determine their Burgers vectors.

The lower carbon (<0.5 wt% C) martensites on the whole exhibit only dislocations. At higher carbon levels very fine twins (5-10 nm wide) commonly occur (Fig. 5.10c). The twinning plane is $\{112\}_{\alpha'}$, derived from $\{110\}_{\gamma}$, and the twinning direction is $\langle 11\bar{1}\rangle_{\alpha'}$, corresponding to the $\langle 1\bar{1}0\rangle_{\gamma}$ direction. Under favourable circumstances, the twins can be observed in the optical microscope, but the electron microscope allows the precise identification of twins by the use of the selected area electron diffraction technique. Thus, the twin shears can be analysed precisely and have provided good evidence for the correctness of the crystallographic theories discussed above. However, twinning is not always fully developed and even within one plate some areas are often untwinned. The phenomenon is sensitive to composition.

The evidence suggests that deformation by dislocations and by twinning are alternative methods by which the lattice invariant deformation occurs. From a thermodynamic point of view, twinning is never the favoured mode due to the cost of the twin interfaces. However, the transformation interface between internally twinned martensite and austenite is likely to be more glissile and hence favoured on a kinetic basis when the transformation is rapid [14].

The morphological and crystallographic characteristics are particularly sensitive to the carbon concentration. Low carbon martensite has a habit plane that is close to $\{111\}_{\gamma}$ and is found in plain carbon and low-alloy steels containing up to about 0.5 wt% carbon. The morphology is lath- or plate-like (Fig. 5.12), where the laths are long and about 0.5 μm wide. These are grouped together in packets, with low-misorientation boundaries between each lath, although a minority of laths are separated by large-misorientation boundaries.

At higher carbon concentrations, the plates of martensite become more lenticular with habit planes close to $\{225\}_{\gamma}$, $\{3\,10\,15\}_{\gamma}$ or $\{259\}_{\gamma}$ (Fig. 5.11). The plates

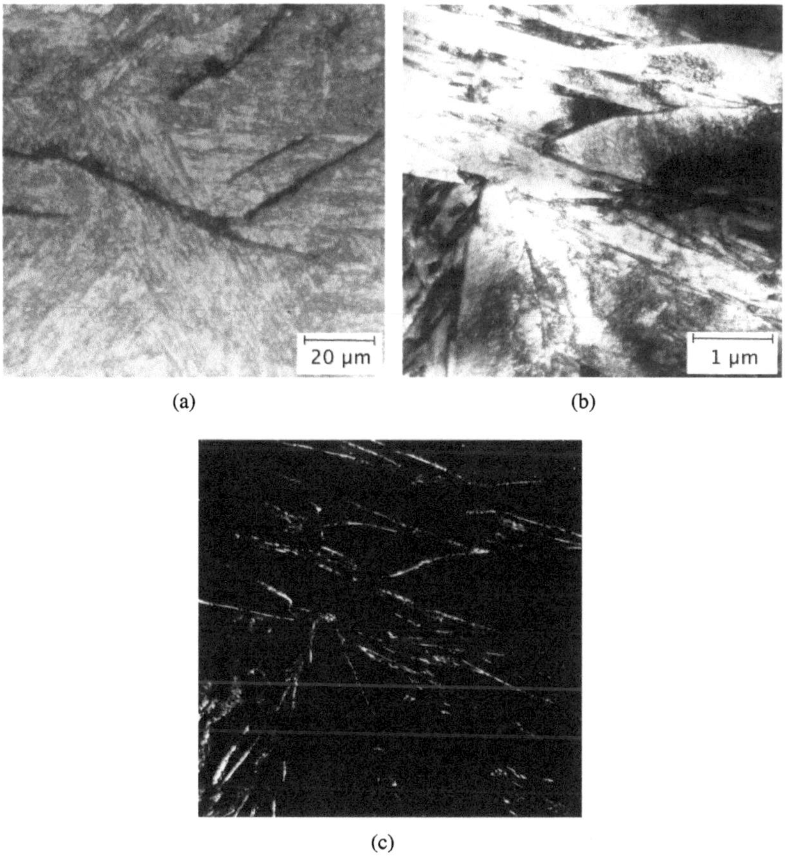

Figure 5.12 An Fe-4Mo-0.2C wt% steel austenitised and quenched into a martensitic microstructure. (a) Optical micrograph of a sample containing cracks that propagate undeviated across packets of plates that are similarly oriented. (b) Bright field transmission electron micrograph. The mottled contrast is due to the presence on a large density of dislocations in the martensite. (c) Corresponding dark-field image showing films of retained austenite around the martensite laths (after Bhadeshia [13]).

are rather well defined because their shape strains are essentially elastically accommodated at the low M_S temperatures where the parent phase is stronger. Fig. 5.13 shows how the amount of lath martensite decreases relative to the plate form as the carbon concentration is increased and as the transformation temperature is reduced. The amount of retained austenite obviously increases with the carbon concentration, so the plates of martensite are well-defined when observed using optical microscopy. This is in contrast to the laths in high M_S alloys, where the transformation leaves little retained austenite (Eq. (5.2)). As a consequence, the individual laths of martensite end up with shapes that are space-filling, determined by physical impingement with other laths.

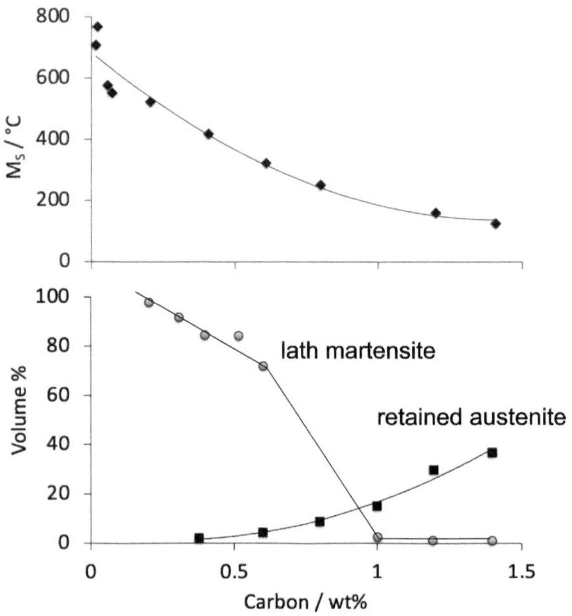

Figure 5.13 Effect of carbon content on the type of martensite and the amount of retained austenite in Fe-C alloys. The plate martensite is simply the total less the laths and retained austenite (adapted from Speich and Leslie [15]).

5.6 Kinetics of martensitic transformation

Martensitic transformations are usually described as athermal, since transformation commences at a well-defined temperature M_S, but for transformation to continue, the temperature must continue to fall until M_F is reached, when the reaction is considered complete. However, there are martensitic reactions, which can proceed at constant temperature (e.g., Fig. 5.1).

5.6.1 Nucleation of martensite

Phase and chemical composition fluctuations can occur as random events due to the thermal vibration of atoms. An individual fluctuation may or may not be associated with a reduction in free energy, but it can only survive and grow if that reduction exceeds its embryonic size. There is a cost associated with the creation of a new phase, the interfacial energy, a penalty which becomes smaller as the particle surface to volume ratio decreases. In a metastable system, this leads to a critical size of fluctuation beyond which growth is favoured.

Consider the homogeneous nucleation of martensite from the parent austenite by the classical heterophase fluctuation mechanism. For a spherical particle with radius r with an isotropic interfacial energy $\sigma_{\alpha'\gamma}$, the change in free energy as a function of radius is:

$$\Delta G = \frac{4}{3}\pi r^3 \Delta G_{\text{chem}} + \frac{4}{3}\pi r^3 \Delta G_{\text{strain}} + 4\pi r^2 \sigma_{\alpha'\gamma}, \tag{5.6}$$

where $\Delta G_{\text{chem}} = G_V^{\alpha'} - G_V^{\gamma}$, G_V is the Gibbs free energy per unit volume of α' and G_{strain} is the strain energy per unit volume of α'. The activation barrier and critical size obtained using Eq. (5.6) are given by:

$$G^* = \frac{16\pi\sigma^3_{\alpha'\gamma}}{3(\Delta G_{\text{chem}} + \Delta G_{\text{strain}})^2} \quad \text{and} \quad r^* = -\frac{2\sigma_{\alpha'\gamma}}{\Delta G_{\text{chem}} + \Delta G_{\text{strain}}}. \tag{5.7}$$

The important outcome is that in classical nucleation the activation energy varies inversely with the square of the driving force. And the mechanism involves random phase fluctuations. It is questionable whether this applies to cases where thermal activation is scarce. In particular, the activation barrier must be very small indeed if the transformation is to occur at a proper rate at low temperatures – and martensite can form at incredibly low temperatures.

One mechanism by which the barrier becomes sufficiently small involves the spontaneous dissociation of specific dislocation defects already present in the parent phase [16,17]. The dislocations are glissile, so the mechanism does not require diffusion. The only barrier is the resistance to the glide of the dislocations. Nucleation cannot occur until the undercooling is sufficient to support the faulting and strains associated with the dissociation process that leads to the creation of the new crystal structure, Fig. 5.14.

The free energy per unit area of fault plane is:

$$G_F = n_\gamma \rho_A (\Delta G_{\text{chem}} + G_{\text{strain}}) + 2\sigma_{\alpha'\gamma}, \tag{5.8}$$

where n_γ is the number of close-packed planes participating in the faulting process, ρ_A is the spacing of the close-packed planes on which the faulting is assumed to occur. The fault energy can become negative when the austenite becomes metastable.

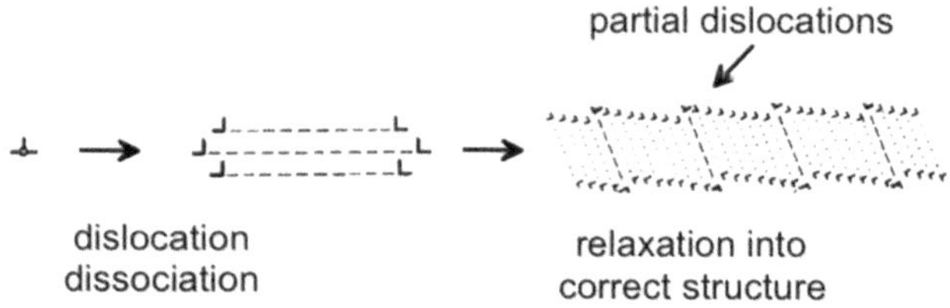

Figure 5.14 Olson and Cohen model for the nucleation of α' martensite. A perfect dislocation $\frac{a_\gamma}{2}\langle 110\rangle$ undergoes three dimensional dissociation over a set of three close-packed planes. This faulted structure is not yet that of martensite. The fault set then relaxes to a body-centred cubic structure with the introduction of partial dislocations in the interface. Additional lattice-invariant deformations, not illustrated here, occur when the nucleus grows sufficiently large so that it becomes necessary to cancel the long range strain fields of the partial dislocations.

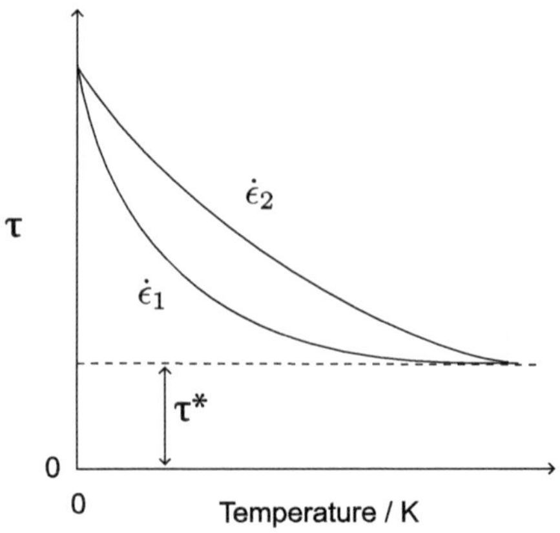

Figure 5.15 Temperature dependence of the applied stress necessary to move a dislocation at two different strain rates ($\dot{\epsilon}_2 > \dot{\epsilon}_1$). τ^* is the athermal resistance which never vanishes (after Conrad [18]).

For a fault bounded by an array of n_γ dislocations, each with a Burgers vector of magnitude b, the force required to move a unit length of dislocation array is $n_\gamma \tau_o b$. τ_o is the shear resistance of the lattice to the motion of the dislocations. It is G_F that provides the stress to overcome this resistance via the chemical free energy so the defect becomes unstable, i.e., nucleation occurs when

$$G_F = -n_\gamma \tau_o b, \tag{5.9}$$

but in general the applied stress τ may be greater than τ_o, so we need to consider also a strain rate that imposes a rate of growth on the fault.

From the theory of dislocations, the temperature dependence of the applied stress needed to achieve two rates of plastic deformation is illustrated in Fig. 5.15, consisting of a temperature-independent τ^* and then a component that depends on temperature and strain rate. Take the energy barrier between adjacent equilibrium positions of a dislocation to be G_o^*. An applied shear stress τ has the effect of reducing the height of this barrier [18,19]:

$$G^* = G_o^* - (\tau - \tau^*)v^*, \tag{5.10}$$

where v^* is an activation volume.

In the context of nucleation, the stress τ is not externally applied but comes from the chemical driving force through $G_F = -n_\gamma \tau b$ (Eq. (5.9)). Therefore, a combination of Eqs. (5.8), (5.9) and (5.10) gives

$$G^* = G_o^* + \left[\tau^* + \frac{\rho_A}{b} G_{\text{strain}} + \frac{2\sigma}{n_\gamma b}\right] v^* + \frac{\rho_A v^*}{b} \Delta G_{\text{chem}}. \tag{5.11}$$

It follows that in this model of nucleation, the activation energy G^* decreases linearly as the magnitude of the driving force ΔG_{chem} increases. This direct proportionality contrasts with the inverse square relationship of classical theory.

5.6.2 Growth of martensite

Martensitic transformation is diffusionless. We have seen that the interface between martensite and austenite has sufficient of the right kind of coherency to enable it to move without the need for diffusion (section 5.2.3). In other words, the dislocations in the interface can glide without a need for climb. As for the ordinary dislocations involved in slip deformation, the $\alpha'\gamma$ interface can in principle glide at speeds limited by that of sound in the metal. The growth of martensite plates has been measured in an Fe-Ni alloy to occur at speeds in excess of $1000\,\text{m}\,\text{s}^{-1}$ [5]. It is emphasised, however, that growth need not occur so rapidly if processes such as plastic relaxation damp the growth, or during stress-induced transformations where the driving force for transformation can be small.

The rate at which the transformation interface moves is given by [20,21]

$$v_i = v_0 \exp\left\{-\frac{Q}{kT}\right\}, \tag{5.12}$$

where the pre-exponential factor v_0 is a limiting velocity when the activation energy Q is zero and k is the Boltzmann factor. The interface is defined by its dislocation structure and these dislocations encounter resistance to their motion such that

$$Q = Q_0\left(1 - \sqrt{\frac{G_{\text{id}}}{G^*_{\text{id}}}}\right), \tag{5.13}$$

where Q_0 is that activation free energy necessary to overcome the resistance to dislocation motion with the help of an interfacial driving force G_{id}. The chemical driving force for transformation is illustrated in Fig. 5.16 and the magnitude of G_{id} will be less than that of $\Delta G^{\gamma\rightarrow\alpha} = G^{\alpha} - G^{\gamma}$ to allow for the stored energy of martensite, which consists of strain energy due to the accommodation of the shape deformation and that due to defects created during the transformation. G^*_{id} represents the maximum glide resistance presented by obstacles to dislocation motion. The velocity of the α'/γ interface will clearly be greater when the driving force is large. In some cases, the martensite forms so rapidly that recalescence occurs. The resulting heating of the sample then reduces the driving force, leading to a corresponding decrease in the interfacial velocity. Since the lattice invariant deformation favoured at high velocities is twinning, the early portion of the martensite plate has twins but this transitions into slip as the plate grows and interfacial velocity decreases due to recalescence, Fig. 5.17 [22].

A typical elastic strain in a typically stressed metal has a magnitude of 10^{-3}, whereas the transformation strain when a plate of α'-martensite grows in steel is as large as $s = 0.26$ and $\zeta = 0.03$. Given such large strains, the plates of martensite can only be accommodated elastically if the following conditions are satisfied:

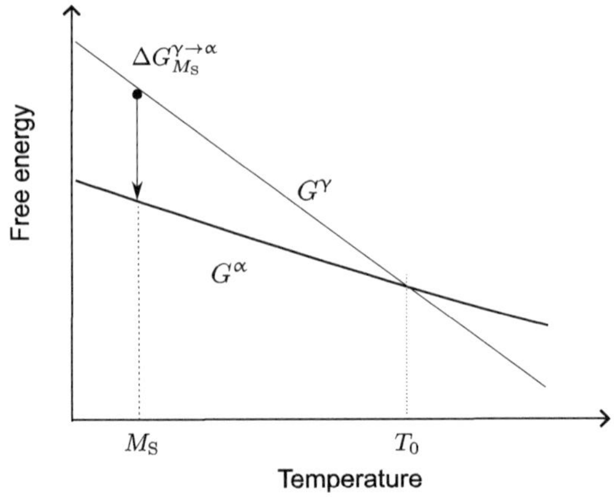

Figure 5.16 An illustration of the driving force $G^{\alpha} - G^{\gamma}$ for the transformation of austenite into ferrite of the same chemical composition. This driving force must reach a critical value $\Delta G_{M_S}^{\gamma\rightarrow\alpha}$ at the martensite-start temperature. T_0 is the temperature at which the austenite and ferrite of identical composition have the same free energy. $G^{\alpha} - G^{\gamma}$ is negative below T_0 and positive above it.

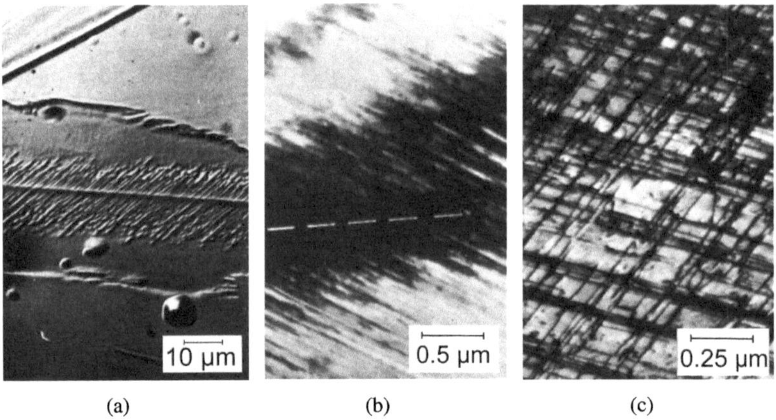

Figure 5.17 The change in the mode of lattice invariant deformation from twinning to slip as a plate of martensite heats up due to the rapid release of the heat of transformation. (a) Optical micrograph showing the twin emanating from the central spine of the plate. (b) Transmission electron micrograph of transformation twins in the central region of the plate. (c) Transmission electron micrographs showing the slip dislocations at the peripheries of the plate. After Patterson and Wayman [22].

a) The austenite is mechanically strong, i.e. has a high yield strength.

b) The austenite has a low elastic modulus because this would permit a larger elastic strain to be accommodated prior to yielding. In some materials, the elastic modulus softens dramatically as the martensite-start temperature is approached.

c) The thickness to length ratio (c/r) of the plate is small because the elastic strain energy per unit volume of martensite scales with this ratio [23]:

$$G_{\text{strain}} \approx \mu \frac{c}{r}(s^2 + \zeta^2). \tag{5.14}$$

In general, ordinary steels do not satisfy these criteria because the M_S temperature tends to be well above ambient, so that the austenite is mechanically weak. There will then be a plastic zone around the martensite plate [24,25]. A plate of martensite lengthens until it hits a barrier such as an austenite grain boundary, and then continues to thicken until G_{strain} can no longer be supported by the chemical driving force. The plastic zone in the austenite typically is greater in extent than the thickness of the plate, as illustrated in Fig. 5.18.

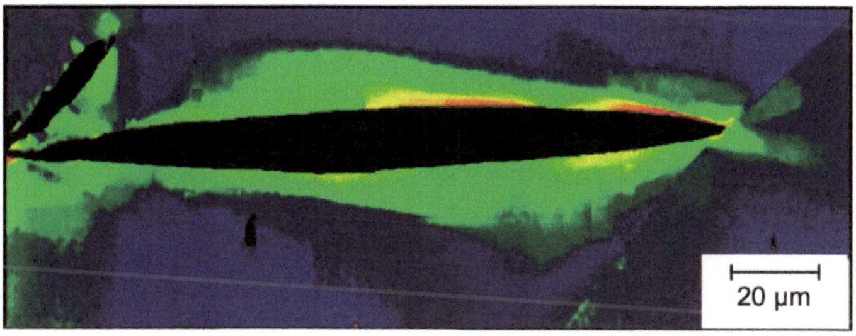

Figure 5.18 A plate of martensite (black) and the crystallographic orientation changes caused in its surrounding austenite due to the plastic accommodation of the shape deformation. The region of austenite far from the plate (blue) is unaffected (after Miyamoto et al. [25], reproduced with permission of Elsevier).

5.6.3 Overall athermal-transformation kinetics

Looking at the kinetics of martensite formation in broad terms, there are three different types of behaviour which can take place (Fig. 5.19). The first type involves normal athermal transformations with a sigmoidal type of curve, where the fraction of austenite transformed is a function solely of the temperature (Fig. 5.19a). This follows approximately the Koistinen and Marburger equation (5.2), the fundamental basis for which was derived by Magee [26] on the basis of experimental observations that the change in the number of new martensite plates per unit volume of austenite number is proportional to the imposed change in the driving force for transformation:

$$dN_V^{\gamma} = -\beta \, d(|\Delta G^{\gamma \rightarrow \alpha}|), \tag{5.15}$$

where β is a proportionality constant. The resulting change in the volume fraction transformed is therefore

$$dV_V^{\alpha'} = \overline{V} \, dN_V, \tag{5.16}$$

where $dN_V = (1 - V_V^{\alpha'})dN_V^{\gamma}$ is the change in the number of plates per unit volume of sample and $\overline{V}$ is the volume per plate of martensite (assumed constant). It follows that

$$dV_V^{\alpha'} = -\overline{V}(1 - V_V^{\alpha'})\beta \frac{d|\Delta G^{\gamma \to \alpha}|}{dT} dT. \tag{5.17}$$

When integrated between appropriate limits, this gives the familiar form of the Koistinen and Marburger equation (5.2),

$$1 - V_V^{\alpha'} = \exp\left\{\beta \overline{V}\left(\frac{d|\Delta G^{\gamma \to \alpha}|}{dT}\right)(M_S - T_Q)\right\} \quad \text{with} \quad T_Q \leq M_S. \tag{5.18}$$

This analysis gives insight into why this kind of transformation is essentially athermal – the time dependence of rapid growth is neglected by assuming that each plate transforms a fixed amount of austenite. Similarly, there is no nucleation rate as such, rather the number of new plates stimulated is related to the undercooling below M_S. This indicates that there is a potency distribution of embryos that eventually evolve into plates [27,28]. The embryos usually represent different configurations of dislocation clusters.

The second type also involves athermal transformation, but the reaction commences suddenly with a burst phenomenon which effectively causes a proportion of the austenite to transform isothermally (Fig. 5.19b). This occurs because the formation of an initial plate stimulates many others by a process known as *autocatalysis*. The stimulated plates are those that are oriented such that the stress induced by the original plate aids its formation [29]. One such example is illustrated in Fig. 5.11, where the zigzag array of mutually accommodating plates is clear. Further transformation is again athermal in character.

Finally, in true isothermal transformation (Fig. 5.19c), the proportion of austenite transformed is related to time at a given temperature. The rate of isothermal reaction can be described using the overall transformation kinetics theory outlined in section 3.6.5 if the nucleation and growth functions are known. The isothermal form

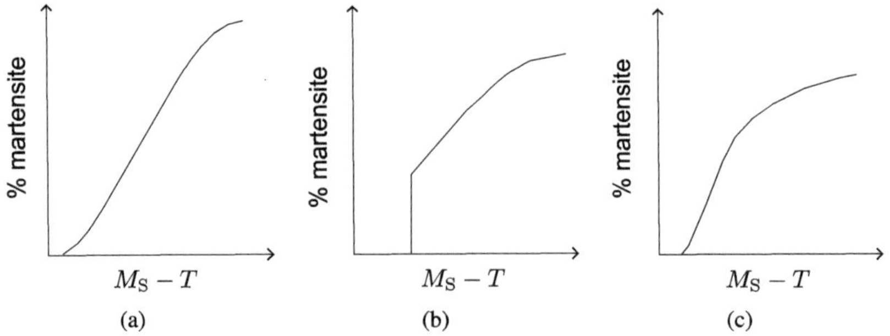

Figure 5.19 Transformation curves for martensite: (a) athermal transformation; (b) athermal with bursts; (c) isothermal transformation (adapted from Christian [12]).

of martensite is most frequently observed in interstitial free iron alloys (e.g., [30]) but also in steels (e.g., [31]).

5.6.4 Effect of alloying elements

Most alloying elements which enter into solid solution in austenite lower the M_S temperature, with the exception of cobalt and aluminium. However, the interstitial solutes carbon and nitrogen have a much greater effect than the metallic solutes. The effect of carbon on both M_S and M_F is shown in Fig. 5.20, from which it can be seen that 1 wt% of carbon lowers the M_S by over 300°C. Since above 0.7 wt% C the M_F temperature is below room temperature, steels with higher carbon concentrations retain austenite, the amount increasing as the martensite-start temperature is reduced; the fraction of retained austenite can be estimated using Eq. (5.2).

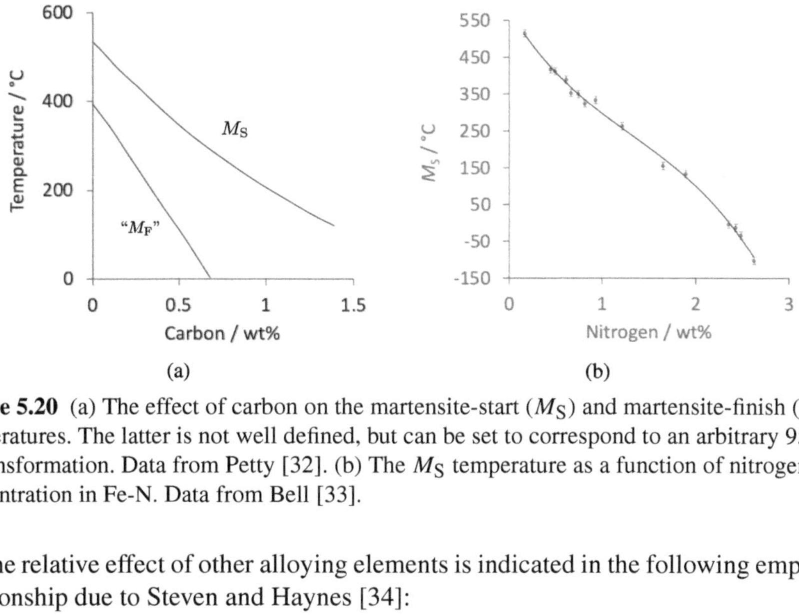

Figure 5.20 (a) The effect of carbon on the martensite-start (M_S) and martensite-finish (M_F) temperatures. The latter is not well defined, but can be set to correspond to an arbitrary 95% of transformation. Data from Petty [32]. (b) The M_S temperature as a function of nitrogen concentration in Fe-N. Data from Bell [33].

The relative effect of other alloying elements is indicated in the following empirical relationship due to Steven and Haynes [34]:

$$M_S^{\alpha'} / °C = 561 - 474w_C - 33w_{Mn} - 17w_{Ni} - 17w_{Cr} - 21w_{Mo} \tag{5.19}$$

	C	Mn	Ni	Cr	Mo
Minimum/wt%	0.10	0.20	0.0	0.0	0.0
Maximum/wt%	0.55	1.70	5.0	3.5	1.0

where w_i represents the weight percent of the solute identified in the subscript. The effect of nitrogen in this context is given by $\Delta M_S = -245w_N$ [33].

The uncertainty with relationships such as Eq. (5.19) is about ±20°C, which is associated mainly with the noise in conducting experiments to measure the M_S tem-

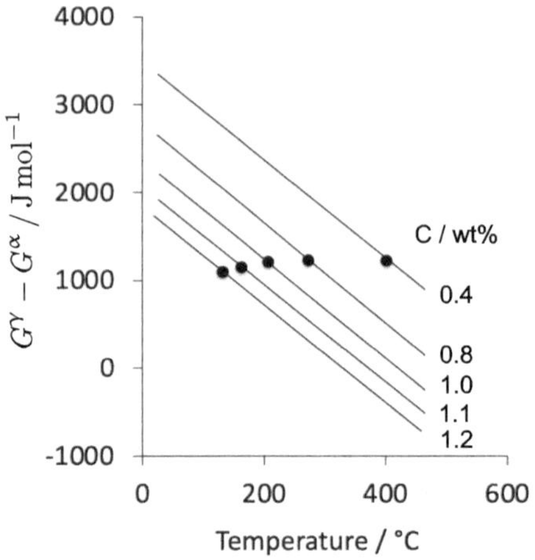

Figure 5.21 Free energy change for the austenite-martensite reaction as a function of temperature and carbon content. The points represent the measured M_S temperatures. Data from Kaufman and Cohen [39].

perature [1].[1] The assumption that M_S is a linear function of solute concentrations may not be justifiable – this can be seen also from the non-linear trends illustrated in Fig. 5.20. More complex expressions that involve products of concentrations are certainly arbitrary; neural network methods are much better in this respect because the non-linear functions appropriate to represent variations in data are deduced rather than assumed [1]. It is not surprising therefore, that the gradient of the curve in Fig. 5.20 is different from that implied by Eq. (5.19), since the relationship between M_S and carbon is not independent of the other solutes.

A better approach is to express M_S in a thermodynamic framework, since the influence of solutes on the chemical driving force $\Delta G^{\gamma \rightarrow \alpha}$ is well established and easily calculated using standard thermodynamic databases. Martensite is said to be triggered when this driving force reaches a critical value $\Delta G_{M_S}^{\gamma \rightarrow \alpha}$ (Fig. 5.16). This typically is -1000 J mol^{-1}, as illustrated in Fig. 5.21 for iron-carbon alloys. More detailed studies have shown that depends also on the substitutional solute content [35,36], via the dependence of the strength of austenite on its total solute content [37]. The reason why a large driving force is required before martensite forms is partly the stored energy that has to be accounted for, consisting of the following terms:

- The strain energy due to the shape deformation, about 600 J mol^{-1}. There is also dislocation debris created by the plastic accommodation of the shape deformation, but since the defects are created by the deformation, this is not an additional term to be added to the 600 J mol^{-1} [38].
- The energy due to the creation of twin interfaces, about 100 J mol^{-1};

[1] Assuming chemically homogeneous samples, the M_S temperature is also dependent on the sensitivity of the measuring technique. Thus, acoustic emission will in general give a greater M_S than dilatometry.

- The α'/γ interfacial energy per unit volume of a plate of martensite, which is a large term at the nucleation stage but becomes negligible at about $1\,\mathrm{J\,mol^{-1}}$ for a fully established plate.

5.6.5 Stress-induced transformation

Martensitic transformation is also a deformation, described accurately as an invariant-plane strain with a shear s on the habit plane, and a dilatation ζ normal to that plane, Fig. 5.22. Given the orthonormal coordinate system Z defined by the unit vectors $\hat{\mathbf{z}}_1$ parallel to the direction of shear, and $\hat{\mathbf{z}}_3$ normal to the habit plane, the deformation matrix $\mathbf{P}$ becomes[2]:

$$(\mathrm{Z}\ \ \mathbf{P}\ \ \mathrm{Z}) = \begin{pmatrix} 1 & 0 & s \\ 0 & 1 & 0 \\ 0 & 0 & 1+\zeta \end{pmatrix}. \tag{5.20}$$

The effect of the deformation on any vector $\mathbf{u}$ to produce a resultant vector $\mathbf{v}$ is then given by

$$(\mathrm{Z}\ \ \mathbf{P}\ \ \mathrm{Z})\,[\mathrm{Z};\mathbf{u}] = [\mathrm{Z};\mathbf{v}]. \tag{5.21}$$

The two vectors are generally not parallel, but a comparison of the magnitudes gives an impression of the strain to be expected due to martensitic transformation.

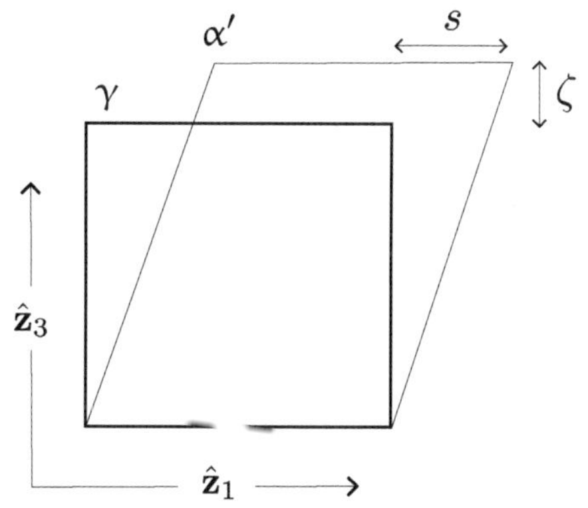

Figure 5.22 Coordinate axes defined by the unit vectors $\hat{\mathbf{z}}_i$ $(i = 1, 2, 3)$ for the derivation of the deformation matrix representing the macroscopic deformation associated with martensitic transformation (α') from austenite (γ). The habit plane contains $\hat{\mathbf{z}}_1$ and $\hat{\mathbf{z}}_2$.

This calculation of strain is along a specific direction $\mathbf{u}$, due to the shape deformation associated with the formation of a plate of martensite. In dealing with TRIP steels,

[2] The notation and method here is due to Bowles and MacKenzie [40,41]. The deformation matrix in Eq. (5.20) is easily derived by inspection of Fig. 5.22. For example, the third column of the matrix represents what happens to the components of the vector $\hat{\mathbf{z}}_3$ as a consequence of the deformation.

the problem needs to be posed somewhat differently, i.e., what is the strain along a particular direction when a stress is applied to induce martensitic transformation in otherwise stable austenite. It is necessary therefore to consider the thermodynamics of stress-affected martensitic transformation. Fig. 5.23 shows that in the absence of external stress ($\sigma = 0$), the martensite-start temperature M_S is defined by the temperature at which the free energy change $\Delta G^{\gamma\to\alpha}$ when austenite decomposes into ferrite of the same composition, reaches a critical value $\Delta G^{\gamma\to\alpha}_{M_S}$.

In contrast, when transformation occurs under the influence of a stress, the latter interacts with the shape deformation and the interaction energy U is given by [42] to be:

$$U = \tau_r s + \sigma_N \zeta, \tag{5.22}$$

where τ_r is the shear stress on the habit plane and σ_N the stress normal to that plane. Notice that the strains involved are plastic, so the interaction energy is given simply by the product of the stress and strain, rather than half that value as is sometimes assumed on the basis of elastic strains. When the stress is such that it favours the formation of martensite, U supplements $\Delta G^{\gamma\to\alpha}$ and the martensite-start temperature is raised to M_S^σ, which is above ambient temperature, so that stress-induced martensitic transformation becomes feasible (Fig. 5.23).

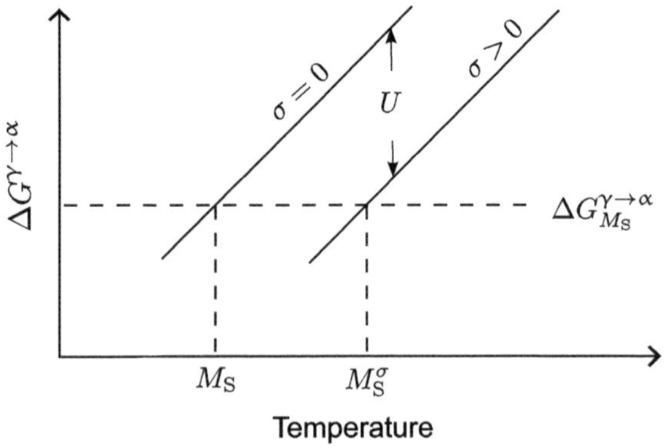

Figure 5.23 Plot of the chemical driving force for martensitic transformation against temperature, with σ representing the applied stress.

Each single crystal of austenite can, in principle, transform into 24 different crystallographic variants of α'-martensite. Each of these variants is associated with a particular value of U, depending on its orientation relative to the applied stress. Therefore, those variants with the largest interaction with the stress, i.e., which transform in a manner that relieves the stress, are favoured. This process is known as *variant selection,* so that stress-induced martensite results in a biased microstructure with reduced variety. This is illustrated in Fig. 5.24, where the martensite is generated by applying a tensile stress to polycrystalline metastable austenite, resulting in plates that are approximately at 45° to the tensile axis.

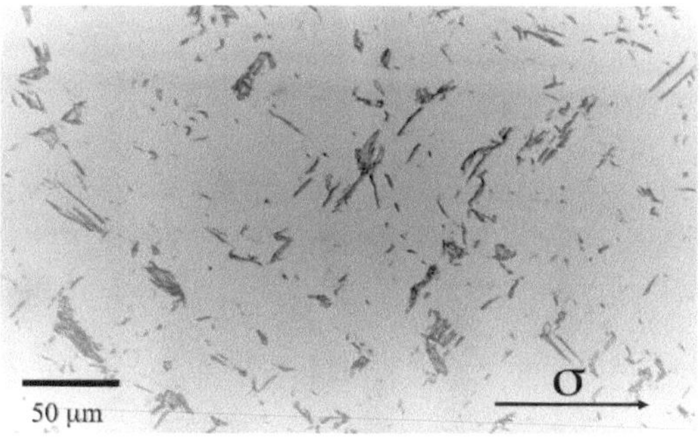

Figure 5.24 A non-random distribution of martensite habit-plane orientations produced by stress-induced martensitic transformation at a temperature between M_S and M_S^σ. The sample is polycrystalline austenite with a grain size of about 30 μm.

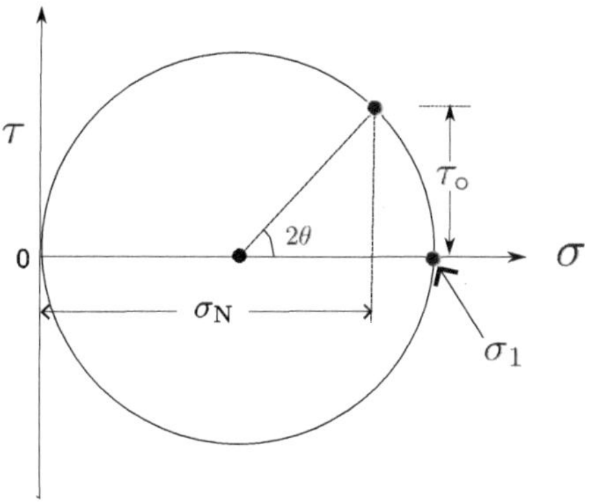

Figure 5.25 Mohr's circle representation of the shear and normal stresses on a habit plane normally inclined at θ to the tensile stress σ_1.

Assuming that a tensile stress σ_1 is applied, inclined at an angle θ to the habit plane normal, with the stress axis in the plane containing $\mathbf{z_1}$ and $\mathbf{z_3}$, then from Eq. (5.22) and Mohr's circle representation in Fig. 5.25,

$$U = \underbrace{\frac{\sigma_1}{2}\sin 2\theta}_{\tau_o} \times s + \underbrace{\frac{\sigma_1}{2}(1+\cos 2\theta)}_{\sigma_N} \times \zeta,$$

$$\frac{dU}{d\theta} = \frac{\sigma_1}{2}[2s\cos 2\theta - 2\zeta \sin 2\theta]. \tag{5.23}$$

Setting the differential to zero to obtain the maximum value of U gives $\tan 2\theta_{max} = s/\zeta$, which for typical values of $s = 0.26$ and $\zeta = 0.03$ gives $\theta_{max} = 41.7°$. Given that there are 24 habit plane orientations within a single austenite grain, it is likely that one close to this value forms first, hence explaining the observation of aligned plates in Fig. 5.24, even though the sample has many orientations of austenite grains.

The tensile axis can be represented as a unit vector

$$[Z; \hat{\mathbf{u}}] = [\sin\theta_{\max} \quad 0 \quad \cos\theta_{\max}],$$

and using $\theta = \theta_{\max}$, the elongation obtained along the tensile axis when a single crystal of austenite transforms completely into the most favoured orientation of martensite, using Eq. (5.21), is given by:

$$[Z; \mathbf{v}] = [(\sin\theta_{\max} + s\cos\theta_{\max}) \;\; 0 \;\; (1+\zeta)\cos\theta_{\max}]$$

$$\%\ \text{elongation} = \left(\frac{|\mathbf{v}|}{|\hat{\mathbf{u}}|} - 1\right) \times 100.$$

This assumes reasonably that $|\mathbf{v}| > 1$, since the unit vector $\hat{\mathbf{u}}$ is parallel to the tensile axis, and makes the approximation that $\mathbf{v}$ is parallel to $\hat{\mathbf{u}}$, neglecting a correction if $\mathbf{v}$ is rotated from $\hat{\mathbf{u}}$. The elongation due to phase transformation is therefore calculated to be 15.3% [43]. This impressive value of elongation due to phase transformation alone supplements that due to ordinary dislocation plasticity, which can be a significant boon to the design of strong steels which usually suffer from early plastic instabilities. However, steels which are fully austenitic at ambient temperature can be expensive in the context of iron-based alloys. Commercially available TRIP steels are based on ingenious alloy designs that avoid the expense but still rely on the consequences of deformation-induced martensite, as discussed in Chapter 10.

5.6.6 Effect of austenite grain size

It is observed experimentally that a reduction in the austenite grain size causes the depression of the martensite-start temperature (Fig. 5.26). There are two factors that govern this relationship:

- The size of the largest martensite plate that can form is related to the austenite grain size as defined by its mean lineal intercept $\overline{L}_\gamma$.
- Each experimental technique has a detection limit that defines the martensite-start temperature.

Relatively few martensite plates therefore need to form before transformation is detected when the austenite grain size is large. Based on this concept, the Koistinen and Marburger equation (5.2) becomes [44]:

$$M_S^o - M_S = \frac{1}{b}\ln\left[\frac{1}{\overline{L}_\gamma^3}\left\{\exp\left(-\frac{\ln(1-V_V^{\alpha'})}{m}\right) - 1\right\} + 1\right], \tag{5.24}$$

where $b = 0.2689$, $\overline{L}_\gamma$ is in mm, $m = 0.05$ is the aspect ratio of martensite plates and M_S is defined when a fraction $V_V^{\alpha'} = 0.01$ of martensite is obtained. The term M_S^o is defined as a fundamental martensite-start temperature for an austenite grain size large enough that the formation of just one martensite plate is detectable using routine methods. It is given by the point where $\Delta G^{\gamma\rightarrow\alpha} = G_\alpha - G_\gamma = -700\,\text{J}\,\text{mol}^{-1}$, i.e., equal in magnitude to the stored energy of martensite due to the shape deformation and twin interfaces. M_S^o is therefore purely a thermodynamic quantity without consideration of kinetic effects.

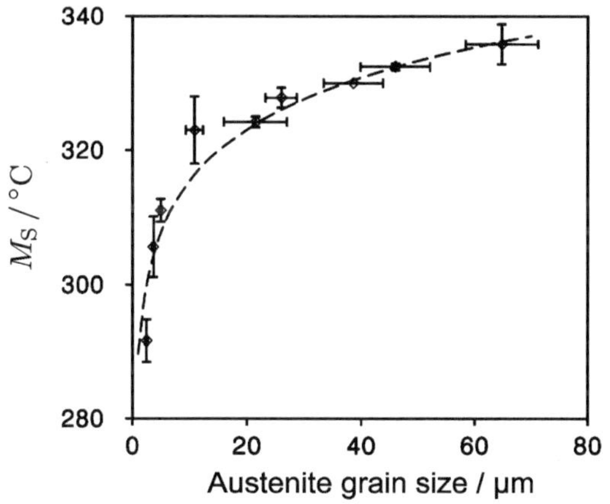

Figure 5.26 Measured variation in the martensite-start temperature as a function of austenite grain size, determined from dilatometric data, for Fe-0.13C-5Ni-2.3Mn wt% steel with $M_S^o = 364°C$. (after Yang et al. [44], where the exact chemical composition of the steel can be found).

5.6.7 *Effect of plastic strain on martensitic transformation*

One effect of plastic strain prior to transformation is to alter $\overline{L}_\gamma$ through a change in the shape of the austenite grains. This problem has been dealt with [45,46] for a large variety of deformations in which an equiaxed grain is strained, but for uniaxial tension or compression the result is:

$$\frac{\overline{L}_0}{\overline{L}} = \frac{(1+3\sqrt{3})S_{11}^{0.5} + 3(S_{11}^3+2)^{0.5}S_{11}^{-1} + (2+2S_{11}^3)^{0.5}S_{11}^{-1}}{3(2\sqrt{3}+1)}, \tag{5.25}$$

where $S_{11} = \exp\{\pm\epsilon\}$, where $+\epsilon$ and $-\epsilon$ are the plastic strains in tension or in compression, respectively, and $\overline{L}_0$ and $\overline{L}$ are the lineal intercepts in the undeformed and deformed states, respectively.

The second effect of plastic strain is through mechanical stabilisation, where martensitic transformation is suppressed when the deformed austenite contains a large density of dislocations (small plastic strains enhance the nucleation rate of martensite). These dislocations impede the motion of glissile martensite interfaces, thus retarding transformation, as illustrated in Fig. 5.27 [47]. The effect is considered by calculating the additional driving force ΔG_{STA} needed in order for the interface to overcome the dislocation density (ρ) created by strain prior to transformation [48,49]:

$$\Delta G_{STA} = \frac{\mu b}{8\pi(1-\nu)}\rho^{0.5} \qquad \mathrm{J\,m^{-3}}$$

$$\text{with} \qquad \rho \approx 2 \times 10^{13} + 2 \times 10^{14} \epsilon_{\mathrm{p}} \qquad \mathrm{m}^{-2}, \tag{5.26}$$

where μ is the shear modulus of austenite at 80 GPa, $b = 0.252$ nm is the magnitude of the Burgers vector of the dislocations and the Poisson's ratio $\nu = 0.27$. To summarise, the martensite–start temperature, independent of grain size effect, is obtained from

$$\Delta G^{\gamma \rightarrow \alpha} + U < \Delta G_{M_{\mathrm{S}}}^{\gamma \rightarrow \alpha} - \Delta G_{\mathrm{STA}}, \tag{5.27}$$

and this result is then corrected for a grain size effect using Eq. (5.24).

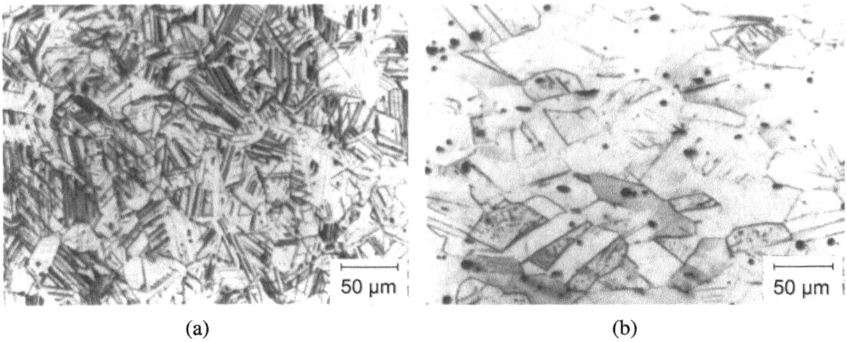

Figure 5.27 Fe-25Mn wt% alloy in which austenite transforms into ε-martensite. (a) Sample in which the austenite was not in a plastically deformed state prior to transformation. (b) The austenite was mechanically stabilised by sufficient plastic deformation, so the amount of martensite obtained is much reduced relative to (a). Images from Tomota et al. [47], with the permission of Springer Nature.

5.6.8 *Thermal stabilisation*

Thermal stabilisation is said to occur when the cooling of a steel is arrested in the $M_{\mathrm{S}} - M_{\mathrm{F}}$ range. When cooling is resumed by lowering the temperature, it does not result in as complete a transformation to martensite as would have been the case if no isothermal pause had occurred. At the chosen delay temperature, the degree of stabilisation increases to a maximum with time, and as the temperature approaches M_{F}, the extent of stabilisation increases. It appears that stabilisation is at a minimum when only a small amount of martensite is present in the matrix.

The explanation for these complex effects lies in the fact that the formation of martensite plates leads to accommodating plastic deformation in the surrounding matrix, which can result in high concentrations of dislocations in the austenite. Interaction of some of these dislocations with the glissile dislocations in the martensite plate boundary then cause it to become immobile, so that the plate cannot grow further. Any phenomena which help to encourage this process lead to stabilisation. Resting at an intermediate temperature gives time for plastic relaxation, i.e., movement of dislocations, as well as the locking of interfacial dislocations by carbon atoms.

Halting the transformation prior to M_{F} can lead to the partitioning of carbon from the martensite into the surrounding austenite [50], which may also explain thermal stabilisation. This concept is now widely exploited in the design of the quench and partitioning process, where the amount of retained austenite in an otherwise martensitic steel is enhanced by partial transformation into martensite followed by an increase in temperature in order to allow carbon to diffuse into the austenite, stabilising it for further transformation [51,52].

5.7 Strength of martensite

The high hardness and brittleness of rapidly quenched steels is the result of the formation of martensite, yet many shear transformations in non-ferrous alloy systems do not produce this dramatic hardening. Indeed, if carbon is eliminated from the steel, the resulting hardness is very much lower and the toughness enhanced. Fig. 5.28a shows the large effect of carbon content on the hardness of martensite compared with the relatively small consequence on the strength of austenite, retained to room temperature by the addition of nickel.

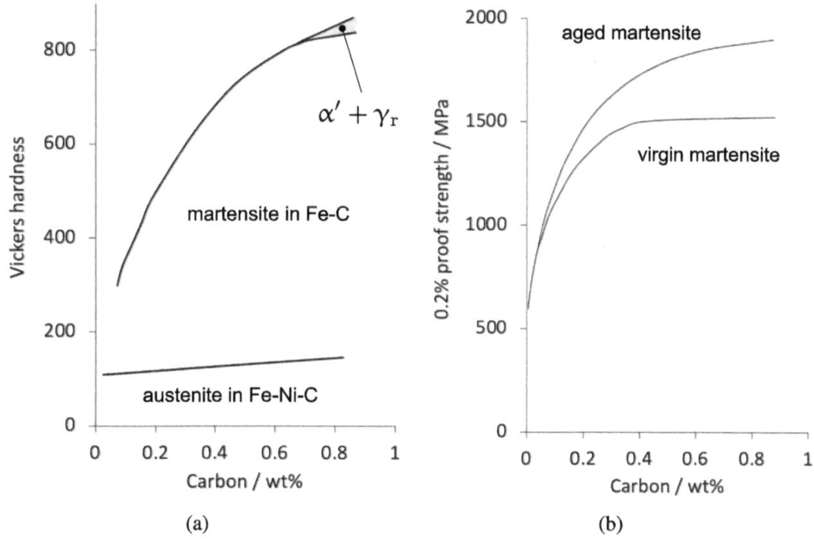

Figure 5.28 (a) The effect of carbon on the hardness of martensite and austenite. The shaded area is the hardness where significant amounts of retained austenite exist in the otherwise martensitic microstructure. (b) Ageing of martensite at 0°C in Fe-Ni-C alloys. Data from Winchell and Cohen [53].

The strength levels reached depend also on the detailed structure of the martensite, e.g., whether it remained stable during quenching and testing at room temperature. By addition of nickel to iron carbon alloys, Winchell and Cohen depressed the M_S temperature to -35°C, so that martensite formed only at low temperatures and autotempering was eliminated (Chapter 8). In addition, the samples were deformed at 0°C, with the results shown in Fig. 5.28b, indicating that the flow stress of martensite increases with carbon content up to about 0.5 wt% C. Allowing the martensite to rest for 3 h at 0°C gives the upper curve (Fig. 5.28b), demonstrating that martensite can harden at ambient temperature or below.

The question of the origin of the high strength of martensite is a difficult one, complicated by the complexity of the structure. It is a tetragonal lattice with interstitial carbon in solid solution formed by shear, which can lead to high density of dislocations and fine twins. As a result, there are several possible strengthening mechanisms:

1) substitutional and interstitial solid solution;
2) dislocation strengthening, i.e., work hardening;
3) fine twins;
4) grain size;
5) segregation of carbon atoms;
6) precipitation of iron carbides.

The interstitial solid solution of carbon, which results in the tetragonality of martensite, is a prime candidate for the role of major strengthening factor. The work of Winchell and Cohen enabled the determination of the yield stress as a function of carbon content under conditions when the carbon atoms were unable to diffuse to form atmospheres and precipitates. The flow stress was shown to vary as $c^{1/3}$, where c is the carbon concentration, but it was found later that the strength could be shown equally well to vary as $c^{1/2}$.

Fleischer [54] examined the situation theoretically with a model of a dislocation bending away from interstitial solute atoms with short range interactions and using a parameter $\Delta\varepsilon$, the difference in longitudinal and transverse lattice strain caused by a tetragonal distortion due to an interstitial carbon atom in martensite ($\Delta\varepsilon \simeq 0.38$). He found the following expression for the flow stress τ:

$$\tau \propto \frac{2\mu\Delta\varepsilon c^{1/2}}{3}, \tag{5.28}$$

predicting that the flow stress should be proportional to $c^{1/2}$. Other experiments on martensites with low M_S temperatures support the $c^{1/2}$ relationship, with little difference between the dislocated and twinned martensite (Fig. 5.29a).

The proposal that the fine twins characteristic of higher carbon martensites make a major contribution to proof strength is not supported by evidence. Certainly there is not a large increase in strength in the transition from dislocated martensite to twinned martensite takes place. However, the high dislocation densities of twin-free martensite must make some contribution to strength, estimated to be no greater than 300 MN m^{-2}, and there is reason to believe that the fine twinning makes a similar but not additive contribution.

The austenite grain-size limits the maximum size of a martensite plate, so some dependence of strength on grain size might be expected. In fact, a Petch-type plot has been found for several alloy steels of different austenitic grain sizes tested in the martensitic condition (Fig. 5.29b). However, when the fine structure of martensite is examined, much smaller sizes than $\overline{L}_\gamma$ operate and dominate the overall strength. Firstly, there is the packet size in lath martensite, or the individual plate in lenticular martensite, and beyond these there is the lath substructure, which is usually well below

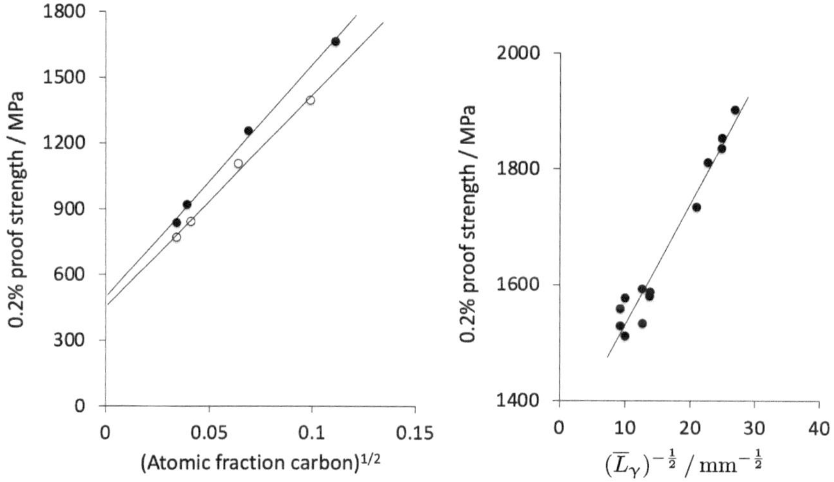

(a) (b)

Figure 5.29 (a) Effect of carbon on the strength of martensite. Data from Chilton and Kelly [55]. (b) Plot of the prior austenite grain size versus the strength of martensite. The observed effect reflects the role of the austenite grain size on the size and distribution of martensite plates, rather than a Hall-Petch effect. The alloy is Fe-0.4C-0.7Mn-0.8Cr-1.5Ni-0.25Mo wt%. Data from Grange [56].

1 μm in thickness. While many of these boundaries may represent small misorientations, they significantly hinder dislocation motion, so the number density of obstacles becomes much greater than expected from $\overline{L}_\gamma$.

It is to be expected that carbon atoms segregate to the dislocation populations typical of martensite, bearing in mind the strong interactions found in the case of ferrite. Internal friction measurements by Kurdjumov and co-workers have revealed the well-defined temperature-dependent peak, the Snoek peak (section 1.2), which occurs as a result of the stress-induced movement of carbon atoms in ferrite and martensite. Fig. 5.30a shows that the Snoek peak is much lower in a 0.026 wt% C martensite than in ferrite of the same composition. This is a direct result of the reduction in free carbon atoms in the martensite structure due to pinning by its relatively large density of dislocations. These pinned carbon atoms cannot contribute directly to the Snoek peak, the height of which is proportional to the concentration of free carbon atoms in the lattice. In contrast, ferrite has only a small dislocation density and so exhibits a more pronounced Snoek peak (Fig. 5.30a) because the carbon atoms are free between the octahedral sites. Fig. 5.30b shows an atom-probe tomograph where the segregation of carbon atoms at dislocations and boundaries in martensite is revealed.

Work on the flow stress of martensite in Fe-Ni-C alloys have shown a strong temperature dependence, together with a peak in the curve associated with serrated flow in the stress-strain curve (Fig. 5.31). Like the development of the yield point in α-iron, this is attributed to the Cottrell–Bilby interaction of carbon with dislocations.

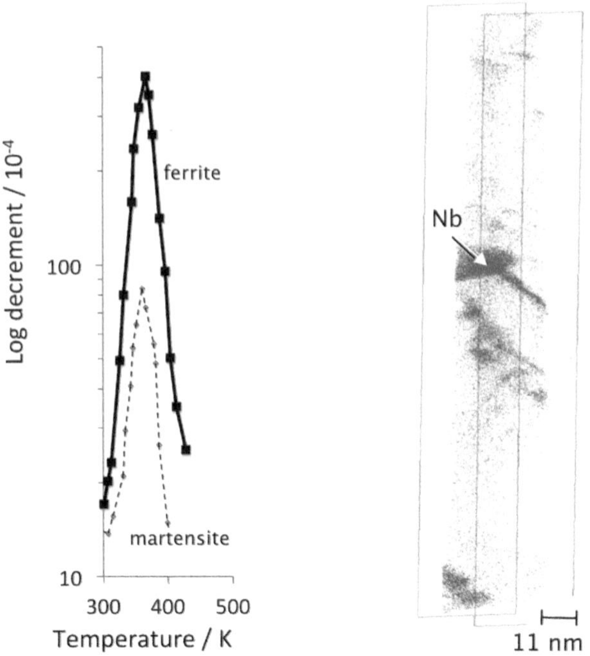

(a) (b)

Figure 5.30 (a) Comparison of the internal friction behaviour of low carbon martensite with that of ferrite in the same steel containing 0.026 wt% with the different structures produced by quenching from 1000°C or 710°C. Data from Speich and Leslie [15]. (b) Atom probe tomograph showing the distribution of carbon in a 0.12 wt% C martensitic steel. The carbon is mostly segregated at lath boundaries and dislocations. The niobium-rich area illustrated is a NbC particle that did not dissolve during austenitisation; it is not relevant to the present discussion. After Hutchinson et al. [57], reproduced with the permission of Elsevier.

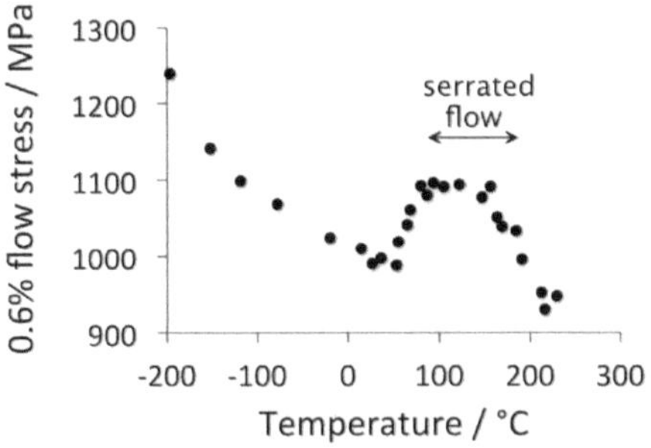

Figure 5.31 Temperature dependence of the flow stress of Fe-21Ni-0.08C wt% martensite tested at a strain rate of $8.3 \times 10^{-4}\,s^{-1}$ (After Owen and Roberts [58]).

The precipitation of iron carbide on the dislocations responsible for the increase in strength shown by martensite aged at room temperature or just above. Martensites with relatively high M_S temperatures may form cementite dispersions during the quench (auto-tempering) that contribute to the observed strength.

The yield strength of martensite, like that of ferrite, is markedly temperature dependent, but this dependence is little affected by the presence or absence of precipitates or by the amount of carbon in solution. It is therefore likely that the temperature dependence arises from the basic resistance of the lattice to dislocation movement, i.e., it is a result of the temperature dependence of the Peierls–Nabarro force.

5.8 Shape memory effect

The shape deformation accompanying martensitic transformation can be reversed by transforming back to the parent phase. Suppose a crystal of austenite is cooled to form many variants of martensite in such a way that they accommodate and hence the overall shape is unaffected by transformation. When a stress is applied, the favoured variant of martensite grows, leading to a shape change (Fig. 5.32). When heated, the shape change is reversed so that the original shape is restored. This is the basis of the shape memory effect. The memory can be lost by introducing defects during transformation, e.g., by repeated cycling. Excessive deformation, beyond that required to produce a single martensite variant by the application of stress will result in irreversible strain and a loss of memory.

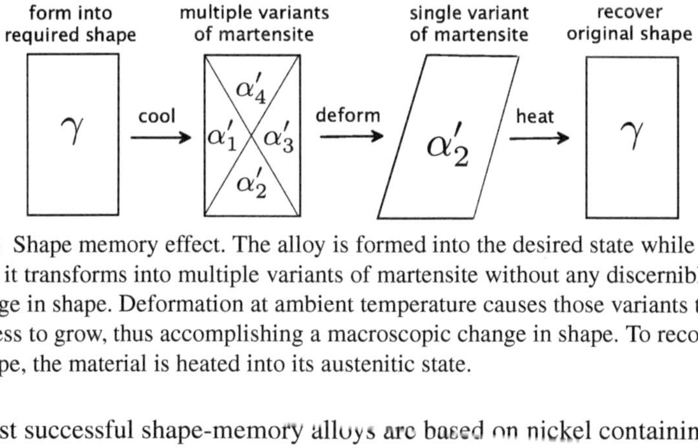

Figure 5.32 Shape memory effect. The alloy is formed into the desired state while austenitic. On cooling, it transforms into multiple variants of martensite without any discernible macroscopic change in shape. Deformation at ambient temperature causes those variants that comply with the stress to grow, thus accomplishing a macroscopic change in shape. To recover the original shape, the material is heated into its austenitic state.

The most successful shape-memory alloys are based on nickel containing titanium and aluminium. A large variety of iron-based shape-memory alloys exist, but their recoverable strains are relatively small and less reversible. They have large cost-advantages and occasionally find engineering applications such as pipe-couplings, prestressing of steel plates or concrete, seismic devices, etc., where the memory effect needs to operate only once to make an integral joint. Some alloys are produced and available in large quantities for such applications [59]. Fig. 5.33 shows how an iron-based shape memory alloy was used to strengthen a century-old bridge. The alloy, in the form of plates, was first deformed to a 2.5% plastic strain, then anchored to the steel beam using bolts, and finally heated using ceramic pads to reverse the strain and place the steel beam into compression (−74 MPa).

Figure 5.33 The installation of the FeSMA (iron-based shape memory alloy) on the lower flange of the bridge girder. Image reproduced under the CC-BY license from [60], copyright 2021.

Some of these alloys exploit the $\gamma \rightarrow \alpha'$ transformation, whereas others rely on $\gamma \rightarrow \epsilon$ martensite. There are even alloys in which the austenite transforms into face-centred tetragonal martensite. The chemical compositions of some iron-based shape memory alloys are listed in Table 5.3.

Table 5.3 Compositions of some iron-based shape memory alloys. The reversible strain and cycle data, where available, are approximate to illustrate that iron-based shape memory alloys are limited compared to nickel–titanium variants.

Composition	$\epsilon_{\text{reversible}}$		**Reference**
Fe-31Ni-0.4C wt%	$\approx 2\%$	single cycle	[61]
Fe-27Ni-0.8C wt%	$\approx 2\%$	single cycle	[61]
Fe-31Ni-10Co-3Ti wt%	$\approx 3\%$	single cycle	[62]
Fe-20Mn-2Si-7Cr-1Cu wt%	$\approx 6\%$	few cycles	[63]
Fe-17Mn-5Si-10Cr-1(V,C) wt%	$\approx 4\%$	single cycle	[64]
Fe-30Mn-6.5Si wt%	$\approx 2\%$	tens of cycles	[65]
Fe-30Pd at%			[66]
Fe-34Mn-7.5Ni-13.5Al at%	$\approx 10\%$	few cycles	[67]

5.9 Summary

Martensitic transformation is probably the most understood of all the phase transformations found to occur in the solid state. Virtually every aspect of the phase can be expressed reliably in a quantitative framework that permits the design of alloys. There are two reasons for this. First, it is a simple transformation involving no change in the chemical composition and second, it involves a rigorously disciplined motion of atoms so the crystallographic characteristics are entirely reproducible. The transformation benefits from dislocation theory because it is, after all, a mechanism for deformation that happens to change the crystal structure at the same time. Just as in slip, where the slip system is defined by a slip plane and Burgers vector, the deformation system for martensite is its habit plane and shape strain.

The role of substitutional solutes is easy to understand and predict, since they only influence the relative thermodynamic stabilities of the parent and product phases. Solution hardening has a small role in changing dislocation dynamics and hence the critical driving force for triggering martensitic transformation.

References

[1] H.-S. Yang, H.K.D.H. Bhadeshia, Uncertainties in the dilatometric determination of the martensite-start temperature, Materials Science and Technology 23 (2007) 556–560.
[2] D.P. Koistinen, R.E. Marburger, A general equation prescribing the extent of the austenite-martensite transformation in pure iron-carbon alloys and plain carbon steels, Acta Metallurgica 7 (1959) 59–60.
[3] G.V. Kurdjumov, O.P. Maksimova, Kinetics of the transformation of austenite into martensite at low temperatures, Doklady Akademii Nauk SSSR 61 (1948) 83–86.
[4] K. Ullakko, M. Nieminen, J. Pietikainen, Prevention of martensitic transformation during rapid cooling, Materials Science Forum 56 (1989) 225–228.
[5] R.F. Bunshah, R.F. Mehl, Rate of propagation of martensite, Transactions of the AIME 193 (1953) 1251–1258.
[6] B.T. Bassler, W.H. Hofmeister, R.J. Bayuzick, The solidification velocity of pure nickel, Materials Science & Engineering A 342 (2003) 80–92.
[7] J.H. Perepezko, Growth kinetics and mechanism of the massive transformation, Metallurgical Transactions A 15 (1984) 437–447.
[8] L. Cheng, A. Böttger, T.H. de Keijser, E.J. Mittemeijer, Lattice parameters of iron-carbon and iron-nitrogen martensites and austenites, Scripta Metallurgica et Materialia 24 (1990) 509–514.
[9] J.C. Fisher, J.H. Hollomon, D. Turnbull, Kinetics of the austenite to martensite transformation, Metals Transactions 185 (1949) 691–700.
[10] M. Cohen, The strengthening of steel, Transactions of the Metallurgical Society of AIME 224 (1962) 638–657.
[11] J.W. Christian, Tetragonal martensites in ferrous alloys – a critique, Materials Transactions, JIM 33 (1992) 208–214.
[12] J.W. Christian, Basic crystallography and kinetics, in: E.R. Petty (Ed.), Martensite Fundamentals and Technology, Longman Group, London, U.K., 1970, pp. 11–41.
[13] H.K.D.H. Bhadeshia, Theory and significance of retained austenite in steels, Ph.D. thesis, University of Cambridge, Cambridge, U.K., 1978.
[14] G.B. Olson, M. Cohen, Theory of martensitic nucleation: a current assessment, in: M.S.H.I. Aaronson, D.E. Laughlin (Eds.), Proceedings of International Conference on Solid→Solid Phase Transformations, TMS-AIME, Warrendale, Pennsylvania, USA, 1981, pp. 1209–1213.
[15] G.R. Speich, W.C. Leslie, Tempering of steel, Metallurgical Transactions 3 (1972) 1043–1054.
[16] J.W. Christian, A theory for the transformation in pure cobalt, Proceedings of the Royal Society of London A 206A (1951) 51–64.
[17] G.B. Olson, M. Cohen, A general mechanism of martensitic nucleation, parts I–III, Metallurgical Transactions A 7A (1976) 1897–1923.
[18] H. Conrad, Thermally activated deformation of metals, Journal of Metals 145 (1964) 582–588.

[19] J.E. Dorn, Low temperature dislocation mechanisms, in: A.R. Rosenfield, G.T. Hahn, A.L. Bement, R.I. Jaffee (Eds.), Dislocation Dynamics, McGraw Hill, New York, 1968, p. 27.
[20] M. Grujicic, G.B. Olson, W.S. Owen, Mobility of martensitic interfaces, Metallurgical & Materials Transactions A 16 (1985) 1713–1722.
[21] G.B. Olson, H.K.D.H. Bhadeshia, M. Cohen, Coupled diffusional-displacive transformations, Acta Metallurgica 37 (1989) 381–389.
[22] R.L. Patterson, C.M. Wayman, The crystallography and growth of partially-twinned martensite plates in Fe-Ni alloys, Acta Metallurgica 14 (1966) 347–369.
[23] J.W. Christian, Accommodation strains in martensite formation, the use of the dilatation parameter, Acta Metallurgica 6 (1958) 377–379.
[24] A. Vevecka, H. Ohtsuka, H.K.D.H. Bhadeshia, Plastic accommodation of martensite in disordered and ordered iron-platinum alloys, Materials Science and Technology 11 (1995) 109–112.
[25] G. Miyamoto, A. Shibata, T. Maki, T. Furuhara, Precise measurement of strain accommodation in austenite matrix surrounding martensite in ferrous alloys by electron backscatter diffraction analysis, Acta Materialia 57 (2009) 1120–1131.
[26] C.L. Magee, The nucleation of martensite, in: H.I. Aaronson, V.F. Zackay (Eds.), Phase Transformations, ASM International, Materials Park, Ohio, USA, 1970, pp. 115–156.
[27] R.E. Cech, D. Turnbull, Heterogeneous nucleation of the martensite transformation, Journal of Metals 206 (1956) 124–132.
[28] G.B. Olson, M. Cohen, in: F.R.N. Nabarro (Ed.), Dislocation Theory of Martensitic Transformations, Elsevier Science Publishers Ltd., North-Holland, 1986, pp. 297–407, chap. 37.
[29] J.C. Bokros, E.R. Parker, The mechanism of the martensite burst transformation in Fe-Ni single crystals, Acta Metallurgica 11 (1963) 1291–1301.
[30] S. Kajiwara, Morphology and crystallography of the isothermal martensite in Fe Ni Mn alloys, Philosophical Magazine 43 (1981) 1483–1503.
[31] K. Ullakko, B. Sundqvist, J. Pietikainen, Isothermal martensite transformation of FeNiC alloys as a function of hydrostatic pressure, Materials Science Forum 56–58 (1990) 197–200.
[32] E.R. Petty (Ed.), Martensite: Fundamentals and Technology, Longman Group, London, U.K., 1970.
[33] T. Bell, Martensite transformation start temperature in iron-nitrogen alloys, Journal of the Iron and Steel Institute 206 (1968) 1017–1021.
[34] W. Steven, A.G. Haynes, The temperature of formation of martensite and bainite in low alloy steels, Journal of the Iron and Steel Institute 183 (1956) 349–359.
[35] H.K.D.H. Bhadeshia, The driving force for martensitic transformation in steels, Metal Science 15 (1981) 175–177.
[36] H.K.D.H. Bhadeshia, Thermodynamic extrapolation and the martensite-start temperature of substitutionally alloyed steels, Metal Science 15 (1981) 178–180.
[37] G. Ghosh, G.B. Olson, Computational thermodynamics and kinetics of martensitic transformation, Journal of Phase Equilibria 22 (2001) 199–207.
[38] J.W. Christian, Thermodynamics and kinetics of martensite, in: G.B. Olson, M. Cohen (Eds.), International Conference on Martensitic Transformations ICOMAT '79, Alpine Press, Massachusetts, USA, 1979, pp. 220–234.
[39] L. Kaufman, M. Cohen, Thermodynamics and kinetics of martensitic transformation, Progress in Metal Physics 7 (1958) 165–246.
[40] J.S. Bowles, J.K. Mackenzie, The crystallography of martensite transformations, part I, Acta Metallurgica 2 (1954) 129–137.
[41] H.K.D.H. Bhadeshia, Geometry of Crystals, 2nd edition, Institute of Materials, 2001.

[42] J.R. Patel, M. Cohen, Criterion for the action of applied stress in the martensitic transformation, Acta Metallurgica 1 (1953) 531–538.
[43] H.K.D.H. Bhadeshia, TRIP-assisted steels?, ISIJ International 42 (2002) 1059–1060.
[44] H.S. Yang, H.K.D.H. Bhadeshia, Austenite grain size and the martensite-start temperature, Scripta Materialia 60 (2009) 493–495.
[45] S.B. Singh, H.K.D.H. Bhadeshia, Topology of grain deformation, Materials Science and Technology 15 (1998) 832–834.
[46] J.Y. Chae, R.S. Qin, H.K.D.H. Bhadeshia, Topology of the deformation of a non-uniform grain structure, ISIJ International 49 (2009) 115–118.
[47] Y. Tomota, M. Strum, J.W. Morris Jr., Tensile deformation behavior of mechanically stabilized Fe-Mn austenite, Metallurgical Transactions A 19 (1988) 1563–1568.
[48] S. Chatterjee, H.S. Wang, J.R. Yang, H.K.D.H. Bhadeshia, Mechanical stabilisation of austenite, Materials Science and Technology 22 (2006) 641–644.
[49] M. Maalekian, E. Kozeschnik, S. Chatterjee, H.K.D.H. Bhadeshia, Mechanical stabilisation of eutectoid steel, Materials Science and Technology 23 (2007) 610–612.
[50] S. Matas, R.F. Hehemann, Retained austenite and the tempering of martensite, Nature 187 (1960) 685–686.
[51] J. Speer, D.K. Matlock, B.C.D. Cooman, J.G. Schroth, Carbon partitioning into austenite after martensite transformation, Acta Materialia 51 (2003) 2611–2622.
[52] J.G. Speer, E. de Moor, A.J. Clarke, Critical assessment 7: quenching and partitioning, Materials Science and Technology 31 (2015) 3–9.
[53] P.G. Winchell, M. Cohen, The effect of carbon on the hardness of martensite and austenite, Transactions of the Metallurgical Society of AIME 224 (1962) 638–657.
[54] R.L. Fleischer, Solution hardening by tetragonal dist ortions: application to irradiation hardening in fcc crystals, Acta Metallurgica 10 (1962) 835–842.
[55] J.M. Chilton, P.M. Kelly, The strength of ferrous martensite, Acta Metallurgica 16 (1968) 637–656.
[56] R.A. Grange, Strengthening steel by austenite grain refinement, Transactions of the ASM 59 (1966) 26–48.
[57] B. Hutchinson, J. Hagström, O. Karlsson, D. Lindell, M. Tornberg, Microstructures and hardness of as-quenched martensites (0.1-0.5%C), Acta Materialia 59 (2011) 5845–5858.
[58] M.J. Roberts, W.S. Owen, Solid solution hardening and thermally activated deformation in iron-nickel-carbon martensites, Journal of the Iron and Steel Institute 206 (1968) 375–384.
[59] M. Shahverdi, S. Raza, E. Ghafoori, C. Czaderski, J. Michels, M. Motavalli, Recent advancements in development and application of an iron-based shape memory alloy at EMPA, Chimia 76 (2022) 242–248.
[60] J.V. Jtěch, P. Ryjáček, J. Campos Matos, E. Ghafoori, Iron-based shape memory alloy for strengthening of 113-year bridge, Engineering Structures 248 (2021) 113231.
[61] S. Kajiwara, Nearly perfect shape memory effect in Fe Ni C alloys, Transactions of the Japan Institute of Metals 26 (1985) 595–596.
[62] T. Maki, S. Furutani, I. Tamura, Shape memory effect related to thin plate martensite with large thermal hysteresis in ausaged Fe-Ni-Co-Ti alloy, ISIJ International 29 (1989) 438–445.
[63] D.P. Dunne, H. Li, The mechanism of thermomechanical training of a newly developed Fe-Mn-Si-Cr-Cu shape memory alloy, Journal de Physique. Colloque 5 (1995) C8–415–C8-420.
[64] Z. Dong, U.E. Klotz, C. Leinenbach, A. Berfamini, C. Czqderski, M. Motavall, A novel Fe-Mn-Si shape memory alloy with improved shape recovery properties by VC precipitation, Advanced Engineering Materials 11 (2009) 40–44.

[65] J. Robinson, P. McCormick, Factors influencing shape memory behaviour in an Fe-Mn-Si alloy, Scripta Metallurgica 23 (1989) 1975–1978.
[66] R. Oshima, K. Tanaka, A. Taniyama, F.E. Fujita, Study of bct martensite of Fe-Pd alloys, Materials Science Forum 56–58 (1990) 175–180.
[67] W. Abuzaid, Y. Wu, R. Sidharth, F. Brenne, S. Alkan, M. Vallmer, P. Krooß, T. Niendorf, H. Shitoglu, FeMnNiAl iron-based shape memory alloy: promises and challenges, Shape Memory and Superelasticity 5 (2019) 263–277.

Bainite

6

Abstract

There are high temperatures, where all atoms are mobile, and low temperatures, where none of the atoms can diffuse during the evolution of microstructure. However, in an intermediate temperature range, the logic becomes fuzzy – it may be that the carbon diffuses during the transformation, even if the substitutional lattice is displaced into position. This is the regime where bainite forms, and therefore represents probably the most difficult solid-state phase transformation to understand. We describe in this chapter, the experimental observations and how they can be used to define the atomic mechanism of transformation. This understanding is the basis of the design of some of the most innovative steels today, as discussed in the remainder of this book.

6.1 Introduction

Solid-state transformations are interesting precisely because it ordinarily is difficult to appreciate how the arrangement of atoms can change in a substance that after all, is solid to the touch. Steels are doubly fascinating in this respect, since they contain both large and very small atoms. There are circumstances where the small atoms zip through the space between the large and immobile atoms of iron. We have seen (Chapter 3) that all atoms must diffuse during a reconstructive transformation, but it is possible at lower temperatures to obviate the need for large atoms to diffuse. Such a case is typified by the displacive growth of Widmanstätten ferrite at a rate controlled by the diffusion of carbon in the austenite ahead of the interface – interestingly, in vanadium, a hydride precipitates by this displacive mechanism, but with hydrogen partitioning into the hydride [1]. However, further lowering the transformation temperature, but still above M_S, leads to the possibility that neither carbon nor the large atoms move during the time scale of the evolution of the new phase. It is this regime between the high-temperature transformations (α, P, α_W) and martensite in which the bainite transformation occurs.

Fig. 6.1 illustrates schematically the sequence of transformations expected as a function of the isothermal transformation temperature. Bainite typically occurs in the range 250–550°C, with the lower limit determined by the onset of martensitic transformation. It is now known that there is no lower limit as long as a gap can be maintained between the bainite-start (B_S) and martensite-start temperature, with the possibility that bainite can actually form at room temperature [2–4]; this will be discussed in more detail in Chapter 14.

Bainite is named after Edgar Bain, who with Davenport first isolated these structures during their pioneering and systematic studies of the isothermal decomposition of austenite [6,7]. It consists of fine aggregates of ferrite plates and cementite particles. The nature of bainite changes as the transformation temperature is lowered. Two main forms can be identified: *upper* and *lower* bainite.

Steels. https://doi.org/10.1016/B978-0-44-318491-8.00012-5

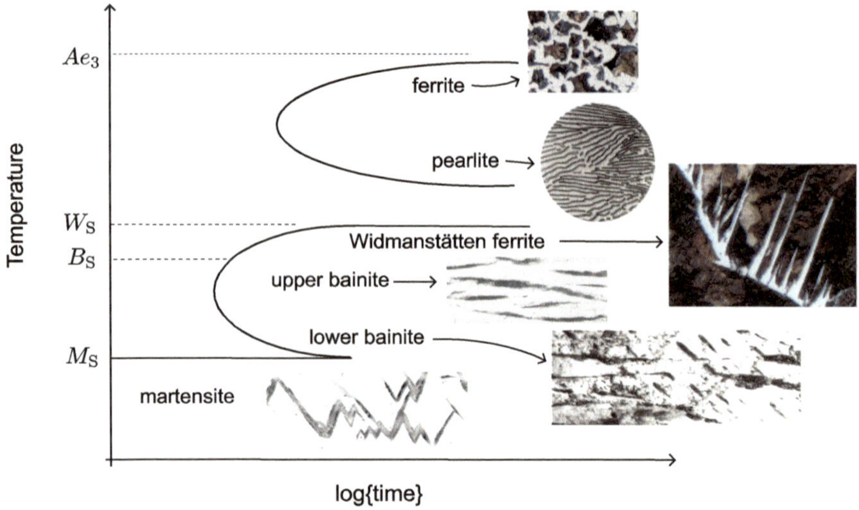

Figure 6.1 An illustration of the variety of microstructures that in principle can be generated during the isothermal transformation of austenite. The pearlite image is from Bain [5], with the original book reproduced with the permission of ASM International. The Widmanstätten ferrite image courtesy of Rolando M. Núñez of the Pontifical Catholic University of Peru.

6.2 Upper bainite (≈ 400 – 550°C)

The microstructure of upper bainite consists of fine plates of ferrite, each of which is about 0.2 μm thick and about 10 μm long. The plates grow in clusters called sheaves, Fig. 6.2a. Within each sheaf, the plates are parallel and of identical crystallographic orientation, each with a well-defined crystallographic habit. The individual plates in a sheaf often are called the sub-units of bainite [8], Fig. 6.2b. They usually are separated by low-misorientation boundaries, cementite particles or retained austenite.

Upper bainite evolves in distinct stages, beginning with the nucleation of ferrite plates at the austenite grain boundaries. The growth of each plate is accompanied by a change in the shape of the transformed region (Fig. 6.3a), a change, which can be described precisely as an invariant-plane strain (IPS) with a large shear component virtually identical to that observed during martensitic transformation [9]. However, bainite grows at relatively high temperatures when compared with martensite. The large strains associated with the shape change cannot be sustained by the austenite, the strength of which decreases as the temperature rises. These strains are relaxed by the plastic deformation of the adjacent austenite. The local increase in dislocation density caused by the yielding of the austenite blocks the further movement of the glissile transformation interface, Fig. 6.3b [10]. This localised plastic deformation therefore halts the growth of the ferrite plate, so that each sub-unit only achieves a limited size, usually much less than the size of an austenite grain.

As with martensite, the shape change implies that the mechanism of growth of bainitic ferrite is displacive. It is the minimisation of the strain energy associated with the displacements that ensures that bainite grows in the form of thin plates. Since the

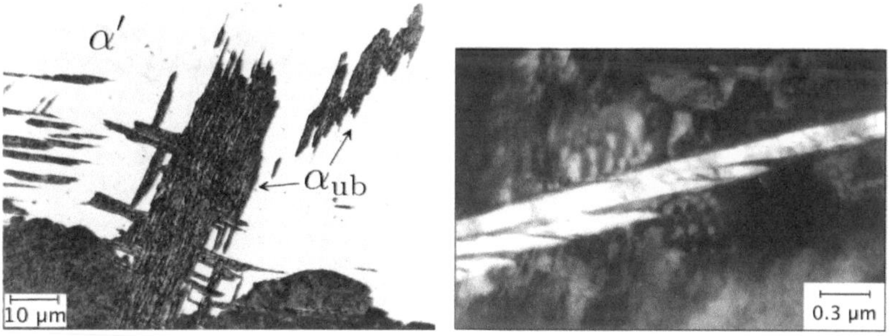

(a) (b)

Figure 6.2 Upper bainite. (a) An optical micrograph of upper bainite (α_{ub}) generated in a Fe-0.8C wt% steel by isothermal transformation at 400°C for 20 s followed by quenching so that the remaining austenite transforms into martensite (α'). Micrograph courtesy of Y. Ohmori. (b) Transmission electron micrograph to show the sub-unit structure of a sheaf of bainite.

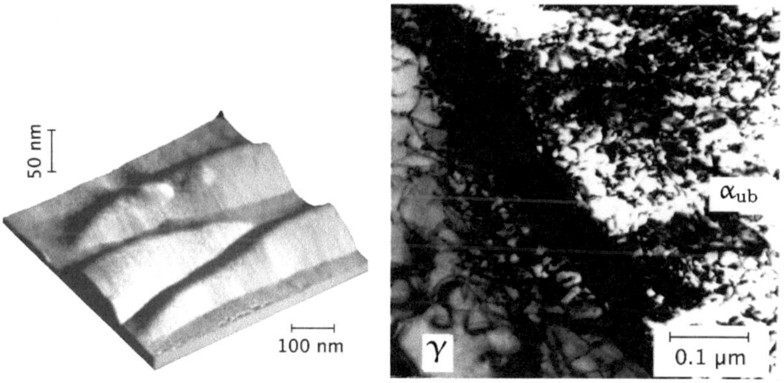

(a) (b)

Figure 6.3 Upper bainite. (a) Atomic force micrograph showing the deformation caused by the formation of α_{ub} when a pre-polished sample of austenite is transformed. Image courtesy of M. Peet. (b) Intense dislocation arrays at and near the α_{ub}/γ interface.

crystal structure of bainite is generated by a coordinated movement of atoms, it follows that there must exist an orientation relationship between the austenite and the bainite. This relationship is found experimentally to be of the type where a pair of the most densely packed planes of the two lattices is approximately parallel, as is a corresponding pair of close-packed directions within those planes. This is loosely described by a Kurdjumov–Sachs type orientation relationship.

Bainite forms on particular crystallographic planes, but the indices of the habit plane show considerable scatter (Fig. 6.4). This is because most of the measurements are made using light microscopy, in which case the habit plane determined is not that of an individual sub-unit. It corresponds instead to some average value depending on the number, size and distribution of sub-units within the sheaf. All these factors can vary with the transformation temperature, time and chemical composition.

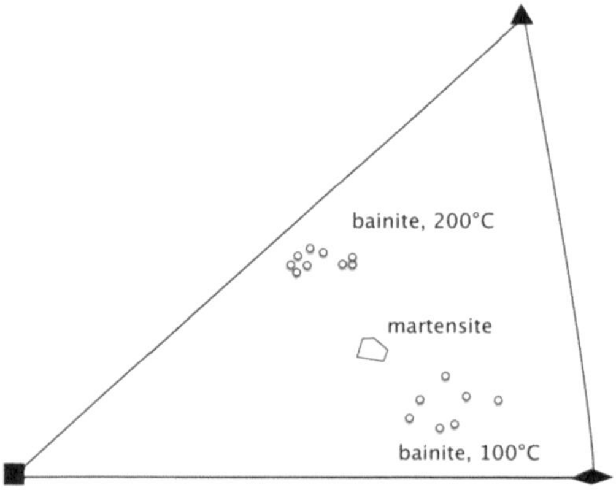

Figure 6.4 Stereographic triangle showing the habit plane of bainite determined using optical microscopy techniques compared with that of martensite in the same steel. Data from Greninger and Troiano [11].

It has already been emphasised that upper bainite forms in two distinct stages, the first involving the formation of bainitic ferrite that eventually retains little carbon in solid solution (<0.02 wt%). The remaining austenite therefore becomes enriched in carbon. Cementite then precipitates from the enriched austenite layers between the ferrite sub-units. The amount of cementite depends on the carbon concentration of the alloy. Large concentrations lead to microstructures in which the ferrite platelets are separated by continuous layers of cementite. Small, discrete particles of cementite form when the alloy carbon concentration is low.

The cementite (*Pnma*) particles have a Pitsch orientation relation to the austenite from which they precipitate:

$$[100]_\theta \parallel [110]_\gamma$$

$$[010]_\theta \parallel [\bar{2}25]_\gamma$$

$$[001]_\theta \parallel [5\bar{5}4]_\gamma.$$

Many variants of carbide may precipitate from the austenite, each particle being indirectly related to the ferrite via the ferrite/austenite orientation relationship.

If sufficient quantities of alloying elements (such as Si or Al) which retard the formation of cementite are added to the steel, then it is possible to suppress the formation of cementite altogether. An upper bainite microstructure consisting of just bainitic ferrite and carbon-enriched retained austenite is obtained instead (Fig. 6.3). The microstructure may also contain martensite if the residual austenite decomposes on cooling to ambient temperature.

6.3 Lower bainite (≈ 250 – 400°C)

Lower bainite has a microstructure and crystallographic features very similar to those of upper bainite. The major distinction is that cementite particles also precipitate inside

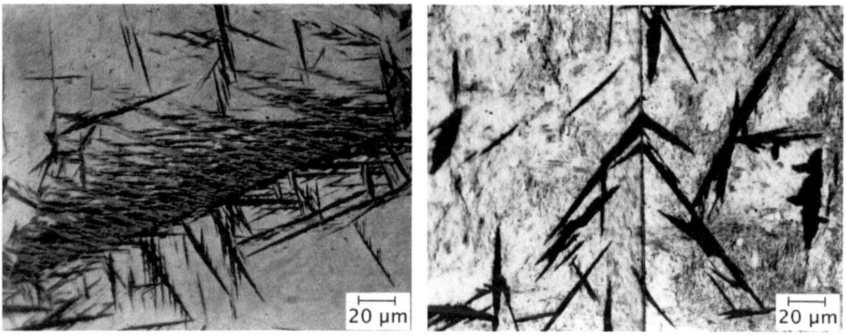

(a) (b)

Figure 6.5 Microstructure of lower bainite. (a) Optical micrograph, Fe-0.8C wt% steel transformed at 300°C, showing sheaves of lower bainite. The bainite etches dark relative to the surrounding untempered martensite. (b) Two-surface composite micrograph. Courtesy of Ohmori.

the plates of ferrite (Fig. 6.5). There are, therefore, two kinds of cementite precipitates: those which grow from the carbon-enriched austenite that separates the platelets of bainitic ferrite, and others which appear to precipitate from supersaturated ferrite. These latter particles exhibit the tempering orientation relationship, which is found when carbides precipitate during the heat treatment of martensite [12] and often described as the Isaichev orientation relationship [13]:

$$\begin{aligned} \langle 1\,0\,0\rangle_\theta \quad &\| \langle 1\,\bar{1}\,\bar{1}\rangle_\alpha \\ \{0\,3\,1\}_\theta &\approx\| \{1\,0\,1\}_\alpha. \end{aligned} \tag{6.1}$$

The carbides in the ferrite need not always be cementite. Depending on the chemical composition and the transformation temperature, other transition carbides may precipitate first. For example, in high-carbon steels containing more than about 1 wt% silicon (which retards cementite formation), epsilon carbide is commonly observed to precipitate in the bainitic ferrite.

In contrast to tempered martensite, the cementite particles in lower bainite frequently precipitate in just one variant of the orientation relationship (Fig. 6.6a), such that they form parallel arrays inclined to the axis of the bainite plate. In tempered martensite, the carbides tend to precipitate in Widmanstätten arrays.

However, these general observations are not always true. Widmanstätten arrays of cementite are also found in lower bainite when the latter forms in high-carbon steels or when the transformation occurs at low temperatures (Fig. 6.6b). Similarly, martensite in low-carbon steels exhibits only a single variant of carbide on tempering. This is because the carbide precipitation is influenced by the stresses associated with the displacive growth of lower bainite or martensite – those variants of cementite that best comply with the stress are dominant. If the driving force for precipitation is large (i.e., the carbon concentration inherited by the bainite is large), then multiple variants, including those which do not comply with the stress, are able to precipitate [14].

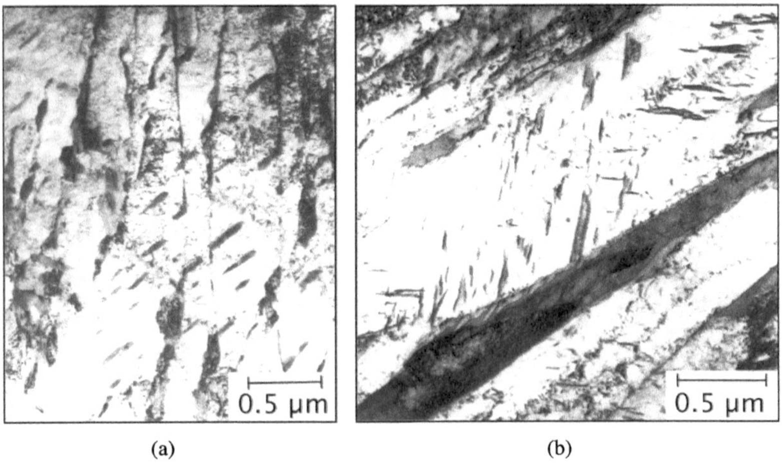

Figure 6.6 (a) Single variant of cementite in lower bainite, Fe-0.3C-4Cr wt%, transformed isothermally at 435°C. (b) Multiple variants of cementite in lower bainite, Fe-0.4C-2Si-3Mn wt%, transformed isothermally at 300°C.

The carbides in the lower bainite are extremely fine, just a few nanometres thick and about 500 nm long. Because they precipitate within the ferrite, a smaller amount of carbon is partitioned into the residual austenite. This in turn means that fewer and finer cementite particles precipitate between the ferrite plates compared to an upper bainitic microstructure. An important consequence is that lower bainite is usually found to be much tougher than upper bainite, in spite of the fact that it also tends to be stronger. The coarse cementite particles in upper bainite are notorious in their ability to nucleate cleavage cracks and voids.

The observed orientation relationships of the cementite in both upper and lower bainite, i.e., Pitsch or Isaichev, are such that a coherent line exists between the parent and product lattices [15,16]. Such a line is one of the necessary conditions (Chapter 5) for displacive transformation to be possible. Combined with the fact that the precipitation occurs at low temperatures, the evidence suggests that the cementite forms without the diffusion of large atoms by paraequilibrium growth from the supersaturated parent phase.

6.4 The shape deformation

The IPS surface relief caused by the growth of bainitic ferrite has a large shear strain component of 0.24 in addition to the volume strain (0.03) on transformation. There is, therefore, a coordinated movement of atoms as the transformation occurs. Consistent with this, the iron and substitutional solutes such as Mn, Si, Ni, Mo and Cr, have been demonstrated using high-resolution techniques to be frozen into position during transformation, Fig. 6.7 [17,18]. The change in crystal structure is therefore achieved by a deformation of the austenite crystal. If the strain is elastically accommodated, then the

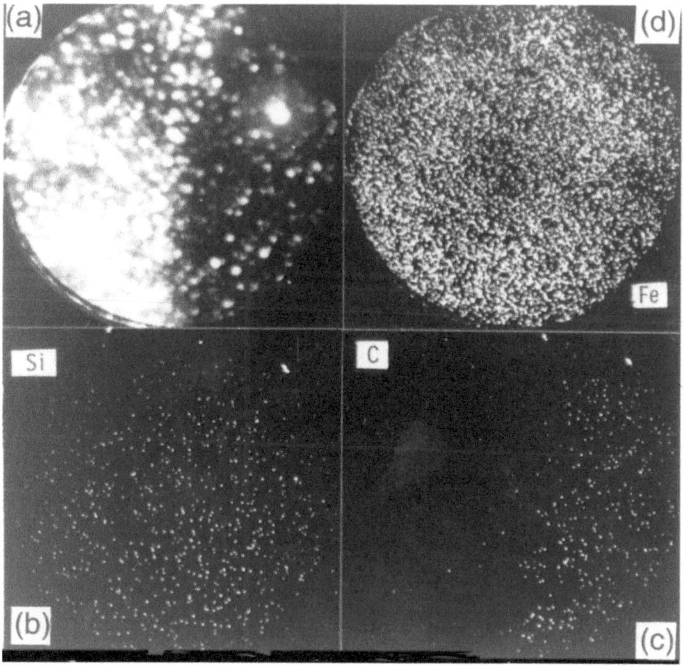

Figure 6.7 Imaging atom-probe micrographs across an austenite-bainitic ferrite interface in a Fe-C-Si-Mn alloy. Substitutional atoms clearly do not diffuse during transformation. (a) Field ion image; each dot corresponds to an atom. The interface is vertical in the image, the austenite located on the right-hand side. (b) Fe atom map. (c) Corresponding Si atom map, showing a uniform distribution. (d) C atom map [17].

strain energy of bainitic ferrite amounts to about $400\,\mathrm{J\,mol^{-1}}$. Some of the strain can be relaxed by plastic deformation in the adjacent austenite.

The movement of interstitial atoms during the change in crystal structure has no influence on the development of surface relief [19]. Conversely, the observation of relief cannot yield information on whether or not carbon diffuses during transformation.

6.5 Carbon in bainite

It is simple to establish that martensitic transformation is diffusionless by measuring the local compositions before and after transformation. Bainite forms at somewhat higher temperatures where the carbon can escape out of the plate within a fraction of a second. Its original composition cannot therefore be measured directly.

There are three possibilities. The carbon may partition during growth so that the ferrite may never contain any excess carbon. On the other hand, growth may be diffusionless, with carbon being trapped by the advancing interface. Finally, there is an intermediate case in which some carbon may diffuse with the remainder being trapped to leave the ferrite partially supersaturated. It is therefore much more difficult to determine the precise role of carbon during the growth of bainitic ferrite than in martensite.

Diffusionless growth requires that transformation occurs at a temperature below T_0, when the free energy of bainite becomes less than that of austenite of the same composition. A locus of the T_0 temperature of the function of the carbon concentration is called the T_0 curve, an example of which is plotted on the Fe-C phase diagram in Fig. 6.8. Growth without diffusion can occur only if the carbon concentration of the austenite lies to the left of the T_0 curve. For example, the transformation of austenite of composition "a" to ferrite of the same composition "b" would lead to an increase in free energy; in contrast, the transformation "c"→"d" without a composition change, would lead to a decrease, as illustrated on Fig. 6.8.

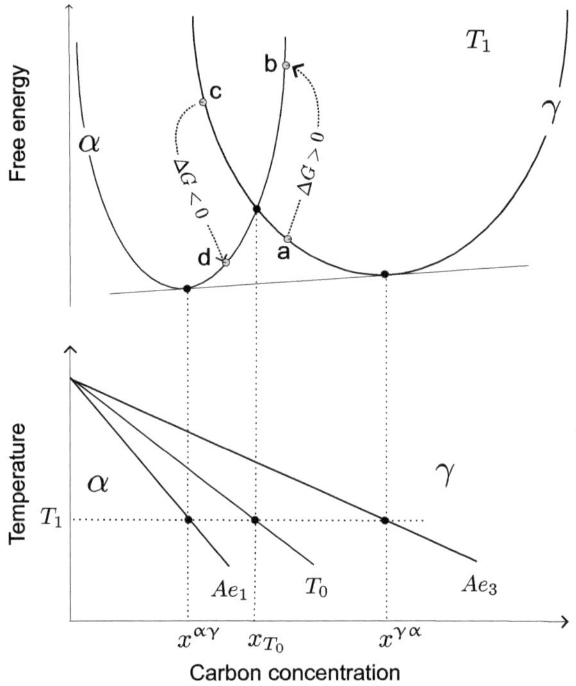

Figure 6.8 Schematic illustration of the origin of the T_0 construction on the Fe-C phase diagram. Austenite with a carbon concentration to the left of the T_0 boundary can, in principle, transform without diffusion. Diffusionless transformation is thermodynamically impossible if the carbon concentration of the austenite exceeds the T_0 curve.

Suppose that the plate of bainite forms without diffusion, but that any excess carbon is soon afterwards rejected into the residual austenite. The next plate of bainite then has to grow from carbon-enriched austenite (Fig. 6.9a). This process must cease when the austenite carbon concentration reaches the T_0 curve. The reaction is said to be incomplete, since the austenite has not achieved its equilibrium composition (given by the Ae_3 curve) at the point the reaction stops. On the other hand, if the ferrite grows

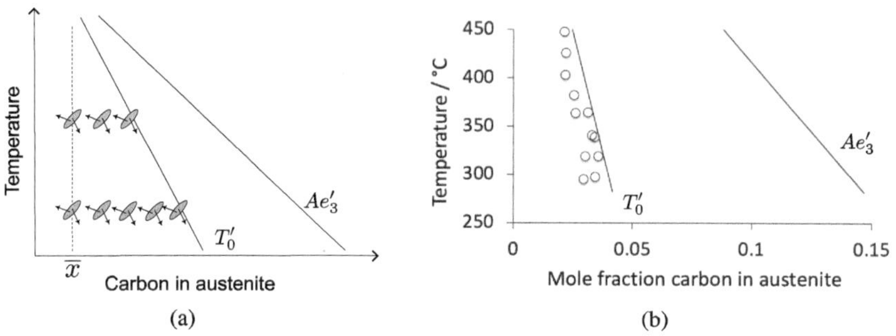

Figure 6.9 (a) Illustration of the incomplete-reaction phenomenon. During isothermal transformation, a plate of bainite grows without diffusion and then partitions its excess carbon into the residual austenite. The next plate therefore has to grow from carbon-enriched austenite. This process continues until diffusionless transformation becomes impossible, when the austenite composition eventually reaches the T_0' boundary, which, unlike T_0, includes the stored energy of transformation. The Ae_3' curve is for paraequilibrium since substitutional solutes are not partitioned during the bainite reaction. (b) Experimental data showing that the growth of bainite stops when the austenite carbon concentration reaches the T_0' curve (Fe-0.43C-3Mn-2.12Si wt% alloy).

with an equilibrium carbon concentration then the transformation should cease when the austenite carbon concentration reaches the Ae_3 curve.

It was found experimentally that the transformation to bainite indeed stops at the T_0 boundary (Fig. 6.9b). The balance of the evidence is that the growth of bainite below the B_S temperature involves the successive nucleation and martensitic growth of sub-units, followed in upper bainite by the diffusion of carbon into the surrounding austenite. The possibility that a small fraction of the carbon is nevertheless partitioned during growth cannot entirely be ruled out, but there is little doubt that the bainite is at first substantially supersaturated with carbon.

These conclusions are not significantly modified when the strain energy of transformation is included in the interpretation of the mechanism, because the difference between the T_0 and Ae_3' carbon concentrations is much larger.

There are two important features of bainite which can be shown by a variety of techniques, e.g., dilatometry, electrical resistivity, magnetic measurements and by metallography. Firstly, there is a well-defined temperature B_S above which bainite does not form, which has been confirmed for a wide range of alloy steels. The amount of bainite formed increases as the transformation temperature is reduced below the B_S temperature. The fraction increases during isothermal transformation as a sigmoidal function of time, reaching an asymptotic limit which does not change on prolonged heat treatment, even when substantial quantities of austenite remain untransformed. Transformation in fact ceases before the austenite achieves its equilibrium composition, so that the effect is dubbed the *incomplete-reaction phenomenon*. These observations are understood when it is realised that growth must cease if the carbon concentration in the austenite reaches the T_0 curve of the phase diagram.

Since this condition is met at ever-increasing carbon concentrations when the transformation temperature is reduced, more bainite can form with greater undercoolings below B_S. But the T_0 restriction means that equilibrium, when the austenite has a composition given by the Ae_3 phase boundary, can never be reached, as observed experimentally. A bainite-finish temperature B_F is sometimes defined, but this clearly cannot have any fundamental significance.

6.6 Kinetics

The rate of the bainite reaction needs to be considered in terms of a number of distinct events (Fig. 6.10). A sub-unit nucleates at an austenite grain boundary and lengthens at a certain rate before its growth is stifled by plastic deformation within the austenite. New sub-units then nucleate at its tip and the sheaf structure develops as this process continues. The overall lengthening rate of a sheaf is therefore lower than that of an individual sub-unit because there is an interval between the formation of successive sub-units [20]. The volume fraction of bainite depends on the totality of sheaves growing from different regions in the sample. Carbide precipitation events also influence the kinetics, primarily by removing carbon from either the residual austenite or the supersaturated ferrite.

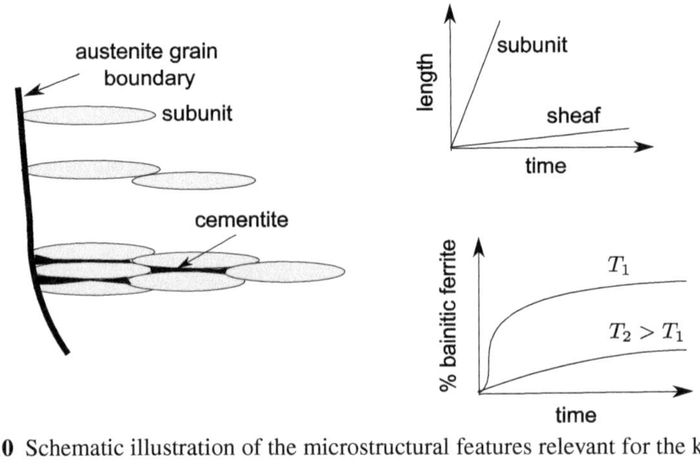

Figure 6.10 Schematic illustration of the microstructural features relevant for the kinetic description of a bainitic microstructure.

Little is known about the nucleation of bainite except that the activation energy for nucleation is directly proportional to the driving force for transformation [21]. This is consistent with the theory for martensite nucleation (Equation 5.11). However, unlike martensite, carbon must partition into the austenite during bainite nucleation, although the nucleus then develops into a sub-unit which grows without diffusion.

The scale of individual plates of ferrite is too small to be resolved adequately using optical microscopy, which is capable only of revealing clusters of plates. Using higher-resolution techniques such as photoemission electron microscopy (Fig. 6.11), it was possible to study directly the progress of the bainite reaction [22]. Not surprisingly,

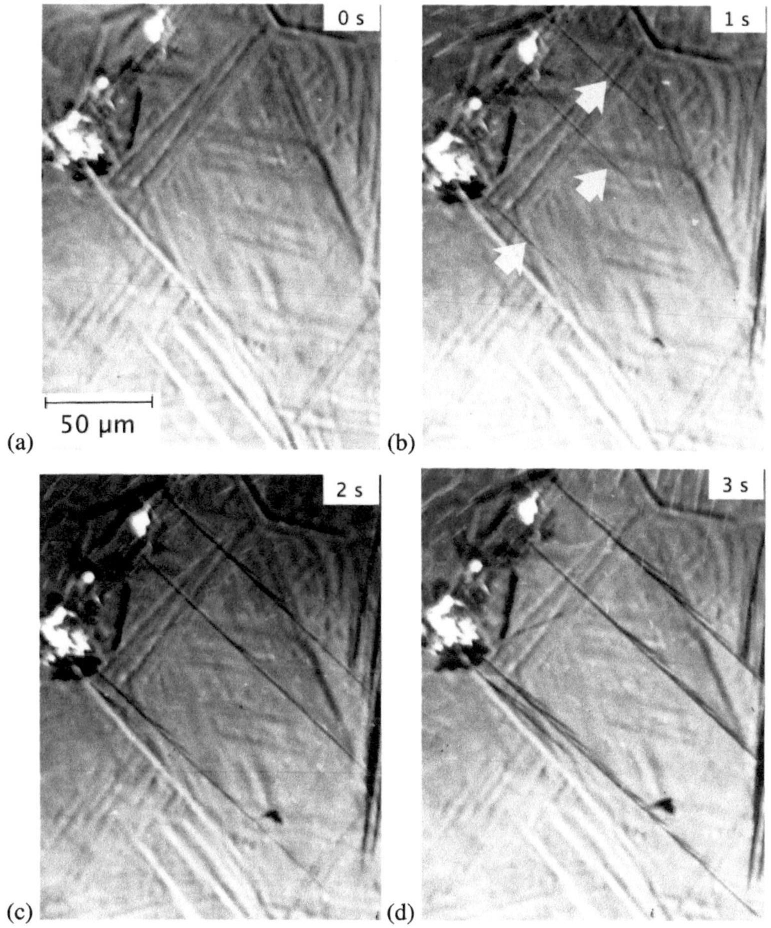

Figure 6.11 Photoemission electron microscope observations on the growth of individual sub-units in a bainite sheaf. The pictures are taken at 1 s intervals [22].

the lengthening of individual bainite platelets is much faster than would be expected from a diffusion-controlled process. The growth rate is nevertheless much lower than that of martensite because the driving force for bainite formation is lower due to the higher transformation temperatures involved. The platelets tend to grow at a constant rate, but usually are stifled before they can traverse the austenite grain.

The lengthening rate of a sheaf is slower still, because of the delay caused by the need to repeatedly nucleate new sub-units. Nevertheless, sheaf lengthening rates are generally found to be about an order of magnitude higher than expected from carbon diffusion-controlled growth. Low-resolution measurements have also been made of the thickening of bainite sheaves, a process which appears to be discontinuous, presumably because of the sub-unit mechanism of growth. The thickening process therefore depends on the rate at which sub-units are nucleated in adjacent locations within a sheaf.

The bainitic reaction has several of the recognised features of a nucleation and growth process. It takes place isothermally, starting with an incubation period during which no transformation is detected (within the precision of the measuring equipment), followed by an increasing rate of transformation to a maximum and then a gradual slowing down. These features are illustrated in the dilatometric results of Fig. 6.12a, for three transformation temperatures in the bainitic range for a Fe-1Cr-0.4C wt% steel, the extent of transformation increasing with decreasing temperature. For this steel, the reaction stops at 510°C after about 1 h, and the remaining austenite is stable at this temperature for a long time.

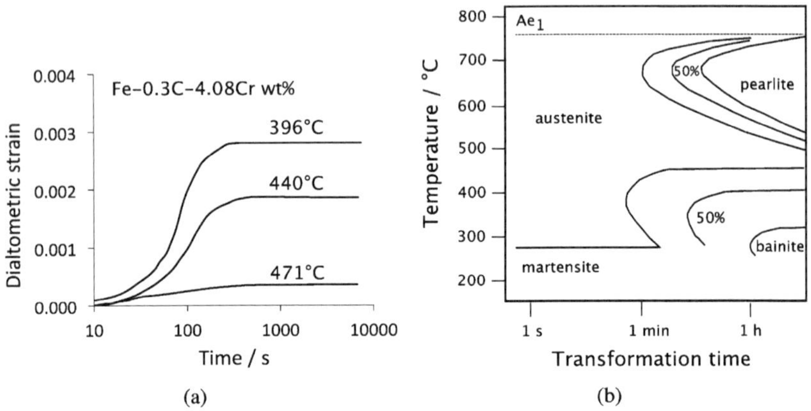

Figure 6.12 (a) Isothermal reaction curves for the formation of bainite, illustrating the incomplete reaction phenomenon. (b) TTT curves for a Fe-3Cr-0.5C wt% steel, adapted from Thelning [23].

These overall transformation characteristics, i.e., the change in the fraction of bainite with time, temperature, austenite grain structure and alloy chemistry, are therefore best considered in terms of a TTT diagram (Fig. 6.12b). A simplified view is that the TTT diagram consists of two separable C-shaped curves. The one at higher temperatures describes the evolution of diffusional transformation products such as ferrite and pearlite, whereas the lower C-shaped curve represents displacive reactions such as Widmanstätten ferrite and bainite. In lean steels, which transform rapidly, these two curves overlap so much that there is apparently just one curve which is the combination of all reactions. As the alloy concentration is increased to retard the decomposition of austenite, the two overlapping curves begin to become distinct, and a characteristic "gap" develops at about the B_S temperature in the TTT diagram. This gap is important in the design of some high-strength (ausformed) steels, which have to be deformed in the austenitic condition at low temperatures before the onset of transformation.

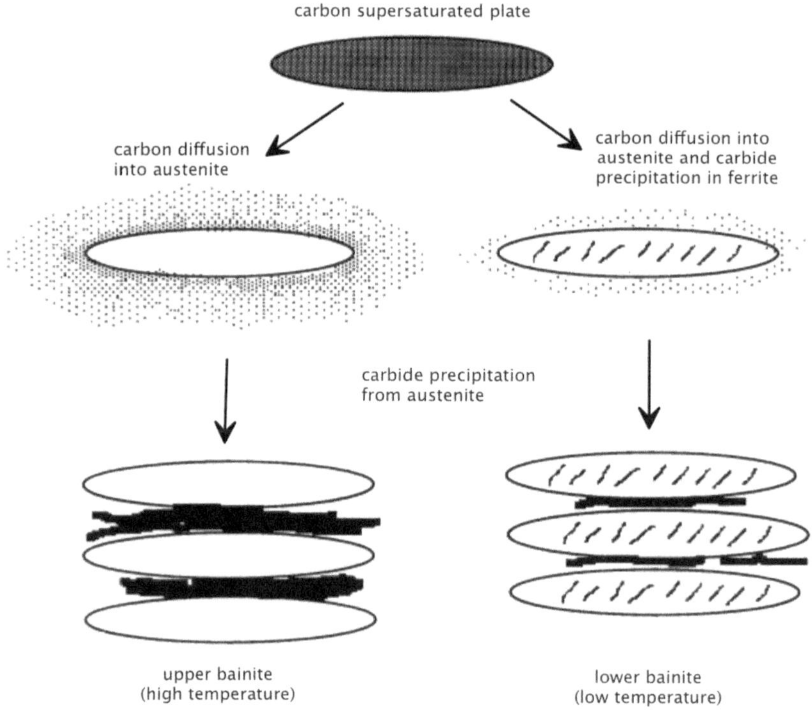

Figure 6.13 Schematic representation of the transition from upper to lower bainite.

6.7 Transition from upper to lower bainite

As the isothermal transformation temperature is reduced to below B_S, lower bainite is obtained in which carbides precipitate in the ferrite, with a correspondingly reduced amount of precipitation from the austenite between the ferrite. This transition from upper to lower bainite can be explained in terms of the rapid tempering processes that occur after the growth of a supersaturated plate of bainite, Fig. 6.13 [24]. Excess carbon tends to partition into the residual austenite by diffusion, but the supersaturation may also be reduced by precipitation in the ferrite.

The time required for a supersaturated plate of ferrite to decarburise by diffusion into austenite is illustrated in Fig. 6.14 for a typical steel. At elevated temperatures, the diffusion is so rapid that there is no opportunity to precipitate carbides in the ferrite, giving rise to an upper bainitic microstructure. Cementite eventually precipitates from the carbon-enriched residual austenite.

As the transformation temperature is reduced and the time for decarburisation increases, some of the carbon has an opportunity to precipitate as fine carbides in the ferrite, whereas the remainder partitions into the austenite, eventually to precipitate as inter-plate carbides. This is the lower bainite microstructure. Because only a fraction of the carbon partitions into the austenite, the inter-plate carbides are much smaller than those associated with upper bainite. For this reason, the lower bainite, with its highly refined microstructure, is always found to be much tougher than upper bainite, even though it usually has a much higher strength.

A corollary to the mechanism of the transition from upper to lower bainite is that in steels containing high concentrations of carbon, only lower bainite is ever obtained. The large amount of carbon trapped in the ferrite by transformation simply cannot escape fast enough into the austenite, so that precipitation from ferrite is unavoidable. Conversely, in very low-carbon steels, the time for decarburisation is so small that only upper bainite is obtained by transformation at all temperatures between the pearlite-finish and the martensite-start temperatures.

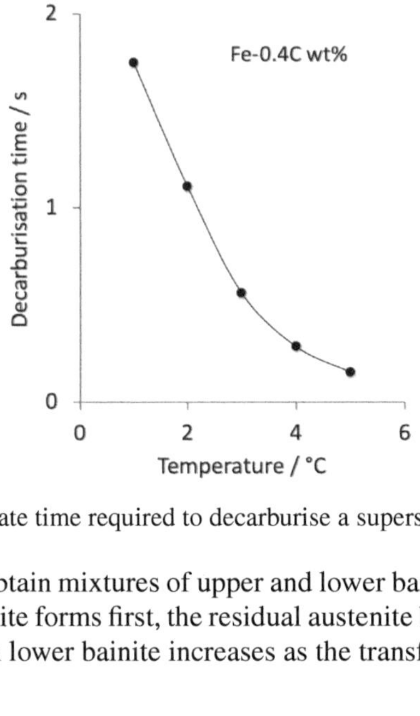

Figure 6.14 The approximate time required to decarburise a supersaturated plate of bainite.

It also is possible to obtain mixtures of upper and lower bainite by isothermal transformation. As upper bainite forms first, the residual austenite becomes richer in carbon and the tendency to form lower bainite increases as the transformation progresses.

6.8 Granular bainite

Granular bainite (Fig. 6.15) describes the bainite morphology generated during continuous cooling transformation. This terminology is used widely in industry, where most steels undergo non-isothermal heat treatments. A good example is the energy generation industry, where large Cr-Mo steel components are allowed to cool naturally from the austenitic state to generate bainitic microstructures.

Granular bainite cannot readily be distinguished from ordinary bainite when examined using transmission electron microscopy, because its mechanism of formation is not different. However, because the microstructure forms gradually during cooling, the sheaves of bainite can be rather coarse. The optical microstructure then gives the appearance of blocks of bainite and austenite, so it may be appropriate to use the adjective “granular”.

A characteristic (though not unique) feature of granular bainite is the lack of carbides in the microstructure. Instead, the carbon that is partitioned from the bainitic

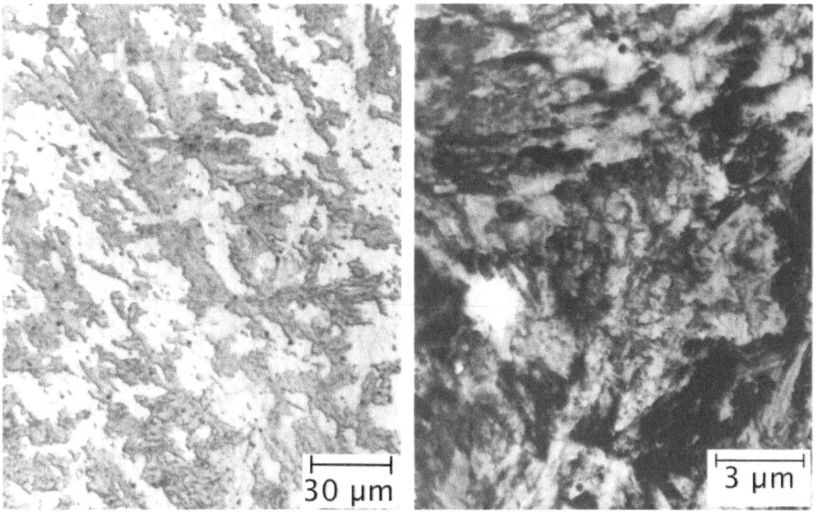

(a) (b)

Figure 6.15 Granular bainite in a Fe-0.15C-2.25Cr-0.5Mo wt% steel of the kind used extensively in the energy generation industry. (a) Light micrograph. (b) Corresponding transmission electron micrograph (Courtesy of B. Joseffson).

ferrite stabilises the residual austenite, so that the final microstructure contains both retained austenite and some carbon-rich martensite in addition to the bainitic ferrite.

6.9 Tempering of bainite

The extent and the rate of change of the microstructure and properties during tempering must depend on how far the initial sample deviates from equilibrium. The behaviour of bainite during tempering is therefore expected to be different from that of martensite.

Unlike martensite, bainitic ferrite usually contains only a slight excess of carbon in solution. Most of the carbon in a transformed sample of bainite is in the form of cementite particles, which in turn tend to be coarser than those associated with tempered martensite. The effects of tempering heat treatments are therefore always milder than in annealing martensite in the same steel.

Bainite forms at relatively high temperatures where some recovery occurs during transformation. Consequently, when low-carbon bainitic steels are annealed at temperatures as high as 700°C (1 h), there are only minor microstructural changes in recovery, morphology or carbide particles. Rapid softening occurs only when the plate-like structure of ferrite changes into equiaxed ferrite. Associated with this change is the spheroidisation and coarsening of cementite. Further tempering has minimal effects.

In marked contrast with martensitic steels, small variations in the carbon concentration (0.06–0.14 wt%) have little effect on the tempering of bainite. Carbon has a very potent solid solution strengthening effect. Thus, the strength of martensite drops sharply as the carbon precipitates during tempering. In bainite, the carbon is present mostly as coarse carbides, which contribute little to strength. It is not therefore surprising that the tempering response is rather insensitive to the bulk carbon concentration.

Many bainitic microstructures contain appreciable quantities of retained austenite. Tempering, usually at temperatures in excess of 400°C, induces the decomposition of this austenite into a mixture of ferrite and cementite. However, the cementite precipitates at the original α_{ub}/γ boundaries, forming intense arrays that pin the α/α boundaries that remain after the austenite has decomposed. This greatly retards recrystallisation, so much so, that the hardness initially increases and then hardly decreases until the tempering temperature is raised to > 550°C, Fig. 6.16. This phenomenon naturally is particularly prominent in carbon-rich steels.

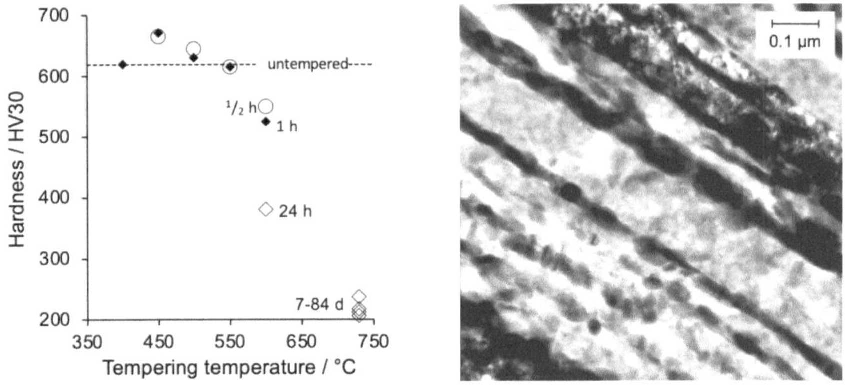

(a) (b)

Figure 6.16 (a) Hardness of a mixture of bainitic ferrite and retained austenite ($V_V^\gamma = 0.3$) after heat treatment as a function of tempering temperature and time, Fe-1C-1.5Si-1.9Mn-1.3Cr-0.3Mo-0.1V wt%. (b) Sample tempered at 500°C for $\frac{1}{2}$ h, showing linear arrays of cementite particles where the retained austenite preceded. Data and image from adapted from [25]. These particles pin the bainitic ferrite boundaries, preventing the structure from recrystallising.

Bainitic steels containing strong carbide-forming elements such as Cr, V, Mo and Nb undergo secondary hardening during annealing at high temperatures. Secondary hardening occurs when fine (more stable) alloy carbides form at the expense of cementite (Chapter 9). Because the cementite in bainite is coarse, the secondary hardening reaction tends to be sluggish compared to martensite.

There is interest in the use of copper-bearing bainitic steels for applications in heavy engineering. Tempering induces the formation of fine particles of copper, which contribute to strength without jeopardising toughness.

To summarise, there are significant differences in the tempering behaviour of bainite and martensite, the most prominent being that there is little carbon in solid solution in bainite. This has the consequence that bainitic microstructures are much less sensitive to tempering, since there is hardly any loss of strength due to the removal of the small quantity of dissolved carbon. Major changes in strength occur only when the bainite plate microstructure coarsens or recrystallises into one consisting of equiaxed grains of ferrite. Minor changes in strength are due to cementite particle coarsening and a general recovery of the dislocation substructure. Bainitic steels containing strong carbide-forming elements tend to exhibit secondary hardening phenomena rather like those observed in martensitic steels, which depends on the precipitation of fine alloy carbides.

6.10 Role of alloying elements

Carbon

Carbon has a large effect on the range of temperature over which upper and lower bainite occur. The B_S temperature is depressed by many alloying elements, but carbon has the greatest influence, as indicated by the following empirical equation [26]:

$$B_S(°C) = 830 - 270w_C - 90w_{Mn} - 37w_{Ni} - 70w_{Cr} - 83w_{Mo}, \tag{6.2}$$

where the concentrations w_i are all in wt% and the equation has 90% confidence limits of ±25°C, over the range of compositions summarised below.

	C	Si	Ni	Cr	Mo
Minimum / wt%	0.10	0.20	0.00	0.00	0.00
Maximum / wt%	0.55	1.70	5.00	3.50	1.00

The method based on the mechanism of the bainite transformation, in the context of the other displacive transformations, is described in [21] and implemented in freely available software [27,28].

Carbon has a much larger solubility in austenite than in ferrite and is a very powerful austenite stabiliser, which leads to a general retardation of reaction kinetics. The fraction of carbides in the final microstructure increases in proportion to the carbon concentration, so that the concentration must be kept below about 0.4 wt% to ensure reliable mechanical properties. We have seen already that an increase in carbon makes it easier for lower bainite to form because it becomes more difficult for plates of supersaturated bainitic ferrite to decarburise before the onset of cementite precipitation.

Other alloying elements

In plain carbon steels, the bainitic reaction is kinetically shielded by the reconstructive ferrite and pearlite reactions, which commence at higher temperatures and shorter times (Fig. 4.2), so in continuously cooled samples bainitic structures are difficult to obtain. Even using isothermal transformation, difficulties arise if, e.g., the ferrite reaction is particularly rapid. As explained in Chapter 4, the addition of metallic alloying elements usually results in the retardation of the ferrite and pearlite reactions. In addition, the bainite reaction is depressed to lower temperatures. This often leads to a greater separation of the reactions, and the TTT curves for many alloy steels show much more clearly separated C-shaped curves for the pearlite and bainitic reactions (Fig. 4.2). However, it still is difficult to obtain a fully bainitic microstructure because of its proximity to the martensite reaction.

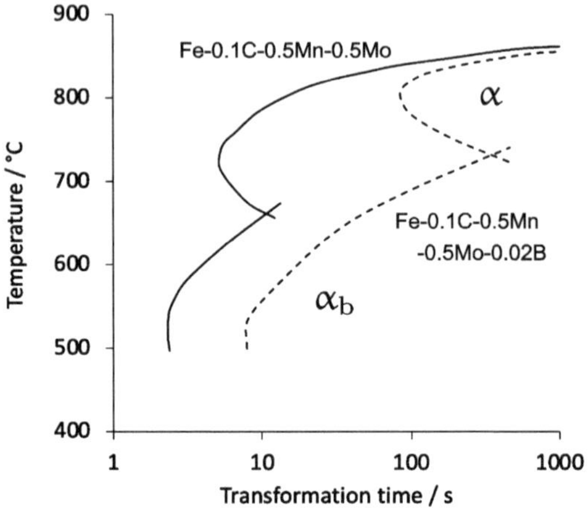

Figure 6.17 Effect of boron on the bainite reaction TTT curves for the initiation of transformation. Adapted from Irvine and Pickering [29].

An effective means of isolating the bainite reaction in low-carbon steels is by adding about 0.002 wt% soluble boron to a Mo-containing steel. While the straight molybdenum steel encourages the bainite reaction (Fig. 6.17), the boron markedly retards the ferrite reaction by preferential segregation to the austenite boundaries, thus reducing their potency as heterogeneous nucleation sites. This permits the bainite reaction to be accessed without interference from the higher temperature transformations. At the same time, the bainite C-shaped curve is hardly affected by the boron addition, so that martensite formation is not enhanced. Consequently, by using a range of cooling rates, fully bainitic steels can be obtained in large components.

6.11 Use of bainitic steels

By far the largest market for steels is at strength levels well below 1000 MPa, and the total alloy concentration rarely exceeds 2 wt%. Bainitic steels are well suited for applications within these constraints. However, alloy design must be carefully selected to obtain the right microstructures. Steels with inadequate hardenability tend to transform into mixtures of allotriomorphic ferrite and bainite. Attempts to improve hardenability usually lead to partially martensitic microstructures. The solution, therefore, lies in low-alloy, low-carbon steels, containing small amounts of boron and molybdenum to suppress allotriomorphic ferrite formation. Boron increases the bainitic hardenability. Other solute additions can, in the presence of boron, be kept at

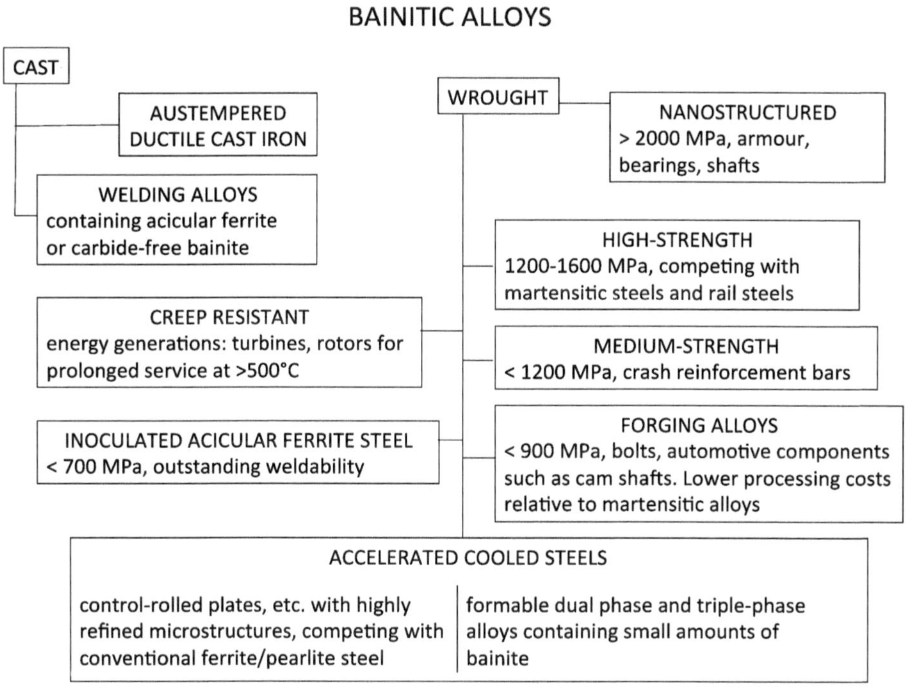

Figure 6.18 Bainitic alloys currently available commercially.

sufficiently low concentrations to avoid the formation of martensite. A typical composition might be Fe-0.1C-0.25Si-0.50Mn-0.55Mo-0.003B wt%. Steels like these are found to transform into virtually fully bainitic microstructures with very little martensite using normalising heat treatments.

The range of bainitic alloys available commercially is summarised in Fig. 6.18 and some typical alloy compositions are stated in Table 6.1. The most modern bainitic steels designed for structural engineering, have much reduced carbon and other alloying element concentrations. They are then processed using accelerated cooling in order to obtain the necessary bainitic microstructure. The reduced alloy concentration not only gives better weldability, but also a larger strength due to the refined bainitic microstructure.

The ultra-high-strength steels consist of mixtures of bainite ferrite, martensite and retained austenite. They have an enhanced hardenability using manganese, chromium and nickel and usually also contain a high silicon concentration (~2 wt%) in order to prevent the formation of cementite. High-strength steels are made with very low impurity and inclusion concentrations, so the dominant fracture initiator becomes cementite particles, which therefore have to be avoided or refined.

Medium strength steels with the same microstructure but somewhat reduced alloy content have found applications in the automobile industry as crash reinforcement bars to protect against sidewise impact. Another major advance in the automobile industry

Table 6.1 Chemical composition, wt%, of typical bainitic steels.

Alloy	C	Si	Mn	Ni	Mo	Cr	V	B	Nb	Other
Early bainitic steel	0.10	0.25	0.5	-	-	0.003	-	-	-	
Ultra-low carbon	0.02	0.20	2.0	0.3	0.30	-	-	0.010	0.05	
Tough pipeline steel	0.05	0.12	1.55	0.13	-	0.23	-	-	0.10	0.011 Ti
Ultra-high strength	0.20	2.00	3.0	-	-	-	-	-	-	
Creep resistant	0.15	0.25	0.50	-	1.00	2.30	-	-	-	
Forging alloy	0.10	0.25	1.00	0.50	1.00	-	-	-	0.10	
Inoculated	0.08	0.20	1.40	-	-	-	-	-	0.10	0.012 Ti
Bearing alloy	1.0	0.30	0.30	-	-	1.50	-	-	-	
Nanostructured bainite	1.0	1.50	1.90	-	0.26	1.26	0.1	-	-	

has been in the application of bainitic forging alloys to the manufacture of components such as camshafts. These were previously made of martensitic steels by forging, hardening, tempering, straightening and finally stress-relieving. All of these operations are now replaced by controlled cooling from the die forging temperature, to generate the bainitic microstructure, with cost savings which on occasions have made the difference between profit and loss for the entire unit.

Creep-resistant bainitic steels have been used successfully in the power generation industry since the early 1940s. Their hardenability has to be such that components as large as 1 m in diameter can be cooled continuously to generate a bainitic microstructure throughout the section. The alloys utilise chromium and molybdenum, which serve to enhance hardenability, but also cause the precipitation of alloy carbides during subsequent heat-treatment, which greatly improve the creep resistance.

By inoculating molten steel with controlled additions of non-metallic particles, bainite can be induced to nucleate intragranularly on the inclusions rather than from the austenite grain surfaces. This intragranularly nucleated bainite is called *acicular ferrite* (Chapter 7). It is a much more disorganised microstructure with a larger ability to deflect cracks. Inoculated steels are now available commercially and are used in demanding structural applications such as the fabrication of oil rigs for hostile environments.

Advances in rolling technology have led to the ability to cool the steel plate rapidly during the rolling process, without causing undue distortion. This has led to the development of accelerated cooled steels, which have a bainitic microstructure, can be highly formable and compete with conventional control-rolled steels.

A much more complete description of the design and exploitation of bainitic steels can be obtained in [4].

6.12 Summary

Bainite can be regarded to form exactly as does martensite, but that it tempers rapidly because of the higher transformation temperatures involved. Thus, excess carbon is able to partition into the residual austenite immediately after transformation, where it either precipitates as cementite or remains in solid solution within the austenite, which is then retained to room temperature. The partitioning process is slower when the transformation temperature is reduced, so some carbon has an opportunity to precipitate with the bainitic ferrite to form the lower bainite microstructure.

Bainitic steels offer the ability to make strong and large components, such as power-plant rotors or wind-turbine bearings, without the need to quench rapidly or richly-alloy the material for the sake of hardenability. The structure can even be generated in automotive steels where the rate at which the rapidly moving steel-sheet experiences transformation is large.

References

[1] J.S. Bowles, B.C. Muddle, C.M. Wayman, The crystallography of the precipitation of beta vanadium hydride, Acta Metallurgica 25 (1977) 513–520.

[2] H.K.D.H. Bhadeshia, Hard bainite, in: J.M. Howe, D.E. Laughlin, J.K. Lee, U. Dahmen, W.A. Soffa (Eds.), Solid-Solid Phase Transformations, vol. 1, TME-AIME, Warrendale, USA, TMS-AIME, Warrendale, Pennsylvania, USA, 2005, pp. 469–484.

[3] H.K.D.H. Bhadeshia, The first bulk nanostructured metal, Science and Technology of Advanced Materials 14 (2013) 014202.

[4] H.K.D.H. Bhadeshia, Bainite in Steels: Theory and Practice, 3rd ed., Maney Publishing, Leeds, U.K., 2015.

[5] E.C. Bain, Alloying Elements in Steel, American Society of Materials, Cleveland, Ohio, USA, 1939, https://www.phase-trans.msm.cam.ac.uk/2004/Bain.Alloying/ecbain.html.

[6] E.S. Davenport, E.C. Bain, Transformation of austenite at constant subcritical temperatures, Transactions of the American Institute of Mining and Metallurgical Engineers 90 (1930) 117–154.

[7] H.K.D.H. Bhadeshia, A personal commentary on transformation of austenite at constant subcritical temperatures, Metallurgical and Materials Transactions A 41 (2010) 1351–1390.

[8] R.F. Hehemann, The bainite transformation, in: H.I. Aaronson, V.F. Zackay (Eds.), Phase Transformations, American Society of Materials, Materials Park, Ohio, USA, 1970, pp. 397–432.

[9] E. Swallow, H.K.D.H. Bhadeshia, High resolution observations of displacements caused by bainitic transformation, Materials Science and Technology 12 (1996) 121–125.

[10] H.K.D.H. Bhadeshia, D.V. Edmonds, The bainite transformation in a silicon steel, Metallurgical Transactions A 10A (1979) 895–907.

[11] A.B. Greninger, A.R. Troiano, Kinetics of the austenite to martensite transformation in steel, Transactions of the ASM 28 (1940) 537–563.

[12] Y. Ohmori, The crystallography of the lower bainite transformation in a plain carbon steel, Transactions of the ISIJ 11 (1971) 95–101.

[13] I.V. Isaichev, Orientation of cementite in tempered carbon steel (Orientatsiya tsementita v otpushchennoi uglerodistoi stali), Zhurnal Tekhncheskoi Fiziki 17 (1947) 835–838.

[14] J.W. Stewart, R.C. Thomson, H.K.D.H. Bhadeshia, Cementite precipitation during tempering of martensite under the influence of an externally applied stress, Journal of Materials Science 29 (1994) 6079–6084.
[15] H.K.D.H. Bhadeshia, Solution to the Bagaryatskii and Isaichev ferrite-cementite orientation relationship problem, Materials Science and Technology 34 (2018) 1666–1668.
[16] H.K.D.H. Bhadeshia, Theory of Transformations in Steels, CRC Press, Taylor and Francis Group, London, U.K., 2021.
[17] H.K.D.H. Bhadeshia, A.R. Waugh, Bainite: an atom probe study of the incomplete reaction phenomenon, Acta Metallurgica 30 (1982) 775–784.
[18] H.K.D.H. Bhadeshia, A.R. Waugh, An atom-probe study of bainite, in: H.I. Aaronson, D.E. Laughlin, R.F. Sekerka, C.M. Wayman (Eds.), Solid-Solid Phase Transformations, TMS-AIME, Warrendale, Pennsylvania, USA, 1982, pp. 993–998.
[19] J.W. Christian, Military transformations: an introductory survey, in: Physical Properties of Martensite and Bainite, Special Report 93, Iron and Steel Institute, London, U.K., 1965, pp. 1–19.
[20] H. Matsuda, H.K.D.H. Bhadeshia, Kinetics of the bainite transformations, Proceedings of the Royal Society of London A 460 (2004) 1710–1722.
[21] H.K.D.H. Bhadeshia, Rationalisation of shear transformations in steels, Acta Metallurgica 29 (1981) 1117–1130.
[22] H.K.D.H. Bhadeshia, Solute-drag, kinetics and the mechanism of the bainite transformation, in: A.R. Marder, J.I. Goldstein (Eds.), Phase Transformations in Ferrous Alloys, TMS-AIME, Ohio, USA, 1984, pp. 335–340.
[23] K.E. Thelning, Steel and Its Heat Treatment, Butterworths, London, U.K., 1975.
[24] M. Takahashi, H.K.D.H. Bhadeshia, Model for transition from upper to lower bainite, Materials Science and Technology 6 (1990) 592–603.
[25] C. Garcia-Mateo, M. Peet, F.G. Caballero, H.K.D.H. Bhadeshia, Tempering of a hard mixture of bainitic ferrite and austenite, Materials Science and Technology 20 (2004) 814–818.
[26] W. Steven, A.G. Haynes, The temperature of formation of martensite and bainite in low alloy steels, Journal of the Iron and Steel Institute 183 (1956) 349–359.
[27] H.K.D.H. Bhadeshia, Software for transformations in steels, https://www.msm.cam.ac.uk/map/steel/programs/mucg46-b.html, 1982.
[28] M. Peet, H.K.D.H. Bhadeshia, Software for transformations in steels, http://www.msm.cam.ac.uk/map/steel/programs/mucg83.html, 1982.
[29] K.J. Irvine, F.B. Pickering, Low carbon bainitic steels, Journal of the Iron and Steel Institute 187 (1957) 292–309.

Acicular ferrite

7

Abstract

Non-metallic inclusions are anathema when it comes to the design of strong steels because they become the initiation sites for fracture. Huge efforts have been made devoted to making clean steels – the oxygen concentration of a hard bearing steel is routinely less than 10 ppm. However, there are other structural steels that have to be welded where the localised heat input generates microstructures in the heat-affected zone that are undesirable. The alloys that are used to deposit the weld must have good properties in the as-cast state. In both of these circumstances, specific non-metallic inclusions are advantageous because they provide substrates for the intragranular nucleation of bainite. As a consequence, highly organised sheaves of bainite are altered into a more chaotic arrangement that frequently deflects propagating cracks, thus increasing toughness. This is the so-called acicular ferrite that is the subject of this chapter.

7.1 Introduction

Highly organised microstructures can often be found in steels, e.g., ferrite can grow in the form of packets containing parallel plates which share a crystallographic orientation (Fig. 7.2a). This can be harmful to mechanical properties, as cleavage cracks or deformation processes can extend readily across the packets (Fig. 5.12). The effects of the individual plates within these packets then have a minimal effect on the mechanical properties that determine structural integrity. Indeed, it is now common in studies relating structure to toughness, to discover relationships with the crystallographic grain size, which describes regions that have a fairly homogeneous crystallographic orientation, rather than the conventional grain size that is determined using imaging alone [1,2]. Fig. 7.1 shows the coarse crystallographically homogeneous regions that contain fine structure normally used to define a grain size. Thus, the toughness of the steel was compromised since it is the coarse regions that determine how far a cleavage crack can propagate without deviation.

Some of the counterintuitive developments in wrought and welded steel technology have involved *acicular ferrite* [4,5]. Far from being organised, this microstructure is better described as chaotic. The plates of acicular ferrite nucleate heterogeneously on small non-metallic inclusions and radiate in many different directions from these "point" nucleation sites (Fig. 7.2b). It is believed that propagating cleavage cracks are deflected frequently as they cross an acicular ferrite microstructure with its many different orientations. This gives rise to superior mechanical properties, especially a reduced tendency to cleave. Pure, entirely brittle cleavage, defined simply as the separation of atomic planes, is not representative of the macroscopically-brittle fracture that occurs in steels because there always is a limited amount of localised plasticity associated with the fracture process. This is why the surface energy associated with this kind of cleavage is higher than the expected $1\,\mathrm{J\,m^{-2}}$. For a microstructure containing 70% and 96% acicular ferrite, the effective surface energy increased from 37 and $74\,\mathrm{J\,m^{-2}}$, respectively [6].

Steels. https://doi.org/10.1016/B978-0-44-318491-8.00013-7

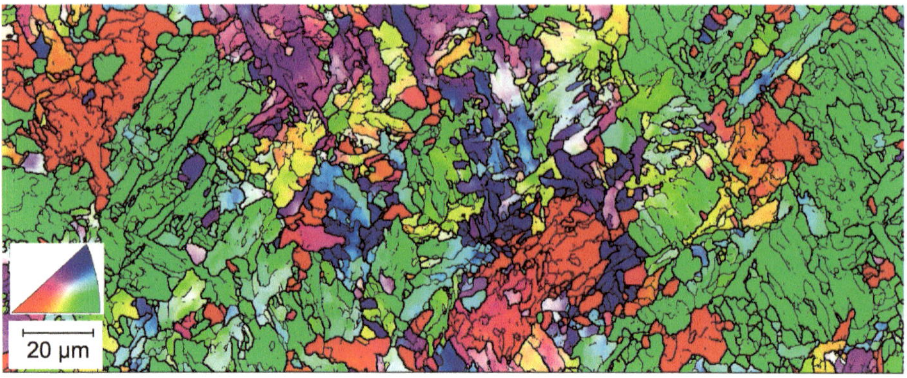

Figure 7.1 High-frequency induction-welded pipe, ≈ Fe-0.05C-1.5Mn-0.2Si-0.05Nb wt%. The orientation image shows the relationship between the crystallographic grain size (homogeneously coloured) and the much finer grain size determined from the black-boundaries. After [3].

Acicular ferrite is recognised, with some justification, to be a desirable microstructure that is generated routinely in steel weld deposits where inclusions inevitably are present and can be manipulated by adjusting the composition of the weld metal and associated flux (e.g., [7–9]). The structure can also be promoted in cast steel containing intentional additions of solutes that generate intragranular heterogeneous-nucleation sites, such as the alloy used in the manufacture of cast high-speed railway brake discs [10]. This chapter deals with the mechanism by which it forms and with the role of inclusions in stimulating its formation.

7.2 Microstructure

The term *acicular* means shaped and pointed like a needle, but it generally is recognised that acicular ferrite in three dimensions has the morphology of thin, lenticular plates (Fig. 7.3). In two-dimensional sections, the acicular ferrite always appears like a plate rather than a section of a rod. Serial sectioning experiments with a depth resolution of about 0.5 μm have confirmed that the shape is plate-like, with the length, width and thickness normally less than about 36, 6 and 3 μm, respectively [11]. There are difficulties with such experiments. When the images are combined to produce a three-dimensional shape, there is a further loss of resolution. Many of the published images show features such as steps that are known to be artifacts of the reconstruction process.

An optimal method now available to reveal the true shape of fine crystals in a polycrystalline sample involves observation simultaneously on two different surfaces, an experiment that is the high-resolution equivalent of conventional two-surface analysis using optical microscopy, for example, the classic work of Srinivasan and Wayman [12]. The technique uses a scanning electron microscope equipped with a focused ion-beam facility that is used to cut a sample of specified shape with incredible accuracy. Fig. 7.4 shows pillars that have been machined out of a weld from two regions of the

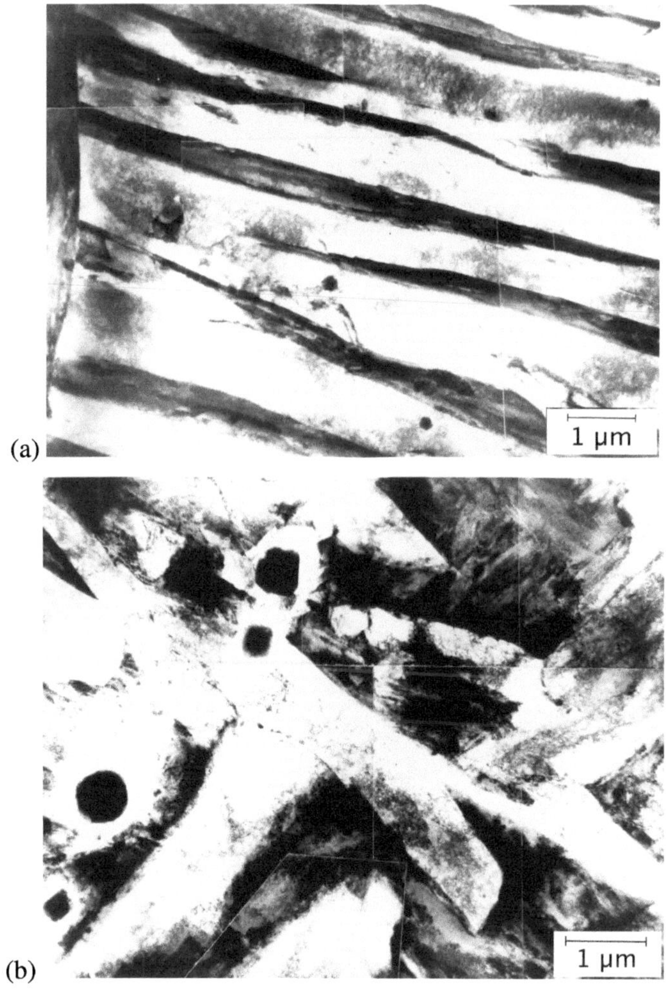

Figure 7.2 Transmission electron micrographs of samples transformed at the same temperature but with different austenite grain sizes. (a) Small austenite grain size leading to plates of ferrite that nucleate at the γ-boundary on the left, growing in parallel formations while sharing a common orientation in space. (b) Large austenite grain size with plates of ferrite nucleating intragranularly on non-metallic inclusions and growing along many crystallographic variants of the habit plane (courtesy of J. R. Yang). Note that the nucleating inclusions are not visible in all plates, partly because the micrograph is a two-dimensional projection of a thin foil, but also because some plates are stimulated by the stress field of prior plates that initiated on inclusions.

same chemical composition, one showing the parallel formations of bainite plates and the other the more disorganised arrays of acicular ferrite plates [13]. The individual plates can be followed across the edges between surfaces to verify unambiguously that the shapes are plates in three dimensions.

(a)

(b)

Figure 7.3 (a) Optical micrograph showing multiple plates of acicular ferrite emanating from inclusions. (b) Montage of transmission electron micrograph showing the inclusion as the nucleating point from which the ferrite plates grow (courtesy of G. Barritte).

Although the plates are nucleated heterogeneously on non-metallic inclusions, the chance of observing an inclusion in any given plate is rather small. If the acicular ferrite is approximated as a square plate of side 10 μm and thickness $t = 1$ μm, containing an inclusion of radius $r = 0.2$ μm, the ratio of the mean linear intercepts of the two phases is given by $4r/6t$ [14]. If every plate contains an inclusion, some 13% will show the nucleating particle in a plane section that is large enough [15]. It also is likely that once a plate forms on a particle, it stimulates the nucleation of others, an effect known as autocatalysis [16–19]. Plates generated in this way may not contain any inclusions except those engulfed accidentally by the growing ferrite.

7.3 Mechanism of transformation

Acicular ferrite and bainite are in many respects similar in their transformation mechanisms. Their microstructures differ in detail, as bainite sheaves grow as a series of

Figure 7.4 Images of samples ion beam machined from a weld, illustrating the three-dimensional shapes of (a) bainite and (b) of acicular ferrite in the same steel – reproduced from Costin et al. [13] with permission of Elsevier.

parallel platelets emanating from austenite grain surfaces, whereas acicular ferrite platelets nucleate intragranularly at point sites so that parallel formations of plates cannot develop. The nucleation site in the latter case is smaller than the thickness of the plate, so that the inclusion normally is engulfed by the plate of ferrite which it stimulates.

The growth of both bainite and acicular ferrite causes an invariant-plane strain shape deformation with a large shear component (Fig. 7.5). Consequently, plates of acicular ferrite cannot cross austenite grain boundaries, because the coordinated movement of atoms implied by the shape change cannot in general be sustained across grains in different crystallographic orientations. The lattice of the acicular ferrite is therefore generated by a homogeneous deformation of the austenite, so that the iron and substitutional solutes are unable to diffuse during the course of transformation. It is not therefore surprising that the concentrations of substitutional solutes are unchanged during the growth of acicular ferrite.

The macroscopic deformation accompanying the growth of acicular ferrite from austenite occurs on particular planes and directions, so the ferrite structure and orientation are related intimately to that of the austenite. It follows that plates of acicular ferrite, like bainite, must without exception have an orientation relationship with the austenite. This is not necessarily the case when a transformation occurs by a diffusion mechanism, because a grain of ferrite can grow easily into any adjacent grain of austenite with which it happens to come into contact.

During isothermal transformation, the acicular ferrite reaction stops when the carbon concentration of the remaining austenite makes diffusionless transformation impossible. This implies that the plates of acicular ferrite grow supersaturated with carbon, but the excess carbon is shortly afterwards rejected into the remaining austenite. This is, of course, the incomplete reaction phenomenon described in Chapter 6 for bainite, where the austenite never reaches its equilibrium composition since the reaction stops at the T_0' curve of the phase diagram, Fig. 7.6 [20]. The obvious conclusion is that acicular ferrite cannot form at temperatures above the bainite-start temperature, and this is indeed the case in practice.

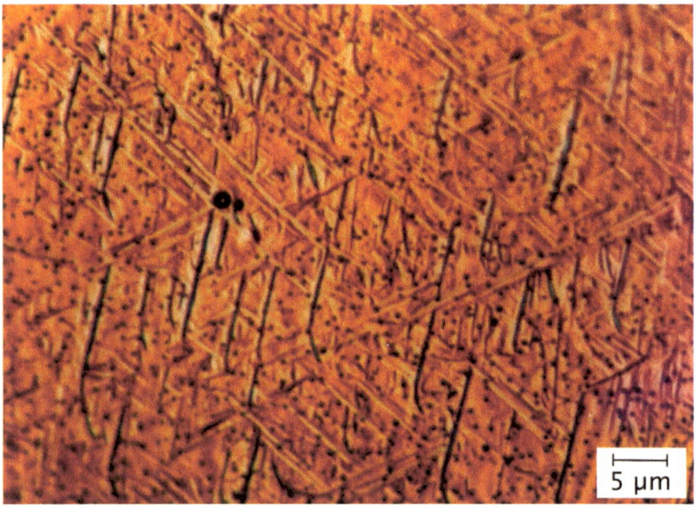

Figure 7.5 Interference contrast micrograph showing the surface relief caused when a metallographically polished sample of steel is transformed to acicular ferrite (courtesy of M. Strangwood).

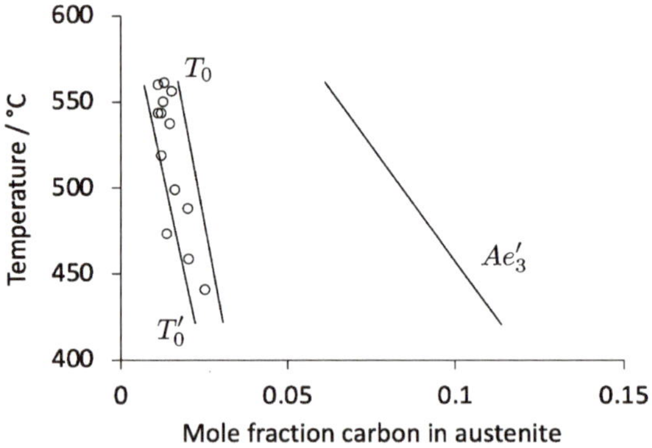

Figure 7.6 Data from experiments in which the austenite is transformed isothermally to acicular ferrite, showing that the reaction stops when the carbon concentration of the austenite reaches the T_0' curve. Adapted using data from Strangwood and Bhadeshia [20].

There are many other correlations which reveal the analogy between acicular ferrite and bainite. For example, the removal of inclusions by vacuum arc melting, without changing any other features, causes an immediate change in the microstructure from acicular ferrite to bainite [21]. The same effect can be obtained by increasing the number density of austenite grain nucleation sites relative to intragranular sites. This can be done by refining the austenite grains to obtain a transition from an acicular ferrite microstructure to one which is predominantly bainitic (Fig. 7.7).

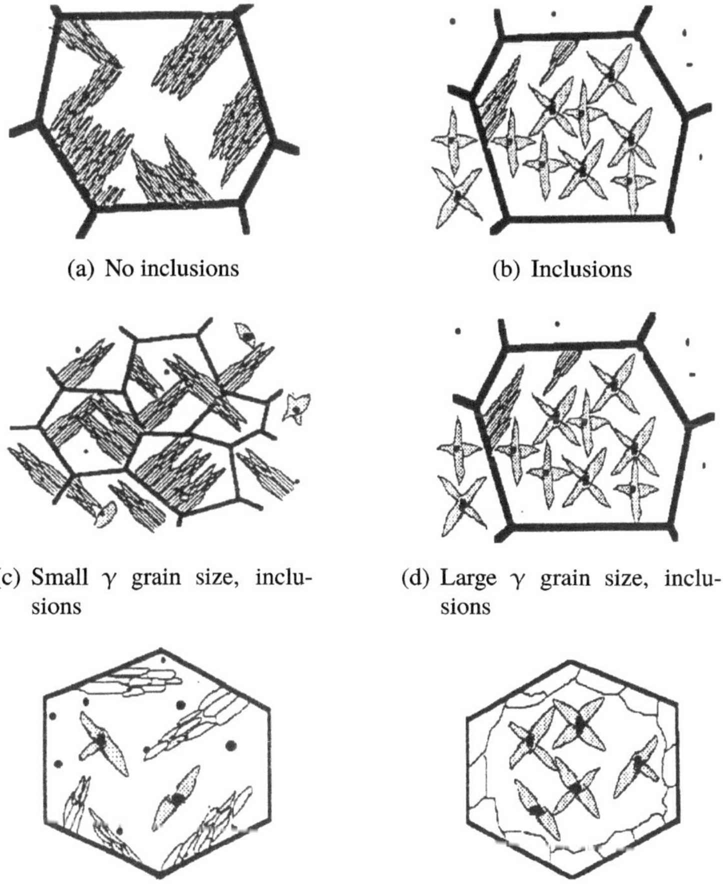

(a) No inclusions

(b) Inclusions

(c) Small γ grain size, inclusions

(d) Large γ grain size, inclusions

(e) No allotriomorphic ferrite, inclusions

(f) Austenite grain boundary covered by allotriomorphic ferrite, inclusions

Figure 7.7 The transition from bainite to acicular ferrite because of the change in the ratio of the number density of nucleation sites at austenite grain boundaries to inclusions. (a)→(b) acicular ferrite promoted by introducing inclusions. (c)→(d) acicular ferrite promoted by an increase in the austenite grain size. (e)→(f) acicular ferrite promoted by saturating austenite grain boundaries with allotriomorphic ferrite. Figures courtesy of S. S. Babu.

(a)

(b)

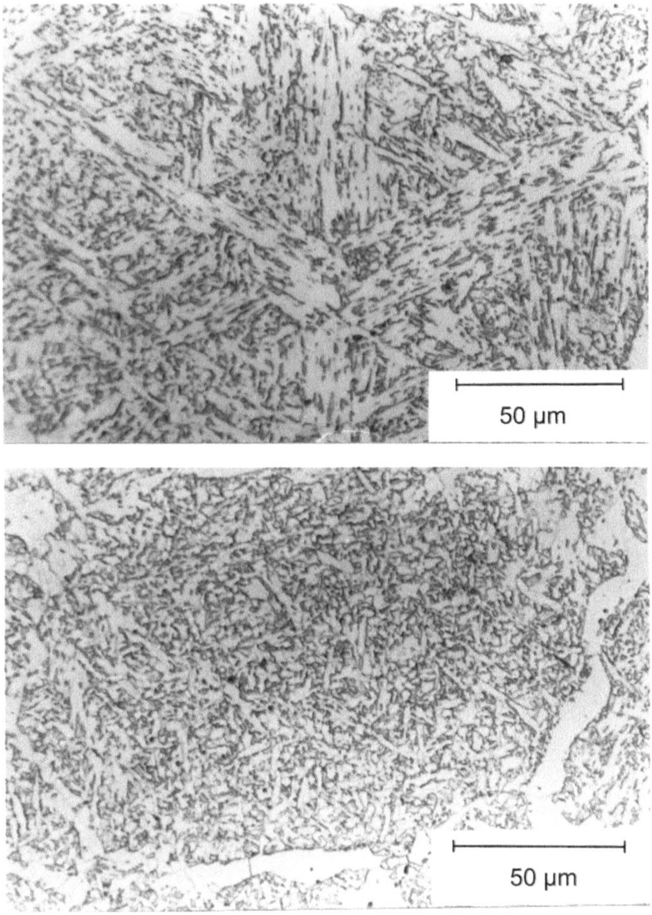

Figure 7.8 The change from a bainitic (a), to an acicular ferrite (b) microstructure, when the austenite grain boundaries are eliminated as nucleation sites, by decoration of the boundaries with inert layers of allotriomorphic ferrite (courtesy of S. S. Babu). The steel is the same for both micrographs.

The opposite phenomenon, in which an inclusion-containing steel with bainite can be induced to transform into an acicular ferrite microstructure, is also observed. This can be done by rendering the austenite grain surfaces ineffective as nucleation sites, either by decorating the boundaries with a thin layer of inert allotriomorphic ferrite (Fig. 7.8) or by adding a small amount of boron (30 ppm). The boron segregates to the γ-boundaries, thereby reducing their energy and making them less favourable sites for heterogeneous nucleation [22]. In general, any method that increases the number density of intragranular nucleation sites relative to austenite grain boundary sites promotes the acicular ferrite microstructure.

7.4 Inclusions as heterogeneous nucleation sites

Many experiments show that inclusions rich in titanium are most effective in acicular ferrite production (Fig. 7.9). A number of different mechanisms have been proposed. It is rare, however, that the specific titanium compound responsible for the observed effects is identified. This is because many of the compounds have similar crystal structures and lattice parameters. When microanalysis is used, elements such as C, N and O are either undetectable or cannot be estimated with sufficient accuracy to determine the stoichiometric ratio with respect to Ti. In reality, the non-metallic inclusions tend to consist of many crystalline and amorphous phases, so that it becomes difficult to identify the particular component responsible for nucleation of acicular ferrite.

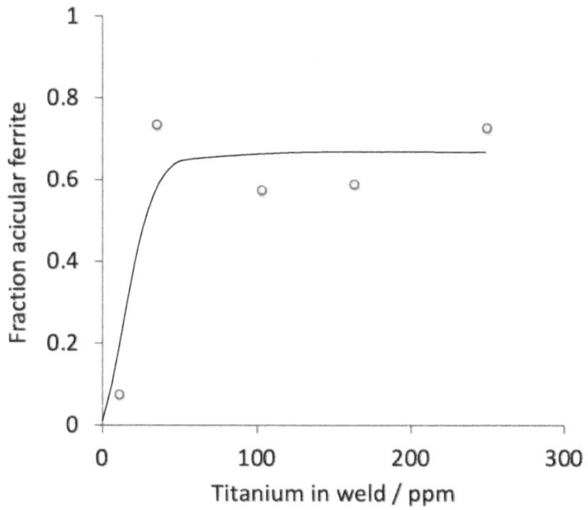

Figure 7.9 Large change in the acicular ferrite content as titanium is introduced into a welding alloy. Data from [23] for manual metal arc weld metal of composition about Fe-1.8Mn-0.3Si-0.08C wt%. The sharp initial rise in acicular ferrite content can be attributed directly to the potency of Ti-rich inclusions in stimulating intragranular nucleation, with subsequent smaller changes dependent also on the evolution of other microstructural constituents.

There are now many results which prove that the inclusions responsible for the heterogeneous nucleation of acicular ferrite are themselves inhomogeneous, as illustrated in Fig. 7.10. The microstructure of the inclusions is particularly important from the point of view of developing a clear understanding of their role in stimulating the nucleation of ferrite. As an example, it is sometimes found that the non-metallic particles in some submerged arc weld deposits consist of titanium nitride cores surrounded by a glassy phase containing manganese, silicon and aluminium oxides, with a thin layer of manganese sulphide (and possibly titanium oxide) partly covering the surface of the inclusions. The inclusions may therefore be a wide variety of oxides or other compounds, but some can influence the development of microstructure during cooling.

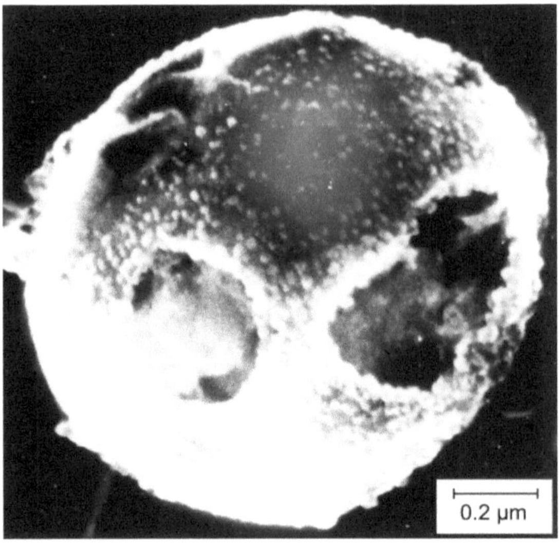

Figure 7.10 Scanning transmission electron micrograph of a non-metallic inclusion in a steel weld metal. The inclusion surface is rather irregular and it features many phases (courtesy of G. Barritte).

7.5 Nucleation of acicular ferrite

It has been demonstrated, assuming classical nucleation theory, that inclusions are less effective in nucleating ferrite compared to austenite grain surfaces. Experiments confirm this, as ferrite formation first begins at the austenite grain boundaries. Furthermore, larger inclusions are expected to be more effective since the curvature of the inclusion/ferrite interface is then smaller. This is confirmed by experimental observations.

7.5.1 Lattice matching theory

Inclusions have long been used to control solidification in aluminium alloys. The aluminium melts are inoculated with particles in order to increase the solid nucleation rate and hence produce a refined grain structure in the fully solidified condition. It is found that inclusions whose lattices match well with solid aluminium are quite effective nucleating agents. This idea has been extrapolated to solid-state transformations in steels, where it is argued that those inclusions which show the best lattice matching with ferrite are most effective in nucleating the ferrite.

The lattice matching κ is expressed in terms of a mean percentage planar misfit [24]. To calculate this, it is assumed that the inclusion is faceted on a plane $(h\,k\,l)_{\mathrm{I}}$, and that the ferrite deposits epitaxially with its plane $(h\,k\,l)_{\alpha}||(h\,k\,l)_{\mathrm{I}}$, with the corresponding rational directions $[u\,v\,w]_{\mathrm{I}}$, and $[u\,v\,w]_{\alpha}$ are inclined at an angle ϕ to each other. The interatomic spacings d along three such directions ($j = 1, 2, 3$) within the plane of epitaxy are examined to obtain:

$$\kappa = \frac{100}{3}\sum_{j=1}^{3}\frac{|d_j^{\mathrm{I}}\cos\phi - d_j^{\alpha}|}{d_j^{\alpha}}. \tag{7.1}$$

Data calculated in this manner for a variety of inclusions phases are presented in Table 7.1.

Table 7.1 Some misfit values between different substrates and ferrite. The data are from a more detailed set published in reference [25] and include all cases where the misfit is less than 5%. The inclusions all have a cubic-F lattice and the ferrite is body-centred cubic (cubic-I). The plane of epitaxy has identical indices in the ferrite and inclusion.

Inclusion	Orientation	Plane of epitaxy	Misfit (%)
TiO	Bain	{1 0 0}	3.0
TiN	Bain	{1 0 0}	4.6
γ-alumina	Bain	{1 0 0}	3.2
Galaxite	Bain	{1 0 0}	1.8
CuS	Cube	{1 1 1}	2.8

To enable the lattice matching concept to be compared against experiments, it is necessary not only to obtain the right orientation relationship, but the inclusion must also be faceted on the correct plane of epitaxy. Many compounds, including some of the titanium oxides, show good matching with ferrite and indeed seem effective in nucleating ferrite. However, there are other compounds, such as γ-alumina, which show good fit but are ineffective nucleants. It is likely that there is more than one mechanism, which helps make a nonmetallic phase a potent heterogeneous nucleation site.

7.5.2 Other possibilities

Other ways in which inclusions may assist the formation of acicular ferrite include stimulation by thermal strains or by the presence of chemical heterogeneities in the vicinity of the inclusion/matrix interface. Alternatively, the inclusions may simply act as inert sites for heterogeneous nucleation. Chemical reactions also are possible at the inclusion matrix interface (Table 7.2). Those minerals that are natural oxygen sources are found to be very effective in stimulating nucleation, probably by inducing decarburisation in the adjacent steel. This effect seems to be independent of the crystallographic nature of the mineral, except in the ability of the mineral to tolerate oxygen vacancy defects, or to thermally decompose. Ti_2O_3 has the ability to cause a dramatic reduction in the manganese concentration of the adjacent steel, and this in turn stimulates nucleation since manganese is an austenite stabiliser. TiO is puzzling in the sense that it is an effective nucleant and yet does not cause any pronounced modification of the adjacent steel. It does have a good lattice match with ferrite, but so does TiN, which is not an effective nucleant.

Table 7.2 List of ceramics tested for their potency in stimulating the nucleation of ferrite plates [26].

Effective: oxygen sources	Effective: other mechanisms	Ineffective
TiO_2, SnO_2	Ti_2O_3	TiN, $CaTiO_3$
MnO_2, PbO_2	TiO	$SrTiO_3$, α-Al_2O_3
KNO_3		NbC

7.6 Summary

Bainite and acicular ferrite have essentially the same transformation mechanism, but their microstructures differ in detail because the former nucleates at grain surfaces and hence grows in the form of sheaves of parallel platelets. Acicular ferrite, on the other hand, nucleates intragranularly on non-metallic inclusions, which are in effect point nucleation sites. The platelets of acicular ferrite therefore radiate from the individual inclusions, thus generating a microstructure, which is much more disorganised, with adjacent platelets pointing in different directions. There are many kinds of non-metallic inclusions which are effective in stimulating intragranular nucleation, but some titanium compounds are found to be particularly potent. The exact mechanism of nucleation remains to be resolved.

Acicular ferrite grows without diffusion, but the excess carbon is not retained in the supersaturated ferrite. It is partitioned into the residual austenite shortly after growth. The transformation is accompanied by shear and rather smaller dilatational displacements, which, together with the reproducible orientation relationship, plate shape and lack of chemical composition change, fit a displacive mechanism of transformation.

Associated with the topic of acicular ferrite is the subject of inclusion design and control in the context of steels. This involves the study of reactions that occur as the liquid steel cools, the role of alloying elements such as aluminium and titanium, and the complexity of the multiphase compounds that then become embedded in the solid steel. The liquid steel can be manipulated to induce a greater number density of fine inclusions. After all, coarse particles can be detrimental to the mechanical properties of the solidified steel. For example, TiN is known to heterogeneously nucleate on $MgO \cdot Al_2O_3$ spinel, so the injection of magnesium vapour into the molten steel during continuous casting has been proven to lead to a refinement of the TiN dispersion [27]. The low-carbon steel used in these experiments has the composition Fe-0.15C-0.16Si-1.5Mn-0.04Al-0.026Ti-0.0044N wt%, but magnesium is also effective in stimulating the nucleation of TiN on MgO in Fe-1C-1.5Cr-0.3Mn wt% bearing steel [28]. What these studies illustrate is there is a vast subject not covered here in detail, on the inclusion control.

We have now dealt with all the major solid-state transformations that occur in steels and their characteristics can be summarised as in Table 7.3 [29].

Table 7.3 Key transformation characteristics in steels. Martensite α', lower bainite α_{lb}, upper bainite α_{ub}, acicular ferrite α_{a}, Widmanstätten ferrite α_{w}, allotriomorphic ferrite α, idiomorphic ferrite α_{i}, pearlite P, substitutional solutes designated X. Consistency of a comment with the transformation concerned is indicated by =, inconsistency by $\neq$; a bullet • identifies the case where the comment is only sometimes consistent with the transformation. The term *parent* γ implies the γ grain from which the product phase grows. IPS is the abbreviation for invariant-plane strain. Adapted from [29].

Characteristic	α'	α_{lb}	α_{ub}	α_{a}	α_{w}	α	α_{i}	P
Nucleation and growth reaction	=	=	=	=	=	=	=	=
Plate shape	=	=	=	=	=	$\neq$	$\neq$	$\neq$
IPS shape change with large shear	=	=	=	=	=	$\neq$	$\neq$	$\neq$
Lattice correspondence during growth	=	=	=	=	$\neq$	$\neq$	$\neq$	$\neq$
Co-operative growth of ferrite and cementite	$\neq$	$\neq$	$\neq$	$\neq$	$\neq$	$\neq$	$\neq$	=
High dislocation density	=	=	=	=	•	$\neq$	$\neq$	$\neq$
Necessarily has a glissile interface	=	=	=	=	=	$\neq$	$\neq$	$\neq$
Always has an orientation within the Bain region	=	=	=	=	=	$\neq$	$\neq$	$\neq$
Grows across austenite grain boundaries	$\neq$	$\neq$	$\neq$	$\neq$	$\neq$	=	=	=
High interface mobility at low temperatures	=	=	=	=	=	$\neq$	$\neq$	$\neq$
Acoustic emissions during transformation	=	=	=					
Reconstructive diffusion during growth	$\neq$	$\neq$	$\neq$	$\neq$	$\neq$	=	=	=
Bulk redistribution of X atoms during growth	$\neq$	$\neq$	$\neq$	$\neq$	$\neq$	•	•	•
Displacive transformation mechanism	=	=	=	=	=	$\neq$	$\neq$	$\neq$
Reconstructive transformation mechanism	$\neq$	$\neq$	$\neq$	$\neq$	$\neq$	=	=	=
Diffusionless nucleation	=	$\neq$	$\neq$	$\neq$	$\neq$	$\neq$	$\neq$	$\neq$
Only carbon diffuses during nucleation	$\neq$	=	=	=	=	$\neq$	$\neq$	$\neq$
Reconstructive diffusion during nucleation	$\neq$	$\neq$	$\neq$	$\neq$	$\neq$	=	=	=
Often nucleates intragranularly on defects	=	$\neq$	$\neq$	=	$\neq$	$\neq$	=	$\neq$
Diffusionless growth	=	=	=	=	$\neq$	$\neq$	$\neq$	$\neq$
Local equilibrium at interface during growth	$\neq$	$\neq$	$\neq$	$\neq$	$\neq$	•	•	•
Local paraequilibrium at interface during growth	$\neq$	$\neq$	$\neq$	$\neq$	=	•	•	$\neq$
Diffusion of carbon during transformation	$\neq$	$\neq$	$\neq$	$\neq$	=	=	=	=
Carbon diffusion-controlled growth	$\neq$	$\neq$	$\neq$	$\neq$	=	•	•	•
Incomplete reaction phenomenon	$\neq$	=	=	=	$\neq$	$\neq$	$\neq$	$\neq$

References

[1] A.F. Gourgues, H.M. Flower, T.C. Lindley, Electron backscattering diffraction study of acicular ferrite, bainite, and martensite steel microstructures, Materials Science and Technology 16 (2000) 26–40.

[2] L. Rancel, M. Gomez, S.F. Medina, I. Gutierrez, Measurement of bainite packet size and its influence on cleavage fracture in a medium carbon bainitic steel, Materials Science & Engineering A 530 (2011) 21–27.

[3] P. Yan, O.E. Güngör, P. Thibaux, M. Liebeherr, H.K.D.H. Bhadeshia, Tackling the toughness of steel pipes produced by high frequency induction welding and heat-treatment, Materials Science & Engineering A 528 (2011) 8492–8499.

[4] D.J. Abson, The role of inclusions in controlling weld metal microstructures in C-Mn steels, Research Report 69/1978/M, The Welding Institute, Abingdon, U.K., 1978.
[5] K. Nishioka, H. Tamehiro, High-strength Ti-oxide bearing line pipe steel for low temperature service, in: Microalloying'88, ASM International, Ohio, U.S.A., 1988, pp. 1–9.
[6] G. Jovanović, D. Glišić, S. Dikić, N. Rdović, A. Patarić, Cleavage fracture of the air cooled medium carbon microalloyed forging steels with heterogeneous microstructures, Materials 15 (2022) 1760.
[7] X. Qi, X. Wang, H. Di, X. Shen, Z. Liu, P. Huan, L. Chen, Effect of ti content on the inclusions, microstructure and fracture mechanism of x100 pipeline steel laser-mag hybrid welds, Materials Science & Engineering A 831 (2022) 142207.
[8] L. Cho, A. Tselikova, K. Hotgrewe, E. de Moor, R. Schmidt, K.O. Findley, Critical assessment 42: acicular ferrite formation and its influence on weld metal and heat-affected zone properties of steels, Materials Science and Technology 38 (2022) 1425–1433.
[9] Z. Wang, J. Gao, M. Zhong, S. Basu, D. Wang, Z. Qu, C. Wang, Revealing charpy impact toughness variations of EH36 shipbuilding steel weld metals processed by CaF2-Al2O3-TiO2 fluxes under high heat input submerged arc welding, Journal of Materials Engineering and Performance (2023) 1–8.
[10] H.S. Kim, M. Jo, J.Y. Park, B.J. Kim, H.C. Kim, D. Nam, B. Kim, Y.S. Ahn, Effects of Ti microalloying on the microstructure and mechanical properties of acicular ferrite casting steel for high-speed railway brake discs, Materials Science & Engineering A 857 (2022) 144125.
[11] K.M. Wu, Three-dimensional analysis of acicular ferrite in a low-carbon steel containing titanium, Scripta Materialia 54 (2006) 569–574.
[12] G.R. Srinivasan, C.M. Wayman, Isothermal transformation in an Fe-7.9Cr-1.1C alloy, Transactions of the Metallurgical Society of AIME 242 (1968) 78–81.
[13] W.L. Costin, O. Lavigne, A. Kotousov, A study on the relationship between microstructure and mechanical properties of acicular ferrite and upper bainite, Materials Science & Engineering A 663 (2016) 193–203.
[14] E.E. Underwood, in: R.T. DeHoff, F.N. Rhines (Eds.), Quantitative Microscopy, McGraw Hill, New York, USA, 1968, pp. 77–125, Chap. 4.
[15] H.K.D.H. Bhadeshia, Bainite in Steels: Theory and Practice, 3rd ed., Maney Publishing, Leeds, U.K., 2015.
[16] E.S. Machlin, M. Cohen, Burst phenomenon in the martensitic transformation, Transactions of the Metallurgical Society of AIME 191 (1951) 746–754.
[17] C.H. Shih, B.L. Averbach, M. Cohen, Some characteristics of the isothermal martensitic transformation, Journal of Metals 203 (1955) 709–710.
[18] J.C. Bokros, E.R. Parker, The mechanism of the martensite burst transformation in Fe-Ni single crystals, Acta Metallurgica 11 (1963) 1291–1301.
[19] A. Entwisle, The kinetics of martensite formation in steel, Metallurgical Transactions 2 (1971) 2395–2407.
[20] M. Strangwood, H.K.D.H. Bhadeshia, Mechanism of acicular ferrite formation in alloy steel weld depos, in: S.A. David (Ed.), Advances in Welding Technology and Science, ASM International, Materials Park, Ohio, USA, 1987, pp. 209–213.
[21] P.L. Harrison, R.A. Farrar, Influence of oxygen-rich inclusions on the $\gamma \rightarrow \alpha$ transformation, Journal of Materials Science 16 (1981) 2218–2226.
[22] H. Hatano, Effects of Nb and Mo on microstructure and toughness of simulated HAZ in 590 MPa class low carbon bainitic steels, Tetsu-to-Hagané 91 (2005) 875–881.
[23] G.M. Evans, The effect of Ti on the microstructure and properties of manganese containing MMA weld deposits, Oerlikon Swhweissmitt 50 (1992) 19–34.

[24] B.L. Bramfitt, The effect of carbide and nitride additions on the heterogeneous nucleation behaviour of liquid iron, Metallurgical Transactions 1 (1970) 1987–1995.
[25] A.R. Mills, G. Thewlis, J.A. Whiteman, Nature of inclusions in steel weld metals and their influence on the formation of acicular ferrite, Materials Science and Technology 3 (1987) 1051–1061.
[26] J.M. Gregg, H.K.D.H. Bhadeshia, Solid-state nucleation of acicular ferrite on minerals added to molten steel, Acta Materialia 45 (1997) 739–748.
[27] H. Murakami, M. Numata, Heterogeneous nucleation in carbon steel by magnesium vapor injection during continuous casting, ISIJ International 61 (2021) 2524–2533.
[28] Y.Y. Xiao, L. Cao, G.D. Wang, J.F. Wei, S. Sridhar, Formation and precipitation mechanism of TiN inclusion in Mg-treated GCr15 bearing steel, Metallurgical & Materials Transactions B 53 (2022) 916–930.
[29] H.K.D.H. Bhadeshia, J.W. Christian, The bainite transformation in steels, Metallurgical & Materials Transactions A 21A (1990) 767–797.

Heat treatment of steels: hardenability

8

Abstract

A heat treatment that causes steel to harden is so much more than the mere plunging of hot metal into a fluid that is often a liquid. The initial red-hot state represents the austenitic condition and the subsequent cooling results in a variety of transformations that depend on the chemical composition of the steel. If the intention is to produce a martensitic structure, then the constituents of the steel must be such that the phase is obtained over the depth required. This leads to the definition of hardenability, but it is necessary to describe the complex phenomena involved in the quench, which in turn depend on the properties of the quenchant and whether it is agitated or not. Heat treatment leads to non-uniform microstructures although the extent of the variation can be controlled. But in addition, there are more subtle effects that leave the final component in a state of stress that can sometimes be beneficial. If the residual stress is allowed to relax, then the component may distort. These and other complexities that form the science of heat treatment are introduced in this chapter.

8.1 Introduction

The traditional route to strong steels is by quenching to form martensite, which subsequently is reheated or *tempered* at an intermediate temperature below that at which austenite can form (Ae_1) to optimise the combination of toughness and strength in the context of the requirements of a specific application. After quenching, the strength achieved is greatest when all the austenite is converted into martensite. The steel must therefore be quenched sufficiently quickly to avoid the formation of such products as ferrite, pearlite and bainite during cooling to the martensite-start temperature. The efficacy of the quench in a given medium depends primarily on two factors: the geometry of the specimen, and the chemical composition of the steel. For the same cross-sectional area, a rectangular strip will cool faster than a round bar if the cooling conditions are identical. A steel that transforms slowly because of its alloy content can be cooled at a slower rate to obtain a fully martensitic state than a lean steel.

With the exception of cobalt and aluminium, the addition of common alloying elements to a steel usually moves the time-temperature-transformation curve to longer times, thus making it easier to pass the nose of the curve during a quenching operation, i.e., there is a reduction in the critical rate of cooling needed to make a steel specimen fully martensitic. If this critical cooling rate is not achieved across the sample, then martensite forms in the outer regions which cool faster, but as the core experiences slower cooling, may transform into bainite, ferrite and pearlite depending on the exact circumstances.

The *hardenability* of a steel is often defined as the distance below the surface where 50% martensite is achieved in a steel rod of standard size, after a standard quenching treatment.

Steels. https://doi.org/10.1016/B978-0-44-318491-8.00014-9

8.2 Use of TTT and continuous cooling diagrams

Time–temperature transformation diagrams provide a good starting point for an examination of hardenability, but as they are statements of the kinetics of transformation of austenite carried out isothermally, they can only be a rough guide. To take one example, the effect of increasing molybdenum, Fig. 8.1 shows the TTT diagrams for a Fe-0.4C-0.2Mo wt% steel and another that has the composition Fe-0.3C-2Mo wt%. The 0.2 wt% Mo steel begins to transform in about 1 s at 550°C, but on increasing the molybdenum concentration to 2 wt% the C-shaped curves for reconstructive and displacive transformations separate and the reactions are substantially slowed. The high-Mo steel clearly has a significantly higher hardenability than that of the 0.2 wt% Mo steel.

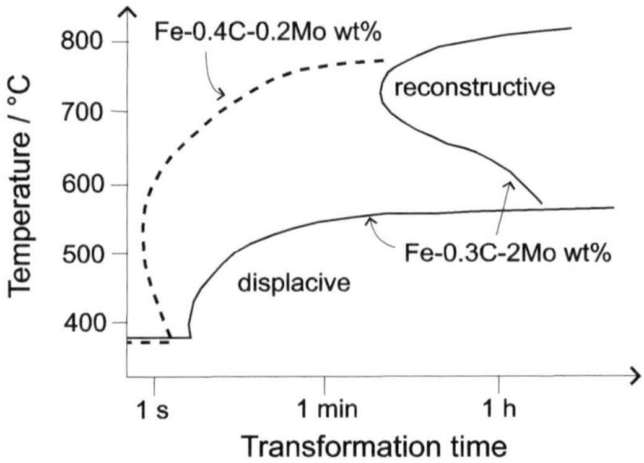

Figure 8.1 TTT diagrams for two steels, one of which is richly alloyed with molybdenum. For clarity, only the curves for the initiation of transformation (5%) are shown. Adapted from [1].

Also noticeable is the fact that in the high-hardenability alloy the curves for the reconstructive and displacive transformations are well separated, but because of the rapidity of the reactions in the low-molybdenum steel, there is an overlap of information resulting in an apparently single C-curve over the whole transformation-temperature range (cf. Fig. 4.2). If the experimental techniques used have sufficient time resolution, then even the diagram for the low-molybdenum steel shows the two C-curves [2].

The obvious limitations of using isothermal diagrams for situations involving a variety of cooling rates through the transformation temperature range have led to efforts to develop more representative diagrams, i.e., continuous cooling transformation (CCT) diagrams. These diagrams record the progress of the transformation with falling temperature for a series of cooling rates. They are determined using cylindrical rods subjected to different rates of cooling, and the onset of transformation is detected by dilatometry, magnetic permeability or some other physical technique. The products of the transformation, whether ferrite, pearlite or bainite, are assessed partly by reference to isothermal diagrams, but confirmed by metallographic examination. The results are then plotted on a temperature/cooling time diagram in which records, e.g., the time to reach the beginning of the pearlite reaction over a range of cooling rates.

There is an approximate relationship between TTT and CCT curves based on the so-called additive reaction rule due to Scheil [3]. In this, the cooling curve is treated as a series of isothermal steps. Referring to Fig. 8.2, if the ith step has anisothermal interval Δt_i at a temperature T_i, and if the time taken to achieve $\xi = 0.05$ of isothermal transformation at T_i is t_i, then the same degree of reaction is obtained during continuous cooling when a temperature is reached at step j where

$$\sum_{i=1}^{j} \frac{\Delta t_i}{t_i} = 1, \tag{8.1}$$

with the summation beginning as soon as the parent phase cools below the equilibrium temperature Ae_3. The method may fail when the transformation rate is determined by multiple functions that vary differently with temperature.

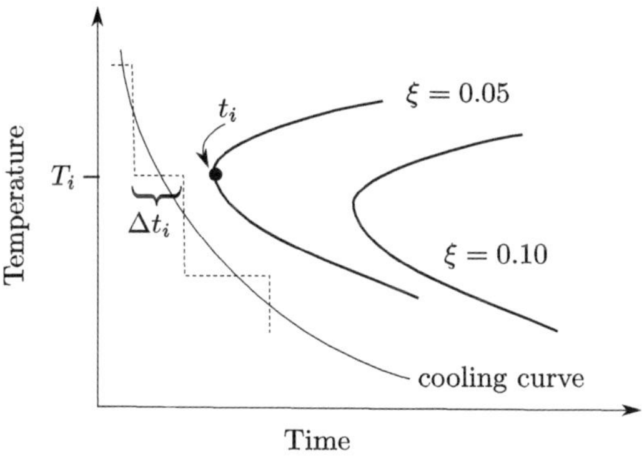

Figure 8.2 The TTT C-curves for particular fractions ξ of isothermal transformation. Superimposed is a cooling curve that is then represented as a series of isothermal steps, each of interval Δt_i, whereas t_i is the time required to achieve a specific value of ξ at a particular temperature. That fraction is said to have been achieved when the sum of all $\Delta t_i / t_i$ equals unity. This is the Scheil method for relating isothermal and anisothermal transformation data.

Fig. 8.3a,b show the TTT and corresponding CCT diagrams for a $2\frac{1}{4}$Cr1Mo steel used typically as a creep-resistant alloy in power plant or petroleum engineering. Such applications require large components, which cannot be transformed isothermally. The thick cylindrical pipe shown in Fig. 8.3c following austenitisation and cooling in air for one hour contains large temperature variations from the core to the surface. The surface experiences air cooling, leading eventually to a bainitic microstructure, whereas the core is akin to furnace cooling and contains primarily allotriomorphic ferrite and pearlite.

Fig. 8.3d shows an interesting superposition of the TTT and CCT diagrams. Continuous cooling at slow rates in general accelerates transformation because the incubation of transformation begins immediately the material is cooled below the Ae_3 temperature, whereas in the isothermal case, the steel is cooled too rapidly to the transformation temperature for its cooling history to have any influence.

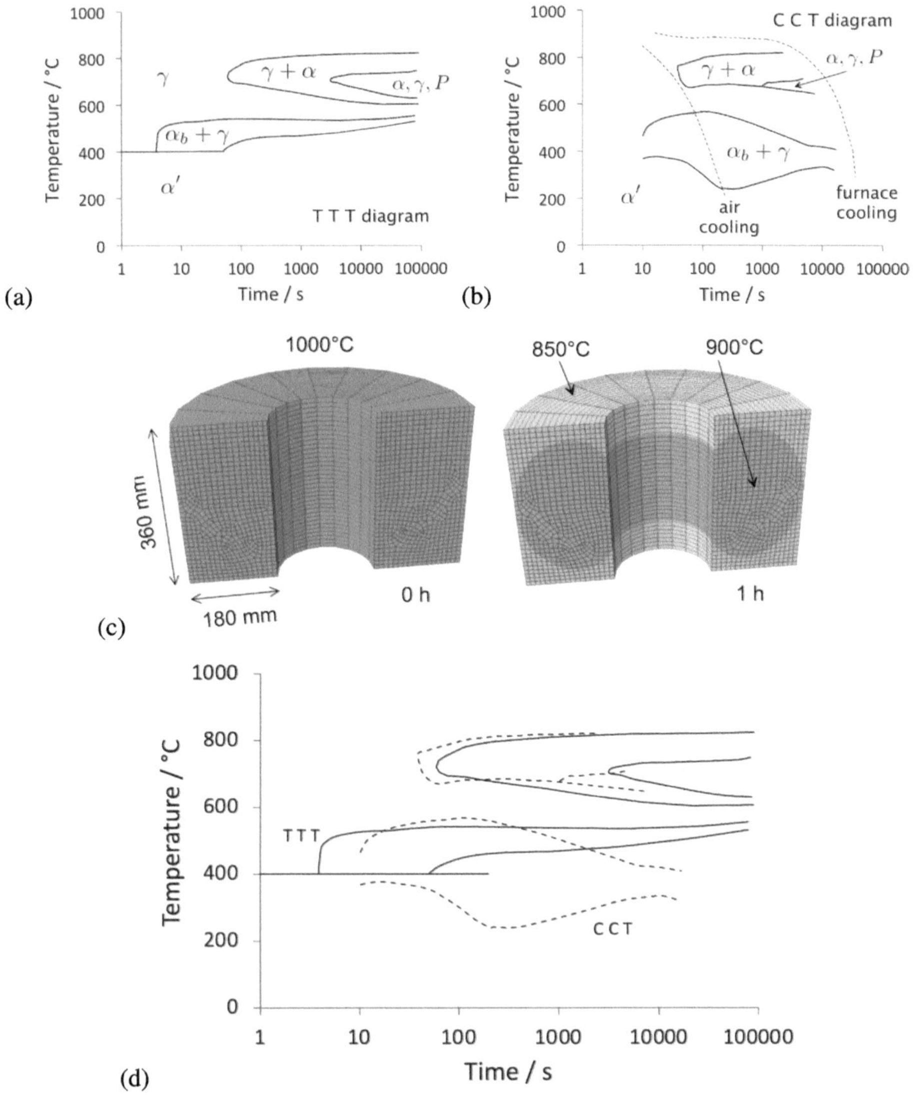

Figure 8.3 $2\frac{1}{4}$Cr1Mo steel. (a) TTT diagram. (b) Corresponding CCT diagram. (c) Large components do not cool uniformly. Here, the core of a $2\frac{1}{4}$Cr1Mo steel cylinder lags the surface temperature by 50°C. The internal diameter of the cylinder is 180 mm (calculations courtesy of S. W. Ooi). (d) Superimposed CCT (dashed lines) and TTT diagrams. The data for plotting the transformation diagrams are from Lundin et al. [4].

8.3 Hardenability testing

The rate at which austenite decomposes to form ferrite, pearlite and bainite is dependent on the composition of the steel as well as other factors such as the austenite grain size, and the degree of homogeneity in the distribution of the alloying elements.

It is now possible to estimate hardenability using pragmatic versions of phase transformation theory, but for generic use there are useful practical tests that allow the hardenability of any steel to be determined experimentally.

8.3.1 The Grossman test

Much of the earlier systematic work on hardenability was done by Grossman and co-workers who developed a test involving the quenching, in a particular cooling medium, of cylindrical bars of different diameters, of the steel under consideration. Hardness measurements on transverse sections of the bars then reveal the effect of hardenability. Fig. 8.4 shows such data for bars of different diameters, oil-quenched from the austenitisation temperature. The greatest hardness at the centre of the bar is achieved in the smaller sections, which can dissipate heat more rapidly.

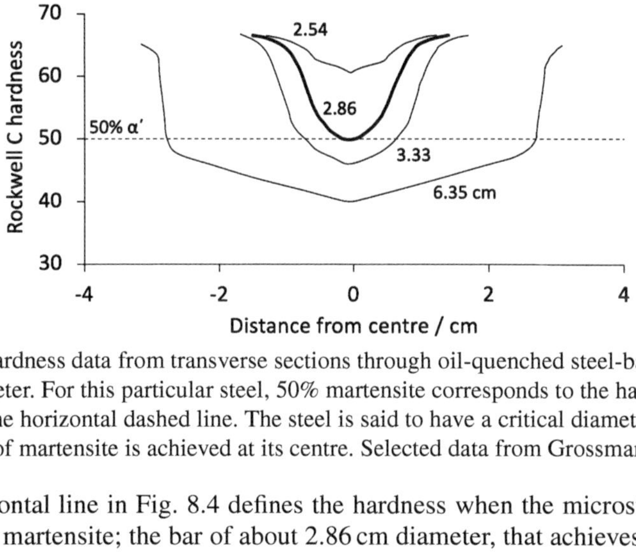

Figure 8.4 Hardness data from transverse sections through oil-quenched steel-bars of increasing diameter. For this particular steel, 50% martensite corresponds to the hardness value indicated by the horizontal dashed line. The steel is said to have a critical diameter ≈2.86 cm because 50% of martensite is achieved at its centre. Selected data from Grossman [5].

The horizontal line in Fig. 8.4 defines the hardness when the microstructure contains 50% of martensite; the bar of about 2.86 cm diameter, that achieves this level of martensite at its core is said to have a critical diameter $d_{50\%\alpha'}$. However, this critical diameter depends on the severity of the quench and therefore is not a complete representation of hardenability. Steel quenched in water has a different $d_{50\%\alpha'}$ than the same steel quenched in oil.

If the efficacy of a water quench is represented by a coefficient $H = 1$, then a comparison can be made with other quench media (Table 8.1), with a smaller value indicating a slower quench and *vice versa*. Using this concept, it is possible to define an ideal diameter $d^{\mathrm{I}}_{50\%\alpha'}$ of a steel bar that would achieve 50% martensite when quenched at an infinitely rapid rate, i.e., when $H = \infty$. This ideal diameter therefore becomes a unique representation of the hardenability of the steel.

Suppose that an experiment is done using a quenching medium with $H = 0.8$ and a bar diameter of $d_{50\%\alpha'} = 10\,\mathrm{mm}$ yields 50% martensite at the centre, then Fig. 8.5 indicates that the ideal diameter $d^{\mathrm{I}}_{50\%\alpha'} = 25\,\mathrm{mm}$ if the quench coefficient is infinite. Thus, in general, $d^{\mathrm{I}}_{50\%\alpha'} > d_{50\%\alpha'}$, although these quantities would be equal

Table 8.1 H-coefficients of quenching media.

Agitation	Cooling medium		
	Oil	**Water**	**Brine**
None	0.25-0.30	1.0	2.0
Moderate	0.35-0.40	1.2-1.3	
Violent	0.8-1.1	4.0	5.0

if $H = \infty$. The relationship between the ideal critical diameter and austenite grain size over the range 20-300 μm has been established empirically as $d^{\mathrm{I}}_{50\%\alpha'} \approx (\overline{L}_\gamma)^{0.22}$, where the grain size is expressed in micrometres whereas the diameter in centimetres.

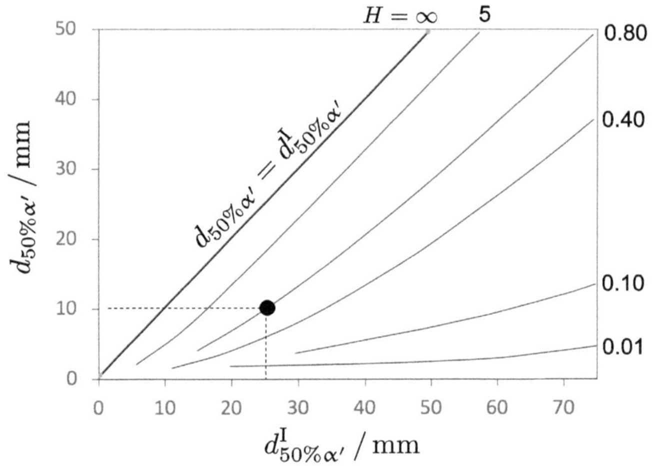

Figure 8.5 Chart for relating the ideal diameter $d^{\mathrm{I}}_{50\%\alpha'}$, the critical diameter $d_{50\%\alpha'}$ and the severity of quench (H) for carbon and medium alloy steels (after Grossman and Bain [6]).

8.3.2 The Jominy end quench test

While the Grossman approach to hardenability is reliable, other less elaborate tests have been devised to indicate trends and to serve in quality control. Foremost amongst these is the Jominy test, in which a standardised round bar (25.4 mm diameter, 102 mm long) is heated to the austenitising temperature and then placed on a rig in which one end of the rod is quenched by a standard jet of water (Figs. 8.6a). This results in a progressive decrease in the rate of cooling along the bar from the quenched end. The temperature as a function of time and the distance (z) from the quenched end is described roughly by the equation [7]:

$$T - T_{\mathrm{o}} = (T_\gamma - T_{\mathrm{o}})\,\mathrm{erf}\frac{z}{2\sqrt{D_{\mathrm{th}}t}}, \qquad (8.2)$$

where T_γ is the austenitisation temperature, T_o that of the quenchant and D_{th} the thermal diffusivity (radiation heat losses and recalescence effects due to phase change are neglected). The equation assumes conduction heat transfer and that the end of the sample away from the quenchant stays at T_γ, i.e., the sample is semi-infinite in length. Fig. 8.6b shows a finite element calculation of the temperature contours following one minute since the beginning of the water quench. This method avoids most of the approximations associated with the analytical equation for the dissipation of heat. For example, the contour at the top of the sample reflects the loss of heat through radiation effects.

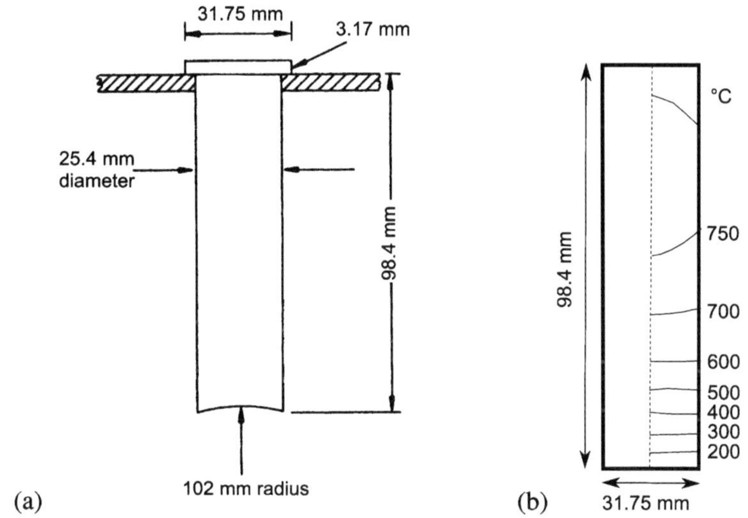

Figure 8.6 (a) The Jominy sample. Following austenitisation, a water jet is directed upward at the lower end of the sample. (b) Finite element calculation of the temperature profile following one minute of operation of the water jet. The contours are symmetrical about the centreline (adapted from Buchmayr and Kirkaldy [8]).

The effects of the variations in cooling rate along the length of the test sample generate corresponding variations in microstructure and hardness, as illustrated in Fig. 8.7. The two steels illustrated have different hardenabilities, despite their identical carbon concentration, because one of them contains a high nickel concentration, which enhances the stability of the austenite. Therefore, only martensite and bainite are obtained to a depth of some 50 mm, making the steel suitable for relatively deep hardening applications. In any steel that is quenched, the appearance of ferrite and pearlite corresponds to a large reduction in hardness, as expected from the following empirical equations that have been derived from a large experimental database [10]:

$$\mathrm{HV}_{\alpha'} = 127 + 949w_{\mathrm{C}} + 27w_{\mathrm{Si}} + 11w_{\mathrm{Mn}} + 8w_{\mathrm{Ni}} + 16w_{\mathrm{Cr}} + 21\log \dot{T}$$

$$\begin{aligned}\mathrm{HV}_{\alpha_b} = &-323 + 185w_{\mathrm{C}} + 330w_{\mathrm{Si}} + 153w_{\mathrm{Mn}} + 65w_{\mathrm{Ni}} + 144w_{\mathrm{Cr}} + 191w_{\mathrm{Mo}}\\ &+ (89 + 53w_{\mathrm{C}} - 55w_{\mathrm{Si}} - 22w_{\mathrm{Mn}} - 10w_{\mathrm{Ni}} - 20w_{\mathrm{Cr}} - 33w_{\mathrm{Mo}}) \times \log \dot{T}\end{aligned}$$

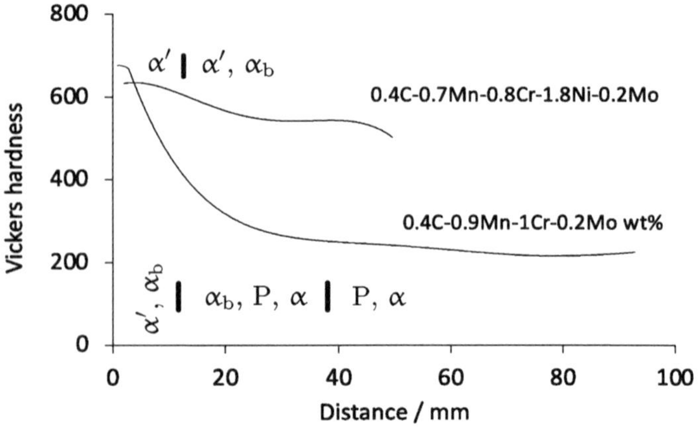

Figure 8.7 Jominy sample hardness data for two steels whose approximate chemical compositions are stated in wt% (adapted using selected data from Buchmayr and Kirkaldy [8] and Kandpal et al. [9]).

$$\begin{aligned} \mathrm{HV}_{\alpha/\mathrm{P}} &= 42 + 223w_{\mathrm{C}} + 30w_{\mathrm{Mn}} + 12.6w_{\mathrm{Ni}} + 7w_{\mathrm{Cr}} + 19w_{\mathrm{Mo}} \\ &\quad + (10 - 19w_{\mathrm{Si}} + 4w_{\mathrm{Ni}} - 8w_{\mathrm{Cr}} + 130w_{\mathrm{V}}) \times \log \dot{T}, \end{aligned} \tag{8.3}$$

where $\dot{T}$ is the cooling rate in °C h^{-1}. The equations apply over the range $0.1 < w_{\mathrm{C}} < 0.5$, $w_{\mathrm{Si}} < 1$, $w_{\mathrm{Mn}} < 2$, $w_{\mathrm{Ni}} < 4$, $w_{\mathrm{Cr}} < 3$, $w_{\mathrm{Mo}} < 1$, $w_{\mathrm{V}} < 0.2$, $(w_{\mathrm{Mn}} + w_{\mathrm{Ni}} + w_{\mathrm{Cr}} + w_{\mathrm{Mo}}) < 5$. w_i represents the wt% of the solute identified in the subscript.

The Jominy test is now used widely to determine the largest diameter round bar that can be fully hardened.

8.4 Effect of grain size and chemical composition on hardenability

Two important variables that influence hardenability are austenite grain size and chemical composition. The hardenability increases with increasing austenite grain size, because the grain boundary area per unit volume decreases. The sites for the nucleation of ferrite and pearlite are then reduced in number, with the result that these transformations are slowed down, so the hardenability increases. Alloying elements, which slow down the ferrite and pearlite reactions, increase hardenability, as described in Chapter 4.

The most economical way to increase the hardenability of a plain carbon steel is to increase the manganese content. Chromium and molybdenum are also very effective and are amongst the cheaper alloying additions per unit of increased hardenability. Boron has a particularly large effect when added to a fully deoxidised low carbon steel, even in concentrations on the order of 0.001 wt%, and would be more widely used if its distribution in steel could be more easily controlled. The role of grain size should not be overlooked, as an increase from 0.02 to 0.125 mm can enhance the hardenability by as much as 50%, which is acceptable provided the toughness is not adversely affected.

High hardenability is not always desirable for many tool and machine parts, where a hard wear-resistant surface is best combined with a tough core. Such shallow hardening situations are additionally preferred because, on quenching, the core develops a tensile internal stress while the surface becomes stressed in compression. This situation is desirable because any fatigue cracks nucleated at surface stress concentrations are more difficult to propagate when a compressive residual-stress is present.

Perhaps an important example in this context is the steel M50 that is used in the manufacture of large bearings for aircraft engines [11]. The composition of the steel is approximately Fe-0.8C-4Cr-4Mo-1V wt%; the TTT curve in Fig. 8.8 shows that large components of the material are easy to harden by martensitic transformation, given the slow rate of transformations above M_S. Aeroengine bearings experience considerable trauma during service: contact stresses between the rolling elements and raceways of ≈ 2 GPa, centrifugal hoop stresses of ≈ 200 MPa, slightly elevated temperatures and aggressive lubricants. It may also be necessary to sustain combined radial and thrust loads. One consequence is that the steel must have a hardness of about 60 HRC, which is straightforward to achieve, but there also is a requirement for toughness, which cannot be achieved for this through hardened steel of this kind. The high solute content in M50 also makes it susceptible to chemical segregation, which causes coarse eutectic-carbides to form that can be damaging to the mechanical properties.

The solution is to avoid through-hardening of the component and limit hard regions to the locations on the component where they are needed. Hence the development of M50-NiL, which has a low carbon concentration (Fe-0.1C-4Cr3.5Ni-4Mo-1.2V wt%) so the steel develops a tough microstructure. When fabricated into bearing-raceways, the surface is locally enriched with carbon, making it hard and the process of case-hardening leaves that surface in a beneficial state of compression.

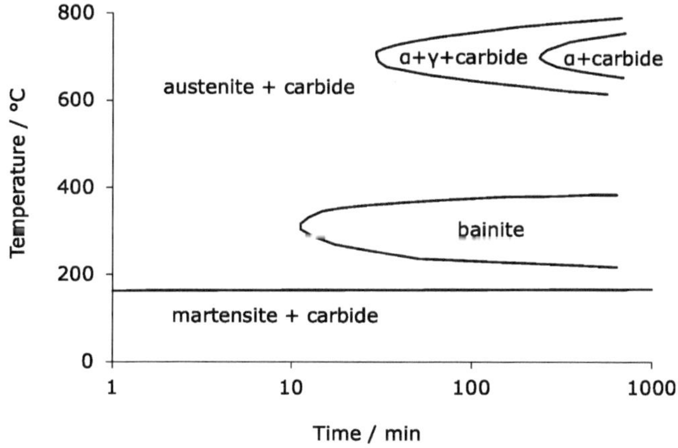

Figure 8.8 Time–temperature transformation diagram of M50 bearing-steel austenitised at 1120°C. Data from [12]. The carbides are not cementite but a mixture of M_6C, MC, $M_{23}C_6$, etc., and some of them are also present in the austenitic condition.

8.5 Quenching stresses and quench cracks

The act of quenching is quite complex because it involves a red-hot sample of steel being plunged into a fluid that usually is at ambient temperature. The instant effect is for the fluid to vaporise. The cooling rate at this stage is slow because the vapour shields the sample from the fluid (Fig. 8.9a). Once the vapour envelope collapses, the heat transfer rate increases dramatically due to the rapid removal of heat as the liquid that touches the sample is boiled off. In the final stage, when the sample has a temperature less than the boiling temperature of the fluid, heat transfer slows down because it occurs primarily by conduction and convection.

It frequently is impossible to achieve uniform cooling at all locations on the sample that is quenched. Even when something as simple as a straight bar is plunged into water, the end that enters the water first will cool more rapidly than the opposite end. Having a complex shape does not help – the shaft illustrated in Fig. 8.9b,c has a keyway parallel to its length, which gives rise to non-uniform cooling. As a result, the shaft becomes distorted on quenching. Distortion must be minimised in production processes for engineering components in order to maintain tolerances and preserve the intended shape of the sample. One method of minimising deformation is to quench more gently, for example in water containing polymeric additives, in oil or by air hardening. All of these lead to more uniform cooling of components.

Quenching processes are in general used when producing strong components based on martensitic microstructures. Strong steels are necessarily less tolerant to plastic relaxation during the quench. As a result of the quench stresses, the component may develop serious quench cracks. Fig. 8.10a shows large cracks in a bearing steel that in the quenched state typically has a hardness in excess of 750 HV. Since the interiors of the prior austenite grains are so hard, the sample gives way at the prior austenite grain boundaries, with the cracks meandering between those boundaries and occasionally extending into the martensitic matrix.

Fig. 8.10b shows a different kind of cracking induced by quenching. Microscopic cracks form through the thickness of untempered, carbon-rich martensite plates. These cracks can be minimised by reducing the austenite grain size, which in turn limits the martensite plate size. Small martensite plates are less prone to cracking [15,16].

To summarise, cracking is a consequence of the internal stresses, which develop during quenching from two sources:

a. *Thermal stresses* arising directly from the different cooling rates experienced by the surface and the interior of the steel.

b. *Transformation stresses* due to the volume changes, which occur when austenite transforms to other phases.

An example of the effect of thermal stresses is given in Fig. 8.11 for a 100 mm diameter steel bar quenched into water from 850°C. The temperature-time relationship for the surface and the core are given in Fig. 8.11a, from which it is seen that the maximum temperature difference occurs after a time t, when it is about 500°C, which could give rise to a stress in excess of 1000 MPa, if no relaxation took place. Under these conditions, the surface stress-time relationship would be that of curve A, Fig. 8.11b. However, the maximum stress level is not sustained because plastic deformation takes place and the stress-time relationship in reality is that indicated by

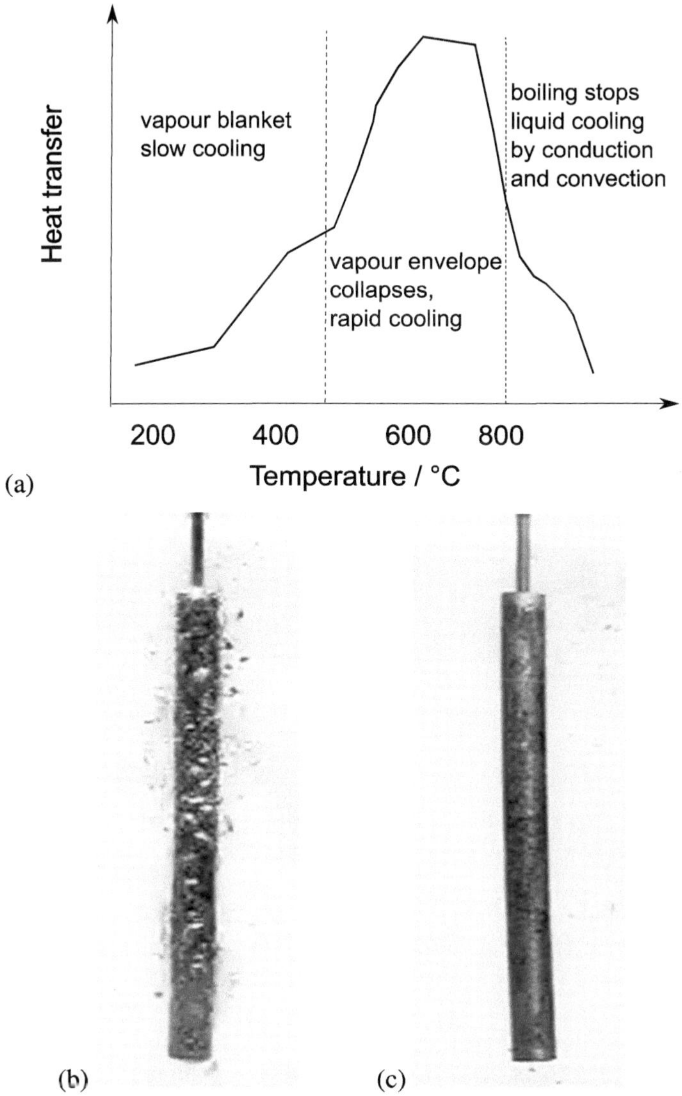

Figure 8.9 (a) Variation in the heat transfer mechanism during a quench (courtesy of Hala Salman Hasan). (b) The moment a red-hot steel shaft containing a keyway parallel to its length, is plunged into water, leading to localised boiling and bubbles. (c) The same shaft at ambient temperature, showing significant distortion. The images are courtesy of M. Narasaki and K. Arimoto of Usunomiya University. The full movie for this can be found on the YouTube channel "bhadeshia123" and more details in Arimoto [13,14].

curve B. The tensile stress in the surface is balanced by a compressive stress in the core, as shown by curve C. At some lower temperature corresponding to t_2 the compressive and tensile stresses drop to zero, but as the temperature drops further to room

(a)

(b)

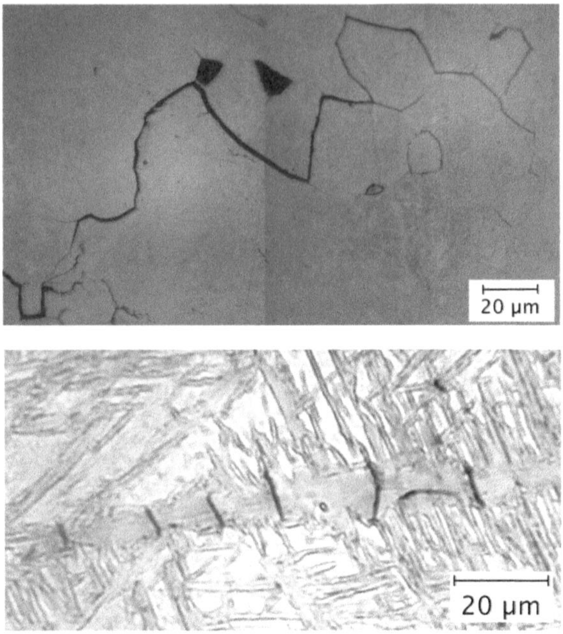

Figure 8.10 (a) Crack in bearing steel containing 1C-1.5Cr wt%, austenitised at 1100°C and quenched into water at 0°C (courtesy of W. Alvarez-Solano). (b) Microscopic cracks in a plate of untempered martensite containing 1C wt% amongst other elements (courtesy of S. Chatterjee).

temperature, the stress situation reverses and the core goes into tension and the surface into compression. Fig. 8.11c shows the stress distribution through the bar at room temperature.

The more rapid the quench, the higher the temperature difference between core and surface during quenching and therefore the greater the resulting stresses at room temperature. In practical terms this means that avoidance of distortion involves the use of less drastic quenching media, e.g., oil instead of water, and consequently adjustments have to be made to the hardenability if full hardening through the section is required.

Transformation stresses arise from the change in volume associated with the formation of a new phase. The expansion coefficient of austenite is much greater than that of ferrite (Fig. 8.12). This means that transformations that occur at high temperatures lead to a smaller volume change than those that occur at lower temperatures. For example, when austenite transforms into martensite in a 1 wt% carbon steel, there is an increase in volume of 4%, while the transformation to pearlite results in a smaller increase of 2.4%. When the austenite is crystallographically textured, the much larger shear strains associated with displacive transformation also manifest [18,19]. But in random microstructures, the shear components cancel, leaving on average only the volume changes.

The effect of these volume changes on the stress pattern developed depends on whether the reactions at the surface and the core start simultaneously, and whether the

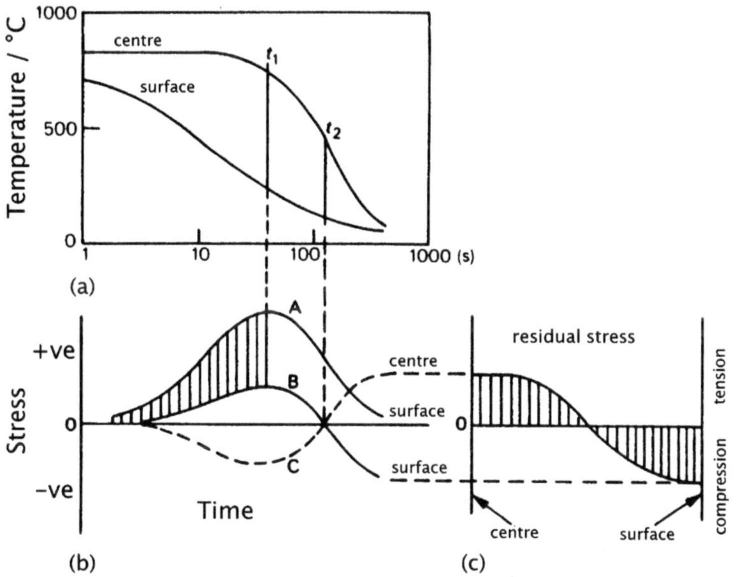

Figure 8.11 Development of thermal stresses during cooling of a 100 mm diameter bar quenched into water from 850°C (after Rose [17]).

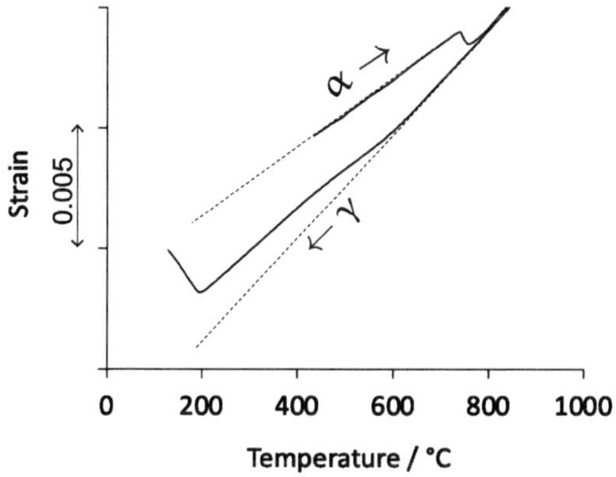

Figure 8.12 A typical heating and cooling curve showing the linear transformation strain, i.e., the vertical distance at any temperature between the dashed lines, as a function of temperature. It is clear that the austenite expansion coefficient (slope of the cooling dashed line) is much greater than that of ferrite. Data courtesy of Nibedita Behara.

hardenability is sufficient to permit full hardening or not. When the martensite reaction starts at the surface, a tensile stress is generated there and a compressive stress occurs in the centre. This situation is accentuated by having the martensite reaction throughout the diameter, i.e., in small sections, or in steels with high hardenability. The presence of a tensile stress in the surface region is clearly undesirable, so high hardenability can sometimes create problems. This can be avoided by the use of steels which provide only a relatively thin hardened layer at the surface that can be maintained in a state of compression. Surface treatment methods such as carburising and nitriding (Section 1.4.3), where the interstitial element concentration is substantially increased by a diffusive process, not only lead to hard, wear-resistant surfaces, but also surfaces which resist crack propagation by being subject to compressive stresses.

Martensite becomes more brittle with increasing carbon content. In higher carbon martensites, which tend to exhibit the burst phenomenon in which individual martensite plates are successively nucleated by previous plates, cracks often are observed in plates at points of impact of later plates upon them. These micro-cracks provide obvious nuclei for the propagation of major cracks. In broader terms, quench cracking is likely to occur when quenching stresses have not been sufficiently released by plastic relaxation at elevated temperatures, and they therefore reach the fracture stress of the steel. As in the case of fatigue cracking, the safest situation is to have the most sensitive region of the steel in a state of compression.

There are some fairly obvious precautions which can be taken to avoid such cracking, including the use of the slowest quench compatible with the achievement of adequate hardness. Also stress concentrations in the form of notches, heavy machining grooves and sudden changes in cross section should be avoided where possible, as these promote the development of hardening cracks.

The composition of the steel is important because the transformation characteristics influence the incidence of cracking. The effect of carbon has already been referred to, but additionally, the M_S temperature decreases with increasing carbon content. Thus, in carbon-rich steels, the quenching stresses are less likely to be relieved than when martensite begins to form at a higher temperature, at which the steel is more able to relieve stresses by flow than by fracture. Further, the lower the M_S temperature the larger the change in volume during the transformation and, therefore, the higher the transformation stresses developed. Metallic alloying elements also depress the M_S, but by substantially increasing the hardenability, they allow the use of less drastic quenching, which greatly reduces the probability of distortion and cracking.

8.6 High-speed, large-scale heat treatment

It is common now for cold-rolled steel strip to be annealed using high-productivity continuous furnaces, where the line speed can be as large as 10-15 $m\,s^{-1}$, Fig. 8.13. There are clever accumulators able to store strip between static coils at each end, allowing for the process to be continuous. When an empty coil is stopped to allow it to be replaced by a new coil welded to the tail end of the original stock, the sheet stored within the entry accumulator can continue to be fed into the heating section, Fig. 8.14a. This is because the rolls inside the accumulator can translate vertically. This can reduce or increase the overall length of the sheet in the accumulator, creating time for coil changes. The accumulator can also help in maintaining the required speed of throughput.

The actual equipment is much more sophisticated than illustrated in Fig. 8.13 [20]. For example, the cleaning of the cold-rolled strip prior to its entry into the furnace consists of dunking, scrubbing, electrolytic action, scrubbing, rinsing and drying. Ra-

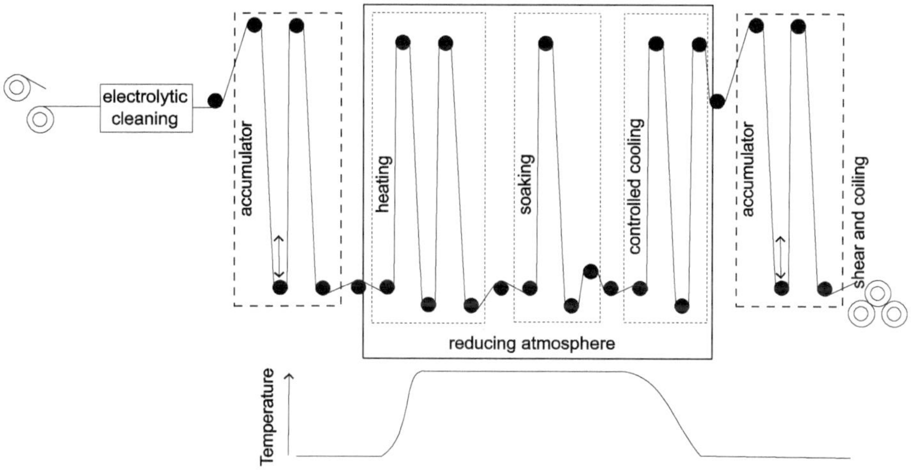

Figure 8.13 A greatly simplified representation of a modern continuous annealing line for the rapid processing of cold-worked steel strip of high quality.

(a)

(b)

Figure 8.14 (a) Accumulator showing the steel strip prior to its entry into the furnace. (b) Batch annealing.

diant heating is implemented to ensure uniformity and avoid oxidation. This usually involves ceramic radiant tubes, which themselves are heated using hot gas. The soaking furnace may be electrically heated.

A variety of cooling procedures also are available. The cooling part contains fans that circulate the water-cooled reducing gas that often is a mixture of a small amount of hydrogen with nitrogen. There are different stages of cooling available, which may include final cooling achieved by spraying and immersion in water. The entire furnace can be about 500 m in length and more than 15 m in height. The production capacity can exceed a million tonnes per year.

An automotive steel produced using continuous annealing is a bake-hardening variety (p. 50) Fe-0.002C-0.25Mn-0.04Al-0.0014N-0.019Ti wt%, which is hot-rolled to

2.3 mm thickness in its austenitic state, coiled, then pickled and cold-rolled down to 0.75 mm thickness. This is followed by continuous annealing in which the steel is heated to 830°C at $8°C\,s^{-1}$, held there for 30 s, slow cooled to 650°C, rapidly cooled at $40°C\,s^{-1}$ to 400°C, held there for 180 s and then allowed to cool to ambient temperature [21]. Attempts have been made to produce dual-phase and TRIP-assisted steels using continuous heat-treatment.

8.7 Large-scale batch heat-treatment

Batch annealing (Fig. 8.14b) is obviously less productive, but is used in circumstances where long annealing times are needed, such as for steels used in transformers where large ferrite grains reduce hysteresis loss. It remains relevant for short production runs or when the carbon concentration is high. Batch annealing can provide a more precise control of the heating and cooling cycles, although the coils have a large thermal mass, so there is a variation in the heat-treatment experienced in different regions of the coil. Coils are stacked inside a column containing inert gas. The column has two walls, one to contain the coils and the other to fire gas, so that heat is transferred to the coils by radiation.

8.8 Cryogenic treatment

Liquid nitrogen is cheap and boils at −196°C so this temperature is in practice the lowest used to implement cryogenic treatments. However, many applications do not require such a low temperature so other fluid mixtures are used when sub-zero temperatures are required.

The major use of such treatments is to control retained austenite content. In components containing high concentrations of alloying elements an excessive amount of retained austenite can reduce dimensional stability if the austenite transforms during service. Rotating bearings in particular require dimensional stability because they are press-fitted on to shafts, a fit that can be compromised if the diameter of the bearing increases when the austenite transforms into martensite [11]. Cooling below ambient temperature can, according to the Koistinen and Marburger Equation (5.2), reduce the amount of retained austenite before the bearing is put into service. Although this equation does not contain time, it is necessary to allow sufficient time permit the sample to reach a homogeneous temperature. In addition, it is possible for martensite to form isothermally at low temperatures (Chapter 5).

Excessive retained austenite can compromise the hardness, so many tool steels are given sub-zero treatments to induce more martensite to form. It also is possible that some transition carbides form during the cryogenic treatment, so the combined effect of martensite and fine precipitates increases the hardness and the tool wear-performance [22]. The formation of carbides at such low temperatures is intriguing and not fully understood.

In the context of carbide-free steels containing only bainitic ferrite and carbon-enriched retained austenite, cryogenic treatment in liquid nitrogen for 2 h resulted in a

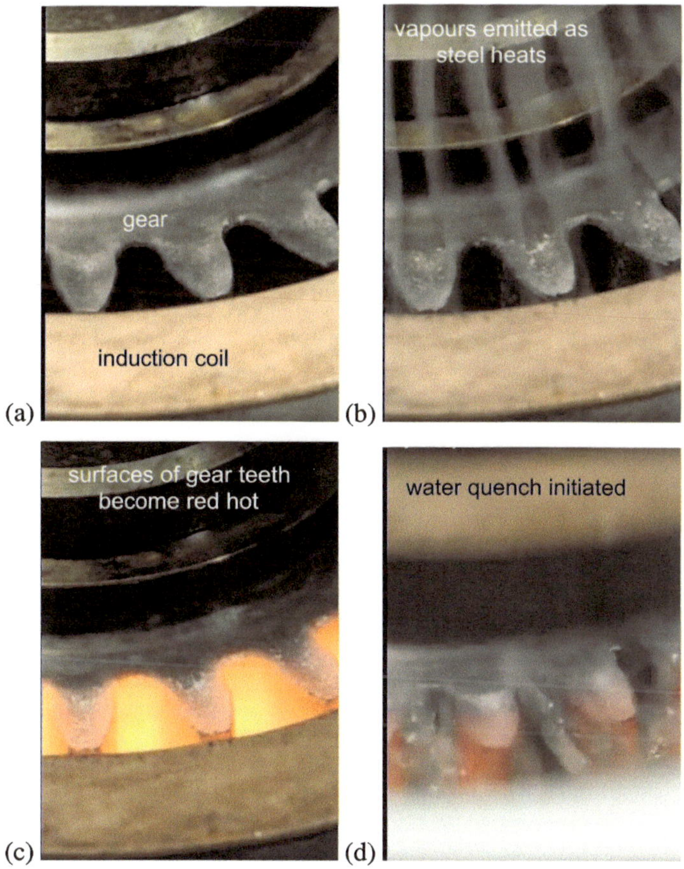

Figure 8.15 Automated inductive hardening of a static gear. The sequence of images is from a movie courtesy of Hans-Werner Zoch of Bremen University. The full movie is available on YouTube channel "bhadeshia123".

general refinement of the austenite regions due to partial transformation into martensite [23]. As a result, the hardness increased from 640 to 670 HV.

8.9 Summary

There are many historical records on the heat treatment of steel dating back to as long ago as the 12th century [24]; some of this relates to mythology, but there are some underlying truths that survive to this day. For example, it was known in the 16th century that quenching in urine is effective in hardening red-hot iron because of the salt it contained. In the modern context, quenching fluids based on water may contain additives to achieve the most effective cooling conditions. Much is understood about the intimately connected roles of steel composition, the quenching medium and the consequences of the process. These consequences go well beyond the hardness of the steel.

The quenching may introduce states of stress in the material that must be accounted for in service. The reproducibility of the process has increased dramatically with the emphasis on automation. Fig. 8.15 illustrates one automated process where carburised gear teeth are heated inductively so that only the regions that require heat treatment experience it, with an automated water quench initiated at the appropriate stage in the sequence.

References

[1] K.E. Thelning, Steel and Its Heat Treatment, Butterworths, London, U.K., 1975.
[2] H.K.D.H. Bhadeshia, A thermodynamic analysis of isothermal transformation diagrams, Metal Science 16 (1982) 159–165.
[3] E. Scheil, Anlaufzeit der Austenitumwandlung, Archiv für das Eisenhüttenwesen 12 (1935) 565–567.
[4] C.D. Lundin, B.J. Kruse, M.R. Pendley, High temperature properties of 2.25Cr1Mo weld metal, Welding Research Council Bulletin 277 (1982) 1–27.
[5] M.A. Grossmann, Elements of Hardenability, American Society of Metals, Cleveland, Ohio, USA, 1952.
[6] M.A. Grossmann, E.C. Bain, Principles of Heat Treatment, 5th ed., ASM, Cleveland, Ohio, USA, 1964.
[7] J.S. Kirkaldy, Prediction of alloy hardenability from thermodynamic and kinetic data, Metallurgical Transactions 4 (1973) 2327–2333.
[8] B. Buchmayr, J.S. Kirkaldy, Modelling of the temperature field, transformation behaviour hardness and mechanical response of low alloy steels during cooling from the austenite region, Journal of Heat Treatment 8 (1990) 127–136.
[9] B.C. Kandpal, A. Chutani, A. Gulia, Harsimran, C. Sadanna, A review of Jominy test and determination of effect of alloying elements on hardenability of steel using Jominy end quench test, International Journal of Advances in Engineering & Technology 1 (2011) 65–71.
[10] R. Blondeau, P. Maynier, J. Dollet, B. Vieillard-Baron, Estimation of hardness, strength and elastic limit of C- and low-alloy steels from their composition and heat treatment, Mémoires Scientifiques de la Revue de Métallurgie 72 (1975) 759–769.
[11] H.K.D.H. Bhadeshia, Steels for bearings, Progress in Materials Science 57 (2012) 268–435.
[12] H.E. Kline, The effect of heat treat variables on the microstructure and high cycle fatigue endurance limit of M50 steel for bearing applications, Technical Report 22-0751, Airsearch Manufacturing Company of Arizona, Arizona, USA, 1978.
[13] K. Arimoto, Thermally-processed steels: residual stresses and distortion, in: G.E. Totten (Ed.), Encyclopedia of Iron, Steel and Their Alloys, Taylor & Francis, Abingdon, U.K., 2015, pp. 3605–3633.
[14] K. Arimoto, A brief review on validation for heat treatment simulation, in: Proceedings of the 31st ASM Heat Treating Society Conference, ASM International, Materials Park, Ohio, USA, 2021, pp. 71–80.
[15] S. Chatterjee, H.K.D.H. Bhadeshia, TRIP-assisted steels: cracking of high carbon martensite, Materials Science and Technology 22 (2006) 645–649.
[16] A.R. Chintha, M. Peet, S. Kundu, H.K.D.H. Bhadeshia, K. Valtonen, V.-T. Kuokkala, Role of fracture toughness in impact-abrasion wear, Wear 428–429 (2019) 430–437.
[17] A. Rose, Eigenspannungen als Ergebnis von Wärmebehandlung und Umwandlungsverhalten, Härterei-technische Mitteilungen 21 (1966) 1–5.

[18] H.K.D.H. Bhadeshia, H. Abreu, S. Kundu, Calculation of crystallographic texture due to displacive transformations, International Journal of Materials Research 99 (2008) 342–346.
[19] H.K.D.H. Bhadeshia, Problems in the calculation of transformation texture in steels, ISIJ International 50 (2010) 1517–1522.
[20] T. Fukushima, Recent technological progress in high speed continuous annealing, Transactions of the Iron and Steel Institute of Japan 25 (1985) 278–293.
[21] W.C. Jeong, Effect of prestrain on aging and bake hardening of cold-rolled, continuously annealed steel sheets, Metallurgical and Materials Transactions A 29 (1995) 463–467.
[22] N.S. Kalsi, R. Sehgal, V.S. Sharma, Cryogenic treatment of tool materials: a review, Materials and Manufacturing Processes 25 (2010) 1077–1100.
[23] F. Hu, P.D. Hodgson, K.M. Wu, Acceleration of the super bainite transformation through a coarse austenite grain size, Materials Letters 122 (2014) 240–243.
[24] D.S. MacKenzie, G. Graham, Beer, blood and urine – mythological quenchants of ancient blacksmiths, in: Proceedings of the 23rd IFHTSE Congress, ASM International, Ohio, U.S.A., 2016, pp. 101–109.

Tempering of martensite

9

Abstract

Martensite in steels can be a very strong, but in its virgin condition, can be brittle. It is then necessary to ameliorate this difficulty by heat treatment in the range 150-700°C. This process is called *tempering*, which in plain English means *to moderate*. Tempering compromises the strength, so the art is to obtain a good combination of strength and toughness. The mechanism of tempering involves the thermally activated approach to equilibrium, for example, by the redistribution of carbon or its precipitation, the rearrangement or elimination of crystal defects and the relief of stresses. These and other aspects of the approach to equilibrium are described in this Chapter. In particular, it may be unnecessary to temper martensite in high-carbon steels if the structure is engineered correctly.

9.1 Introduction

Most of the steel that is in use has internal structures that are not in equilibrium, but rather are engineered to achieve particular combinations of properties. It may not matter that they are not in equilibrium as long as the temperature at which they are applied does not induce significant changes within the steel, over the lifetime of the component of interest. After all, engineering design in general assumes a fixed set of properties. For service at elevated temperatures, for example, in power plants where the steam can be as hot as 630°C, calculations of safe service-life have to account for changes within the steel that accumulate over time periods as long as 40 years. However, the focus here is on inducing changes that help moderate the properties of strong martensitic steels prior to service without causing the formation of austenite.

Tempering is used to optimise properties for particular applications, but it is driven entirely by the fact that martensite is not an equilibrium phase, which, of course, is why it does not appear in equilibrium phase diagrams. These diagrams nevertheless define the ultimate state that can be achieved during prolonged tempering. The tendency of the microstructure to react to a tempering heat treatment depends on how far it deviates from equilibrium. The data in Table 9.1 show the components of the excess free energy of martensite in a typical low alloy steel with chemical composition Fe-0.2C-1.5Mn wt%. The reference state is the equilibrium mixture of ferrite, graphite and cementite, defined here as having a zero stored energy. Graphite precipitates incredibly slowly in steels and is almost never observed during tempering - not surprising given that the reduction in free energy when it forms from a mixture of ferrite and cementite is small (70 J mol^{-1}, Table 9.1). Preventing the substitutional solute, manganese, from partitioning between the ferrite and the austenite adds a substantial amount of energy, but the greatest stored energy increase comes from the trapping of carbon in supersaturated ferrite. Martensite is in Table 9.1 distinguished from supersaturated ferrite by including the strain and interfacial energies due to its mechanism of transformation.

Steels. https://doi.org/10.1016/B978-0-44-318491-8.00015-0

Table 9.1 Stored free energies of a variety of phase-mixtures in their undeformed states. The stored energy of the equilibrium mixture of ferrite, graphite and cementite is set to zero. The supersaturated ferrite and martensite have the same chemical composition but different mechanisms of transformation. Adapted from [1].

Phase mixture in Fe-0.2C-1.5Mn wt% at 300 K	Stored energy / J mol^{-1}
Ferrite, graphite and cementite	0
Ferrite and cementite	70
Para-equilibrium ferrite and para-equilibrium cementite	385
Supersaturated ferrite	1214
Martensite	1914

The general trend during the tempering of martensite therefore begins with the rejection of excess carbon to precipitate carbides, but the substitutional solutes do not diffuse during this process. The end result of tempering is a dispersion of coarse carbides in a ferritic matrix, which bears little resemblance to the original martensite.

In most steels, the martensite reaction does not go to completion on quenching, as is implicit in the Koistinen and Marburger Equation (5.2). Retained austenite is therefore present in the quenched microstructure, but does not necessarily remain stable during the tempering process.

9.2 Tempering: cementite and transition carbides

Carbon can diffuse readily in steel, so its precipitation is the first to occur during tempering, either as cementite or one of the many iron-based transition carbides that are richer in carbon than cementite (Table 9.2). At low temperatures, many of these carbides can form by a displacive transformation mechanism that does not require the diffusion of iron atoms. Indeed, carbide precipitation may occur *during* the quench that generates martensite (Fig. 9.1), in steels that have a high martensite-start temperature; any martensite that forms on cooling below the start-temperature then has an opportunity to temper during the remainder of the quench; this phenomenon is referred to as *auto-tempering*.

The second aspect important in considering the tempering of martensite is that when the latter does not exhibit a shape-memory effect, the martensite can be rich in defect structure, as described in Chapter 5. When the shape deformation of martensite is plastically accommodated, the martensite becomes populated with a dislocation density that sometimes makes it difficult to discern individual dislocations using transmission electron microscopy. A typical dislocation density for a 0.2 wt% carbon steel in its martensitic state is between 0.3 and $1.0 \times 10^{16}\,\mathrm{m}^{-2}$. As the carbon content rises above about 0.3 wt%, fine transformation-twins about 5-10 nm wide may also be observed.

Table 9.2 Variety of iron carbides in the binary Fe-C system. The compositions are approximate – for example, even cementite does not have an exact 3:1 Fe:C ratio. Similarly, the space group of ϵ-carbide changes from $P6_322$ to $P6_3/mmc$ if the carbon atoms are distributed at random in the octahedral interstices. The lattice parameter data are for ambient temperatures and steels containing substitutional solutes can affect not only the parameters but lattice type. More details in [2].

Phase	Lattice type	Space group	Lattice parameters / nm
θ Fe_3C	orthorhombic	*Pnma*	$a = 0.50837$ $b = 0.67475$ $c = 0.45165$
χ Fe_5C_2	monoclinic	$C2/c$	$a = 1.1588$ $b = 0.4579$ $c = 0.5059$ $\beta = 97.75°$
ε $Fe_{2.4}C$	hexagonal	$P6_322$	$a = 0.4767$ $c = 0.4354$
η Fe_2C	orthorhombic	*Pnnm*	$a = 0.4704$ $b = 0.4318$ $c = 0.2830$

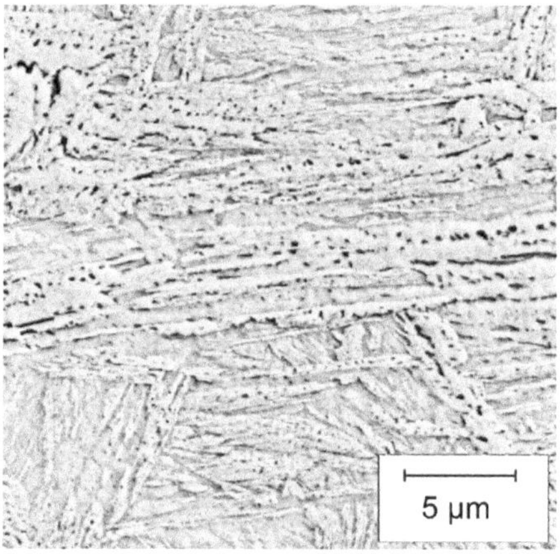

Figure 9.1 A nuclear pressure vessel steel containing just 0.17 wt% carbon, showing autotempered martensite that has formed below the measured M_S temperature of 410°C. The fine particles are cementite (courtesy of H. Pous-Romero). For more details, see [3].

9.3 Stages of tempering

On reheating as-quenched martensite, the tempering can be considered to occur in four stages, but the categorisation though convenient, is not rigorous. The temperatures quoted for the different stages of tempering can depend on the heat-treatment time and steel composition – for example, a large concentration of silicon retards tempering reactions to greater temperatures [4,5]. The categorisation is therefore loose:

Stage 1 up to 250°C: precipitation of ε-iron carbide or other transition carbides; partial loss of tetragonality in the martensite. Excess carbon in the martensite may partition slowly into residual austenite.

Stage 2 between 200°C and 300°C: decomposition of retained austenite.

Stage 3 between 200°C and 350°C: replacement of transition carbides by cementite; martensite loses tetragonality.

Stage 4 above 350°C: cementite coarsens and spheroidises; recrystallisation of ferrite.

9.3.1 Tempering: stage 1

Martensite, formed rapidly by diffusionless transformation from austenite, can undergo changes in the distribution of carbon within its lattice under ambient conditions, with any rearrangement of the carbon being accompanied by changes in tetragonality [6]. Observations using the atom-probe show that the carbon atoms tend to segregate to defects and form small clusters, at temperatures as low as 22°C [7]. Fig. 9.2a shows that the segregation leads to a steady reduction in the carbon that remains in solid solution, as some of the total amount is attracted to defects or into clusters. Direct observations of the heterogeneous distribution of carbon in quenched martensite are shown in Fig. 9.2b.

As the tempering temperature is raised towards 250°C, transition iron-carbides such as ε precipitate in the martensite (Fig. 9.3). This carbide has a hexagonal crystal structure and precipitates as narrow platelets in the matrix, with a well-defined and reproducible orientation relationship [8]:

$$(011)_{\alpha'} \parallel (0001)_{\varepsilon},$$
$$[11\bar{1}]_{\alpha'} \parallel [\bar{1}2\bar{1}0]_{\varepsilon},$$
$$(2\bar{1}1)_{\alpha'} \parallel (10\bar{1}0)_{\varepsilon}.$$

This orientation relationship may be optimal, but nevertheless represents poor fit with the martensite lattice, explaining why ε-carbide is observed only as extremely fine particles visible using transmission electron microscopy [2, p. 431]. It is possible that the carbon content of the ε-carbide is less than $Fe_{2.4}C$ during the early stages of formation [9]; since the lattice parameters and crystallography would therefore be affected, it might be reasonable to speculate that fit with the matrix may be better than expected at the nucleation stage.

Heat treatment in the range 50-100°C hardly affects the hardness, possibly because there is a new contribution from the precipitation of fine carbides that compensates for the reduced amount of carbon in solid solution, Fig. 9.4. At the end of stage 1, the martensite still possesses a tetragonality. We note that although martensite when it first forms in steel must have an ordered arrangement of carbon atoms as a consequence of the Bain strain, it is possible with low-carbon martensites for the carbon to disorder subsequently if it forms at temperatures greater than that associated with Zener ordering (p. 143).

There are other transition carbides that may form, depending on the carbon concentration and tempering conditions. The tempering sequence at temperatures where the iron carbides dominate may be as follows [13], although it is emphasised that the relative thermodynamic stabilities of these carbides are not established [14]:

$$\alpha'\text{martensite} \rightarrow \underbrace{\varepsilon + \alpha' \rightarrow \eta + \alpha' \rightarrow \chi + \alpha' \rightarrow \theta + \alpha}_{\text{reduction in dissolved carbon in } \alpha \;\Longrightarrow}. \tag{9.1}$$

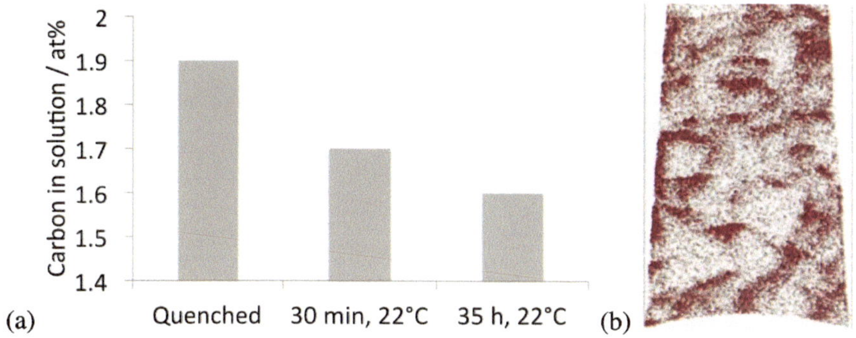

Figure 9.2 (a) Atom-probe observations on a Fe-Ni-C alloy containing 2.08 at% of carbon, with $M_S = -50°C$, aged at room temperature. The carbon dissolved in the matrix is depleted with time as segregation and clustering occurs. Data from Miller et al. [7]. (b) Atom probe three-dimensional image reconstructed from sections, of a quenched Fe-C-Si alloy, showing the heterogeneous distribution of carbon in the martensite. The horizontal extent of the image is about 60 nm and each dot represents the position of a carbon atom. After Sherman et al. [10], with permission of Springer Nature.

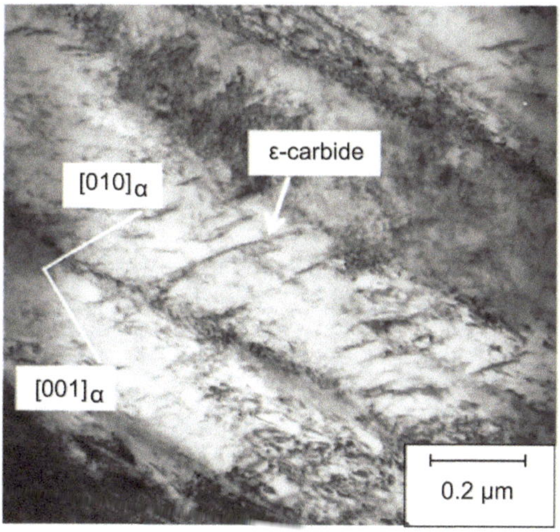

Figure 9.3 ε-carbide in a quenched and tempered steel of composition Fe-0.3C-1.6Si-3.5Mn wt% (after Hajyakbary et al. [11], reproduced with permission of Elsevier).

It follows that the transition carbides should eventually be replaced by cementite – that this is correct in the Fe-C system, is indicated on Fig. 9.5, where the Larson-Miller time-temperature parameter is used to indicate the potency of the heat-treatment by combining the effects of temperature and time. Transition carbides occur only because they are easier to nucleate than cementite. They never are thermodynamically favoured with respect to cementite. They can form only when the driving force for carbon precipitation is large enough to accommodate transition phases that lead to a

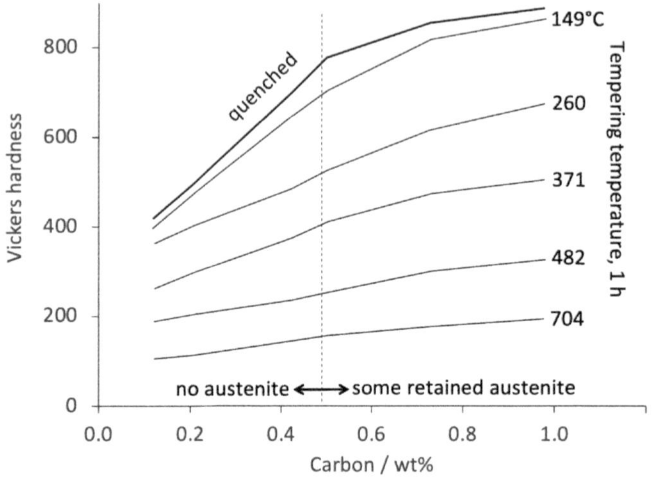

Figure 9.4 The hardness of martensite in binary Fe-C steels as a function of the carbon concentration and tempering temperature. At the larger carbon concentrations on the right of the dashed line, there are increasing amounts of retained austenite left in the microstructure following quenching to ambient temperature, which is why the net hardness of the mixture becomes less sensitive to the carbon concentration. Selected data from Grange et al. [12].

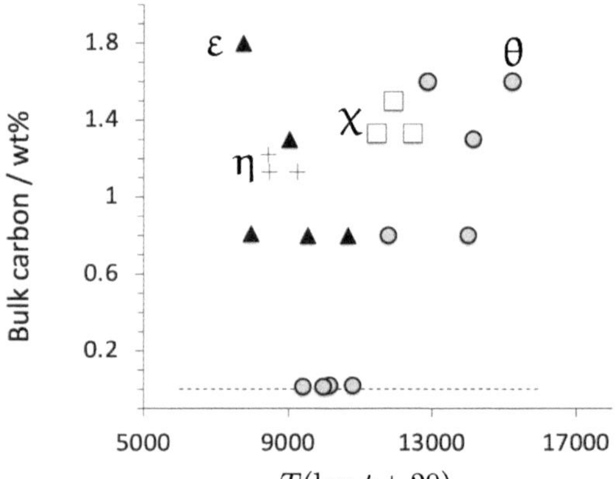

Figure 9.5 The Larson–Miller parameter $T(\log t + 20)$, where T is the absolute tempering-temperature and t the tempering time in hours. The data represent the evolution of carbides in Fe-C binary steels containing martensite or supersaturated ferrite as a function of the potency of the tempering treatment [14].

smaller free energy reduction than cementite. For this reason, tempering supersaturated ferrite that contains only 0.02 wt% of carbon directly precipitates cementite, as shown in Fig. 9.5. The driving force for precipitation in this case is so low that the transition iron-carbides cannot form at all.

The crystallographic orientation between the η-carbide (orthorhombic) and martensite is approximately as follows [15]:

$$(0\,1\,0)_\eta \parallel \{0\,1\,1\}_{\alpha'} \qquad (0\,0\,1)_\eta \parallel \{1\,0\,0\}_{\alpha'} \qquad [1\,0\,0]_\eta \parallel \{0\,\bar{1}\,1\}_{\alpha'},$$

and other features such as the habit plane are consistent with the crystallographic theory of martensite, because the carbide has been shown to form without the diffusion of iron atoms. The displacive transformation mechanism leads to well-defined lenticular plates of η-carbide in tempered martensite [16]. The orientation relationship adopted by χ-carbide (monoclinic) with martensite is [17]:

$$(1\,0\,0)_\chi \parallel \{1\,1\,\bar{2}\}_{\alpha'} \qquad (0\,1\,0)_\chi \parallel \{1\,1\,0\}_{\alpha'} \qquad [0\,0\,1]_\chi \parallel \{1\,\bar{1}\,1\}_{\alpha'}.$$

χ-carbide follows η-carbide and then transforms into cementite. The mechanism of this latter change involves the intergrowth of the two phases [18–21], i.e., interpenetrating layers of cementite and χ, as illustrated in Fig. 9.6.

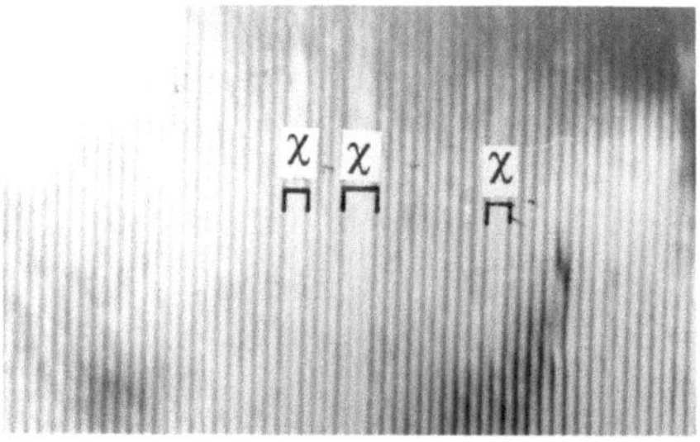

Figure 9.6 Lattice image showing the intergrowth of layers of (010) cementite (*Pnma* setting, spacing 0.67 nm) and (200) χ-carbide (spacing 0.57 nm) formed during the tempering of cementite. Image courtesy of Professor Y. Ohmori, further details in [22].

It has been shown that during the tempering of martensite in steel, some carbon may partition from the supersaturated martensite into the adjacent austenite which acts "as an effective sink to drain the excess carbon from the martensitic matrix" [23]. Later work confirmed this observation using the atom probe technique [24,25]. The partitioning occurs because the carbon has a lower free energy (lower chemical potential) within the austenite, where its concentration is far less than demanded by equilibrium with tempered martensite or ferrite.

A modern variant of this interstitial-atom partitioning, which exploits the process to stabilise the austenite to improve properties, is known as the *quench and partitioning* method [26–29]. The steel is transformed partly into martensite by quenching to a temperature between M_S and M_F (Fig. 9.7). The resulting mixture of martensite and residual austenite is then heated to a higher temperature where some of the excess carbon in the martensite precipitates as carbides, but a significant portion diffuses into the austenite. The enriched austenite is then retained on cooling to ambient temperature and confers a work-hardening capacity to the steel due to deformation-induced martensitic transformation.

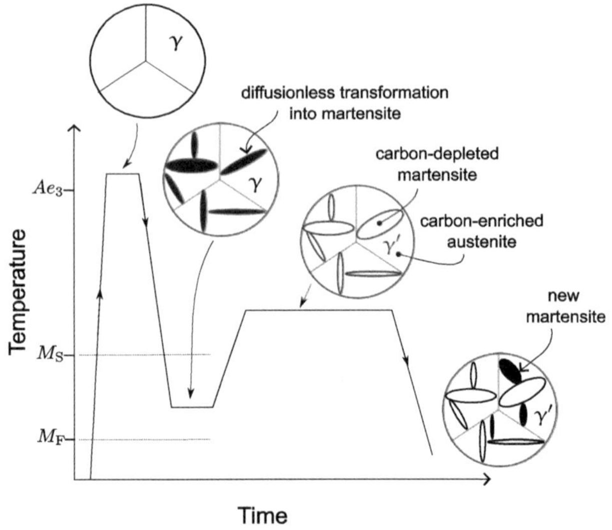

Figure 9.7 The heat treatment used in the quenching and partitioning process. After austenitisation the steel is quenched to produce a mixture of untempered martensite and austenite, which is then heated to allow some of the excess carbon in the martensite to partition into the austenite. The enriched-austenite is referred to as γ'. On further cooling, a considerable amount of austenite is retained, the remainder decomposing to untempered martensite.

9.3.2 Tempering: stage 2

During stage 2, austenite retained during quenching tends to decompose, usually in the temperature range 230-300°C. Cohen and coworkers were able to detect this stage by X-ray diffraction measurements as well as dilatometric and specific volume measurements. Experiments using Mössbauer spectroscopy reveal the increase in cementite associated with the decomposition of austenite, Fig. 9.8.

In martensitic plain carbon steels with less than 0.5 wt% carbon, the amount of retained austenite is often below 2%, rising to around 6% at 0.8 wt% C and over 30% at 1.25 wt% C. The available evidence suggests that in the range 230-300°C, retained austenite decomposes into bainitic ferrite and cementite. Fig. 9.9b,c show how retained austenite films decompose on tempering into discrete particles of cementite, which can be detrimental to mechanical properties (Chapter 11). When thin films of retained austenite decompose, it is natural for the crystallographic orientation of the resulting ferrite to be identical to one of the adjacent plates of martensite. Nucleation is not required, the process can simply involve the epitaxial growth of ferrite. Any carbide must, of course, nucleate, with multiple nucleation events necessary at the transformation front, since the small volume fraction of carbide cannot form a continuous film there (Fig. 9.9c).

Although the focus here is on the precipitation of $\theta + \alpha$ during tempering, prolonged tempering at low temperatures (300°C, 9 h) of carbon-rich retained austenite

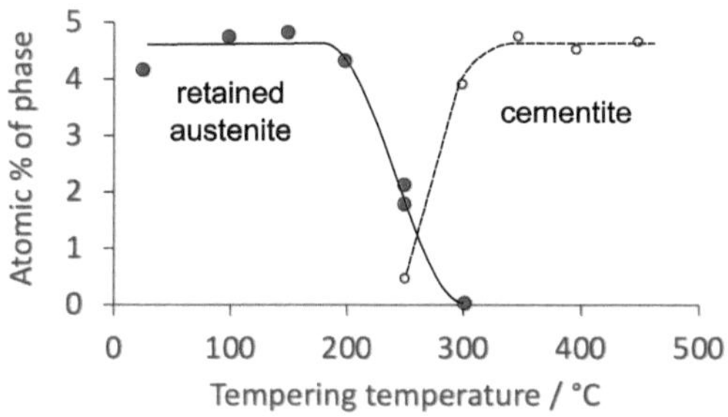

Figure 9.8 Fe-0.4C-0.8Mn-0.3Si-1.8Ni-0.8Cr-0.2Mo wt% steel quenched to martensite and then tempered at the temperatures indicated, for 1 h for temperatures up to 300°C and 2 h at higher temperatures. The phase quantities were measured using Mössbauer spectroscopy. Selected data from Clarke et al. [30] and Williamson et al. [31].

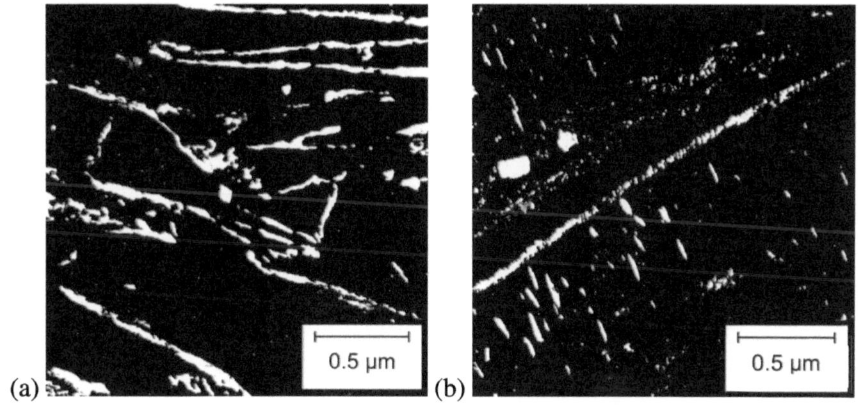

Figure 9.9 (a) Fe-4Mo-0.2C wt% steel. Dark-field transmission electron micrograph showing retained austenite films between martensite plates in the as-quenched condition. (b) Fe-4Mo-0.2C wt% steel. Dark-field transmission electron micrograph showing that the retained austenite has decomposed into an array of discrete particles of cementite at the plate boundaries, following tempering at 295°C for 1 h. After Bhadeshia and Edmonds [32].

($\approx$ 1 wt%) in a TRIP-assisted steel has been shown to decompose into a mixture of $\varepsilon + \alpha$ [33].

9.3.3 Tempering: stage 3

During the third stage of tempering, cementite first appears in the microstructure of the martensite plates, as a Widmanstätten distribution of tiny platelets that have a well-defined orientation relationship with the matrix, which has now lost its tetragonality and become ferrite. The orientation relationship is often stated as that due to Bagaryatskii, but in fact accurate measurements show that it is the Isaichev relationship [34,35]:

$$\langle 1\,0\,0\rangle_{\theta} \;\; \| \; \langle 1\,\bar{1}\,\bar{1}\rangle_{\alpha}$$
$$\{0\,3\,1\}_{\theta} \approx\| \; \{1\,0\,1\}_{\alpha}.$$

It is important to note the approximation sign, because these planes are not exactly parallel, as pointed out in the original paper by Isaichev [36]. There is an angle of some 1.5–2° between $\{0\,3\,1\}_{\theta}$ and $\{1\,0\,1\}_{\alpha}$. The origin of this orientation relationship lies in the fact that it provides the high degree of lattice matching necessary to enable the cementite to grow by displacive transformation without the need for large atoms to diffuse; the Bagaryatskii relationship fails in this respect [37].

This reaction can commence at temperatures as low as 100°C since the driving force for getting rid of excess carbon from the martensite is much greater than when the precipitation occurs from austenite. The precipitation is fully developed at ≈ 300°C (e.g., Fig. 9.9b), with particles up to 200 nm long and ∼15 nm in thickness. Similar structures are often observed in lower-carbon steels as-quenched, as a result of the formation of Fe_3C during the quench. During tempering, the most likely sites for the nucleation of the cementite are the ε-iron carbide interfaces with the matrix (Fig. 9.3), and as the Fe_3C particles grow, the ε-iron carbide particles gradually disappear.

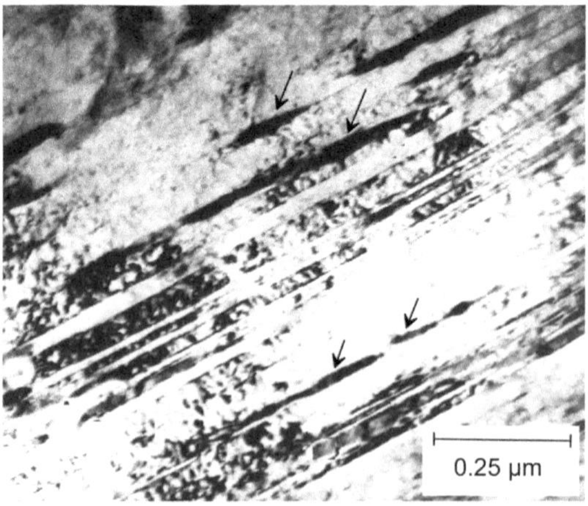

Figure 9.10 Cementite particles (arrows) precipitating on twin boundaries in tempered martensite. After Kraševec et al. [38], with permission from Elsevier.

The twins occurring in the higher carbon martensites are also sites for the nucleation and growth of cementite, which tends to grow along the twin boundaries, Fig. 9.10. This is because the atomic arrangement of iron atoms at the twin boundaries is not dissimilar to that found in cementite [38]. The orientation relationship adopted is (cementite space group *Pnma*)

$$(211)_{\alpha'} \;\; \| \; (01\bar{1})_{Fe_3C},$$
$$[\bar{1}11]_{\alpha'} \approx\| \; [100]_{Fe_3C}.$$

A third site for the nucleation of cementite is the martensite lath boundaries (Fig. 9.11a), and precipitation may also occur at the prior austenite grain boundaries (Fig. 9.11b). The cementite can form as very thin films which are difficult to detect, but gradually spheroidise to give rise to well-defined particles of Fe_3C in the grain boundary regions. There is some evidence that these grain boundary particles or films can adversely affect ductility. However, they can be modified by addition of alloying elements.

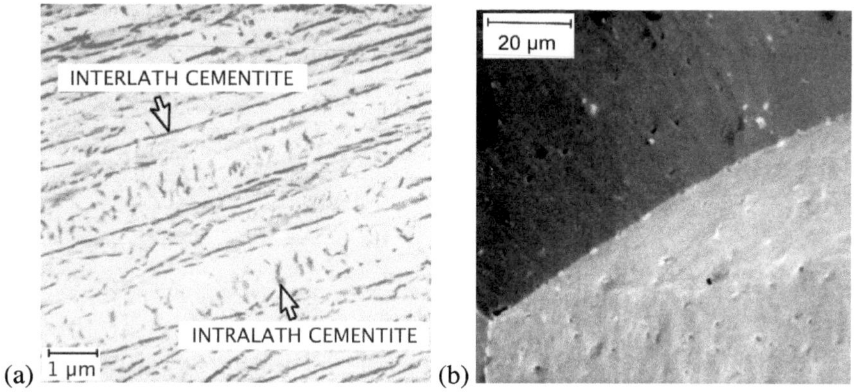

Figure 9.11 (a) Interlath cementite in a quenched and tempered (350°C, 1 h), Fe-0.38C-0.8Mn-0.5Si-1.7Ni-0.8Cr-0.3Mo wt% steel. Adapted from B. C. Kim et al. [39], with permission of Springer. (b) Cementite precipitates at the prior austenite grain boundaries, in a Fe-0.3C-3.5Ni-1.7Cr wt% steel, quenched and tempered at 350°C, 1 h. After Briant and Banerjee [40], with permission of Springer.

During the third stage of tempering, the tetragonality of the matrix disappears and it is then essentially ferrite, not supersaturated with respect to carbon, although the consequences of the shape deformation are not yet eliminated.

Subsequent changes in the morphology of the cementite particles occur by a coarsening process known as Ostwald ripening, where the smaller particles dissolve in the matrix providing carbon for the selective growth of the larger particles. There is a minimal change in the amount of cementite during this process. The mechanism of coarsening can be described as follows, using a simplified scenario where all particles are treated as spherical in shape and with an isotropic interfacial energy per unit area, $\sigma_{\theta\alpha}$.

The free energy per atom of a spherical particle that is small is greater than that of a particle which has a larger radius. If v_a is the volume per atom, then when an atom is added to a spherical particle of radius r, its surface area increases by $2v_a/r$. Therefore, the free energy increases by $\Delta G = 2\sigma_{\theta\alpha}v_a/r$. Such an increase is smaller for larger particles, so there is diffusion from the small to the large particle, with the former dissolving and the latter growing. The flux is driven by ΔG, which, as we have seen, is proportional to $1/r$. For a set of spherical particles, the interparticle spacing is also proportional to $1/r$. It follows that the rate of coarsening, $\mathrm{d}r/\mathrm{d}t \propto 1/r^2$, which on integration gives an r^3 dependence on time. A more thorough derivation by Lifshitz and Wagner [41,42] leads to this dependence as follows:

$$\bar{r}_t^3 - \bar{r}_o^3 = \frac{C_3}{RT} D \sigma_{\theta\alpha} t, \tag{9.2}$$

where $\bar{r}_o$ is the mean particle radius in the initial distribution of particles, $\bar{r}_t$ is the corresponding mean particle radius at time t, D is an effective diffusion coefficient in the matrix (a function of the diffusivities of both carbon and iron) and C_3 is a constant.

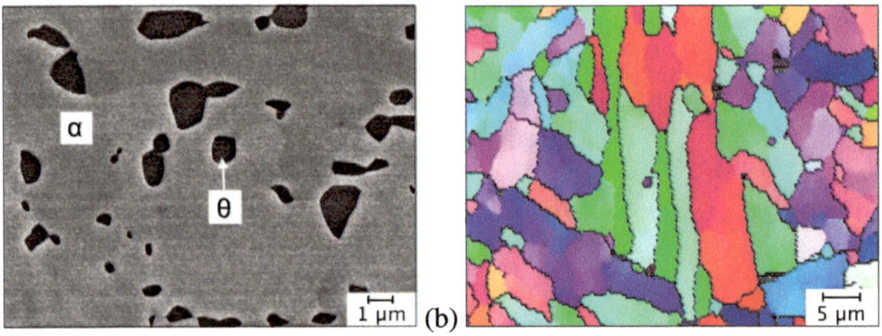

(a) (b)

Figure 9.12 Iron-0.6C wt% steel, quenched to martensite and then tempered at 650°C for 1.5 h. (a) Coarse, spheroidised cementite particles. (b) Electron backscatter diffraction orientation image showing the coarsened state of the original martensite plates, which would typically have been only 0.24 μm thick. After Miyamoto et al. [43], with permission of Elsevier.

9.3.4 Tempering: stage 4

It is useful to define a fourth stage of tempering in which the cementite particles undergo a coarsening process and essentially lose their faceted, crystallographic morphology, becoming spheroidised. The coarsening commences between 300°C and 400°C, while spheroidisation takes place increasingly up to 700°C. At the higher end of this range of temperature, the martensite lath boundaries are replaced by equiaxed ferrite grain boundaries in a process, which is best described as recrystallisation. The final result is an equiaxed array of ferrite grains with coarse spheroidised particles of Fe_3C (Fig. 9.12), partly, but not exclusively, at the grain boundaries.

The spheroidisation of the Fe_3C is encouraged by the resulting decrease in total interfacial energy per unit volume. The particles which preferentially grow and spheroidise are located mainly at interlath boundaries and prior austenite boundaries, although some particles remain in the matrix. The boundary sites are preferred because of the greater ease of diffusion in these regions.

The original martensite lath boundaries remain stable up to about 600°C, but in the range 350–600°C there is considerable rearrangement of the dislocations within the laths and at those lath boundaries which are essentially small-misorientation boundaries. This leads to a marked reduction in the dislocation density and to lath-shaped ferritic grains closely related to the packets of similarly oriented laths in the original martensite. This process, which is essentially one of recovery, is replaced between 600°C and 700°C by recrystallisation, which results in the formation of equiaxed ferrite grains with spheroidal Fe_3C particles in the boundaries and within the grains. This process occurs most readily in low-carbon steels. At higher carbon contents, the increased density of Fe_3C particles is much more effective in pinning the ferrite boundaries, so recrystallisation is much more sluggish. The final process is the continued coarsening of the Fe_3C particles and gradual ferrite grain growth (Fig. 9.13).

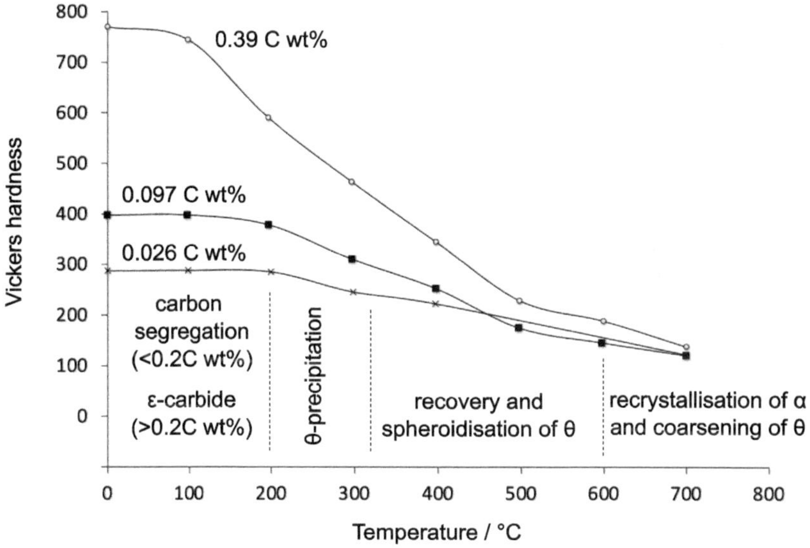

Figure 9.13 Hardness of iron–carbon martensites tempered 1 h at 100-700°C. Adapted from Speich [44].

9.4 Role of carbon content

Carbon has a profound effect on the behaviour of steels during tempering. Firstly, the hardness of the as-quenched martensite is influenced mostly by the carbon content (Fig. 5.28), as is the morphology of the martensite which have an approximately $\{111\}_\gamma$ habit plane up to 0.3 wt% C, changing to $\{225\}_\gamma$ at higher carbon contents. The M_S temperature is reduced as the carbon concentration increases, so the probability of auto-tempering is reduced. During fast quenching in alloys with less than 0.2 wt% C, the majority of the carbon segregates to dislocations and α'-boundaries, but with slower quenching, some precipitation of cementite occurs over the range M_S-M_F. On subsequent tempering of low-carbon steels up to 200°C further segregation of carbon takes place, but no precipitation has been observed. Under normal circumstances, it is difficult to detect any tetragonality in the martensite in steels with less than 0.2 wt% C, perhaps due to segregation and because M_S exceeds the Zener ordering temperature, in which case, the carbon atoms forced into an ordered state by the Bain deformation would tend to disorder amongst the three sets of octahedral interstices. The lattice then has cubic symmetry. The ordering of carbon atoms is determined by a competition between the enthalpy change that favours and the entropy change which opposes ordering. The order–disorder transition occurs at the Zener ordering temperature T_{Zener} [45–47]. The critical ordering temperature is carbon concentration dependent [2, p. 71]:

$$T_{\text{Zener}}\,/\,\text{K} \approx 28080 \times \frac{x_{\text{C}}}{1-x_{\text{C}}}, \tag{9.3}$$

where x_{C} is the mole fraction of carbon.

The hardness changes during tempering are also dependent on carbon content, as shown in Fig. 9.13 for steels up to 0.4 wt% C. Above this concentration, a small increase in hardness has been observed in the temperature range 50-150°C, as ε-carbide precipitation strengthens the martensite. However, the general trend is an overall softening, as the tempering temperature is raised. The diagram indicates the main physical processes contributing to the change in mechanical properties.

A useful empirical representation of the hardness of tempered martensite has been derived using a parameter that accounts for tempering time and temperature, and the chemical composition of the steel [48]. The parameter is defined as

$$\frac{p}{T} = \log\{t\} + 17.4 - 6.7w_{\text{C}} - 1.6w_{\text{Mn}} - 3.4w_{\text{Si}} - 0.25w_{\text{Ni}} - 1.1w_{\text{Cr}} - 4.4w_{\text{Mo}}, \tag{9.4a}$$

where the time has units of seconds, the temperature is in °C. The Vickers hardness (HV) given by

$$\text{HV} = \left(1543 - \frac{25.3}{w_{\text{C}}}\right) \times \exp\{-1.23 \times 10^{-4} p\}, \tag{9.4b}$$

applicable over the ranges listed below.

	w_{C}	w_{Mn}	w_{Si}	w_{Ni}	w_{Cr}	w_{Mo}	T / °C	t / s	HV
Min.	0.12	0.00	0.00	0.00	0.00	0.00	145	30	105
Max.	1.15	1.75	0.90	3.50	2.00	0.50	716	3.5×10^6	899

9.5 Mechanical properties of tempered martensite

The intrinsic mechanical properties of tempered plain carbon martensitic steels are difficult to measure for several reasons. Firstly, the absence of other alloying elements means that the hardenability of the steels is low, so a fully martensitic structure is possible only in thin sections. However, this may not be a disadvantage, where shallow hardened surface layers are all that is required. Secondly, at lower carbon levels, the M_{S} temperature is rather high, so auto-tempering is likely to take place. Thirdly, at the higher carbon levels, the presence of retained austenite will influence the results. Added to these factors, plain carbon steels can exhibit quench cracking, which makes

Table 9.3 Mechanical properties of plain carbon steels, both as-quenched and tempered (after Irvine et al. [49]).

wt% C	Property	Tempered			
		As-quenched	Tempered 7 h at		
			100°C	200°C	300°C
0.2	σ_Y / MPa	1270	1460	1235	1110
0.3		1360	1370	1270	1140
0.4			1670	1410	
0.2	σ_{UTS} / MPa	1470	1690	1450	1340
0.3		1580	1605	1460	1240
0.5				2040	1600
0.2	% elongation	5	6	6	9
0.3		5	7	7	10
0.5				4	7
0.2	HV	446	444	446	357
0.3		564	517	502	420
0.5		680	666	571	470

it difficult to obtain reliable mechanical-test results. This is particularly the case at higher carbon levels, i.e., above 0.5 wt% carbon.

Provided care is taken, good mechanical properties, in particular proof and tensile stresses, can be obtained on tempering in the range 100–300°C. However, the elongation is frequently small and the impact values poor. Table 9.3 shows some typical results for plain carbon steels with between 0.2 and 0.5 wt% C, following tempering at low temperatures.

Plain carbon steels with less than 0.25 wt% are not normally quenched and tempered, but in the range 0.25–0.55 wt% C heat treatment is often used to upgrade mechanical properties. The usual tempering temperature is between 300°C and 600°C allowing the development of tensile strengths between 1700 and 800 MPa, the toughness increasing as the tensile strength decreases. This group of steels is versatile as they can be used for crankshafts and general machine parts as well as hand tools, such as screwdrivers and pliers. But it is emphasised that almost all commercial steels contain solutes other than carbon.

The high-carbon steels (0.5-1.0 wt%) are much more difficult to fabricate and are therefore particularly used in applications where hardness and wear resistance are required, e.g., axes, knives, hammers, cutting tools. Typical mechanical properties as a function of tempering temperature are shown in Fig. 9.14 for a steel at the lower level (0.5 wt% C) of this range. Another important application is for springs, where often the required mechanical properties are obtained simply by heavy cold work, i.e. hard, drawn spring-wire. However, steels in the range 0.5-0.75 wt% C are quenched and then tempered to the required yield stress.

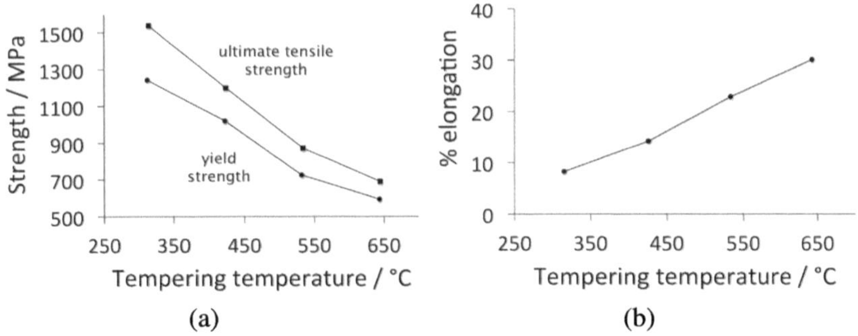

Figure 9.14 Fe-0.52C-0.93Mn wt% steel, quenched into a martensitic state and then tempered at the temperatures indicated, in each case for 1 h. (a) Strength. (b) Ductility. Selected data from the Metals Handbook [50].

9.6 Tough, untempered, high-carbon martensite

The idea that strong high-carbon martensite is brittle when new is embedded firmly in metallurgical convention. The brittleness is apparent in Fig. 8.10b, which illustrates the periodic cracking of a plate of martensite in a quenched steel containing 1C wt% [51]. The cracks are not caused by the application of an external stress, but form spontaneously under the influence of thermal or transformation stresses, as the microstructure evolves.

The tendency for such cracking diminishes as the austenite grain size is reduced (p. 224). This is because it becomes difficult to transfer stress on to an individual martensite plate as its dimensions decrease. Fig. 9.15 shows that in a composite containing short fibres, the stress transferred onto the fibre when the composite is loaded, increases from zero at the end to a maximum at the centre. Therefore, if the cracks on a martensite plate are separated by a distance Δz then making the plates shorter in length than Δz would avoid cracking because of the difficulty of transferring load onto an individual platelet.

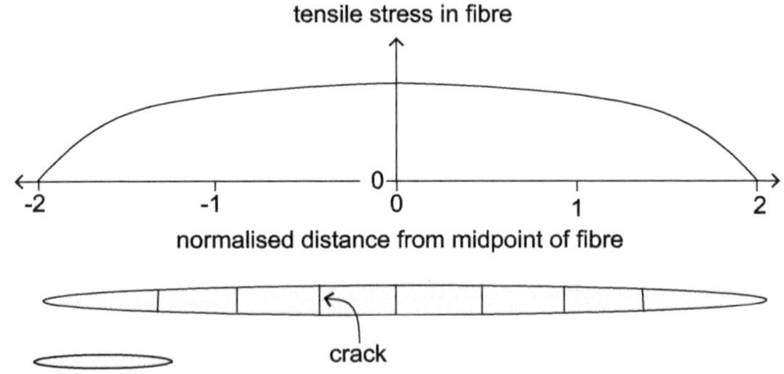

Figure 9.15 Graphs illustrate schematically, the tensile stress transferred onto a short fibre in a composite, assuming some bonding at the fibre/matrix interface. The stress is zero at the ends of the fibre. The long martensite plate is periodically cracked, the smaller one is not.

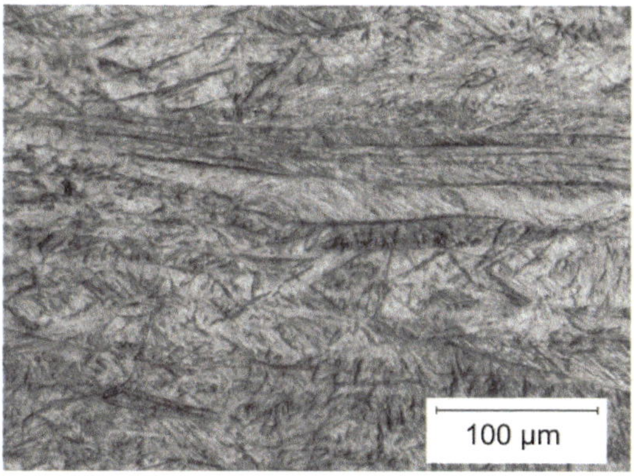

Figure 9.16 Severely pancaked prior austenite grains, also containing intense deformation bands. [52].

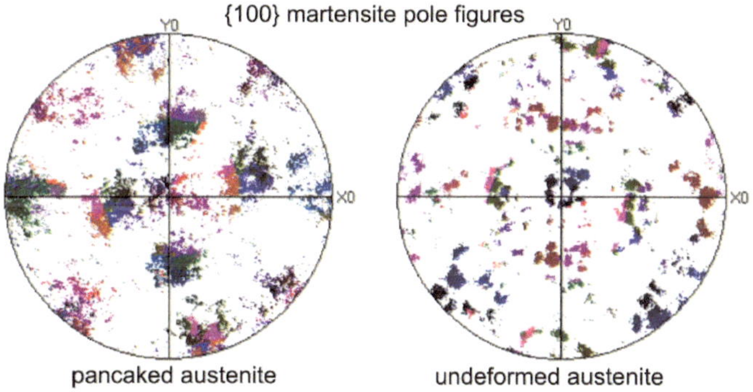

Figure 9.17 {100} martensite pole figures, each from a single austenite grain. On the left, the pole figure representing the martensite that forms in a severely deformed γ-grain. On the right, the corresponding figure where the martensite is from an undeformed parent. [52].

The longest length of a martensite plate that can form is given by the austenite grain size. Therefore, the austenite grain size must be refined to avoid the cracking of new martensite. This can be achieved by severely deforming the austenite prior to its transformation [52]. Fig. 9.16 shows severely elongated prior austenite grains that not only are pancaked, but contain numerous deformation bands, and inevitably, a large defect density. The chemical composition of the steel is Fe-0.34C-1.45Si-1.5Cr-3.53Ni-0.39Mo-0.3V wt% [52]. An *individual* deformed-grain therefore contains large crystallographic orientation gradients, with discontinuities at the bands or at sub-boundaries formed by the rearrangement of dislocations. Therefore, martensite can form in a huge number of orientations in a single austenite grain, unlike the limited number possible in an undeformed parent structure (Fig. 9.17).

Fig. 9.18 shows that there are huge spreads in the crystallographic orientation of martensite plates within a single, deformed prior austenite grain. In contrast, the expected 24 crystallographic variants of martensite are obtained from an undeformed austenite grain. Martensite plates cannot cross large crystallographic discontinuities and therefore are greatly refined in size. The deformed austenite is also mechanically stabilised (p. 165), resulting in a further degree of refinement.

All this means that the untempered martensite has exceptional properties. Fig. 9.18 shows that the steel has a reproducible toughness, $72 \pm 1.5\,\mathrm{MPa\,m^{1/2}}$, much more so than nanostructured bainite at the same ultimate tensile strength. The Charpy properties at $-40°\mathrm{C}$ are impressive and the elongation tolerable. When transformed from undeformed austenite, the Charpy toughness is less than 5 J at $-40°\mathrm{C}$. The steel can also be welded without cracking in the heat-affected zone. It is possible therefore, to use martensitic steels containing substantial quantities of carbon, without tempering as long as the austenite grains prior to transformation are broken into crystallographic domains no bigger than Δz, in this case approximately 0.4 μm.

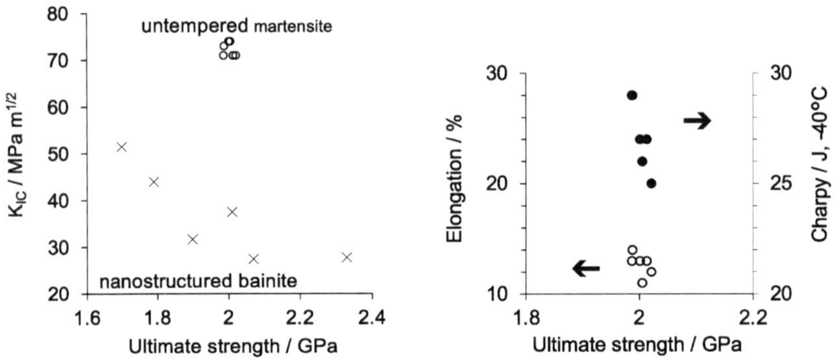

(a) (b)

Figure 9.18 (a) The fracture toughness of untempered, high-carbon martensite compared with the strongest form of bainite (Chapter 14). (b) Elongation and Charpy ($-40°\mathrm{C}$) of untempered, high-carbon martensite.

9.7 Steels with strong carbide-forming elements

The addition of alloying elements to a steel has a substantial effect on the kinetics of the $\gamma \rightarrow \alpha$ transformation and also of the pearlite reaction. Most common solutes move the time-temperature-transformation curves to longer times, with the result that it is much easier to avoid the nose of the C-curve during quenching. This essentially gives greater hardenability, since martensite structures can be achieved at slower cooling rates and, in practical terms, thicker specimens can be made fully martensitic. Alloying elements have also been shown to have a substantial effect in depressing the M_S temperature (Eq. (5.19)). In this section, we will examine the further important effects of solutes during the tempering of martensite, where not only the kinetics of the basic reactions are influenced but also the products of these reactions can be substantially changed, e.g., cementite can be replaced by other carbide phases. Several of the simpler groups of alloy steels are used to provide examples of the general behaviour.

9.7.1 Effect of alloying elements on the formation of iron carbides

The structural changes during the early stage of tempering are difficult to follow. However, it is clear that certain elements, notably silicon, can stabilise the ε-iron carbide to such an extent that it still is present in the microstructure after tempering at 400°C in steels with 1-2 wt% Si, and at even higher temperatures if the silicon concentration is further increased. The evidence suggests that both the nucleation and growth of the carbide are slowed down and that silicon enters into the ε-carbide structure. Thus, the transformation of ε-iron carbide into cementite is delayed considerably.

Fig. 9.19a shows calculated kinetics of cementite precipitation from supersaturated ferrite, assuming that the Fe/X ratio remains constant, as a function of the silicon and carbon concentrations [5]. This condition is a reflection of the displacive mechanism by which cementite precipitates at low temperatures. It is clear that because of the large driving force $\Delta G^{\alpha' \rightarrow \alpha+\theta}$ for the precipitation of carbon from martensite, silicon has a minor effect on the kinetics of cementite formation even when it is forced into the cementite by para-equilibrium transformation, except at low carbon concentrations when $\Delta G^{\alpha' \rightarrow \alpha+\theta}$ too is small in magnitude.

This is reflected in the effect of Si on the precipitation of cementite from austenite, where the magnitude of the driving force $\Delta G^{\gamma' \rightarrow \gamma+\theta}$ is much smaller than the corresponding term for precipitation from supersaturated ferrite. At small silicon concentrations, it still is possible to precipitate cementite from high-carbon austenite under para-equilibrium conditions, but this becomes impossible if the silicon concentration is raised, Fig. 9.19b – only equilibrium precipitation is then possible, albeit at very slow rates. These observations for the precipitation from both supersaturated ferrite and austenite are entirely consistent with experimental measurements [5,53]. And it is noteworthy that at high temperatures, where there is sufficient atomic mobility, the cementite need not grow with the trapping of silicon, such as during the formation of pearlite, whence the rate of transformation can be large – it is only the lack of mobility at low temperatures, which forces the cementite to absorb silicon. The formal definition of *trapping* is that the chemical potential of the species increases on transfer across the interface.

While the tetragonality of martensite disappears by 300°C in plain carbon steels, in steels containing some alloying elements, e.g., Cr, Mo, W, V, Ti, Si, the tetragonal lattice is still observed after tempering at 450°C and even as high as 500°C, Fig. 9.20.

Alloying elements influence the proportion of austenite retained on quenching. Typically, a steel with 4Mo wt%, 0.2C wt%, in the martensitic state contains less than 2 vol.% austenite; about 5 vol.% is detected in a steel with 1V wt% and 0.2C wt%. The amount of austenite retained can be calculated by first estimating the M_S temperature using Eq. (5.19), followed by substitution into the Koistinen and Marburger Equation (5.2).

In lean steels, the austenite can be revealed as a fine network around the martensite laths, by using dark-field electron microscopy (Fig. 9.21). On tempering at 300°C, the austenite decomposes to give thin grain boundary films of cementite, which, in the case of the higher quantities of retained austenite, can be fairly continuous along the

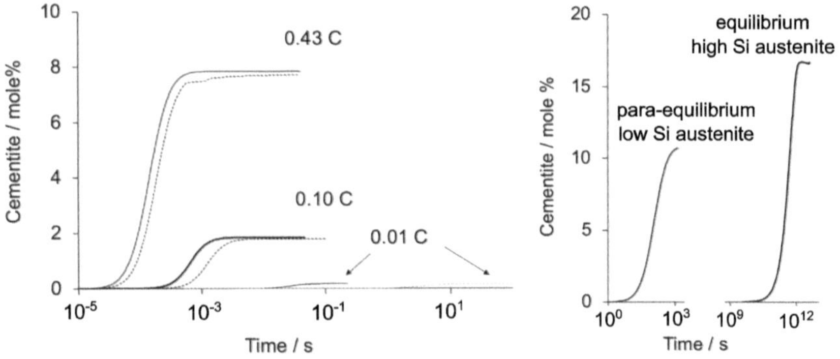

(a) (b)

Figure 9.19 (a) Kinetics of para-equilibrium cementite precipitation from supersaturated ferrite in Fe-0.4C-0.7Mn-0.28Si-0.8Cr-1.8Ni-0.25Mo wt% (continuous curves) and Fe-0.4C-0.7Mn-1.6Si-0.8Cr-1.8Ni-0.25Mo wt% (dotted curves) steels as a function of carbon concentration. (b) Para-equilibrium cementite precipitation from austenite containing Fe-0.8C-0.38Si-1.3Mn wt% held at 370°C, and equilibrium cementite precipitation from austenite containing Fe-1C-1.41Si-1.42Mn wt% held at 360°C. After Kozeschnik and Bhadeshia [5].

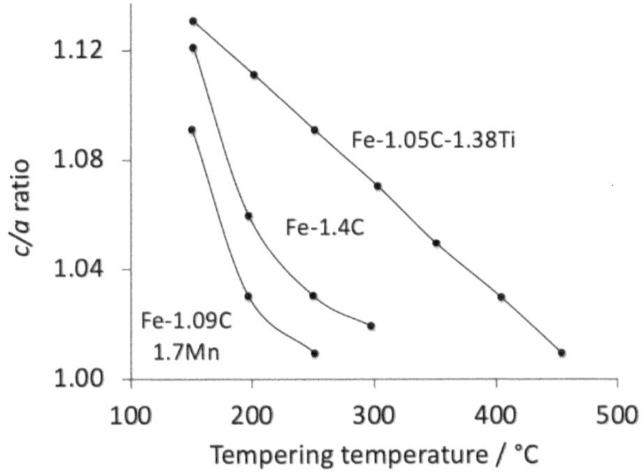

Figure 9.20 Effect of Ti and Mn on the tetragonality of martensite during tempering. Concentrations in wt%. Data from Kurdjumov [54].

lath boundaries. It is likely that this interlath cementite (Fig. 9.9b) is responsible for *tempered martensite embrittlement*, frequently encountered as a toughness minimum in the range 300–350°C, by leading to easy nucleation of cracks, which then propagate across the tempered martensite laths.

Alloying elements can also restrain the coarsening of cementite in the range 400–700°C, a basic process during the fourth stage of tempering. Several alloying elements, notably silicon, chromium, molybdenum and tungsten, cause the cementite to retain its fine Widmanstätten structure to higher temperatures, either by entering into the cementite structure or by segregating at the carbide-ferrite interfaces. Whatever

(a) 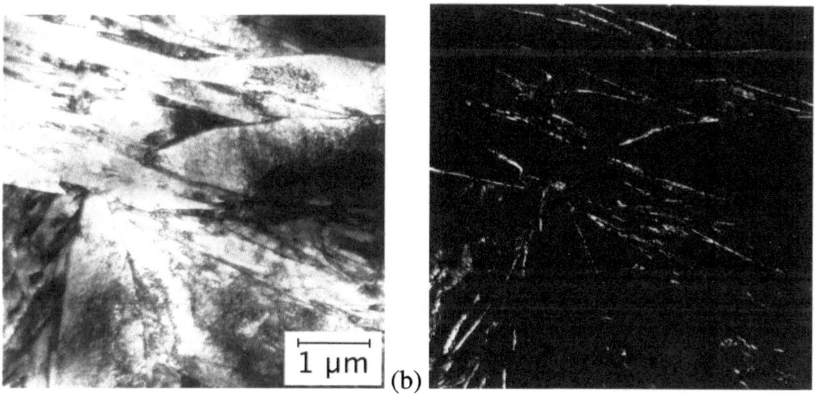

(b)

Figure 9.21 Fe-4Mo-0.2C wt% steel austenitised and quenched to produce martensite. (a) Bright-field image. (b) Corresponding dark-field image showing films of austenite retained between the martensite platelets [32].

the basic cause may be, the effect is to delay significantly the softening process during tempering. This influence on the cementite dispersion has other effects, in so far as the carbide particles, by remaining fine, slow down the reorganisation of the dislocations inherited from the martensite, with the result that the dislocation substructures refine more slowly. The cementite particles are found also on ferrite grain boundaries, where they control the rate at which the α-grains grow.

Recrystallisation and grain growth involve the movement of grain boundaries. The motion is inhibited by second phase particles. The drag on the boundary due to an array of insoluble, incoherent spherical particles is because the grain boundary area decreases when a boundary intersects the particle. Therefore, to move away from the particle requires the creation of new surface. The net drag force on a boundary of energy $\sigma_{\alpha\alpha}$ per unit area due to a particle of radius r is given by (Fig. 9.22)

$$F = \sigma_{\alpha\alpha} \sin\{\theta\} \times 2\pi r \cos\{\theta\}, \tag{9.5}$$

where $\sigma_{\alpha\alpha} \sin\{\theta\}$ is the force per unit length in the direction of boundary motion, and the remaining term on the right represents the length of the intersection of the boundary plane with the particle. By differentiating this, the maximum force occurs at $\theta = 45^\circ$, with $F_{max} = \sigma_{\alpha\alpha} \pi r$.

Suppose now that there is a random array of particles, volume fraction V_V with N_V particles per unit volume then

$$N_V = \frac{V_V}{\frac{4}{3}\pi r^3}. \tag{9.6}$$

Only those particles within a distance $\pm r$ can intersect a plane. The number of particles intersected by a plane of area $1\,m^2$ is

$$n = 2r \times N_V = \frac{3V_V}{2\pi r^2}. \tag{9.7}$$

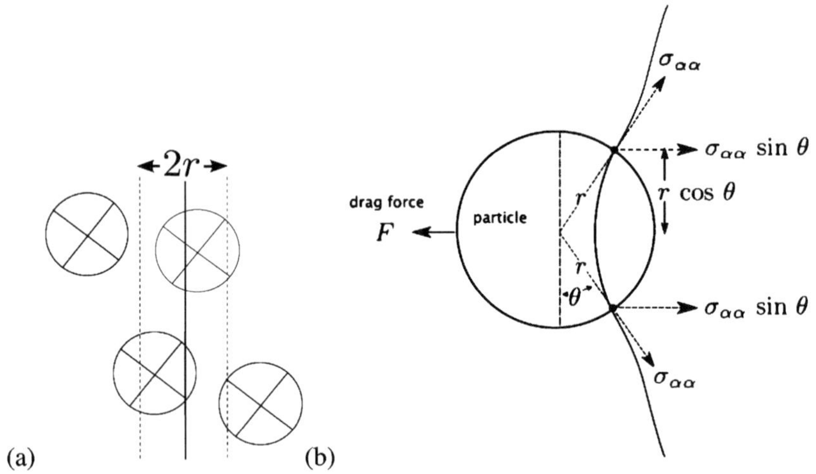

Figure 9.22 (a) Particles whose centres are located a distance r from the boundary intersect it. (b) The pinning geometry when a boundary intersects a particle.

The drag pressure P is then often expressed as

$$P = F_{\text{max}} n = \frac{3\sigma_{\alpha\alpha} V_{\text{V}}}{2r}. \tag{9.8}$$

This may be a significant pressure if the particles are small. Anisotropic particles may have a larger effect if they present a greater surface area for interaction with the boundary.

The pinning force F_{max} opposes the movement of a boundary, whereas the need to minimise the total interfacial energy per unit volume drives the motion. A grain size given by a mean lineal intercept $\overline{L}$ has an amount $S_{\text{V}} = 2/\overline{L}$ of surface per unit volume. It follows that the energy locked as grain boundaries per unit volume is $\sigma_{\alpha\alpha} S_{\text{V}}$ so that the force driving the grain boundary motion is $2\sigma_{\alpha\alpha}/\overline{L}$. Grain growth stops when a limiting size is reached, where the particle pinning force equals this driving force,

$$\frac{3\sigma_{\alpha\alpha} V_{\text{V}}}{2r} = \frac{2\sigma_{\alpha\alpha}}{\overline{L}} \qquad \text{so that} \qquad \overline{L}_{\text{limiting}} = \frac{4r}{3V_{\text{V}}}. \tag{9.9}$$

The theory presented here roughly follows Zener, more sophisticated models are reviewed in [55].

In plain carbon steels, cementite particles begin to coarsen in the temperature range 350–400°C; the addition of chromium, silicon, molybdenum or tungsten delays the coarsening to the range 500–550°C. At first, the only carbides to form are those of iron which may incorporate or reject varying amounts of alloying elements into solid solution during tempering. This is because they do not have the equilibrium composition when they first precipitate. This has inspired an interesting application, where the chemical composition of the cementite is used to estimate the remaining life of power plant steels [1]. These steels have to serve reliably over periods of 30–40 years under conditions where the temperatures are typically in the range 450–600°C so that creep and other damage mechanisms must be assessed carefully in the context of continued service. The problem is that the mechanisms are sensitive to the service temperature, which can vary for a variety of reasons. Therefore, one method to determine an effective service temperature is to non-destructively measure the chemical composition of the cementite, using for example, an equation that relates the enrichment of solute in cementite as a function of diffusivity and other parameters – the time t taken to reach a particular concentration c_θ of a solute such as Cr or Mn is given by [56]

$$t = \frac{\pi[z_\theta(c^\theta - \overline{c})]^2}{16D_\alpha(\overline{c} - c^{\alpha\theta})^2}, \tag{9.10}$$

where $c^{\alpha\theta}$ is the concentration in the ferrite in equilibrium with cementite, $\overline{c}$ is the average concentration, D_α is the diffusivity of the solute concerned in ferrite and z_θ is the thickness of the cementite particle, assumed to be in the form of a platelet. The concentration in the cementite will therefore vary with $\sqrt{t}$ until soft-impingement occurs with the ferrite after long periods of time. Fig. 9.23 illustrates some data from a power plant life-assessment exercise covering about 10,000 h [57], with the calculated curve accounting for any soft-impingement effects.

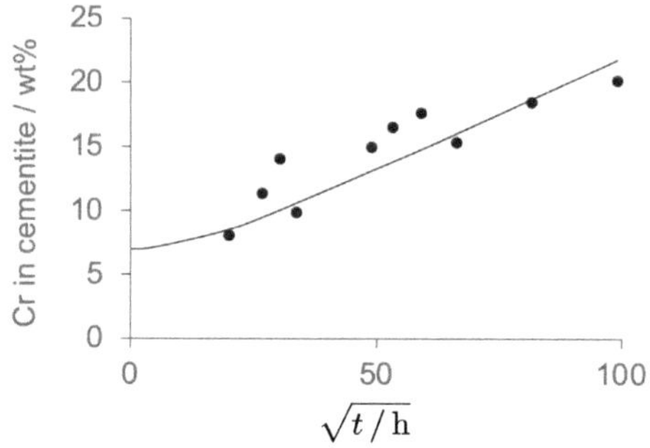

Figure 9.23 Comparison between measured data for the chromium concentration cementite during heat-treatment at 550°C [57], a typical steam temperature encountered in the older power plant, against a line calculated as described in the text [56].

9.7.2 The formation of alloy carbides: secondary hardening

A number of the familiar alloying elements in steels form carbides, which are thermodynamically more stable than cementite. It is interesting to note that this also is true of a number of nitrides and borides. Nitrogen and boron are increasingly used in steels in small but significant concentrations. The thermodynamic stabilities of some of these compounds are shown in Fig. 4.8, in which iron carbide is the least stable compound, situated at the right of the diagram. The alloying elements Cr, Mo, V, W and Ti are all strong carbide-formers, while nickel, cobalt and copper do not form carbide phases in steels, although there are exotic circumstances, unconnected with steels, in which Cu, Co and Ni carbides have been synthesised. Manganese carbide Mn_3C (space group *Pnma*) has the same crystal structure and similar lattice parameters as cementite. Manganese therefore dissolves readily in cementite to form $(Fe,Mn)_3C$, rather than precipitate independently.

When strong carbide-forming elements are present in a steel in sufficient concentration, their carbides would be formed in preference to cementite. Nevertheless, during the tempering of all alloy steels, alloy carbides are unable form until the temperature exceeds about 500°C, because otherwise, the substitutional solutes cannot diffuse sufficiently rapidly. These carbides have distinct chemical compositions that are quite different from that of the ferrite matrix, so long-range diffusion is essential for them to grow, unlike the iron carbides that grow at much lower temperature by a displacive mechanism in which only carbon has to diffuse. The metallic elements diffuse substitutionally, in contrast to carbon and nitrogen which move through the iron lattice interstitially, with the result that the diffusivities of carbon and nitrogen are several orders of magnitude greater in iron than those of the metallic alloying elements (Table 1.3). Consequently, higher temperatures are essential to precipitate the so-called alloy carbides.

The coarsening rate is dependent on the diffusion coefficient of the solute (Eq. (9.2)) and, under the same conditions, at a given temperature, cementite would coarsen at a greater rate than any of the alloy carbides once formed. This occurs in alloy steels in which cementite and an alloy carbide coexist, where the cementite dispersion is always much coarser. It is this ability of certain alloying elements to form and retain a fine alloy carbide dispersion in spite of high temperatures, allows the development of strong, creep-resistant steels. Indeed, the formation of alloy carbides is accompanied by a marked increase in strength, whereas the plain carbon steel shows a continuous decrease in hardness at the tempering temperature is increased (Fig. 9.24). This phenomenon, which is referred to as *secondary hardening* because it leads to a hardness increase that follows the initial decrease due the rapid precipitation of iron carbides. The effect is characteristic of steels containing molybdenum, vanadium, tungsten, titanium and also in chromium steels.

To summarise, during secondary-hardening the cementite dispersion is replaced by a new and much finer alloy carbide dispersion. In Fig. 9.24, the effect of increasing molybdenum content is thus effectively demonstrated in a series of steels containing 0.1 wt% carbon. Significantly, non-carbide-forming elements such as nickel, cobalt and silicon do not give secondary hardening. However, some elements, e.g., silicon, by delaying the coarsening of cementite, lead to a plateau on the tempering curve in the range 300–500°C.

9.7.3 Nucleation and growth of alloy carbides

The dispersions of alloy carbides, which occur during tempering can be complex, but some general principles can be discerned which apply to a wide variety of steels. The alloy carbides can form in at least three ways:

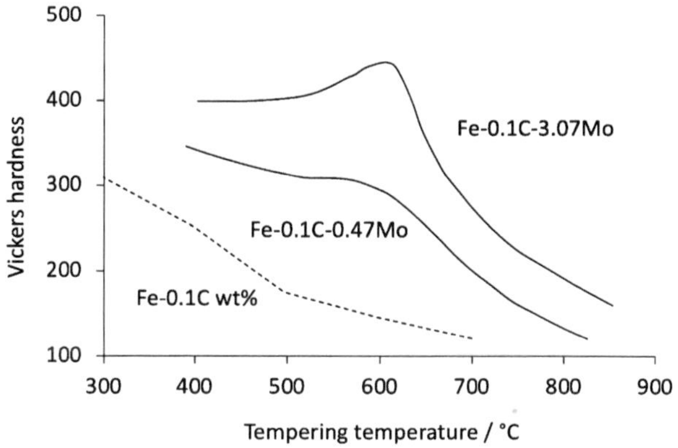

Figure 9.24 The effect of molybdenum on the tempering for 1 h, of quenched 0.1 wt% C steels. Selected data from Irvine and Pickering [58], with the Fe-0.097 wt% data from Speich [44]. The plain carbon steel shows a monotonic decline in hardness with temperature, whereas the Mo-containing steels both show a secondary-hardening peak.

a) *In-situ* nucleation at pre-existing cementite particles – it has been shown that the nuclei form on the interfaces between the cementite particles and the ferrite. As they grow, carbon is provided by the less stable adjacent cementite, which gradually disappears. If carbon is present in quantities greater than can combine with the strong carbide-former, then residual cementite persists in the microstructure.

b) By separate nucleation within the ferrite matrix – usually at dislocations inherited from the martensitic structure.

c) At grain boundaries and sub-boundaries – these include the former austenite boundaries, the original martensitic lath boundaries (now ferrite) and the new ferrite boundaries formed by coalescence of sub-boundaries or by recrystallisation.

In-situ nucleation at pre-existing cementite particles is a common occurrence but because these particles are fairly widely spaced at temperatures above 500°C, the contribution of this type of alloy carbide nucleation to strength is very limited. Fig. 9.25a shows, in a 4 wt% molybdenum steel tempered 4.5 h at 550°C, the relatively coarse Widmanstätten precipitation of Fe_3C, which at this stage has largely transformed to fine Mo_2C particles. These are readily identified by dark field microscopy. On further tempering, the positions of the original cementite particles are indicated by small necklaces of alloy carbides, which tend to be coarser than the matrix precipitation.

The scenario for Fe-Mo-C is somewhat different in high-carbon steel (0.8 wt%), where excess cementite persists even after the early precipitation of Mo_2C at the expense of some of the Fe_3C. At the later stages of heat-treatment, both Fe_2MoC and $(Fe, Mo)_{23}C_6$ form and replace some of the cementite [59]. The Fe_2MoC in particular grows into the cementite; it too has an orthorhombic crystal structure but much larger lattice parameters (≈1.6, 1, 1.1 nm) than cementite.

Fig. 9.25a also illustrates the dislocation network characteristic of tempered steels as inherited from the martensite, although there has been considerable rearrangement

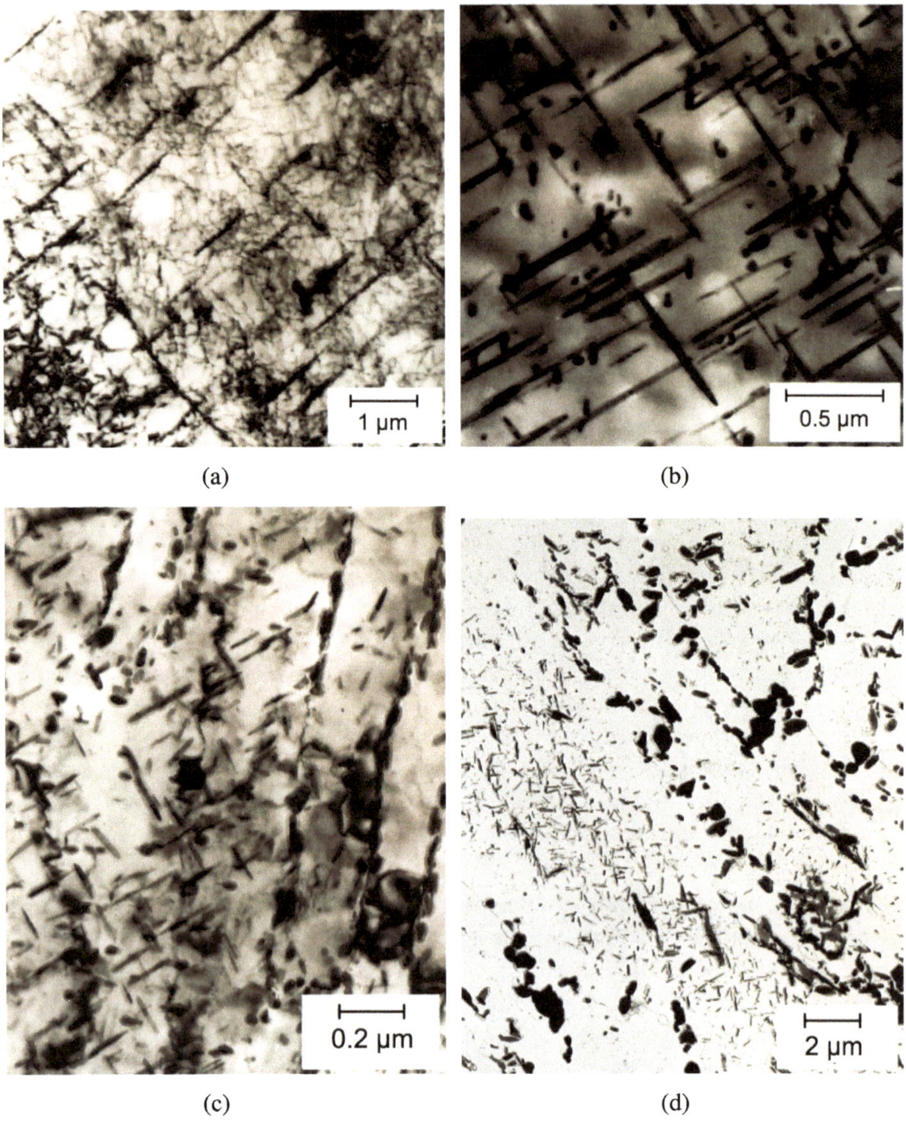

Figure 9.25 Fe-4Mo-0.2C wt% steel, quenched into a martensitic microstructure and then tempered. (a) Tempered at 563°C for $4\frac{1}{2}$ h, showing cementite transforming *in situ* into Mo_2C (courtesy of Raynor – more details in [60].). The structure remains heavily dislocated. (b) Tempered at 675°C for 15 h, showing three variants of Mo_2C needles growing along $\langle 001 \rangle_\alpha$ directions. One of these variants appears in cross-section (courtesy of Raynor). (c) Tempered at 700°C for 30 min showing fine Mo_2C within the laths and M_6C at lath boundaries (courtesy of Irani). (d) Extraction replica image of the sample in (c), showing with greater clarity, the two kinds of precipitates (courtesy of Irani).

and reduction in dislocation density. Dark-field electron microscopy reveals that these dislocations are the sites for very fine precipitation of the appropriate alloy carbide. On further ageing the particles are more readily resolved, e.g., in Mo steels as a Widmanstätten array, comprising Mo_2C rod-lets lying along $\langle 001 \rangle_\alpha$ directions. Fig. 9.25b illustrates this stage in a single martensitic lath. More intense precipitation is evident at the lath boundaries.

The nucleation of carbides at the various types of boundary is to be expected, because these are energetically favourable sites, which also provide paths for relatively rapid diffusion of solute. Consequently the ageing process is usually more advanced in these regions and the precipitate is coarser (Fig. 9.25c). In many alloy steels, the first alloy carbide to form is not the final equilibrium carbide and in some steels as many as three alloy carbides can form successively. Under these circumstances, the equilibrium alloy carbide frequently nucleates first in the grain boundaries, grows rapidly and eventually completely replaces the Widmanstätten non-equilibrium carbide within the grains. This is illustrated in Fig. 9.25c for a 4 wt% Mo steel tempered 30 min at 700°C, in which M_6C equiaxed particles are growing at the grain boundaries, but Widmanstätten Mo_2C is still visible within the grains. It is interesting to note that the structure still possesses the lath-shaped ferrite grains inherited from the martensite. Recrystallisation occurs after longer tempering times at 700°C.

9.7.4 Tempering of steels containing vanadium

Vanadium is a strong carbide former. In steel containing as little as 0.1V wt%, the face-centred cubic vanadium carbide V_4C_3 ($Fm\bar{3}m$) is able to precipitate. Normally, this is the only vanadium carbide formed in steels, so the structural changes observed during the tempering of vanadium steels are relatively simple.

Vanadium carbide forms as small platelets, initially less than 5 nm across and not more than 1 nm thick. These form within the ferrite grains on dislocations (Fig. 9.26a) in the range 550-650°C, and produce a marked secondary-hardening peak. There is a well-defined orientation relationship (Baker/Nutting) with the ferrite matrix: $(100)_\alpha \parallel (100)_{V_4C_3}$ and $[010]_\alpha \parallel [011]_{V_4C_3}$. In the early stages of precipitation at 550°C, the particles are coherent with the matrix, there being only a 3% misfit between $\langle 010 \rangle_\alpha$ and $\langle 110 \rangle_{V_4C_3}$. However, at 700°C, the platelets coarsen rapidly (Fig. 9.26b) and begin to spheroidise; the original martensite laths can still be recognised and are replaced by equiaxed ferrite grains only after long periods at 700°C.

Many steels containing vanadium, e.g., $\frac{1}{2}$Cr$\frac{1}{2}$Mo$\frac{1}{4}$V, 1Cr$\frac{1}{4}$V, 3Cr1Mo$\frac{1}{4}$V, 1Cr1Mo$\frac{3}{4}$V, exhibit extensive vanadium carbide precipitation on tempering, because of the stability of this carbide, not only with respect to cementite, but also the several chromium carbides and molybdenum carbide (Fig. 4.8). Because of its ability to maintain a fine carbide dispersion, even at temperatures approaching 700°C, vanadium is an important constituent of steels for elevated temperature service.

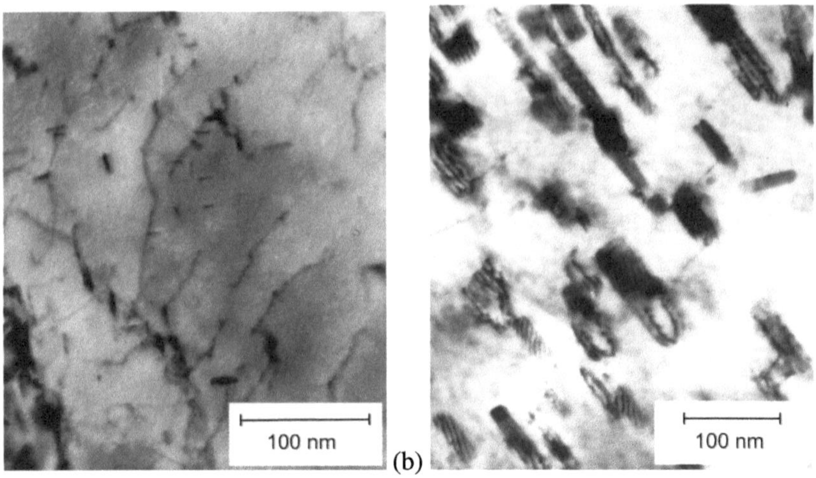

(a) (b)

Figure 9.26 Fe-1V-0.2C wt% quenched and tempered. Thin-foil electron micrographs: (a) 72 h at 550°C, V_4C_3 nucleation on dislocations (courtesy of Raynor); (b) 50 h at 700°C. Plates of V_4C_3 (courtesy of Irani).

9.7.5 Tempering of steels containing chromium

In chromium-containing steels, two chromium carbides are encountered: Cr_7C_3 (orthorhombic *Pmcn*) and $Cr_{23}C_6$ (cubic $Fm\overline{3}m$), although both of these are likely to contain other metallic elements, particularly iron, on the substitutional lattice. The normal carbide sequence during tempering is:

$$\text{Matrix} \rightarrow M_3C \rightarrow M_7C_3 \rightarrow M_{23}C_6,$$

where as usual 'M' represents a mixture of iron and chromium atoms, although the M_7C_3 and $M_{23}C_6$ are rich in Cr. An example for a creep-resistant steel is illustrated in Fig. 9.27. While this sequence occurs in higher-chromium steels, below about 7 wt% Cr, $Cr_{23}C_6$ is absent unless other metals such as molybdenum are present. Chromium is a weaker carbide former than vanadium, illustrated by the fact that Cr_7C_3 does not normally occur until the chromium content of the steel exceeds 1 wt% at a carbon level of about 0.2 wt%.

In steels up to 4 wt% Cr, the transformation from Fe_3C to Cr_7C_3 occurs mainly by nucleation at the Fe_3C/ferrite interfaces. Steels up to 9 wt% Cr do not show secondary-hardening peaks in tempering curves (Fig. 9.28). However, these curves do exhibit plateaux at the higher chromium contents, which are associated with the precipitation of Cr_7C_3. Chromium diffuses rapidly in ferrite and is usually present in relatively large concentrations, with the result that Cr_7C_3 is detected during tempering at temperatures as low as 500°C, and in comparison with vanadium carbide, chromium carbide coarsens rapidly. Thus, in a 2 wt% Cr-0.2C wt% steel, continuous softening will normally occur on tempering between 500°C and 700°C, although addition of other alloying elements, e.g., Mo, can reduce the rate of coarsening of Cr_7C_3.

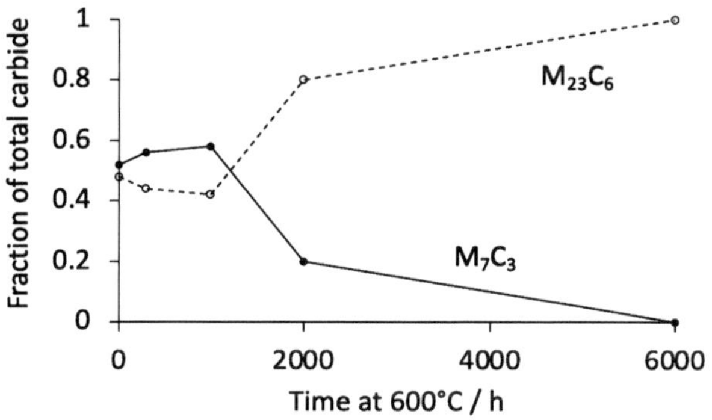

Figure 9.27 Chromium-rich M_7C_3 being replaced progressively by the more stable $M_{23}C_6$ during tempering of Fe-0.1C-4.5Cr-0.5Mo-0.5Mn wt% steel at 600°C. Adapted using data from Saucedo-Muñoz et al. [61].

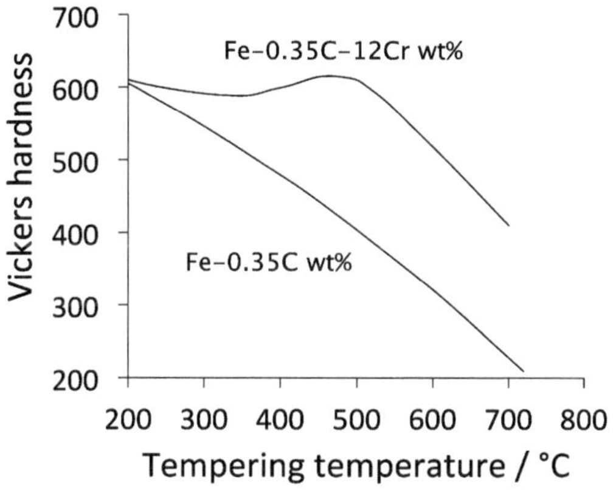

Figure 9.28 The effect of chromium on the tempering of a 0.35 wt% C steel. Selected data from Bain and Paxton [62].

In contrast, a 12 wt% Cr steel exhibits secondary hardening in the same temperature range (Fig. 9.28) due to precipitation of Cr_7C_3. Additionally, $Cr_{23}C_6$ nucleates at about the same time but at different sites, particularly former austenite grain boundaries and at ferrite lath boundaries. This precipitate grows at the expense of the Cr_7C_3, which eventually disappears from the microstructure, at which stage the steel has completely over-aged. This transition from Cr_7C_3 to $Cr_{23}C_6$ in high-chromium steels is by separate nucleation and growth. Further alloying additions can promote one or other of these carbide reactions, e.g., addition of tungsten encourages the formation of $Cr_{23}C_6$ by allowing it to nucleate faster, while vanadium tends to stabilise Cr_7C_3. In doing so, it decreases the rate of release into solution of chromium and carbon needed for the growth of $Cr_{23}C_6$. Clearly, vanadium would be a preferred addition than tungsten, if a fine, stable chromium carbide dispersion is needed in the temperature range 550–650°C.

9.7.6 Tempering of steels containing molybdenum and tungsten

When molybdenum or tungsten is the predominant alloying element in a steel, a number of different carbide phases are possible, but for compositions between 4 and 6 wt% of the element, the carbide sequence is likely to be:

$$Fe_3C \rightarrow M_2C \rightarrow M_6C.$$

The carbides responsible for the secondary hardening, in the case of both tungsten and molybdenum, are the isomorphous hexagonal carbides Mo_2C ($P6_3/mmc$ or $Pbcn$) and W_2C, both of which, in contrast to vanadium carbide, have a well-defined needle morphology, (Fig. 9.29a). This needle-like shape is a consequence of good lattice matching with the ferrite along the $\langle 100 \rangle_\alpha$ direction, but not along the other two axes, which are parallel to $\langle 011 \rangle_\alpha$ [2, p. 477]. When formed in the matrix, M_2C adopts a Widmanstätten distribution of needles lying along $\langle 100 \rangle_\alpha$ directions. On the other hand, Mo_2C grows in equiaxed forms or as faceted plates when unconstrained by the ferrite lattice.

In molybdenum steels, peak hardness occurs after about 25 h at 550°C, when the rods are about 10–20 nm long and 1–2 nm in diameter. The orientation relationship is:

$$(0001)_{M_2C} \parallel (011)_\alpha,$$
$$[11\bar{2}0]_{M_2C} \parallel [100]_\alpha \text{ (rod growth direction).}$$

M_2C also nucleates at former austenite, ferrite lath boundaries, on dislocations in the ferrite, and at the Fe_3C/ferrite interfaces, though the secondary hardening arises primarily from the dislocation-nucleated dispersion of M_2C.

On prolonged tempering at 700°C, the cubic M_6C ($Fd\bar{3}m$) forms predominantly at grain boundaries as massive particles which grow quickly, while the M_2C phase goes back into solution. The equilibrium microstructure is equiaxed ferrite with coarse M_6C in the form of faceted particles at grain boundaries, and plates, illustrated in Fig. 9.29b for a 6 wt% tungsten steel tempered 26 h at 700°C.

For similar atomic concentrations, the secondary hardening response in the case of tungsten steels is less than that of molybdenum steels. The M_2C dispersion in the former case is coarser, probably because the slower diffusivity of tungsten allows a coarsening of the dislocation network prior to being pinned by the genesis of M_2C particles.

At lower concentrations of tungsten and molybdenum (0.5-2 wt%), two other alloy carbides are interposed in the precipitation sequences, i.e., the complex cubic $M_{23}C_6$ and the orthorhombic M_aC_b, probably Fe_2MoC (p. 259). These carbides are found as intermediate precipitates between M_2C and M_6C.

9.7.7 Complex alloy steels

The presence of more than one carbide-forming element can complicate the precipitation processes during tempering. In general terms, the carbide phase, which is the most

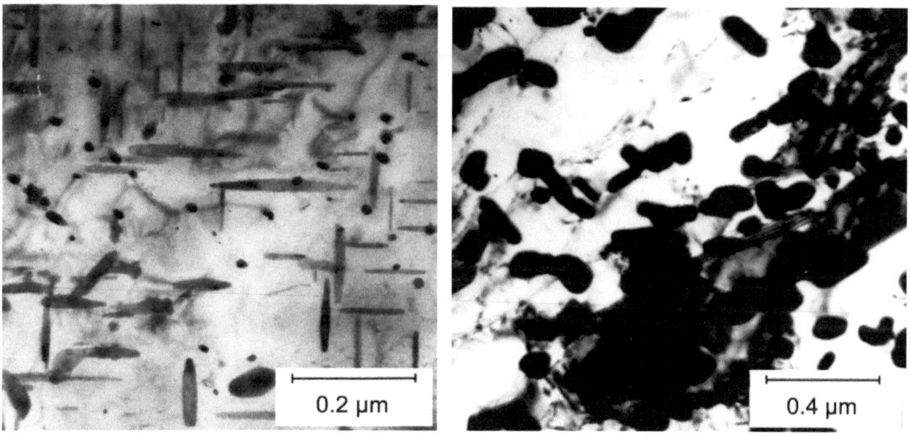

(a) (b)

Figure 9.29 Fe-6W-0.23C wt%, quenched into a martensitic microstructure and then tempered. (a) Tempered for 100 h at 600°C, showing needles of W_2C along the $\langle 100 \rangle_\alpha$ directions. (b) Tempered for 5 h at 700°C, showing coarse precipitates of M_6C. Courtesy of Davenport.

stable thermodynamically, predominate, but this assumes that equilibrium is reached during tempering. This is clearly not so at temperatures below 500–600°C. The use of pseudo-binary diagrams for groups of steels, e.g., Cr-V, Cr-Mo, can be a useful guide to carbide phases likely to form during tempering (Chapter 4, Section 4.3). The sequence of precipitation for a particular composition can be approximated to by drawing a line from the origin of the diagram, e.g., Fig. 4.10, to the composition of interest. The phase fields traversed are usually those that occur during tempering, but the exact conditions cannot be forecast from such data.

Certain strong carbide formers, notably niobium, titanium and vanadium, have effects on tempering out of proportion to their concentration. In concentrations of 0.1 wt% or less, provided the tempering temperature is high enough, i.e., 550–650°C, they combine preferentially with part of the carbon and, in addition to the major carbide phase, e.g., Cr_7C_3, Mo_2C, they form a separate, very much finer dispersion, more resistant to overaging (Fig. 9.30). This secondary dispersion can greatly augment the secondary-hardening reaction, illustrating the importance of these strong carbide-forming elements in achieving high strength levels not only at room temperature but also at elevated temperatures, where creep resistance often is an essential requirement.

9.7.8 Mechanical properties of tempered alloy steels

A wide range of mechanical properties is obtainable by tempering alloy steels between 200°C and 700°C. A typical example is shown in Fig. 9.31 for a steel containing 1.5Ni-1Cr-0.25Mo-0.4C wt%, the tensile strength of which can be varied from 1800

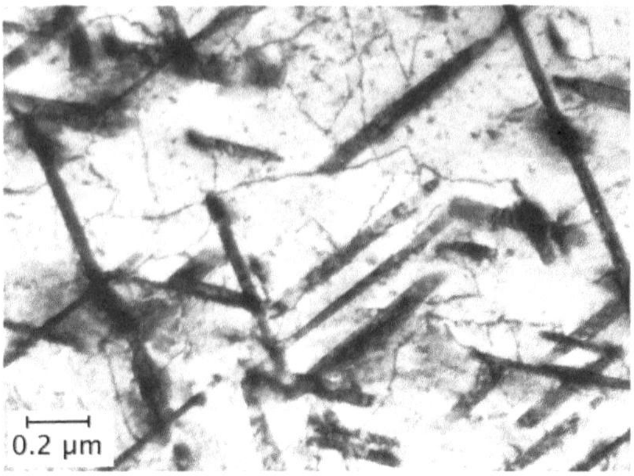

Figure 9.30 Fe-4Mo-0.1Nb-0.2C wt% steel tempered 6 h at 700°C. Coarse needles of Mo_2C in ferrite and fine particles of NbC on dislocations (courtesy of Irani).

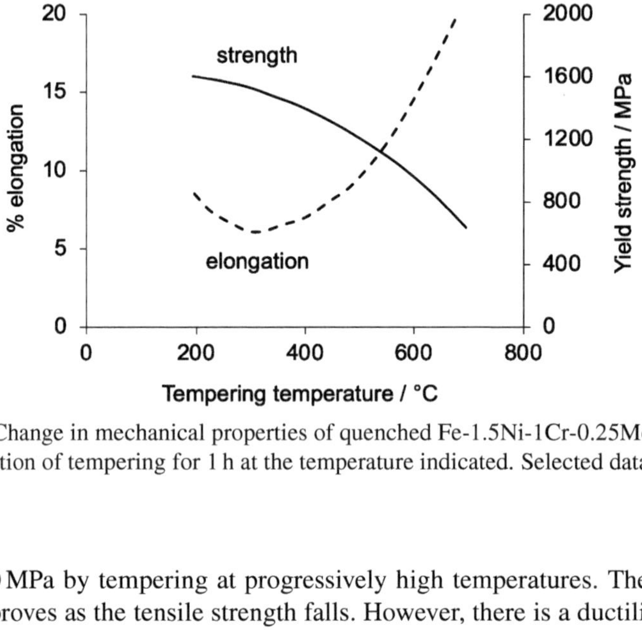

Figure 9.31 Change in mechanical properties of quenched Fe-1.5Ni-1Cr-0.25Mo-0.4C wt% steel as a function of tempering for 1 h at the temperature indicated. Selected data from Thelning [63].

down to 900 MPa by tempering at progressively high temperatures. The ductility of the steel improves as the tensile strength falls. However, there is a ductility minimum around 275–300°C, which often is observed in plain carbon and lower-alloy steels. This is due to the formation of thin cementite films, as a result of the transformation of austenite at the interlath boundaries, a phenomenon known as tempered martensite embrittlement, as discussed in more detail in Chapter 11. At higher temperatures, these films spheroidise and the toughness improves.

To obtain really high strength levels in tempered steels ($\sim$1500 MN m^{-2}), it is usual to temper at low temperatures, i.e., 200–300°C, when the martensite is still heavily dislocated and the main strengthening dispersion is cementite or ε-iron carbide. Alloy steels, when tempered in this range, not only provide high tensile-strengths with some ductility, but are superior to what can be achieved with plain carbon steels, as shown in Fig. 9.32. It is clear from Fig. 9.32a that the carbon content has a large influence on the strength. The alloying elements refine the iron carbide dispersion and as the carbon content is raised, the dispersion becomes more dense and, therefore, more effective. The toughness decreases with increasing strength, as shown in Fig. 9.32b. However, alloying elements very substantially improve the toughness, when compared with plain carbon steels of similar strength levels. When molybdenum is present in the steel, the toughness is increased further, as the scatter bands indicate. This effect of alloying elements is again attributed to the breakdown of carbide films at grain and martensite lath boundaries. These films are particularly less noticeable in steels containing molybdenum.

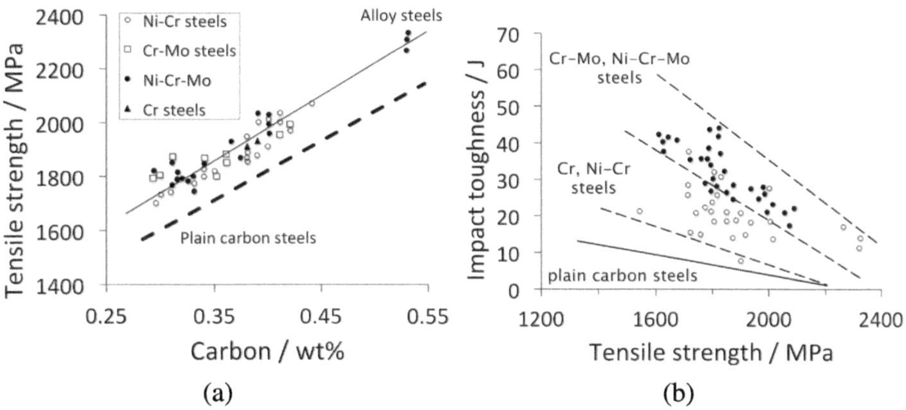

Figure 9.32 Comparison of mechanical properties of plain carbon and alloy steels that are quenched into a martensitic microstructure and then tempered at 200°C. (a) Effect of carbon on tensile strength; (b) relation between tensile strength and ambient temperature Charpy impact toughness. Note the beneficial effect of Mo. After Irvine and Pickering [58].

Alloy steels which exhibit secondary hardening can provide high strength levels on tempering between 500°C and 700°C, with better ductility than that obtained at lower tempering temperatures. However, one of their main advantages is that once a high strength level is reached by alloy carbide dispersion formed between 550°C and 650°C, this structure is relatively stable at temperatures up to 500°C. Therefore, the steels are suitable for use under stress at elevated temperatures. A typical example is given in Fig. 9.33 of a 12Cr-1Ni-0.2C wt% stainless steel, which can be quenched to martensite and then tempered to give a fine dispersion of chromium carbides in a ferritic matrix. The strength is well-maintained up to the secondary-hardening peak at 500°C, and is combined with a reasonable level of ductility. This type of steel is tempered to between 700 and 1000 MPa yield stress and is used frequently in steam and gas turbines, but can also be used for constructional purposes where lower temperatures are involved. Further improvements in mechanical properties at elevated temperatures can be obtained by addition of small concentrations of stronger carbide formers, e.g., molybdenum (2 wt%) and vanadium (0.25 wt%).

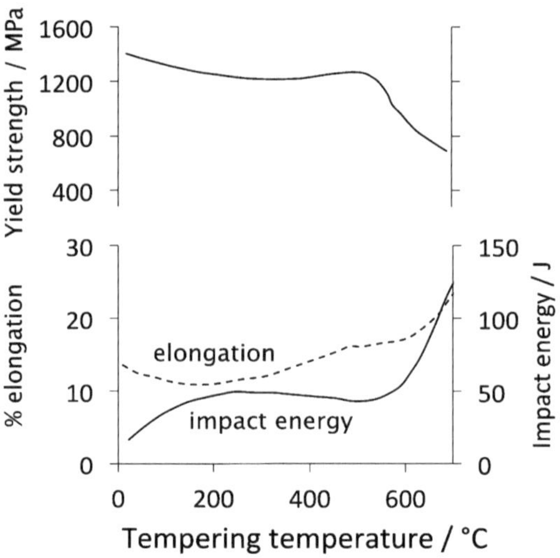

Figure 9.33 Effect of tempering for 1 h on the mechanical properties of a martensitic 12Cr-1Ni-0.2C wt% stainless steel. Selected data from Thelning [63].

9.7.9 Mechanical properties: hydrogen trapping

It is very well established that small concentrations (1 p.p.m.) of atomic, *diffusible* hydrogen can embrittle strong steel [64]. Any method that renders the hydrogen immobile should mitigate its detrimental effects. The strain fields of coherent or usually semi-coherent carbide particles introduced by tempering reactions can attract hydrogen and trap it. The carbide particles that precipitate at temperatures where substitutional solutes such as molybdenum, vanadium, niobium and titanium become mobile over length scales of a few nanometres are particularly interesting from the point of view of hydrogen trapping [65]. With appropriate tempering at temperatures in the range 500–600°C, fine dispersions of alloy carbides can be introduced in the microstructures of strong steel. The nature of the carbide dispersion can be controlled carefully while at the same time selecting tempering conditions that are suitable for the other properties required of the steel [66].

Some typical binding energies of hydrogen with carbides are given in Table 9.4; the higher the binding energy, the more effective the carbide is in preventing the hydrogen from diffusing. Fig. 9.34 shows that the mobility of hydrogen is dramatically reduced in the presence of vanadium carbides, and so is the propensity of hydrogen to embrittle the steel. Such steels are now successfully used as extremely strong bolts in the construction of bridges.

Vanadium-based carbides have long been known to be effective in mitigating hydrogen-induced delayed fracture in strong bolting steels. The binding energy determined using thermal desorption analysis is found to be 33-35 kJ mol^{-1}. It has been argued [74] that the key trap in V_4C_3 is at carbon vacancies in the lattice, but the binding energy calculated for this is not consistent with that measured; furthermore, it has

Table 9.4 Data on trapping (binding) energies E_b for hydrogen or deuterium atoms in ferritic steels. A more comprehensive list is available in [65].

Trap site	$-E_b/\mathrm{kJ\,mol^{-1}}$	Reference
Cementite/α interfaces	11-18	[67,68]
TiC	46-116	[69–71]
V_4C_3	33-35	[72]
Coherent M_2C (Mo-rich needles)	11-12	[73]

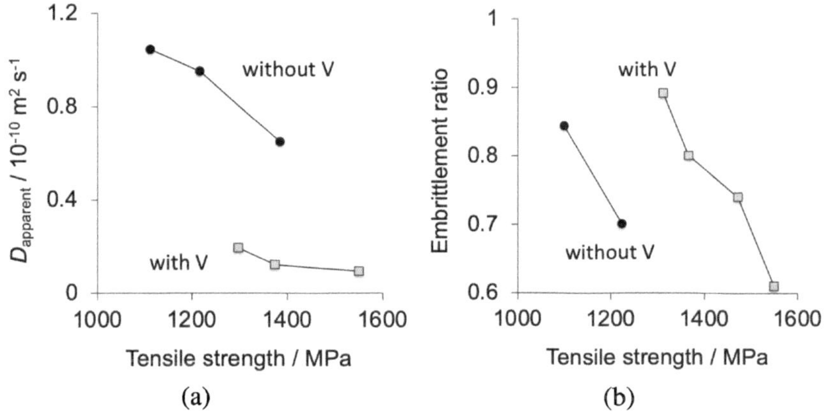

Figure 9.34 Comparison of the susceptibility to delayed fracture of two bolting steels whose detailed chemical composition is given in Table 9.5. The steel with vanadium contains vanadium carbides that serve to trap otherwise diffusible hydrogen, thus dramatically reducing its apparent diffusivity D_{apparent}. Therefore, the embrittlement ratio (fracture stress divided by tensile strength in the absence of hydrogen) is much larger with the vanadium-containing steel, a higher ratio implying that the static fracture strength is less affected by hydrogen. Data from [76].

been demonstrated that the state of coherency with the ferrite influences the hydrogen trapping capacity [75], emphasising the role of the strain fields around the carbides.

9.8 Maraging steels

It has been shown that precipitation of alloy carbides in tempered martensite gives rise to age hardening, usually referred to as secondary hardening. There is no reason why other finely divided phases cannot be used for a similar purpose and in fact an important group of iron-based alloys, the maraging steels [81], reach yield-strength levels as high as 2 GPa by the precipitation of various intermetallic compounds, while exhibiting a toughness of about 70 MPa m$^{1/2}$. When the yield-strength is controlled to 1.5 GPa, the toughness achieved is an extraordinary 170 MPa m$^{1/2}$ [82].

Carbide precipitation is practically eliminated by the use of low-carbon compositions, and the steels contain between 18 and 25 wt% nickel so the hardenability is impressive and even slow-cooling from the austenitic condition leads to martensitic

Table 9.5 Compositions (wt%) of some of the steels containing substitutional solutes to form alloy carbides on tempering, that trap diffusible hydrogen and hence increase the resistance to embrittlement. The enamelling steel also contains 0.3Ti-0.0046N wt%.

Alloy	C	Si	Mn	Mo	Cr	V	Nb	Ref.
Bolt steel, quenched, tempered 550-650°C for 90 min.	0.5		0.3	0.7	1.0	0.3	0.03	[76]
Bolt steel, without vanadium, quenched and tempered at 500-550°C for 90 min.	0.39	0.82		0.16	1.11		0.03	[76]
Bolt steel with molybdenum carbides (NIMS17), quenched, tempered at 570°C, 90 min	0.59	1.98	0.20	0.94	0.99			[77–79]
Enamelling steel	0.048	0.05	0.47					[80]

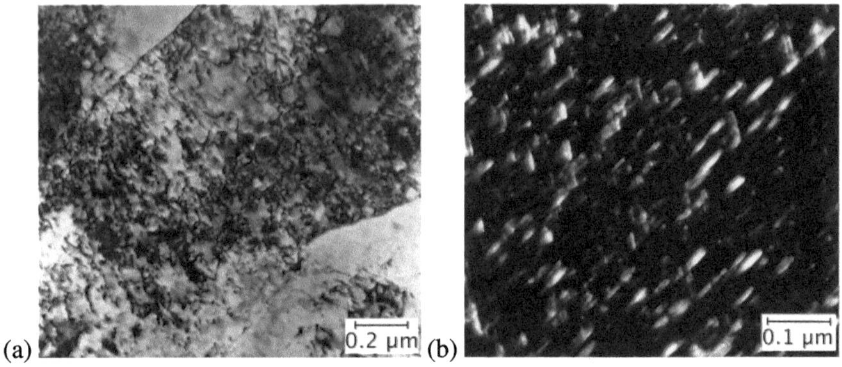

Figure 9.35 Maraging steel of composition Fe-18Ni-4.2Mo-12.5Co-1.7Ti-0.1Al wt%. (a) Martensitic state following air-cooling from 820°C. (b) Dark-field image showing precipitates of Ni_3(Ti,Mo) intermetallic compound following tempering at 510°C for 30 min. After Tewari et al. [83], reproduced with the permission of Elsevier.

transformation with the $M_S \approx 150$°C. The martensite that forms, in the absence of carbon, is soft (≈ 300 HV) but heavily dislocated martensite, Fig. 9.35. The precipitation of strengthening intermetallic compounds can be induced by tempering at about 500°C. Elements such as molybdenum and titanium are necessary additions, which result in the precipitation of Ni_3Mo, Ni_3Ti and the Laves phase, Fe_2Mo (Fig. 9.35b). Cobalt is also a useful alloying element as it reduces the solubility of molybdenum in the matrix and this increases the volume fraction of molybdenum-rich precipitate.

The precipitation reactions can lead to substantial volume fractions (≈ 0.05) of precipitate, and thus to the achievement of high strength levels. For example, a steel with 18-19 Ni, 8.5-9.5 Co, 4.5-5 Mo and 0.5-0.8 Ti wt% can be heat treated to give a yield stress of around 2000 MPa. However, it is useful to reiterate that the strength is accompanied by good ductility and excellent toughness. Combined with their ability to be welded, this combination of properties makes the alloys useful in, for example, rocket motor fabrication, where the casing can burst due to internal pressure, causing stresses approaching the ultimate tensile strength. The alloys have a high strength-to-weight ratio, so are used for aircraft landing gear, helicopter undercarriages, portable bridges and a variety of components in machines where there may be rapid cycling of temperature.

The large nickel concentration of traditional maraging steels limits their elevated temperature capability because of the reversion to austenite during prolonged exposure to high temperatures. Low-pressure turbine shafts in civil jet engines have very demanding requirements so a maraging alloy has been developed [84,85] that avoids the formation of austenite and maintains strength, creep resistance, toughness and corrosion resistance at temperatures up to 450°C. The alloy has been shown to be advantageous relative to conventional technologies for the main shafts of gas turbine engines. It contains a much reduced nickel concentration

wt%	Fe	Cr	Co	Ni	Al	Mo	W
	68	9.9	8	7	1.8	2.8	2.4

and is tempered at ≈540°C for 5 h to generate extremely small precipitates of Laves phase (Mo and Cr-rich) and β-NiAl, Fig. 9.36. A yield strength in excess of 1.8 GPa, ultimate strength of 2 GPa and a tensile ductility of 8% at room temperature, together with a creep rupture life of > 2000 h at 700 MPa and 500°C have been achieved [85].

(a)
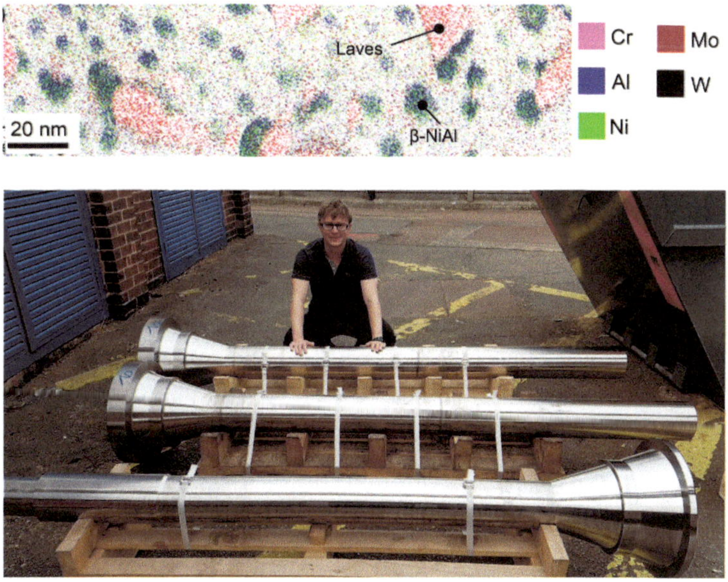

(b)

Figure 9.36 New maraging alloy for main shafts of gas turbine engines. (a) Atom map showing the Laves and β-NiAl precipitates following the tempering heat-treatment. Details in Sun et al. [85]. (b) Actual shaft-blanks made from the alloy prior to machining of intricate features. Image courtesy of Paul Hill (pictured) and Martin Rawson, Rolls-Royce.

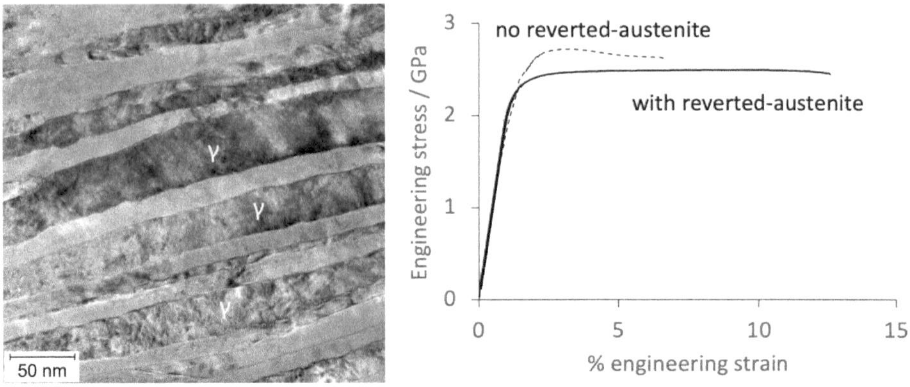

(a) (b)

Figure 9.37 (a) Microstructure consisting of alternating layers of reverted austenite (dark) and precipitation-hardened martensite (light). (b) The dashed curve represents the sample without the reverted austenite because it was not given the 600°C treatment, whereas the continuous curve is for the case where the sample was first aged to precipitate NiAl and then reheated to generate reverted austenite. Both of the materials were cold-deformed prior to heat treatment. Adapted from Yang et al. [87] under CC BY 4.0 license, https://creativecommons.org/licenses/by/4.0/.

9.8.1 TRIP-assisted maraging steels

Whereas the reversion to austenite can compromise the application of maraging steel at elevated temperatures, it may be advantageous to have this phase present for ambient temperature implementations. Thus, a maraging steel of approximate composition Fe-0.03C-8.Ni-13-2.3Mo-1.1Al-2.5Cu wt% aged at 480°C for 4 h martensitic condition led to NiAl precipitation and about 18% retained austenite. The latter contributed significantly to elongation by the TRIP mechanism, leading to a most useful combination of $\sigma_{UTS} = 1650$ MPa with an elongation of $\approx 18\%$ [86].

More astonishing properties have been achieved in Fe-0.35C-8Ni-8Cr-4Mo-2Mn-2Al-0.1V-0.1Nb wt% alloy, containing a significantly enhanced carbon concentration [87]. The material was reduced 90% in thickness by cold-rolling, followed by heat-treatment first at 500°C for 2 h to generate NiAl precipitates, and then at 600°C for 10 min to form reverted austenite. This led to a fine microstructure, Fig. 9.37a, in which the precipitation-hardened martensite plates were separated by reverted austenite. As expected, the austenite was enriched in carbon to almost 1 wt% but oddly, also greatly enriched in chromium and molybdenum concentrations. Fig. 9.37b shows clearly the effect of the reverted austenite on the elongation; a yield strength of 2.2 GPa, ultimate strength of 2.5 GPa and a uniform elongation in excess of 9% were achieved. It is the work-hardening ability of the austenite which prevents the early onset of plastic instability in maraging steels [88] so it is not surprising that the sample in Fig. 9.37b that does not contain the austenite has relatively poor elongation. Elevated temperature properties have not been reported, but it might be speculated that the presence of substantial quantities of austenite than does not contain the NiAl precipitates would be detrimental to creep resistance. In the context of fatigue, the reverted austenite offers a soft region for crack propagation [89].

9.9 Summary

The contents of this chapter are not dissimilar to discussions on precipitation in aluminium or nickel alloys, where allotropic transitions are absent. The tempering of martensite is carried out in a temperature range where austenite does not form – the goal is simply to optimise properties through low-temperature heat treatment, or to introduce precipitate dispersions at temperatures where substitutional solutes are mobile, in a matrix that is toughened by tempering so that strength and toughness can both be accessed. The phenomenon of secondary hardening, either by alloy carbides or intermetallic compounds, accounts for steels that are used in critical scenarios, such as aerospace bearings and rocket motor casings. Secondary hardened steels are the backbone of steam turbines that are responsible for the generation of most of the electricity consumed worldwide.

References

[1] H.K.D.H. Bhadeshia, A. Strang, D.J. Gooch, Ferritic power plant steels: remanent life assessment and the approach to equilibrium, International Materials Reviews 43 (1998) 45–69.

[2] H.K.D.H. Bhadeshia, Theory of Transformations in Steels, CRC Press, Taylor and Francis Group, London, U.K., 2021.

[3] H. Pous-Romero, H.K.D.H. Bhadeshia, Coalesced martensite in pressure vessel steels, Journal of Pressure Vessel Technology 136 (2014) 031402.

[4] W.S. Owen, The effect of silicon on the kinetics of tempering, Transactions of the ASM 46 (1954) 812–829.

[5] E. Kozeschnik, H.K.D.H. Bhadeshia, Influence of silicon on cementite precipitation in steels, Materials Science and Technology 24 (2008) 343–347.

[6] G.V. Kurdjumov, A.G. Khachaturyan, Phenomena of carbon atom redistribution in martensite, Metallurgical Transactions 3 (1972) 1069–1076.

[7] M.K. Miller, P.A. Beaven, S.S. Brenner, G.D.W. Smith, Martensite: an atom probe study of the aging of Fe-Ni-C, Metallurgical Transactions A 14 (1983) 1021–1024.

[8] S. Murphy, J.A. Whiteman, The precipitation of epsilon-carbide in twinned martensite, Metallurgical Transactions 1 (1970) 843–848.

[9] A. Vieweg, E. Provoden-Karadeniz, G. Ressel, P. Prevedel, T. Wojcik, F. Mendez Martin, A. Stark, J. Keckes, E. Kozeschnik, Phase evolution and carbon redistribution during continuous tempering of martensite studied with high resolution techniques, Materials & Design 136 (2017) 214–222.

[10] D.H. Sherman, S.M. Cross, S. Kim, F. Grandiean, G.J. Long, M.K. Miller, Characterization of the carbon and retained austenite distributions in martensitic medium carbon, high silicon steel, Metallurgical & Materials Transactions A 38 (2007) 1698–1711.

[11] F. Hajyakbary, J. Sietsma, G. Miyamoto, T. Furuhara, M.J. Santofimia, Interaction of carbon partitioning, carbide precipitation and bainite formation during the Q&P process in a low C steel, Acta Materialia 104 (2016) 72–83.

[12] R.A. Grange, C.R. Hribal, L.F. Porter, Hardness of tempered martensite in carbon and low-alloy steels, Metallurgical Transactions A 8A (1977) 1775–1785.

[13] M. Dirand, L. Afqir, Identification structural precise des carbures precipites dans lest aciers faiblement allies aux divers stades due revenu. Mechanismes de precipitation, Acta Metallurgica 31 (1983) 1089–1107.
[14] H.K.D.H. Bhadeshia, A.R. Chintha, S. Lenka, Critical assessment 34: are Hägg, eta and epsilon carbides transition-phases relative to cementite in steels?, Materials Science and Technology 35 (2019) 1301–1305.
[15] Y. Imai, Phases in quenched and tempered steels, Transactions of JIM 16 (1975) 721–734.
[16] K.A. Taylor, G.B. Olson, M. Cohen, J. Vander Sande, Carbide precipitation during stage 1 tempering of FeNiC martensite, Metallurgical Transactions A 20 (1989) 2749–2765.
[17] Y. Ohmori, χ-carbide and its transformation into cementite during the tempering of martensite, Transactions of the Japan Institute of Metals 13 (1972) 119–127.
[18] S. Nagakura, S. Suzuki, M. Kusunoki, Structure of the precipitated particles at the third stage of tempering of martensitic iron-carbon steel studied by high resolution electron microscopy, Transactions of the Japan Institute of Metals 22 (1981) 699–709.
[19] Y. Nakamura, T. Mikami, S. Nagakura, In situ high temperature electron microscopic study of the formation and growth of cementite particles in the third stage of tempering, Transactions of the Japan Institute of Metals 26 (1985) 876–885.
[20] G. Ghosh, C.E. Campbell, G.B. Olson, Analytical electron microscopy study of paraequilibrium cementite precipitation in ultra-high strength steel, Metallurgical & Materials Transactions A 30 (1999) 501–512.
[21] G. Ghosh, G.B. Olson, Precipitation of paraequilibrium cementite: experiments, thermodynamic and kinetic modelling, Acta Materialia 50 (2002) 2099–2119.
[22] Y. Ohmori, Precipitation of iron carbides in lower bainite and tempered martensite in Fe-C alloys, in: International Conference on Martensitic Transformations ICOMAT '86, Tokyo, Japan, Japan Institute of Metals, 1986, pp. 587–594.
[23] S. Matas, R.F. Hehemann, Retained austenite and the tempering of martensite, Nature 187 (1960) 685–686.
[24] S.J. Barnard, G.D.W. Smith, M. Sarikaya, G. Thomas, Carbon atom distribution in a dual phase steel, an atom probe study, Scripta Metallurgica 15 (1981) 387–392.
[25] C. Lerchbacher, S. Zinner, H. Leitner, Atom probe study of the carbon distribution in a hardened martensitic hot-work tool steel X38CrMoV5-1, Micron 43 (2012) 818–826.
[26] J. Speer, D.K. Matlock, B.C.D. Cooman, J.G. Schroth, Carbon partitioning into austenite after martensite transformation, Acta Materialia 51 (2003) 2611–2622.
[27] J.G. Speer, D.V. Edmonds, F.C. Rizzo, D.K. Matlock, Partitioning of carbon from supersaturated plates of ferrite, with application to steel processing and fundamentals of the bainite transformation, Current Opinion in Solid State & Materials Science 8 (2004) 219–237.
[28] D.V. Edmonds, K. He, F.C. Rizzo, B.C.D. Cooman, D.K. Matlock, J.G. Speer, Quenching and partitioning martensite – a novel steel heat treatment, Materials Science & Engineering A 438–440 (2006) 25–34.
[29] J.G. Speer, E. de Moor, A.J. Clarke, Critical assessment 7: quenching and partitioning, Materials Science and Technology 31 (2015) 3–9.
[30] A.J. Clarke, J. Klemm-Toole, K.D. Clarke, D.R. Coughlin, D.T. Pierce, V.K. Euser, J.D. Poplawsky, B. Clausen, D. Brown, J. Almer, P.J. Gibbs, D.J. Alexander, R.D. Field, D.L. Williamson, J.G. Speer, G. Krauss, Perspectives on quenching and tempering 4340 steel, Metallurgical & Materials Transactions A 51 (2020) 4984–5005.
[31] D.L. Williamson, R.G. Schupmann, J.P. Materkowski, G. Krauss, Determination of small amounts of austenite and carbide in hardened medium carbon steels by Mössbauer spectroscopy, Metallurgical Transactions A 10 (1979) 379–382.

[32] H.K.D.H. Bhadeshia, D.V. Edmonds, Tempered martensite embrittlement: role of retained austenite and cementite, Metal Science 13 (1979) 325–334.
[33] H.S. Park, J.B. Seol, N.S. Lim, S.I. Kim, C.G. Park, Study of the decomposition behavior of retained austenite and the partitioning of alloying elements during tempering in CMnSiAl TRIP steels, Materials & Design 82 (2015) 173–180.
[34] M.X. Zhang, P.M. Kelly, Crystallography and morphology of Widmanstätten cementite in austenite, Acta Materialia 46 (1998) 4617–4628.
[35] J.X. Zhang, P.M. Kelly, L.K. Bekessy, J.D. Gates, Determination of retained austenite using an X-ray texture goniometer, Materials Characterization 45 (2000) 39–49.
[36] I.V. Isaichev, Orientation of cementite in tempered carbon steel (Orientatsiya tsementita v otpushchennoi uglerodistoi stali), Zhurnal Tekhncheskoi Fiziki 17 (1947) 835–838.
[37] H.K.D.H. Bhadeshia, Solution to the Bagaryatskii and Isaichev ferrite-cementite orientation relationship problem, Materials Science and Technology 34 (2018) 1666–1668.
[38] V. Kraševec, M. Mäček, J. Rodič, On precipitation of cementite along twin boundaries in alloy steel, Acta Metallurgica 28 (1980) 223–229.
[39] B.C. Kim, S. Lee, D.Y. Lee, N.J. Kim, In situ fracture observations on tempered martensite embrittlement in an AlSl 4340 steel, Metallurgical and Materials Transactions A 22 (1991) 1889–1892.
[40] C.L. Briant, S.K. Banerji, Tempered martensite embrittlement in phosphorus doped steels, Metallurgical Transactions A 10 (1979) 1729–1737.
[41] I.M. Lifshitz, V.V. Slyozov, Kinetics of precipitaiton from supersaturated solid solutions, Journal of Physics and Chemistry of Solids 19 (1961) 35–50.
[42] C. Wagner, Theorie der alterung von niederschlägen durch umlösen (Ostwald-Reifung), Zeitschrift für Elektrochemie 65 (1961) 581–591.
[43] G. Miyamoto, H. Usuki, Z.-D. Li, T. Furuhara, Effects of Mn, Si and Cr addition on reverse transformation at 1073K from spheroidized cementite structure in Fe-0.6 mass% C alloy, Acta Materialia 58 (2010) 4492–4502.
[44] G.R. Speich, Tempering of low-carbon martensite, Transactions of the Metallurgical Society of AIME 245 (1969) 2553–2564.
[45] C. Zener, Kinetics of the decomposition of austenite, Transactions of the American Institute of Mining and Metallurgical Engineers 167 (1946) 550–595.
[46] J.C. Fisher, J.H. Hollomon, D. Turnbull, Kinetics of the austenite to martensite transformation, Metals Transactions 185 (1949) 691–700.
[47] D.W. Hoffman, Concerning the elastic free energy of dilute interstitial alloys, Acta Materialia 18 (1970) 819–833.
[48] S. Kang, S.J. Lee, Prediction of tempered martensite hardness incorporating the composition-dependent tempering parameter in low alloy steels, Materials Transactions 55 (2014) 1069–1072.
[49] K.J. Irvine, F.B. Pickering, J. Garstone, The effect of composition on the structure and properties of martensite, Journal of the Iron and Steel Institute 196 (1960) 66–81.
[50] Anonymous, Metals Handbook, 8th ed., American Society for Metals, Metals Park, Ohio, USA, 1961.
[51] S. Chatterjee, H.K.D.H. Bhadeshia, TRIP-assisted steels: cracking of high carbon martensite, Materials Science and Technology 22 (2006) 645–649.
[52] A.R. Chintha, M. Peet, S. Kundu, H.K.D.H. Bhadeshia, K. Valtonen, V.-T. Kuokkala, Role of fracture toughness in impact-abrasion wear, Wear 428–429 (2019) 430–437.
[53] P.J. Jacques, E. Girault, P. Harlet, F. Delannay, Developments of cold-rolled TRIP-assisted multiphase steels. Low silicon TRIP-assisted multiphase steels, ISIJ International 41 (2001) 1061–1067.

[54] G.V. Kurdjumov, Phenomena occurring in the quenching and tempering of steels, Journal of the Iron and Steel Institute 195 (1960) 26–48.

[55] T. Gladman, Physical Metallurgy of Microalloyed Steels, Institute of Materials, London, U.K., 1997.

[56] H.K.D.H. Bhadeshia, Theoretical analysis of changes in cementite composition during the tempering of bainite, Materials Science and Technology 5 (1989) 131–137.

[57] A. Afrouz, M.J. Collins, R. Pilkington, Microstructural examination of 1Cr 0.5Mo steel during creep, Metals Technology 10 (1983) 461–463.

[58] K.J. Irvine, F.B. Pickering, The tempering characteristics of low-carbon low-alloy steels, Journal of the Iron and Steel Institute 194 (1960) 137–153.

[59] D.C. Shtansky, G. Inden, Phase transformation in Fe-Mo-C and Fe-WC steels – I. The structural evolution during tempering at 700°C, Acta Materialia 45 (1997) 2861–2878.

[60] D. Raynor, J.A. Whiteman, R.W.K. Honeycombe, In-situ transformation of Fe_3C to Mo_2C in iron-molybdenum-carbon alloys, Journal of the Iron and Steel Institute 204 (1966) 1114–1116.

[61] M.L. Saucedo-Muñoz, S.I. Komazaki, E.O. Avila-Davila, V.M. Lopez-Hirata, H.J. Dorantes-Rosales, Precipitation characterization and creep strength at 600°C for creep resistant Cr–Mo steel, ISIJ International 60 (2020) 2059–2067.

[62] E.C. Bain, H.W. Paxton, The Alloying Elements in Steel, 2nd ed., ASM, Metals Park, Ohio, USA, 1961.

[63] K.E. Thelning, Steel and Its Heat Treatment, Butterworths, London, U.K., 1975.

[64] W.H. Johnson, On some remarkable changes produced in iron and steel by the action of hydrogen and acids, Proceedings of the Royal Society of London 23 (1875) 168–179.

[65] H.K.D.H. Bhadeshia, Prevention of hydrogen embrittlement in steels, ISIJ International 56 (2016) 24–36.

[66] S. Yamasaki, H.K.D.H. Bhadeshia, Modelling and characterisation of Mo_2C precipitation and cementite dissolution during tempering of Fe-C-Mo martensitic steel, Materials Science and Technology 19 (2003) 723–731.

[67] G.W. Hong, J.Y. Lee, The interaction of hydrogen and the cementite-ferrite interface in carbon steel, Journal of Materials Science 18 (1983) 271–277.

[68] W.Y. Choo, J.Y. Lee, Thermal analysis of trapped hydrogen in pure iron, Metallurgical Transactions A 13 (1982) 135–140.

[69] G.M. Pressouyre, I.M. Bernstein, A quantitative analysis of hydrogen trapping, Metallurgical Transactions A 9 (1978) 1571–1580.

[70] F.G. Wei, T. Hara, K. Tsuzaki, Precise determination of the activation energy for desorption of hydrogen in two Ti-added steels by a single thermal-desorption spectrum, Metallurgical & Materials Transactions B 35 (2004) 587–597.

[71] F.G. Wei, K. Tsuzaki, Quantitative analysis on hydrogen trapping of TiC particles in steel, Metallurgical & Materials Transactions A 37 (2006) 331–353.

[72] H. Asahi, D. Hirakami, S. Yamasaki, Hydrogen trapping behavior in vanadium-added steel, ISIJ International 43 (2003) 527–533.

[73] D. Li, R.P. Gangloff, J.R. Scully, Hydrogen trap states in ultrahigh-strength AERMET 100 steel, Metallurgical & Materials Transactions A 35 (2004) 849–864.

[74] K. Kawakami, T. Matsumiya, Numerical analysis of hydrogen trap state by TiC and V_4C_3 in bcc-Fe, ISIJ International 52 (2012) 1693–1697.

[75] S. Yamasaki, H.K.D.H. Bhadeshia, Precipitation during tempering of Fe-C-Mo-V and relationship to hydrogen trapping, Proceedings of the Royal Society of London A 462 (2006) 2315–2330.

[76] T. Kushida, N. Kuratomi, T. Kudoh, H. Matsumoto, T. Tsumura, F. Nakasato, Delayed fracture and hydrogen absorption of 1.3 GPa grade high strength bolt steel, Tetsu-to-Hagane 82 (1996) 297–302.
[77] Y. Kimura, T. Hara, K. Tsuzaki, Strength and delayed fracture properties of martensitic steels containing Si, Cr and Mo, CAMP-ISIJ 14 (2001) 1307.
[78] S. Li, E. Akyama, K. Yuuji, K. Tsuzaki, N. Uno, B. Zhang, Hydrogen embrittlement property of a 1700-MPa-class ultrahigh-strength tempered martensitic steel, Science and Technology of Advanced Materials 11 (2010) 025005.
[79] E. Akiyama, Evaluation of delayed fracture property of high strength bolt steels, ISIJ International 52 (2012) 307–315.
[80] S. Matera, E. Anelli, TEM-EDS characterization of second phases in ferritic steels, Microscopy Microanalysis Microstructures 6 (1995) 633–646.
[81] R.F. Decker, S. Floreen, Maraging steels - the first 30 years, in: R.K. Wilson (Ed.), Maraging Steels - Recent Developments and Applications, TMS-AIME, Warrendale, Pennsylvania, USA, 1988, pp. 1–38.
[82] Y. Wang, Y. Tomota, T. Ohmura, W. Gong, S. Harjo, M. Tanaka, Continuous and discontinuous yielding behaviors in ferrite-cementite steels, Acta Materialia 196 (2020) 565–575.
[83] R. Tewari, S. Msumder, I.S. Batra, G.K. Dey, S. Banerjee, Precipitation in 18 wt% Ni maraging steel of grade 350, Acta Materialia 48 (2000) 1187–1200.
[84] A.T.W. Barrow, H.K.D.H. Bhadeshia, M. Rawson, p.O. Hill, Alloy steel, U.S. patent no. 217186, G.B. patent no. 1016731.0, Technical Report, U.S. and G.B. Patent and Trademark Offices, Washington D.C. and London, 2015.
[85] L. Sun, T.H. Simm, T.L. Martin, S. McAdam, D.R. Galvin, K.M. Perkins, P.A.J. Bagot, M.P. Moody, S.W. Ooi, P. Hill, M.J. Rawson, H.K.D.H. Bhadeshia, A novel ultra-high strength maraging steel with balanced ductility and creep resistance achieved by nanoscale β-NiAl and Laves phase precipitates, Acta Materialia 149 (2018) 285–310.
[86] L. Tan, D. Li, L. Yan, X. Pang, K. Gao, Simultaneous enhancement of strength-ductility via multiple precipitates and austenite in a novel precipitation-hardened martensitic stainless steel, Materials Science & Engineering A 873 (2023) 145062.
[87] B. Yang, Q. He, H. Wang, C. Wang, F. Guo, R. Hu, C. Wang, G. Fan, Q. Wang, W. Cao, et al., Achieving an extra-high-strength yet ductile steel by synergistic effects of TRIP and maraging, Materials Research Letters 11 (2023) 578–585.
[88] S.W. Ooi, P. Hill, M. Rawson, H.K.D.H. Bhadeshia, Effect of retained austenite and high temperature laves phase on the work hardening of an experimental maraging steel, Materials Science & Engineering A 564 (2013) 485–492.
[89] Z. Zhang, M. Koyama, M. Wang, K. Tsuzaki, C.C. Tasan, H. Noguchi, Microstructural mechanisms of fatigue crack non-propagation in trip-maraging steels, International Journal of Fatigue 113 (2018) 126–136.

Thermomechanical treatment 10

Abstract

The mass production of quality steel by deforming austenite in the presence of small quantities of microalloying additions such as niobium has revolutionised the properties of structural steels. This is because the size of the austenite grains is limited by the compounds of these microalloying additions. The process can be adapted to ensure that the austenite grains are left in a deformed, pancake-like shape prior to the onset of transformation because this again reduces the scale of the structure. The cooling rate during transformation can be controlled precisely using water spray technologies. These factors combine to give a final microstructure that is sufficiently refined to give an excellent combination of strength and toughness in steels that can be welded and fabricated into giant infrastructure projects. Such steels have proven to be so reliable that most people are oblivious of the underlying technologies, and yet, the quality of their lives has been dramatically improved by implementing about 40 billion tonnes of microalloyed steels. Dual phase, TRIP, TWIP and ausformed steels are also introduced.

10.1 Introduction

Thermomechanical processing involves the simultaneous application to a steel of heating and cooling cycles combined with some sort of a deformation, in order to change its shape and ultimately to refine its microstructure for the sake of better mechanical properties. The most common deformation is hot rolling, as this highly-automated process is capable of handling vast quantities of steel and still be subject to precise and seamless control, Fig. 10.1. Continuously cast segments of steel, ranging from 1 to 50 tonnes in weight from a holding furnace, are introduced into the rolling sequence at a temperature typically in the range 1200–1300°C. They are then progressively rolled into a variety of shapes, depending on application.

The use of energy can be optimised by introducing a continuous process in which the hot, solid-steel that emerges from continuous casting is processed immediately by rolling. This eliminates the stage where segments cut from the continuous casting process are held in a holding furnace or reheated to the hot-rolling temperature. In another variant, the steel is cast continuously in the form of thin slab or strip, thereby reducing the effort of rolling, although this may not always lead to the best properties, since deformation has a role in modifying the cast structure.

The deformation breaks down the original coarse microstructure that exists in the cast state by repeated recrystallisation of the steel in the austenitic condition, at the same time gradually reducing the length-scale and magnitude of any chemical segregation. Any non-metallic inclusions, i.e., oxides, sulphides and silicates, are broken up, some deformed and distributed throughout the steel in a more refined and uniform manner.

Hot rolling is not confined to the production of regular plates or sheet. With appropriate roll and rolling sequence design, it can be applied to make rods, rails, beams for infrastructure construction, etc. Since no machining is required following the rolling

Steels. https://doi.org/10.1016/B978-0-44-318491-8.00016-2

Figure 10.1 Thermomechanical processing of steel. (a) A 30 tonne slab emerges from the reheating furnace for hot-rolling. (b) The slab is elongated and thinned by multiple hot-rolling operations, although the width is maintained (plane-strain deformation). (c) Controlled water-cooling. (d) Coiling the long product. The final microstructure may be generated by transformation as the coil cools or the hot-rolled product may be processed further by cold-rolling and heat-treatment.

of rail and beam sections, these processes are superior to those classified as *near net-shape* [1] or *additive manufacturing* [2,3], where significant finishing operations are needed following manufacture. Additive manufacturing is modern jargon for an old process that refers to the layer-by-layer building up of a three-dimensional component using a variety of techniques (Chapter 14).

Essentially the same concepts are applied in the production of specialised objects, for example, the rotors for steam turbines, where large ingots of steel are hot forged into shape and then machined. The nature of chemical segregation is different in ingot castings [4,5] so the forging process must be adjusted to ensure that any detrimental segregate does not occur in a critical region of the final component.

The hot-rolling process is now a finely tuned operation in which more than a billion tonnes of steel are produced annually using automated arrays of equipment, resulting in impressive levels of productivity and reproducibility. The compositions of the low-alloy steels are chosen carefully to provide optimum mechanical properties when the hot deformation and subsequent cooling are complete. This process, in which the rolling parameters (temperature, strain, strain rate, number of rolling passes, finishing temperature, etc.) are predetermined and precisely defined for each steel grade, is called *controlled rolling*. It is now of the greatest importance in obtaining reliable mechanical properties in steels for pipelines, bridges, buildings and a huge variety of other products.

10.2 Controlled rolling of low-alloy steels

10.2.1 General

Before the late 1940s, strength in hot-rolled low-alloy steels was enhanced by the addition of carbon up to 0.4 wt% and manganese up to 1.5 wt%, giving yield stresses of 350–400 MPa. However, their microstructures were relatively coarse ferrite-pearlite aggregates, which compromised toughness and weldability. The toughness, as measured by the ductile/brittle transition, decreases dramatically with carbon concentration, i.e., with increasing proportion of pearlite in the steel [6,7], Fig. 10.2. Furthermore, with the introduction of welding as the main fabrication technique, the high carbon contents led to serious cracking problems that could only be eliminated by the use of lower-carbon steels.

The need for a fine ferrite grain size soon became apparent (Section 2.5), so controlled rolling in the austenitic condition was gradually introduced to achieve this. Fine ferrite grain sizes in the finished steel were found to be expedited by the addition of small concentrations (<0.1 wt%) of grain-refining elements such as niobium, titanium and vanadium, and also aluminium [8]. On adding such elements to steels with 0.03–0.08 wt% C and up to 1.5 wt% Mn, it became possible to produce fine-grained material with yield strengths between 450 and 550 MPa, and with ductile/brittle transition temperatures as low as −70°C. Such steels are now referred to as high-strength low-alloy (HSLA) steels or microalloyed steels. This progress, from the relatively low strength of ordinary mild steel (220–250 MPa) in a period of 20 years represents a major metallurgical development whose importance for engineering applications cannot be overstated.

In the majority of cases, the thermomechanical processing aims to refine the austenite grain size so that the ferrite that forms by transformation after hot-rolling is also

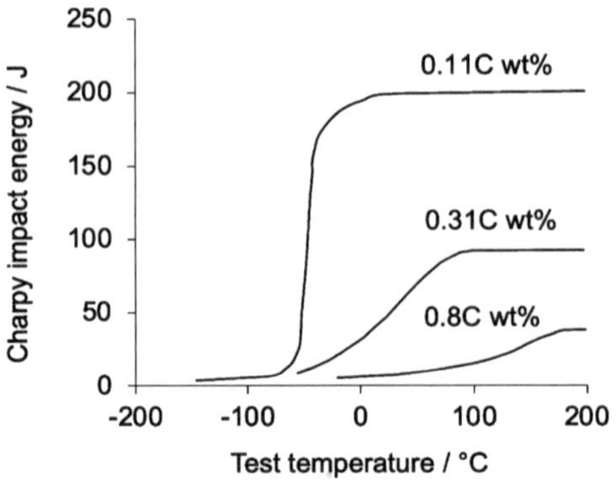

Figure 10.2 Effect of carbon content on the impact transition temperature curves of ferrite–pearlite steels. Selected data from Pickering [6,9].

fine. During hot deformation, the stress required to deform the steel is a function of the plastic strain (ϵ_p), strain rate ($\dot{\epsilon}$) and temperature. Zener and Hollomon proposed [10,11] that the effects of strain rate and temperature can be combined by writing $\sigma = f\{\epsilon_p, Z\}$ where Z, now known as the Zener–Hollomon parameter, is defined as:

$$Z = \dot{\epsilon}_p \exp\left(\frac{Q}{RT}\right) \quad s^{-1}, \tag{10.1}$$

where Q is an unspecified heat of activation, since most rates are associated with an activated event. The material work hardens during hot-rolling, but softens as a critical plastic strain $\epsilon_{c,p}$ is reached that induces the austenite to recrystallise, with

$$\epsilon_{c,p} \propto (\overline{L}_{\gamma_\circ})^{C_4}\, Z^{C_5}, \tag{10.2}$$

where $\overline{L}_{\gamma_\circ}$ is the austenite grain size prior to deformation and C_i are empirical constants.

If the strain $\epsilon_{c,p}$ is reached while the steel is still being rolled, then the process is known as *dynamic recrystallisation*. On the other hand, *metadynamic recrystallisation* is said to occur when recrystallisation occurs immediately after a rolling pass, when the strain retained in the austenite exceeds that needed to induce recrystallisation. The recrystallised austenite grain size is in general smaller than the initial size, and grain growth may follow, as illustrated in Fig. 10.3 and Fig. 10.4 [12]. The full theory of recrystallisation and grain growth is not presented here because it tends to be alloy specific, but can be accessed from extensive literature on the subject [13,14].

The general features of controlled rolling are summarised in Fig. 10.5. Really quite sophisticated process models now exist to treat the entire sequence of rolling [16–18], microstructure [19] and properties [20,21], so much so that some of these are now used in the on-line control [22] of rolling mills using artificial intelligence (i.e., empirical) methods [23] to ensure product uniformity.

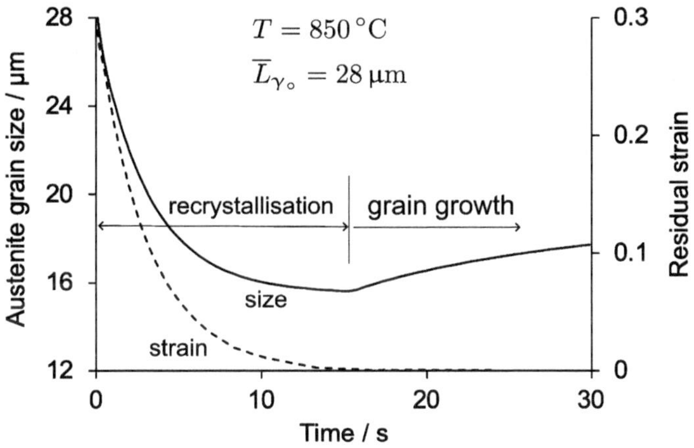

Figure 10.3 The calculated [12] influence of a single pass of rolling deformation to a strain of 0.3, on the evolution of the austenite grain size and residual strain (0.3 multiplied by the fraction of unrecrystallised austenite), as a function of time after the rolling deformation was applied. The residual strain is that accumulated during hot rolling due to incomplete softening.

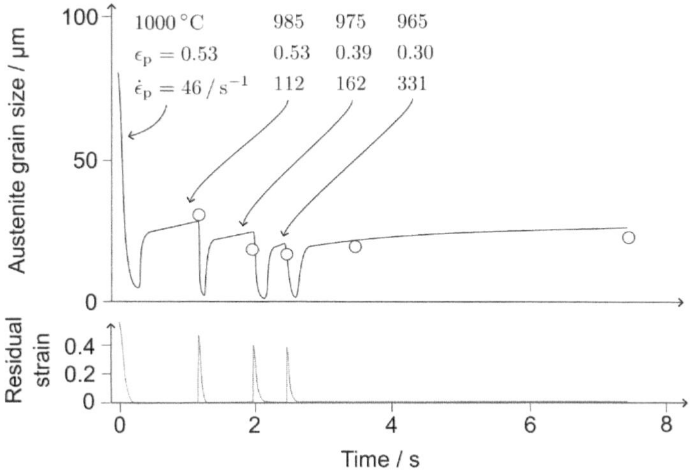

Figure 10.4 The calculated [12] influence of four passes of rolling deformation to a strain of 0.3, on the evolution of the austenite grain size and residual strain. Note that recrystallisation is complete when the residual strain becomes zero, with subsequent grain growth contributing to the increase in grain size. The position where the deformation is applied is indicated by the arrows. The units of $\dot{\epsilon}_p$ are in s^{-1}. The points are actual data from Senuma et al. [15].

10.2.2 Grain size control during controlled rolling

The primary grain refinement mechanism in controlled rolling is the recrystallisation of austenite during hot deformation, referred to earlier as dynamic recrystallisation. This process is influenced by the temperature and the degree of deformation which takes place during each pass through the rolls. However, in austenite devoid of second-

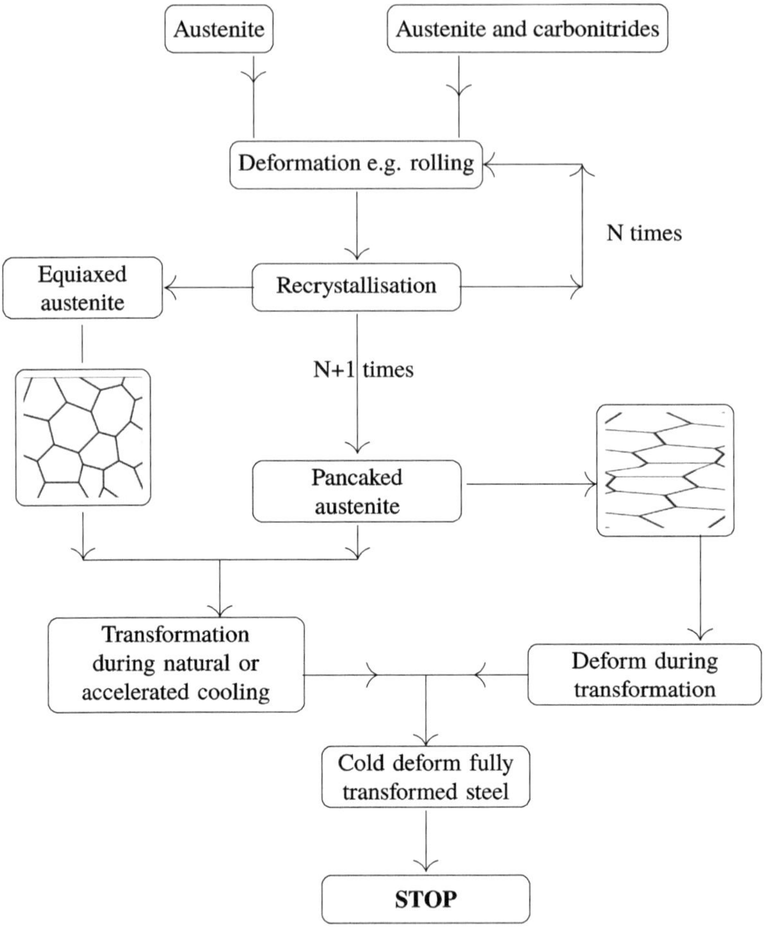

Figure 10.5 The variety of thermomechanical processing routes. Controlled rolling, followed by accelerated cooling is often designated *thermomechanical controlled processing* or TMCP.

phase particles, the high temperatures involved in hot rolling lead to marked grain growth, with the result that grain refinement during subsequent working is limited.

The situation is improved substantially if fine particles are introduced into the austenitic matrix. The particles usually are found on grain boundaries, because an interaction takes place between the particles and the boundary. A short length of grain boundary is replaced by the particle and the interfacial energy ensures a stable configuration. When the grain boundary attempts to migrate away from the particles, the local energy increases and thus a drag is exerted on the boundary by the particles.

The theory of boundary pinning by particles already has been referred to in Chapter 9. Eq. (9.9) defines the critical size of particle below which pinning is effective. Clearly, the control of grain size at high austenitising temperatures requires as fine a grain boundary precipitate as possible, and one which does not dissolve completely in the austenite, even at the highest working temperatures (1200–1300°C). The best grain-refining elements are the strongest of carbide and nitride formers such as niobium, titanium and vanadium, also aluminium, which forms only a nitride. As both carbon and nitrogen are present in control-rolled steels, and as the nitrides can be even more stable than the carbides (Fig. 4.8), it is likely that the most effective grain refining compounds are the respective carbo-nitrides, except in the case of aluminium nitride.

Equally important is the degree of solubility that such stable compounds have in austenite. It is essential that there is sufficient solid solubility at the highest austenitising temperatures to allow fine precipitation to occur during controlled rolling at temperatures, which decrease as rolling proceeds. The solubility products (in atomic per cent) of several relevant carbides and nitrides are shown in Fig. 11.13 as a function of the reciprocal of the temperature. All of these compounds have low but increasing solubility in the critical temperature range ≈900–1300°C (Fig. 10.6). In contrast, the carbides of chromium and molybdenum have much greater solubility, which ensures that they normally go completely into solution in the austenite, if the temperature is high enough, and precipitate only when the temperature is well below the critical range for grain growth. Data from another source [24] have provided the following equations for solubilities expressed in weight per cent as a function of absolute temperature:

$$\begin{aligned}\log_{10}[\mathrm{Al}][\mathrm{N}] &= -6770/T + 1.03,\\ \log_{10}[\mathrm{V}][\mathrm{N}] &= -8330/T + 3.46,\\ \log_{10}[\mathrm{Nb}][\mathrm{C}] &= -6770/T + 2.26,\\ \log_{10}[\mathrm{Ti}][\mathrm{C}] &= -7000/T + 2.75.\end{aligned}$$

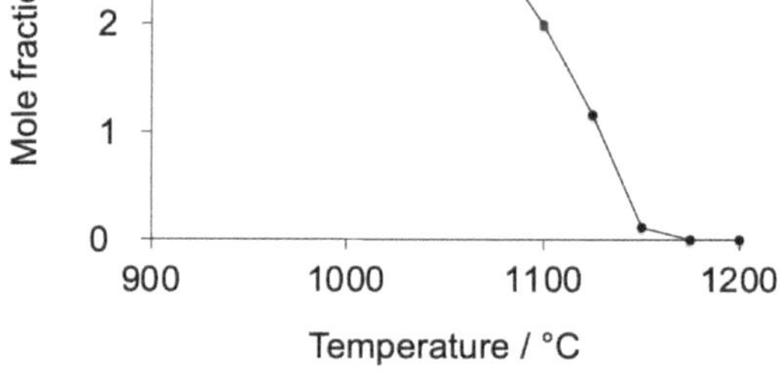

Figure 10.6 Equilibrium solubility curve for NbC in a steel with 0.15C-1.14Mn-0.04Nb wt%. Calculated using MTDATA [25] and the SGTE database. Such calculations should be superior to the use of solubility products, as they are able to account for all the alloying elements present in the steel.

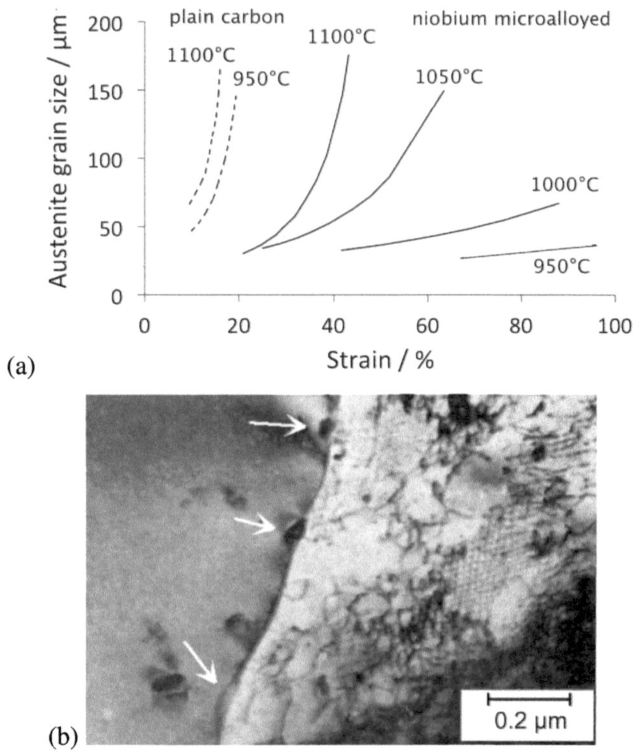

Figure 10.7 (a) Critical strain needed to complete recrystallisation of austenite as a function of deformation temperature and grain size. Comparison of Nb steel with plain carbon steel (adapted from Tanaka et al. [28]). (b) The pinning by niobium carbide precipitates (arrowed) of a recrystallisation front separating deformed austenite on the right from recrystallised austenite on the left. After Jones and Ralph [29], reproduced with permission of Elsevier.

The compositional changes are many, so the discussion will be limited to general principles that apply equally regardless of which compound is the effective grain refiner in a given steel. While grain growth at the highest austenitising temperatures may be restricted to some extent by a residual dispersion, the main refinement is achieved during rolling as the temperature progressively falls, and fine carbo-nitrides are precipitated from the austenite. These new precipitates will:

a) Increase the strain, for a given temperature, at which recrystallisation will commence, Fig. 10.7a.

b) Restrict the movement of recrystallisation fronts, Fig. 10.7b.

Austenite may recrystallise several times during controlled rolling, resulting overall in pronounced austenite grain refinement when the steel reaches the transformation temperature γ/α, (Fig. 10.7). In the later stages of austenite deformation, at lower temperatures, recrystallisation may not occur, with the result that deformed austenite grains elongated and flattened by rolling may transform directly into ferrite (Fig. 10.8).

Such flattened grains are often described as being *pancaked*. The austenite grain surface to volume ratio increases when an equiaxed grain is flattened [26,27], so the number density of ferrite nucleation sites increases, and the deformation structures within the flattened austenite grains themselves contribute as nucleation sites. All this leads to a further refinement of the final microstructure. The effect of the finish-rolling temperature on the ferrite grain size obtained is illustrated in Fig. 10.9a.

In the final stages of controlled rolling, austenite grain growth can be further suppressed by rapid cooling from the finishing temperature, which allows the γ/α transformation to take place sub-critically, i.e., below Ar_1, in austenite that still is deformed. It is sometimes the practice to continue rolling through the γ/α transformation and even into the fully ferritic region. Such treatments lead to finer grain sizes, and greater yield stresses in the finished product, but impose much higher loads on the rolling mills.

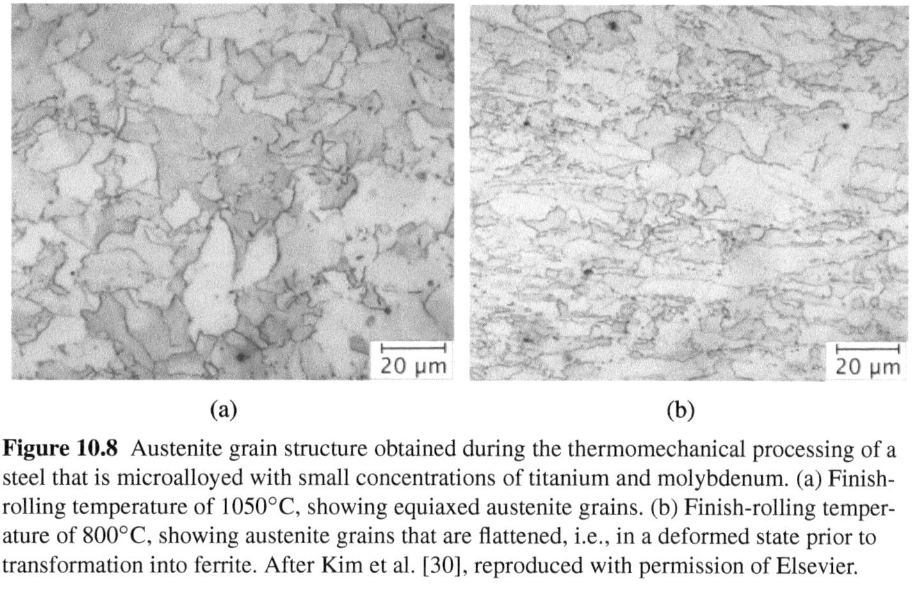

(a) (b)

Figure 10.8 Austenite grain structure obtained during the thermomechanical processing of a steel that is microalloyed with small concentrations of titanium and molybdenum. (a) Finish-rolling temperature of 1050°C, showing equiaxed austenite grains. (b) Finish-rolling temperature of 800°C, showing austenite grains that are flattened, i.e., in a deformed state prior to transformation into ferrite. After Kim et al. [30], reproduced with permission of Elsevier.

As a result of the combined use of controlled rolling and fine dispersions of carbonitrides in low-alloy steels, it has been possible to obtain routinely, ferrite grain sizes between 5 and 10 μm, in commercial practice. Laboratory tests have achieved grain sizes approaching 1 μm, which appears to be a practical limit using this approach. The Hall–Petch relationship between grain size and yield strength, which was discussed in Chapter 2, is very relevant to microalloyed steels, and in fact, linear plots are obtained for the yield stress against $(\overline{L}_\alpha)^{-1/2}$, Fig. 10.9b. Addition of 0.05-0.09 wt% Nb to a plain carbon steel refines the ferrite grain size, allowing it to be reduced to below 5 μm $[(\overline{L}_\alpha)^{-1/2} = 14\,\text{mm}^{-1/2}]$, with a consequent substantial increase in yield strength. The displacement of ~100 MPa between the C-Mn and C-Mn-Nb curves arises from dispersion strengthening due to NbC. This is further illustrated in the two lower curves of Fig. 10.10, which were obtained from specimens austenitised at 950°C

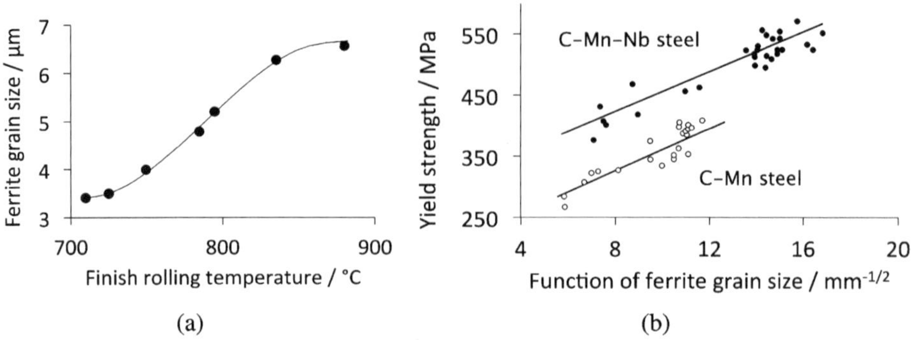

Figure 10.9 (a) Effect of finish rolling temperature on final ferrite grain size of a microalloyed steel (after McKenzie, [31]). (b) Effect of grain size, expressed as $(\overline{L}_\alpha)^{-1/2}$, on yield stress of a carbon-manganese-niobium steel (after Le Bon and Saint Martin, [32]).

prior to air cooling. If, however, progressively higher austenitising temperatures are used, e.g., 1100°C and 1250°C followed by air cooling, the resulting curves, although still linear, have much steeper slopes, indicating a marked increase in yield strength for a particular grain size. This large increment in strength is due to the precipitation of NbC during cooling, following its solution at the higher austenitising temperatures.

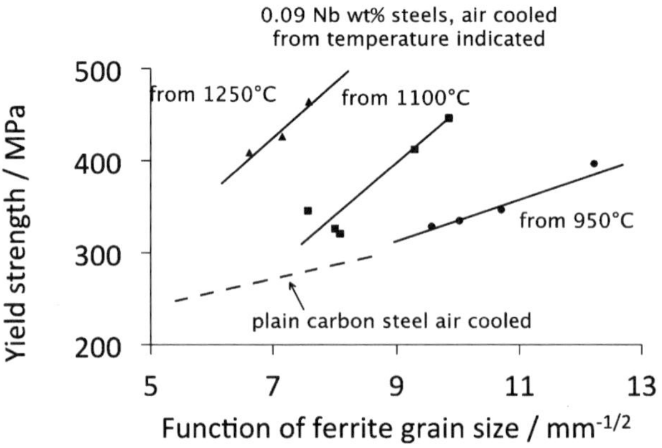

Figure 10.10 Effect of austenitising temperature on the yield strength of a 0.1C- 0.6Mn-0.09Nb wt% steel (adapted from Gladman et al. [33]). The ferrite grain size is expressed as $(\overline{L}_\alpha)^{-1/2}$.

10.2.3 *Niobium atom clusters*

Much of the discussion on the role of microalloying additions naturally focuses on precipitation, but high-resolution techniques have revealed that niobium atoms in particular can cluster within the austenite prior to precipitation proper [34]. The density and size of clusters increases as the temperature is reduced (Fig. 10.11a). The composition of the clusters is not constant, but evolves as a function of temperature. They are said not to significantly affect the recrystallisation of deformed austenite, unless the deformation is conducted at a low temperature.

The mechanism by which such clusters form is not clear, since the iron-niobium solution is of a kind where clustering is not favoured, as shown in Fig. 10.11b and is evident from the fact that the activity coefficient of niobium in austenite is less than one [35]. It is possible, therefore, that the clusters represent nuclei of NbC. The fact that they enrich in composition as they grow may simply be a reflection of the capillarity effect [36], where the interfacial energy changes the equilibrium between the precipitate and matrix due to the very large surface to volume ratio due to the small particle size.

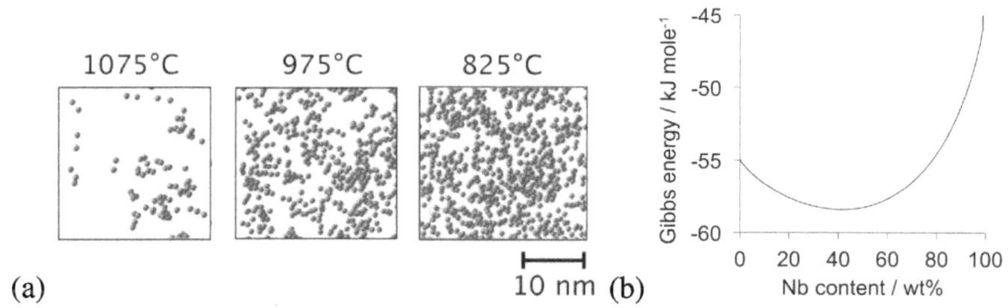

Figure 10.11 (a) Niobium clusters observed using atom-probe tomography. The temperature indicated is that at which the austenite was deformed. The atoms illustrated are only those of niobium. After Kostryzhev et al. [34], reproduced with the permission of Elsevier. (b) Calculated austenite free energy curve for mixtures of iron and niobium at 900°C. The data are generated using MatCalc Version 5.62 with the mc_fe_v2.021.tdb database, courtesy of Arunim Ray.

10.2.4 Minimum achievable grain size

It is interesting to consider what might be the smallest grain size achievable using large-scale thermomechanical processing [37]. A reduction in ferrite grain size (mean linear intercept $\overline{L}$) is equivalent to an increase in the amount of grain boundary surface per unit volume (S_V), since $\overline{L} = 2/S_V$. Grain boundaries have an energy $\sigma_{\alpha\alpha}$ per unit area, so the interfacial energy stored per unit volume of steel is:

$$\sigma_{\alpha\alpha} \times S_V \equiv 2\sigma_{\alpha\alpha}/\overline{L}.$$

This stored energy cannot exceed the magnitude of the free energy change when austenite transforms to ferrite, i.e., $|\Delta G_V^{\gamma\alpha}|$:

$$|\Delta G_V^{\gamma\alpha}| \geq \frac{2\sigma_{\alpha\alpha}}{\overline{L}_\alpha}.$$

It follows that the smallest ferrite grain size that can be achieved is when all of $\Delta G_V^{\gamma\alpha}$ is used up in creating α/α grain boundaries, so that:

$$\overline{L}_\alpha^{\min} = \frac{2\sigma_{\alpha\alpha}}{|\Delta G_V^{\gamma\alpha}|}.$$

Fig. 10.12 shows the variation in this limiting ferrite grain size as a function of $\Delta G_V^{\gamma\alpha}$, together with a compilation of experimental data on the smallest grain size achieved commercially, using thermomechanical processing [37]. The curve indicates that at large grain sizes, $\overline{L}_\alpha^{\min}$ is sensitive to $\Delta G_V^{\gamma\alpha}$ and hence to the undercooling below the equilibrium transformation temperature. However, reductions in grain size in the sub-micrometre range require huge values of $|\Delta G_V^{\gamma\alpha}|$, meaning that the transformations would have to be suppressed to large undercoolings to achieve fine grain size.

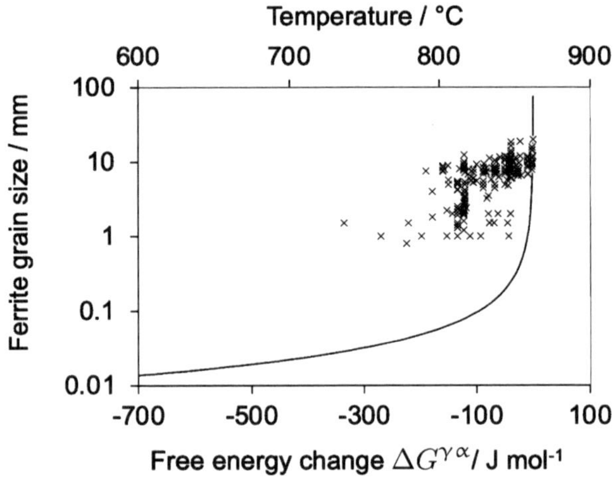

Figure 10.12 Plot of the logarithm of ferrite grain size versus the free energy change available to generate grain boundaries. The curve represents the values of $\overline{L}_\alpha^{\min}$. The free energy change increases in magnitude as the undercooling below the equilibrium transformation temperature is increased. The points are experimental data.

Comparison of the industrial data against the calculated curve indicates that, despite tremendous efforts in developing processing routes, the smallest ferrite grain size obtained commercially using thermomechanical processing is stuck at values greater than 1 μm. The reason is *recalescence*, which is the rise in temperature of the steel caused by release of the latent heat of transformation at a rate which is so high that it cannot easily be dissipated by diffusion. It causes the temperature of the steel to rise, thus reducing $\Delta G_V^{\gamma\alpha}$ and preventing the achievement of ultrafine grain structures. Large-scale thermomechanical processing is therefore limited by recalescence and is unlikely to lead to grain sizes uniformly less than about 1 μm.

10.2.5 Dispersion strengthening during controlled rolling

The solubility data imply that, in a microalloyed steel, carbides and carbo-nitrides of Nb, Ti and V precipitate progressively during controlled rolling as the temperature falls. While the primary effect of these fine dispersions is to control grain size, dispersion strengthening takes place. The strengthening arising from this cause depends on both the particle size r and the interparticle spacing L_P, which is determined by the volume fraction of precipitate (Eq. (2.13)). These parameters depend primarily on the type of compound which is precipitating, and this is determined by the microalloying content of the steel. However, the maximum solution temperature reached and the detailed schedule of the controlling rolling operation are also important variables.

It is now known not only that precipitation takes place in the austenite, but also that further precipitation occurs during the transformation to ferrite. The precipitation of niobium, titanium and vanadium carbides has been shown to take place progressively as the interphase boundaries move through the steel. This is the interphase precipitation discussed in Section 4.5.3. As this precipitation is normally on an extremely fine scale occurring between 850°C and 650°C, it is likely to be the major contribution to the dispersion strengthening. In view of the higher solubility of vanadium carbide in austenite, the effect is most pronounced in the presence of this element, with titanium and niobium in decreasing order of effectiveness. When the rate of cooling through the transformation is large, leading to the formation of supersaturated plates of ferrite, the carbides tend to precipitate within the grains, usually on the dislocations, which are numerous in this type of ferrite.

In determining optimum compositions of microalloyed steels, it should be borne in mind that the maximum volume fraction of precipitate that can be put into solid solution in austenite at high temperatures is achieved by use of stoichiometric compositions. For example, if titanium (atomic weight 47.9) is used, it combines with approximately one quarter its weight of carbon (atomic weight 12), so that for a 0.025 wt% C steel, 0.10 wt% of Ti will provide carbide of the stoichiometric composition. In Fig. 10.13 the stoichiometric line for TiC is shown superimposed on the solubility curves for titanium carbide at 1100°C, 1200°Cand 1300°C. If the precipitation in steels with 0.10 wt% titanium cooled from 1200°C is considered, at low carbon contents, i.e. to the left of the stoichiometric line, the carbide fraction is limited by the carbon content, i.e. zone A, lower diagram. For carbon contents between the stoichiometric line and the solubility line at 1200°C, the full potential volume fraction of fine TiC will form on cooling (zone B). When the carbon content exceeds the solubility limit (>0.10 wt%), the titanium is progressively precipitated at 1200°C as coarse carbide, thus reducing the amount of titanium available to combine with carbon to form fine TiC during cooling. As coarse carbide particles are ineffective in controlling grain growth, it is desirable to have steel compositions, which avoid their formations. It also follows from Fig. 10.13 that high austenitising temperatures are essential to obtain full benefit from the precipitation of finely divided carbide phases.

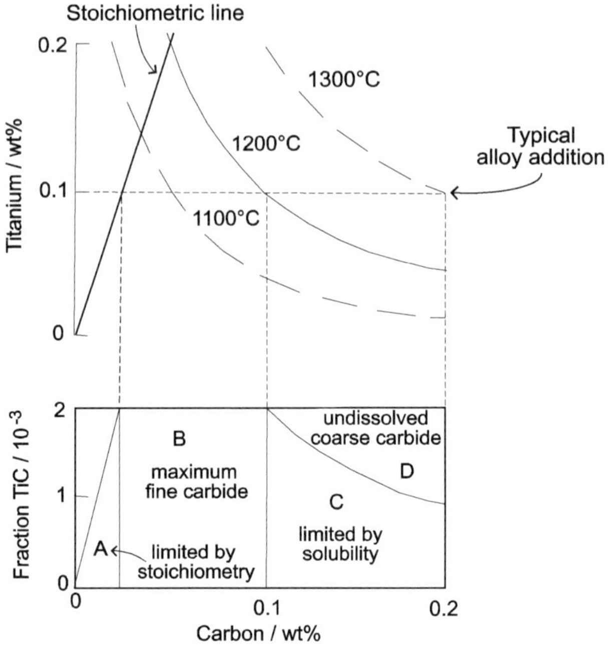

Figure 10.13 Effect of stoichiometry on the precipitation of TiC in a microalloyed steel (after Gladman et al. [33]).

10.2.6 Strength of microalloyed steels: an overall view

In modern controlled microalloyed steels, there are at least three strengthening mechanisms, which contribute to the final strength achieved. The relative contribution from each is determined by the composition of the steel and, equally important, by the details of the thermomechanical treatment to which the steel is subjected. The several strengthening contributions for steels with 0.2 wt% carbon, 0.2 wt% silicon, 0.15 wt% vanadium and 0.015 wt% nitrogen as a function of increasing manganese content are shown schematically in Fig. 10.14. Firstly, there is the solid solution strengthening from manganese, silicon and uncombined nitrogen. Secondly, the grain size contribution to the yield stress is shown to be a substantial component, the magnitude of which is sensitive to the detailed thermomechanical history. Finally, a typical increment for dispersion strengthening is shown. The total result is a range of yield strengths between about 350 and 500 MPa. In this particular example, the steel was normalised (air cooled) from 900°C, but had it been control rolled down to 800°C or even lower, the strength levels would have been raised substantially.

The effect of the finishing temperature for rolling is important in determining the grain size, and therefore, strength level reached for a particular steel. It is now possible to roll through the transformation into the completely ferritic condition, and so obtain fine subgrain structures in the ferrite, which provide an additional contribution to strength. Alternatively, the rolling is finished above the γ/α transformation and the nature of the transformation is altered by increasing the cooling rate. Slow rates

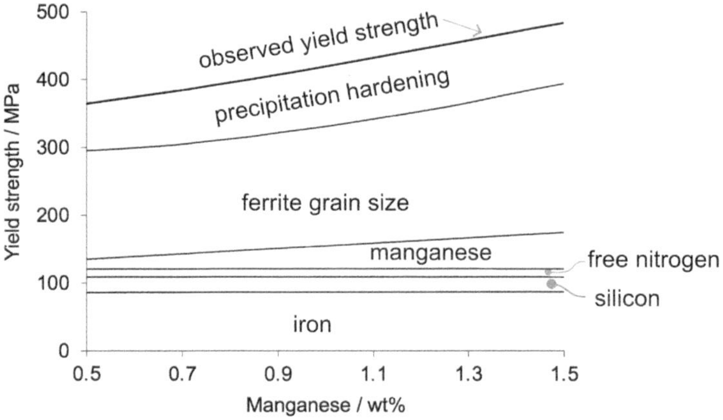

Figure 10.14 The contributions to strength in a 0.2C-0.15V wt% steel as a function of Mn content (after Gladman et al. [33]).

of cooling obtained by coiling at a particular temperature result in lower strengths than rapid rates imposed by water spray cooling following rolling. The latter route can change the ferrite from equiaxed to Widmanstätten with a much higher dislocation density. The result is a steel with improved mechanical properties and, in many cases, the sharp yield point can be suppressed. This has practical advantages in fabrication of sheet steel, e.g., pipe manufacture, where a continuous stress-strain curve is preferred.

10.3 Dual-phase steels

The HSLA steels described in Section 10.2 give improved strength to weight ratios over ordinary steels. However, they are not readily formed, e.g., by cold pressing and related techniques. The worldwide demand for safety and fuel economy in transportation has led to the development of a number of steel types which not only are strong, but have the formability required for mass production of car bodies and components. A measure of formability is the product of strength and uniform elongation.

The *dual-phase* steels are low-alloy steels that meet these requirements by exploiting microstructures in which there are two major phases (Fig 10.15), one of which is soft and the other significantly harder. The ferrite-martensite dual-phase steels typically contain manganese and silicon, and are strong and yet are formable. They exhibit continuous yielding, i.e., no sharp yield point, and a relatively low proof strength (300–350 MPa). The simplest steels in this category contain 0.08–0.2C, 0.5–1.5Mn wt%, but steels microalloyed with vanadium are also suitable, while small additions of Cr (0.5 wt%) and Mo (0.2–0.4 wt%) are used frequently to control the development of microstructure.

The simplest way of achieving a duplex structure is to use intercritical annealing in which the steel is heated into the $(\alpha+\gamma)$ region between the Ae_1 and Ae_3 and held, typically at 790°C for several minutes to allow small regions of austenite to form in

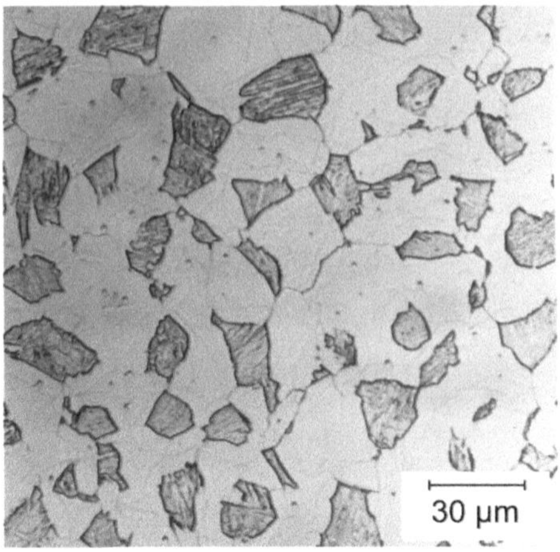

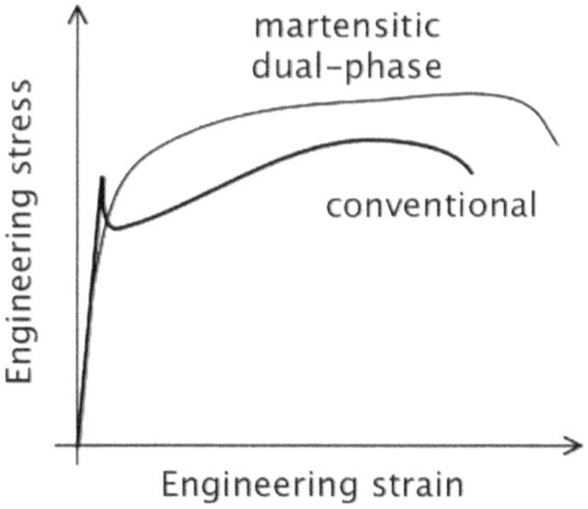

(a) (b)

Figure 10.15 (a) Typical microstructure of dual-phase steel, consisting of a mixture of martensite (dark) and ferrite. (b) Schematic stress-strain curves comparing the behaviour of a conventional automobile steel with that of a dual-phase steel.

the ferrite. As it is essential to transform these regions of austenite into martensite, cooling to ambient temperature must be sufficiently rapid to avoid other intervening transformations. Alternatively, the hardenability of the austenite must be enhanced by adding between 0.2 and 0.4 wt% Mo to a steel already containing 1.5 wt% manganese. The required microstructure can then be obtained by air cooling after intercritical annealing.

To eliminate an extra heat treatment step, dual-phase steels have now been developed, which can be given the required structure during cooling after controlled rolling. Typically, these steels have additions of 0.5Cr and 0.4Mo wt%. After completion of hot rolling around 870°C, the steel forms approximately 80% ferrite on the water-cooled run-out table from the mill. The material is then coiled in the metastable region (510–620°C) below the pearlite/ferrite transformation, and on subsequent cooling the austenite regions transform to martensite.

10.4 TRIP-assisted steels

The steels developed to exploit the properties obtained when the martensite reaction occurs during plastic deformation are known as *transformation-induced plasticity* (TRIP) steels. They are strong and exhibit considerable uniform elongation before failure. There are several varieties of such steels. Those which are made fully austenitic by using large quantities of austenite-stabilising solutes, but transform to martensite when stressed, are simply called the TRIP steels (discussed in Chapter 12). When the austenite is a minor phase in the overall microstructure, but undergoes martensitic transformation during straining, the steels are said to be TRIP assisted and are usually low alloy steels.

Martensitic transformation induced by local stress has the effect of relieving stress concentrations, increasing the work-hardening rate, and promoting homogeneous deformation, with consequent improvements in the strength, ductility and toughness of steels. TRIP-assisted steels are mass produced, made using a complex heat treatment, which is often completed within a short time during the processing of steel strip. Their microstructure consists of allotriomorphic ferrite as the major phase together with a total of 30-40% of harder regions. The latter consist of mixtures of bainite, martensite and carbon-enriched retained austenite. The chemical composition is typically Fe-0.12C-1.5Si-1.5Mn wt%. Some austenite is retained in spite of the low overall solute content because when the bainite forms, the silicon prevents cementite precipitation, thereby enriching the residual austenite with carbon (Chapter 6). The major application of TRIP-assisted steels is in the automobile industries, both for painted surfaces and for enhancing the safety of the passenger compartment in the event of a crash.

There are two kinds of TRIP-assisted steels. In the first, cold-rolled strip is heated rapidly from ambient temperature for intercritical treatment in the $\alpha+\gamma$ phase field between the Ac_1 and Ac_3 temperatures (Fig. 10.16). The intercritical annealing induces partial transformation to austenite and at the same time recrystallises the residual ferrite. The strip is then cooled at a controlled rate during which some of the austenite transforms into allotriomorphic ferrite and at lower temperatures into bainitic ferrite. This latter reaction causes the austenite to become enriched in carbon, allowing it to be retained to ambient temperature (Fig. 10.17).

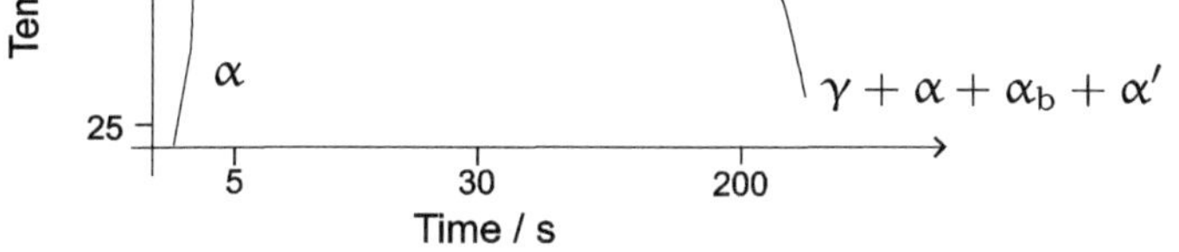

Figure 10.16 The two kinds of heat treatment used to generate the microstructures of TRIP-assisted steels. The terms γ, α, α_b and α' represent austenite, allotriomorphic ferrite, bainite and martensite, respectively. The continuous line represents intercritical annealing to generate the mixture of $\gamma + \alpha$, whereas the dashed line starts with a fully austenitised steel.

The details of the microstructure and mechanical properties can be altered by manipulating the cooling conditions. For example, it is common practice to allow more time in the bainite transformation range than at the higher temperatures where allotriomorphic ferrite grows. Fig. 10.18 shows the effect of holding in the bainite transformation temperature range on the final microstructure. An inadequate amount of bainite leaves the austenite susceptible to martensitic transformation. Similarly, be-

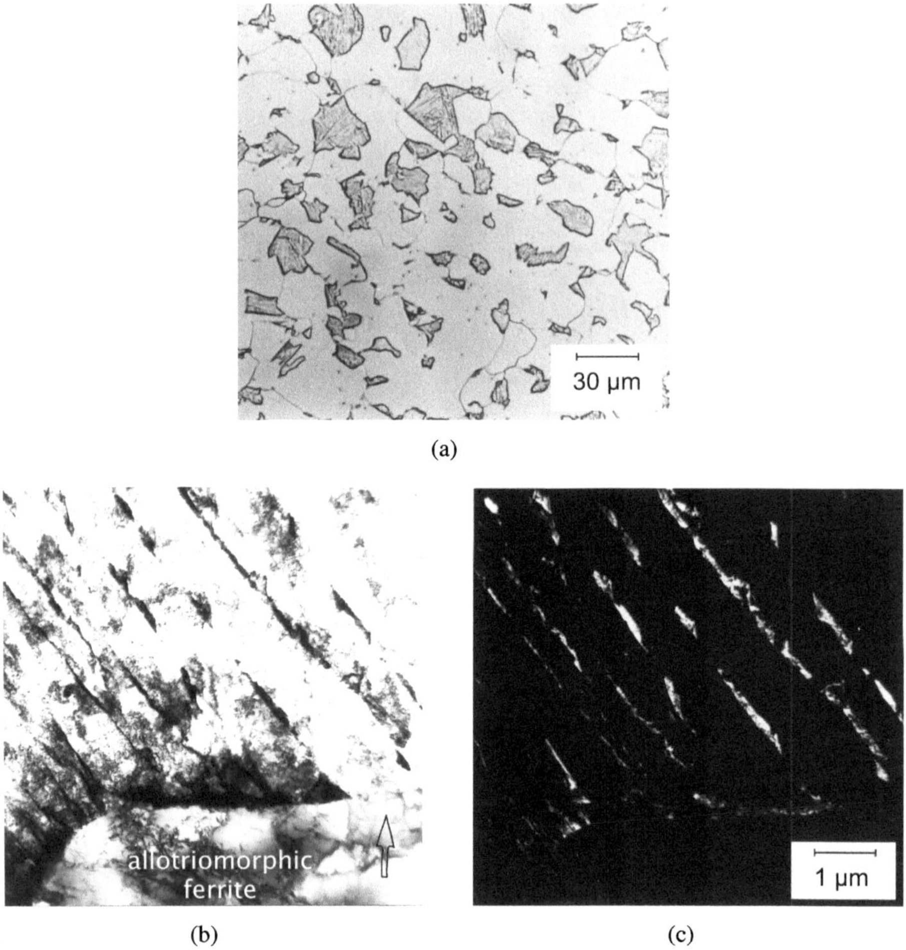

Figure 10.17 TRIP-assisted steel showing a mixed microstructure of allotriomorphic ferrite, bainitic ferrite and retained austenite films. (a) Optical micrograph. (b) Brigh-field transmission electron micrograph. (c) Dark-field image of retained austenite.

cause the carbon concentration in the austenite is limited by the T_0 curve (Chapter 6), transformation to bainite at too high a temperature also renders the austenite unstable.

The second kind of heat treatment starts from a hot-rolled strip which is fully austenitic (Fig. 10.16) and forms both allotriomorphic ferrite and bainite during the cooling part of the thermal cycle. This has the advantage that the microstructure can be produced directly from the hot strip which has been rolled to its final dimensions. The process is cheap since the strip does not have to be heated to the intercritical annealing temperature. However, hot-rolling mills are restricted by rolling loads to strips thicker than about 3 mm, although there are modern mills, which can cope with 1.4 mm thickness. Cold-rolled strips can, on the other hand, be made routinely into thinner gauges.

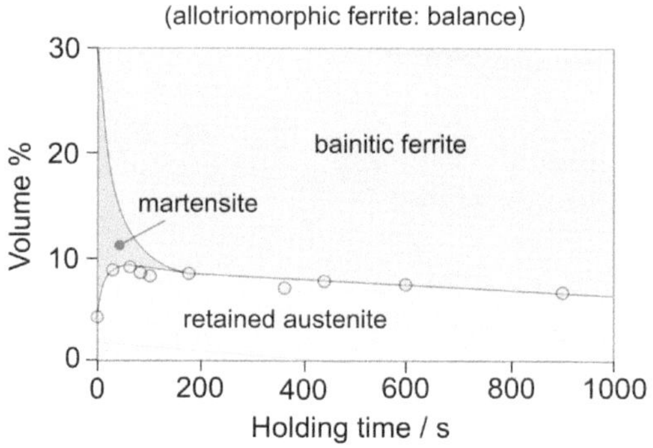

Figure 10.18 Evolution of room temperature microstructure as a function of the time during isothermal transformation to bainite in Fe-0.11C-0.06Si-1.5Al-1.55Mn wt%. Adapted using data from Girault et al. [38].

Hot-rolled strips are preferred for some automobile applications where cost is a prime factor in the choice of materials.

The transformation strain due to the formation of martensite does not account for the observed uniform tensile elongation of some 15–30%. The shape deformation due to martensite (Chapter 5) is at most equivalent to a 2% tensile strain because the amount of austenite available for transformation is quite small in TRIP-assisted steels. The major contributions to uniform elongation arise partly from the enhanced work-hardening coefficient of the material due to the progressive formation of hard martensite during deformation. There is a further significant contribution from dislocations induced into the ferrite by the strains associated with martensitic transformation [39]. These dislocations strengthen the ferrite and are visible in Fig. 10.17b.

The austenite also delays the necking process during a tensile test by transforming into martensite at stress concentrations. It is important therefore to delay the transformation of retained austenite to the late stages of deformation, when significant damage accumulates in the steel. At this time, the TRIP effect can be most beneficial. It is useful therefore to examine further the transformation of austenite as a function of plastic strain.

It is reasonable to assume that the change in the fraction of martensite ($dV_V^{\alpha'}$) obtained for a given increment of plastic strain ($d\epsilon_p$) should be proportional to the fraction of remaining austenite:

$$\frac{dV_V^{\alpha'}}{d\epsilon_p} = k_\gamma V_V^\gamma, \tag{10.3}$$

where k_γ is a function of the steel composition and test temperature; V_V^γ is the fraction of austenite remaining untransformed. If the fraction of austenite at zero strain is $V_V^{\gamma_\circ}$, then $V^{\alpha'} = V_V^{\gamma_\circ} - V_V^\gamma$, and integration of Eq. (10.3) gives:

$$\ln\{V_V^{\gamma_\circ}\} - \ln\{V_V^\gamma\} = k_\gamma \epsilon_p.$$

The form of this equation is illustrated in Fig. 10.19.

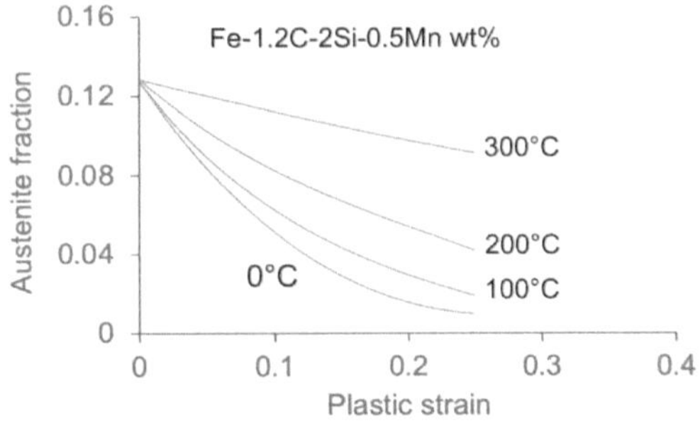

Figure 10.19 Martensitic transformation of retained austenite in a TRIP-assisted steel as a function of deformation temperature and plastic strain. $V_V^{\gamma\circ}$ and $k_\gamma = 0.00446\,\mathrm{mol\,J^{-1}}$. After Sherif et al. [40].

10.4.1 Low- or zero-silicon TRIP-assisted steels

The substantial silicon addition to TRIP-assisted steels leads to the formation of hard, adherent oxide (fayalite, Fe_2SiO_4), which is difficult to remove prior to hot rolling. This leads in turn to a blemish on the surface known as *red oxide* [41–44], which is FeO that has been sequentially oxidised into Fe_3O_4 and Fe_2O_3, with the latter being the coloured oxide from which the name derives, resulting in a poor surface finish [41–43]. Routine descaling operations fail to remove the oxide because the fayalite forms a eutectic with FeO and its liquid phase penetrates both the steel and any prior oxide [43,45]. The resulting convoluted interface with the steel provides a mechanical key that prevents the uniform removal of the oxide. This is not acceptable when the surface finish of the steel is paramount.

Since silicon is the culprit, it is natural to seek low-silicon TRIP steels, as illustrated in Fig. 10.20. The austenite in such alloys is more difficult to retain because of the tendency to precipitate cementite, but this can be minimised with careful heat treatment. When this is done, strengths in excess of 600 MPa with a uniform ductility of 15% have been achieved.

10.4.2 Galvanising of TRIP-assisted steels

There are two basic methods of galvanising, by dipping the steel in liquid zinc or by electrolytically depositing the zinc. The typical concentrations of silicon and manganese in TRIP-assisted steels lead to a stable Mn_2SiO_4 oxide film on the surface during the heat treatment that leads to the desired microstructure. This makes it difficult for the zinc to wet the steel surface, making it necessary to electrolytically galvanise such alloys.

The problem can be alleviated by increasing the humidity in the annealing furnace. The oxide coverage of the surface is then reduced, giving better wetting by zinc. The higher humidity causes the internal oxidation of Mn and Si below the steel surface, thus reducing their availability to form Mn_2SiO_4 at the surface.

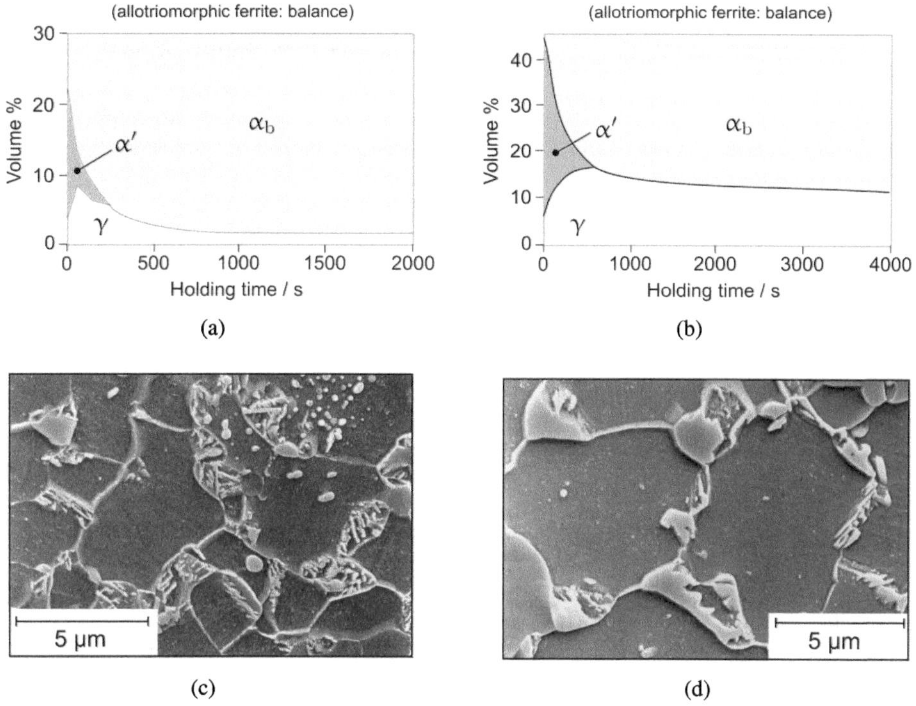

Figure 10.20 Evolution of room-temperature microstructure as a function of time during isothermal transformation to bainite. (a) Low-silicon steel Fe-0.16C-0.38Si-1.3Mn wt%. (b) High-silicon steel Fe-0.29C-1.41Si-1.42Mn wt%. (c) Scanning electron micrograph of low-silicon alloy isothermally transformed to bainite for 1800 s. Much of the austenite has decomposed to bainitic ferrite and cementite. (d) Corresponding micrograph for high-silicon steel transformed to bainite for 900 s with plenty of austenite evident (courtesy of Pascal Jacques).

An alternative approach is to eliminate the silicon and add aluminium to retain the cementite-free microstructure. Aluminium oxidises easily to form alumina by internal oxidation near the surface, again limiting the amount of $FeAl_2O_4$ that can form at the free surface when the humidity in the annealing furnace is low. Such a steel can easily be hot-dip galvanised. At high humidity, MnO begins to cover the surface leading to a deterioration in the ability of the liquid zinc to wet the surface. The aluminium-containing steels are therefore better suited to continuous galvanising lines.

It is clear that silicon and manganese must diffuse to the surface to form oxides. Some of the diffusion flux is via grain boundaries. When phosphorus is present, its segregation to the grain boundaries reduces the boundary diffusion-flux, thereby reducing the extent to which oxides form. Such steels are more amenable to wetting by molten zinc (in contrast, the same effect makes it more difficult to form iron-zinc compounds during the galvannealing process).

10.5 TWIP steels

There are three essential modes by which steels can be permanently deformed at ambient temperature, without recourse to diffusion. Individual dislocations whose Burgers vectors correspond to lattice vectors can glide, leading to a change in shape without altering the crystal structure or volume. In contrast, a displacive transformation (e.g., martensite or bainite) results not only in a plastic strain, but also a change of crystal structure and density. This phenomenon is exploited in the TRIP steels.

The third mode of deformation is mechanical twinning, in which the crystal structure of the steel is preserved, but the twinned region is reoriented in the process. Mechanical twinning results in a much larger shear strain $s = 1/\sqrt{2}$, compared with displacive transformations where s typically is 0.26. There is a particular class of extraordinarily ductile alloys of iron, known as the TWIP steels, which exploit mechanical twinning to achieve their properties.

TWIP stands for twinning-induced plasticity. The alloys are austenitic and remain so during mechanical deformation, but the material is able to accommodate strain via both the glide of individual dislocations and through mechanical twinning on the $\{1\ 1\ 1\}_\gamma \langle 1\ 1\ \bar{2}\rangle_\gamma$ system. The alloys typically contain a large amount of manganese, some aluminium and silicon (e.g., Fe-25Mn-3Si-3Al wt%), with carbon and nitrogen in some versions present essentially as impurities. Larger concentrations of carbon may be added to enhance strength.

At high manganese concentrations, there is a tendency for the austenite to transform into ϵ-martensite (hexagonal close packed) during deformation. ϵ-martensite can form by the dissociation of a perfect $a/2\,\langle 0\ 1\ 1\rangle_\gamma$ dislocation into Shockley partials on a close packed $\{1\ 1\ \bar{1}\}_\gamma$ plane, with a fault between the partials. This faulted region represents a three layer thick plate of ϵ-martensite [46,47]. A reduction in the fault energy therefore favours the formation of this kind of martensite. The addition of aluminium counters this because it raises the stacking fault energy of the austenite. Silicon has the opposite effect of reducing the stacking fault energy, but like aluminium, it leads to a reduction in the density of the steel. The combination of Al and Si at the concentrations used typically reduces the overall density from some 7.8 g cm^{-3} to about 7.3 g cm^{-3}.

The alloys have a rather low yield-strength at 200–300 MPa, but the ultimate tensile strength can be much higher, over 1100 MPa. This is because the strain-hardening coefficient is large, resulting in a great deal of uniform elongation, and a total elongation of some 60–95%. The effect of mechanical twinning is two-fold. The twins add to plasticity, but they also have a powerful effect in increasing the work-hardening rate by subdividing the untwinned austenite into finer regions (Fig. 10.21).

The large strain-hardening capacity of high-Mn steel has been shown to enhance erosion resistance in pipelines that transport mineral slurries in the context of oil sands, which consist of about 85% silica sand, 6-15% bitumen and a residue of clay and water. The bitumen is later converted into crude oil. The big problem, therefore, is the abrasive wear of the steel. Quite dramatic wear results during the transmission of the slurry through steel pipes, typically at a speed of 5.5 m s^{-1}. However, the TWIP steel significantly outperforms ferritic pipeline steels, as established initially in a trial where a 12 m seam-welded pipe segment was inserted into a long production pipeline con-

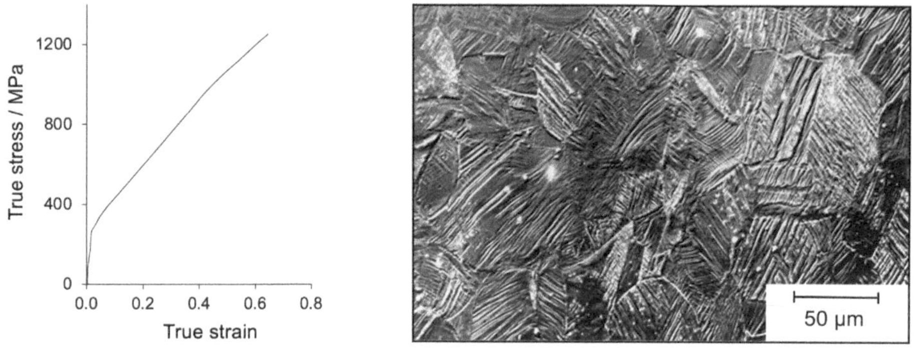

(a) (b)

Figure 10.21 (a) Typical stress-strain curve for a TWIP steel. (b) Optical microstructure of a TWIP steel following deformation, showing profuse twinning (image and data courtesy of Frommeyer, G., Brüx, U. and Neumann, P).

necting the mine to the extraction facility [48]. Some 1.2 km of such slurry pipes have now been installed in Canada.

An advantage of high-Mn steels is that they are austenitic and they maintain attractive properties at cryogenic temperatures (−150°C) and high strain rates, e.g., $10^3\ s^{-1}$. They therefore have great potential in applications wider than just transportation. High-manganese steel has been used in the manufacture of a very large tank for liquefied natural gas [49], but the exact chemical composition and structural changes during deformation have not been published, so it is not clear whether the steel involved exhibits twinning-induced plasticity. The aim is to replace the more expensive 9Ni wt% steels that currently serve that purpose.

10.6 Thermomechanically treated industrial steels

Microalloyed steels produced by controlled rolling are a most attractive proposition in many engineering applications because of their relatively low cost, moderate strength, and very good toughness and fatigue strength, together with their ability to be readily welded. They have, to a considerable degree, eliminated quenched and tempered steels in many applications.

These steels are most frequently available in control-rolled sheet, which is then coiled over a range of temperatures between 750°C and 550°C. The coiling temperature has an important influence as it represents the final transformation temperature, which is reflected in the microstructure. A lower coiling-temperature in general boosts the strength. The normal range of yield strength obtained in these steels varies from about 350 to 550 MPa. The strength is controlled both by the detailed thermomechanical treatment, by varying the manganese content from 0.5 to 1.5 wt%, and by using the micro-alloying additions in the range 0.03 to above 0.1 wt%. Niobium is used alone or with vanadium, while titanium can be used in combination with the other two carbide-forming elements. The interactions between these elements are complex, but in general terms niobium precipitates more readily in austenite than does vanadium as carbide or carbo-nitride, so it is more effective as a grain refiner. The greater solubility of vanadium carbide in austenite underlines the superior dispersion strengthening potential of this element shared to a lesser degree with titanium. Titanium also interacts with sulphur and can have a beneficial effect on the shape of sulphide inclusions. Bearing in mind that the total effect of these elements used in conjunction is not a simple sum of their individual influences, the detailed metallurgy of these steels becomes extremely complex.

One of the most extensive applications is in pipelines for the conveyance of natural gas and oil, where the improved weldability due to the limited alloying content (lower hardenability) and, particularly, the low carbon levels, is a great advantage. Furthermore, as the need for larger diameter pipes has grown, steels of higher yield stress have to be used to avoid excessive wall thicknesses. In practice, wall thicknesses of 10-12.5 mm have been found to be the most convenient. Typical compositions (wt%) to achieve a yield stress of around 410 MPa:

C 0.12   S 0.012   Mn 1.35   Nb 0.03
C 0.12   S 0.006   Mn 1.33   Nb 0.02   V 0.04

for higher yield strengths (450 MPa):

C 0.06   S 0.006   Mn 1.55   Nb 0.05   V 0.10

It should be emphasised that greater strength often is achieved by control of the fabrication variables, such as the temperature at which rolling is finished and the temperature used for coiling the sheet. Nitrogen is often deliberately used as an alloying element. One successful range of steels relies on vanadium to form carbo-nitride precipitates for grain size control and dispersion strengthening. In some steels, rare earth additions are made to control the inclusion shape. Typical compositions at lower and higher strength levels are given in Table 10.1.

Table 10.1 Typical compositions in wt%, of microalloyed vanadium steels in two classes of yield strength (σ_Y).

Alloying element	Steels with $\sigma_Y = 345$ MPa	Steels with $\sigma_Y = 550$ MPa
Carbon	0.08-0.12	0.12-0.17
Manganese	0.75-1.10	1.20-1.55
Phosphorus	0.008-0.013	0.008-0.013
Sulphur	0.007-0.020	0.007-0.020
Silicon	0.05-0.15	0.30-0.55
Aluminium	0.03-0.06	0.03-0.06
Vanadium	0.03-0.07	0.10-0.14
Nitrogen	0.006-0.012	0.015-0.022
Cerium	0.02-0.06	0.02-0.06

At the higher strength levels, microalloyed steels are used for heavy duty truck frames, tractor components, crane booms and lighting standards, etc. The control of sulphide inclusions using rare-earth elements, gives the steels a high degree of formability in the context of cold-fabrication processes. This recent development has allowed the use of HSLA steels for many applications, which previously led to cracking during forming.

10.7 Ausforming

A steel is said to have been ausformed when martensite is produced from plastically deformed austenite. Ausforming has provided some of the strongest, toughest steels so far produced, with the added advantage of good fatigue resistance. However, they usually contain significant concentrations of expensive alloying elements and must be subjected to large deformations that impose heavy work loads on rolling mills. Nevertheless, such steels are particularly useful where a high strength to weight ratio is required and where cost is a secondary factor. Typical applications have included parts for undercarriages of aircraft, special springs and bolts.

The 12 wt% Cr transformable steels respond readily to ausforming to the extent that tensile strengths of over 3000 MPa can be obtained in appropriate compositions. A 0.4C-6Mn-3Cr-1.5Si steel has been ausformed to a tensile strength of 3400 MPa, with an improvement in ductility over the conventional heat treatment. Similar high strength levels with good ductility have been reported for 0.4C-5Cr-1.3Mo-1.0Si-0.5V wt% steel (Fig. 10.22). All of these steels are sufficiently highly alloyed to allow adequate time for substantial deformation in the austenitic bay of the time-temperature transformation curve prior to transformation.

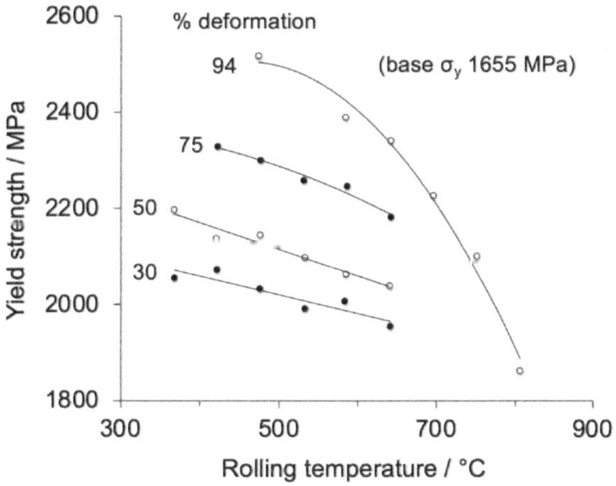

Figure 10.22 Effect of amount and temperature of deformation on the yield strength of 0.4C-5.0Cr-1.3Mo-1.0Si-0.5V wt% steel quenched after ausforming and tempered at 510°C (after Schmatz et al. [50]).

10.8 Summary

There is no need to wax lyrical about the success of high-strength, low-alloy steels. The steels pervade so many aspects of everyday life that we are numbed to their presence. One reason is that the reliability of these steels far exceeds the annoying quality of the software that we are also exposed to in our everyday interactions with technology. It is not necessary, therefore, for the ordinary person to think about the technology behind this wonder material.

The notion that minute quantities of solutes can so dramatically influence the properties of steels is now well-established. When combined with thermomechanical processing, one vision is that the variety of steels that exist today will be greatly reduced to facilitate the recycling of steels.

There will, of course, remain a demand for special steels, which are made in relatively small quantities, for example the TWIP alloys, where the imagination can run wild as to what applications will be possible in the future.

References

[1] D.H. Kirkwood, Semisolid metal processing, International Materials Reviews 39 (2013) 173–189.
[2] S.S. Babu, R. Goodridge, Additive manufacturing, Materials Science and Technology 31 (2015) 881–1010.
[3] T. DebRoy, H.K.D.H. Bhadeshia, Innovations in Everyday Engineering Materials, Springer Nature, Switzerland, 2021.
[4] E.J. Pickering, Macrosegregation in steel ingots: the applicability of modelling and characterisation techniques, ISIJ International 53 (2013) 935–949.
[5] E.J. Pickering, S.S. Al-Bermani, J. Talamantes-Silva, Application of criterion for A-segregation in steel ingots, Materials Science and Technology 31 (2015) 1313–1319.
[6] K.W. Burns, F.B. Pickering, Deformation and fracture of ferrite-pearlite structures, Journal of the Iron and Steel Institute 202 (1964) 889–906.
[7] A.R. Rosenfield, G.T. Hahn, J.D. Embury, Fracture of steels containing pearlite, Metallurgical Transactions 3 (1972) 2797–2804.
[8] B.B. Argent, Research on the metallurgy of niobium at the University of Sheffield, Science Progress 52 (1964) 608–627.
[9] F.B. Pickering, Physical Metallurgy and the Design of Steels, Applied Science Publishers, Essex, U.K., 1978, p. 104.
[10] C. Zener, J.H. Hollomon, Effect of strain rate upon plastic flow of steel, Journal of Applied Physics 15 (1944) 22–32.
[11] C. Zener, J.H. Hollomon, Plastic flow and rupture of metals, Transactions of ASM 33 (1944) 163–215.
[12] D. Bombac, M. Peet, S. Zenitani, S. Kimura, T. Kurimura, H.K.D.H. Bhadeshia, An integrated hot-rolling and microstructure model for dual-phase steels, Modelling and Simulation in Materials Science and Engineering 22 (2014) 045005.
[13] I. Tamura, H. Sekine, T. Tanaka, C. Ouchi, Thermomechanical Processing of High-Strength Low-Alloy Steels, Butterworth & Co., London, U.K., 1988.
[14] T. Gladman, The Physical Metallurgy of Microalloyed Steels, IOM Communications, London, 1996.
[15] T. Senuma, H. Yada, Y. Matsumura, T. Futamura, Structure of austenite of carbon steels in high speed hot working processes, Tetsu-to-Hagane 70 (1984) 2112–2119.

[16] C.M. Sellars, J.A. Whiteman, Principles of hot deformation, in: Controlled Processing of HSLA Steels, vol. Paper 64, British Steel Corporation, Sheffield, U.K., 1976, pp. 1–19.
[17] C.M. Sellars, J.A. Whiteman, Recrystallization and grain growth in hot rolling, Metal Science 13 (1979) 187–194.
[18] C.M. Sellars, Computer modelling of hot working processes, Materials Science and Technology 1 (1985) 325–332.
[19] E.J. Palmiere, C.M. Sellars, S.V. Subramanian, Modelling of thermomechanical rolling, in: Niobium 2001, TMS, Warrendale, Pennsylvania, USA, 2001, pp. 501–526.
[20] R.C. Hwang, Y.J. Chen, H.C. Huang, Artificial intelligent analyzer for mechanical properties of rolled steel bar by using neural networks, Expert Systems with Applications 37 (2010) 3136–3139.
[21] I. Mohanty, S. Sarkar, B. Jha, S. Das, R. Kumar, Online mechanical property prediction system for hot rolled IF steel, Ironmaking and Steelmaking 41 (2014) 618–627.
[22] M. Schlang, B. Feldkeller, B. Lang, T. Poppe, T. Runkler, Neural computation in steel industry, in: Control Conference (ECC), 1999 European, IEEE Computer Society, Washington, D.C., USA, 1999, pp. 2922–2927.
[23] S.B. Singh, H.K.D.H. Bhadeshia, D.J.C. MacKay, H. Carey, I. Martin, Neural network analysis of steel plate processing, Ironmaking and Steelmaking 25 (1998) 355–365.
[24] K.J. Irvine, F.B. Pickering, T. Gladman, Grain refined C-Mn steels, Journal of the Iron and Steel Institute 205 (1967) 161–182.
[25] NPL, MTDATA, Software, National Physical Laboratory, Teddington, U.K., 2006.
[26] I. Czinege, T. Reti, Determination of local deformation in cold formed products by a measurement of the geometric characteristics of the crystallites, in: Eighteenth International Machine Tool Design and Research Conference, Forming, vol. 1, 1977, pp. 159–163.
[27] S.B. Singh, H.K.D.H. Bhadeshia, Topology of grain deformation, Materials Science and Technology 15 (1998) 832–834.
[28] T. Tanaka, N. Tabata, T. Hatomura, C. Shiga, Three stages of the controlled-rolling process, in: M. Korchynsky, L.S. Richardson, G.J. Klems, J.M. Gray (Eds.), Microalloying '75, Union Carbide Corporation, New York, USA, 1975, pp. 107–119.
[29] A.R. Jones, B. Ralph, The influence of recrystallization on carbide particle distributions in a fully stabilized austenitic steel, Acta Metallurgica 23 (1975) 355–363.
[30] J.I. Kim, J.H. Pak, K.S. Park, J.H. Jang, D.W. Suh, H.K.D.H. Bhadeshia, Segregation of phosphorus to ferrite grain boundaries during transformation in Fe-P alloy, International Journal of Materials Research 105 (2014) 1166–1172.
[31] C.R. McKenzie, Unpublished research, in: Rosenhain Centenary Metallurgy Conference: the Contribution of Physical Metallurgy to Engineering Practice, The Royal Society, London, U.K., 1975, p. 293.
[32] A.B. Le Bon, L.N. Saint-Martin, Using laboratory simulation to improve rolling schedules, in: M. Korchynsky, L.S. Richardson, G.J. Klems, J.M. Gray (Eds.), Microalloying '75, Union Carbide Corporation, New York, USA, 1975, pp. 90–99.
[33] T. Gladman, D. Dulieu, I.D. McIvor, Structure/property relationships in high-strength micro-alloyed steels, in: M. Korchynsky, L.S. Richardson, G.J. Klems, J.M. Gray (Eds.), Microalloying '75, Union Carbide Corporation, New York, USA, 1975, pp. 32–55.
[34] A.G. Kostryzhev, A. Al Shahrani, C. Zhu, S.P. Ringer, E.V. Pereloma, Effect of deformation temperature on niobium clustering, precipitation and austenite recrystallisation in a Nb–Ti microalloyed steel, Materials Science & Engineering A 581 (2013) 16–25.
[35] K. Balasubramanian, A. Kroupa, J.S. Kirkaldy, Experimental investigation of the thermodynamics of Fe-Nb-C austenite and nonstoichiometric niobium and titanium carbides ($T = 1273$ to 1473 K), Metallurgical Transactions A 23 (1992) 729–744.

[36] J.W. Christian, Theory of Transformations in Metals and Alloys, Part I, 3rd ed., Pergamon Press, Oxford, U.K., 2003.
[37] T. Yokota, C. Garcia-Mateo, H.K.D.H. Bhadeshia, Formation of nanostructured steel by phase transformation, Scripta Materialia 51 (2004) 767–770.
[38] E. Girault, A. Mertens, P. Jacques, Y. Houbaert, B. Verlinden, J. van Humbeeck, Comparison of the effects of silicon and aluminium on the tensile behaviour of multiphase TRIP-assisted steels, Scripta Materialia 44 (2001) 885–892.
[39] P.J. Jacques, Q. Furnemont, A. Mertens, F. Delannay, On the sources of work hardening in multiphase steels by transformation-aided plasticity, Philosophical Magazine. A 81 (2001) 1789–1812.
[40] M. Sherif, C. Garcia-Mateo, T. Sourmail, H.K.D.H. Bhadeshia, Stability of retained austenite in TRIP-assisted steels, Materials Science and Technology 20 (2004) 319–322.
[41] T. Fukagawa, H. Okada, Y. Maehara, Mechanism of red scale defect formation in Si-added hot-rolled steel sheets, ISIJ International 34 (1994) 906–911.
[42] H. Okada, T. Dukagawa, A. Okamoto, M. Azuma, Y. Matsuda, Prevention of red scale formation during hot rolling of steels, ISIJ International 35 (1995) 886–891.
[43] S. Taniguchi, K. Yamamoto, D. Megumi, T. Shibata, Characteristics of scale/substrate interface area of Si-containing low-carbon steels at high temperatures, Materials Science & Engineering A 308 (2001) 250–257.
[44] E.J. Song, D.W. Suh, H.K.D.H. Bhadeshia, Oxidation of silicon containing steel, Ironmaking and Steelmaking 39 (2012) 599–604.
[45] L. Suárez, J. Schneider, Y. Houbaert, Effect of Si on high-temperature oxidation of steel during hot rolling, Defect and Diffusion Forum 273–276 (2008) 655–660.
[46] J.W. Brooks, M.H. Loretto, R.E. Smallman, In situ observations of martensite formation in stainless steel, Acta Metallurgica 27 (1979) 1829–1838.
[47] J.W. Brooks, M.H. Loretto, R.E. Smallman, Direct observations of martensite nuclei in stainless steel, Acta Metallurgica 27 (1979) 1839–1847.
[48] R. Chung, J. Jiang, C. Pang, B. Yu, R. Eadie, D. Li, Erosion-corrosion behaviour of high manganese steel used in slurry pipelines, Wear 530–531 (2023) 204885.
[49] J.H. Kim, M.-K. Lee, W. Jang, J.-H. Lee, Strain behavior of very new high manganese steel for 200,000 m^3 LNG cryogenic storage tank, Energy 271 (2023) 126889.
[50] D.J. Schmatz, F.W. Schaller, V.F. Zackay, Structural aspects and properties of martensite of high strength, in: National Physical Laboratory Symposium: The Relation Between the Structure and Mechanical Properties of Metals, Her Majesty's Stationary Office, London, U.K., 1963, pp. 613–645.

Embrittlement and fracture

11

Abstract

A brittle material usually is hard and therefore less able to accommodate plasticity, Thus, it can easily be broken. To embrittle means to make something brittle. This is undesirable in most cases. In particular, the nicely engineered properties of steel can be ruined by a number of mechanisms that cause the steel to split or splinter along particular crystallographic planes instead of behaving in a malleable manner. Tiny atoms like hydrogen can permeate into the steel when the latter is exposed to the elements and cause a dramatic drop in ductility. Others that are not well accommodated in the lattice tend to segregate at the interfaces between crystals and cause havoc there. This chapter has a description of the ductile to brittle transition in steels with a focus on mechanisms and methods of preventing the unexpected deterioration in properties that is commonly referred to as *embrittlement.* The role of inclusions and phases that liquefy at a temperature well below that at which iron melts, together with irradiation effects, are also described.

11.1 Introduction

Steels exhibit a wider variety of failure mechanisms. While ultimate failure under excessive loading can reasonably be predicted by appropriate mechanical tests, premature failure is pernicious. Even then, the use of fracture mechanics can allow the quantitative assessment of the evolution of damage, making it possible to assign a safe-life to engineering structures.

There are essentially three fracture modes in steels at ambient temperature. Ductile failure is accompanied by gross plastic deformation, with the work of fracture far exceeding that required to create new surfaces as the steel is parted. Cleavage or intergranular fracture are in contrast, brittle mechanisms because there is minimal absorption of energy associated with the creation of new surfaces. For pure metals, the transition from ductile to brittle fracture is described by the ratio of the modulus μ, which is a measure of the resistance to shear, and the bulk modulus K, since the stress field at a brittle crack tip is similar to pure dilatation [1]. A small value of the ratio μ/K favours ductile fracture because plastic deformation by shear is then relatively easy. For pure α-iron made using electron beam zone-melting under high vacuum, the ratio is 0.33, which is less than the critical value associated with brittle failure. Indeed, high-purity iron is found to be ductile at temperatures as low as 4.2 K [2]. Such behaviour is not expected in commercial steels, which are designed to be strong, i.e., more resistant to plastic deformation.

11.2 Cleavage fracture in iron and steel

Cleavage fracture is familiar in many minerals and inorganic crystalline solids as a mechanism involving little plastic deformation and occurring in a crystallographic fashion along planes of low indices, i.e., relatively large atomic density. At low temperatures, zinc cleaves along the basal plane, but in contrast, body-centred cubic iron

Steels. https://doi.org/10.1016/B978-0-44-318491-8.00017-4

cleaves on $\{100\}_\alpha$, which is not the most densely packed plane (Fig. 11.1). It is all the more surprising then, that the surface energy per unit area increases in the order $\{110\}_\alpha$, $\{100\}_\alpha$ and $\{111\}_\alpha$ [3]. On these grounds alone, it should be easier to cleave on the $\{110\}_\alpha$, in contradiction to experimental evidence. However, cleavage does not simply involve the parting of atoms under the influence of a stress normal to the plane of interest. Crack growth phenomena, even during cleavage, involve features such as dislocation emission and twinning. Atomistic simulations indicate that these processes are more complex when propagation is on the $\{110\}_\alpha$ and $\{111\}_\alpha$ planes [4], which may account for the more brittle behaviour of the $\{100\}_\alpha$ planes. Mechanical twinning, as illustrated in Fig. 11.1, is sometimes observed in association with cleavage, but is predicted by the simulations to occur at a relatively late stage in the growth of cleavage cracks, as a mechanism that relieves stress.

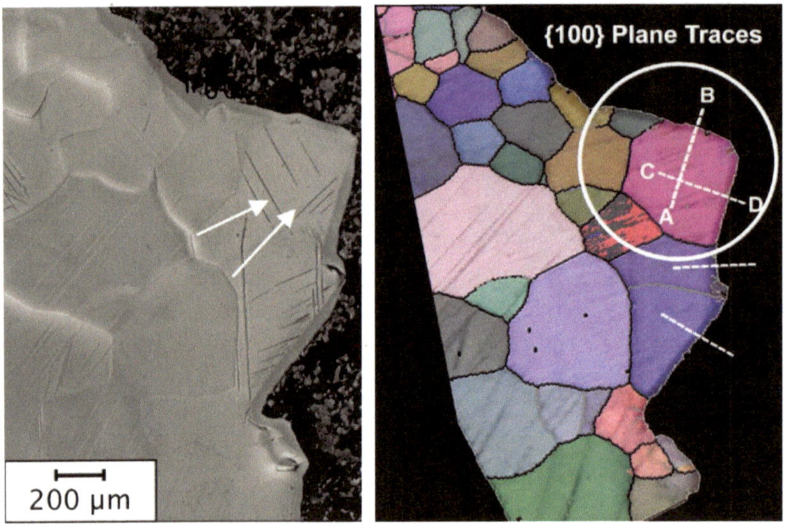

(a) (b)

Figure 11.1 Cleavage fracture in commercially pure iron. (a) Cross-section of fracture surface; the arrows point to mechanical twins on $\{112\}_\alpha$ planes. (b) Electron backscatter diffraction crystallographic-orientation image, with dashed lines parallel to the traces of $\{100\}_\alpha$ planes. After Ayer et al. [5], reproduced with permission of Elsevier.

Fig. 11.1 also illustrates that not all the grains at the fracture surface have clear facets consistent with cleavage; the macroscopic fracture of a polycrystalline material must involve a mixture of fracture mechanisms dependent on the distribution of grain orientations relative to the applied stresses. If a greater fraction of cleavage planes from different grains are aligned along the fracture surface, toughness is expected to be reduced. Fig. 11.2 illustrates how the toughness dramatically decreases in the heat-affected zone of a weld, as large blocks of similarly oriented grains increase the tendency for the cleavage planes of ferrite to align along the macroscopic fracture plane [6]. This effect can be designated *texture embrittlement*, which manifests particularly in steels that are thermomechanically processed. Such steels do not contain random arrangements of polycrystals, but exist in a state best described as between a single crystal and random array of crystals, i.e., the steel is crystallographically textured. In such steels, the texture is a principal reason for the observed anisotropy of toughness [7–9].

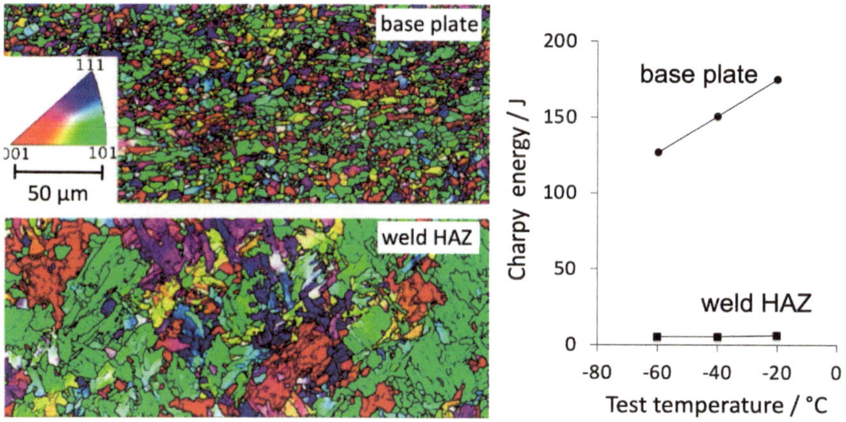

(a) (b)

Figure 11.2 (a) Orientation images of the base plate, and of the heat-affected zone of a weld in the same material. (b) Charpy toughness data for the two regions illustrated in (a). A Charpy test is an empirical measure of the ability of a material to absorb energy during fracture. After Yan et al. [6], reproduced with the permission of Elsevier.

This fact that body-centred cubic iron cleaves when tested in regimes that limit plasticity would appear to be an intrinsic characteristic of iron, but it has been shown that iron, highly purified by zone refining and containing minimal concentrations of carbon, oxygen and nitrogen, is very ductile even at extremely low temperatures. For example, at 4.2 K reductions in area in tensile tests of up to 90% have been observed with iron specimens of the highest available purity [2]. The temperature and strain-rate sensitivity of the flow stress is an intrinsic function of the lattice [10], a Peierls–Nabarro force due to the non-planar decomposition of dislocations parallel to $\langle 111 \rangle_\alpha$ into stacking faults intersecting on the $\{11\bar{2}\}_\alpha$ planes [11]. Since pure iron behaves in a ductile manner at cryogenic temperatures, cleavage is expected when the resistance to dislocation motion becomes relatively large.

As the carbon and nitrogen concentrations of the iron are increased, the transition from ductile to brittle cleavage behaviour takes place at increasing temperatures, until in some steels this can occur at ambient and above-ambient temperatures. Clearly, the significant variables for such a transition are of great basic and practical importance.

Even if there is a tiny bit of plasticity involved with the emission of dislocations during the propagation of a cleavage crack in iron, the work done in extending such a crack is far less than associated with the growth of a macroscopically ductile crack. This is easily shown by carrying out impact tests in a pendulum apparatus, where a notched sample is broken and the energy absorbed measured over a range of temperature. The presence of the notch provides a particular triaxial state of stress which makes brittle failure more likely. In the Charpy test, a square sectioned (10 mm × 10 mm) piece 55 mm long with a central 2 mm deep notch with a tip radius of 0.25 mm on one side is machined.

The energy absorbed by the specimen from the pendulum, when plotted as a function of temperature, usually exhibits a substantial change in slope (Fig. 11.3a) as the mode of fracture changes from ductile to brittle with the lowering of the test temperature. Since the test is essentially empirical, parameters such as the energy absorbed and the transition temperature cannot be used in design, but nevertheless serve in quality control and in the ranking of different steels and heat treatment processes. Completely ductile fracture begins usually with the nucleation, growth and linking of voids and defines the upper shelf region of the impact transition curve. At a sufficiently low temperature, the steel fails by cleavage, and in between the two failure modes is the so-called transition temperature T_c. The ductile-brittle transition can be dramatic, as illustrated in Fig. 11.3a, but the difference between the upper and lower shelf energies diminishes as the strength of the steel increases, and it becomes difficult to identify a clear transition temperature. An alternative strategy is to define a temperature at which 50% of the fracture surface exhibits the ductile mode of failure (Fig. 11.3b).

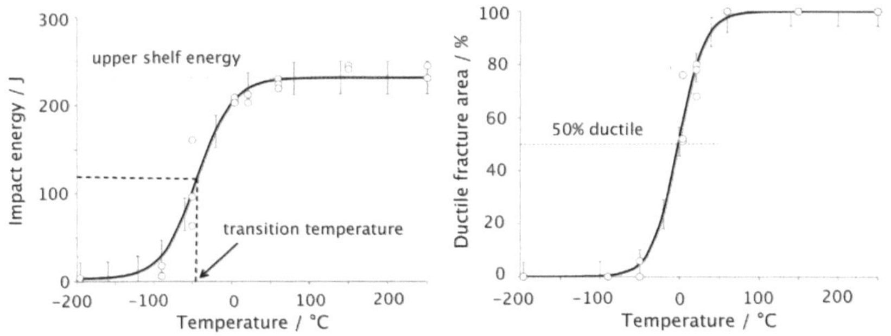

(a) (b)

Figure 11.3 Charpy impact toughness data for a nuclear pressure vessel steel. The points represent experimental data and the curves are fitted with error bars indicating the local level of uncertainty. (a) Impact energy absorbed during the test versus test-temperature. (b) The portion of the resulting fracture surface that is ductile, versus test-temperature. Courtesy of Hector Pous Romero [12].

The impact transition curve has a shape well represented by a hyperbolic tangent function [12]:

$$\text{energy absorbed} = C_7 + C_8\left[1 + \tanh\left(\frac{T - C_9}{C_{10}}\right)\right], \tag{11.1}$$

where C_i are fitting constants and T is the test temperature. One advantage of fitting such an equation is to estimate the uncertainties in the experimental data. Real materials or test samples are not homogeneous and hence exhibit scatter that can be expressed quantitatively.

The failure stress is easy to define for a homogeneous, geometrically smooth sample. However, the presence of a crack or sharp geometrical feature locally intensifies the applied stress to magnitudes greater than expected in the absence of such features. The term *stress intensity* is used to define the concentration of stress at, for example, the sharp tip of a crack.

Certain tests reasonably assume the presence of cracks where stress is intensified. Tests in which the toughness measured is a material parameter rather than a function of sample size and geometry are particularly useful in the safe-design of engineering components and structures. One such test measures the critical value of the stress intensity beyond which a sharp crack propagates rapidly and leads to failure. The stress intensity when a sample is loaded in a direction normal to the crack face is defined as:

$$K_I = \sigma(\pi c)^{1/2} \tag{11.2}$$

for a body of infinite extent containing a through-thickness crack of length $2c$ normal to the applied stress σ. Fracture propagates rapidly when the stress intensity reaches the critical value K_{IC}, so the tolerable defect size for a given applied stress can readily be calculated.

11.3 Factors influencing the onset of cleavage fracture

There are several factors, some interrelated, which play an important part in the initiation of cleavage fracture:

1. The temperature dependence of the yield stress.
2. The development of a sharp yield point.
3. Nucleation of cracks at twins.
4. Nucleation of cracks at carbide particles.
5. Grain size.

All bcc metals, including iron, show a marked temperature dependence of the yield stress, even when the interstitial impurity content is very low, i.e., the stress necessary to move dislocations, the Peierls–Nabarro stress, is strongly temperature dependent. This means that as the temperature is lowered, the first dislocations move more rapidly as the velocity is proportional to the stress, and so the chances of forming a crack nucleus, e.g., by dislocation coalescence, increase. Fig. 11.4 shows schematically two ways in which dislocation pile-ups could nucleate cracks.

The interstitial atoms, carbon and nitrogen cause the steel to exhibit a sharp yield point (Chapter 2), either by the catastrophic break-away of dislocations from their interstitial atom atmospheres, or by the rapid movement of freshly generated dislocations. In either case, the conditions are suitable for the localised, rapid movement of dislocations as a result of high stresses, thus promoting the nucleation of cracks by dislocation coalescence.

The flow stress of iron increases rapidly with decreasing temperature (Fig. 2.3) to a point where deformation twinning is stimulated, which then becomes a significant deformation mechanism. It has been shown that cracks are preferentially nucleated at various twin configurations, e.g., at twin intersections [13] and at points where twins contact grain boundaries, so that, under the same conditions, crack propagation is more likely in twinned iron. The temperature dependence of the flow stress makes plastic

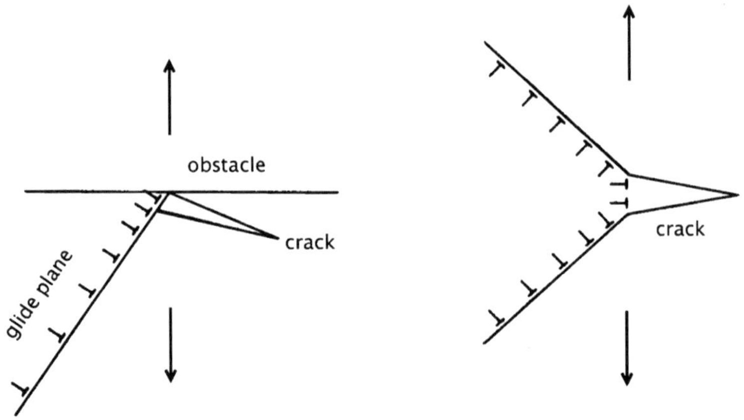

Figure 11.4 Schematic diagram of dislocation mechanisms for crack nucleation.

deformation more difficult at the tip of a moving crack, so less plastic blunting of the crack tip occurs at low temperatures, thus aiding propagation.

So far, we have discussed crack nucleation mechanisms associated with a single-phase material, but in the presence of a brittle phase such as cementite it becomes easier to nucleate cracks. Plastic deformation can crack grain boundary cementite particles or cementite lamellae in pearlite, producing microscopic cracks (Fig. 11.5a), which in turn can propagate to cause catastrophic cleavage failure (Fig. 11.5b). Brittle inclusions such as alumina particles or various silicates found in steels can also be a source of crack nuclei.

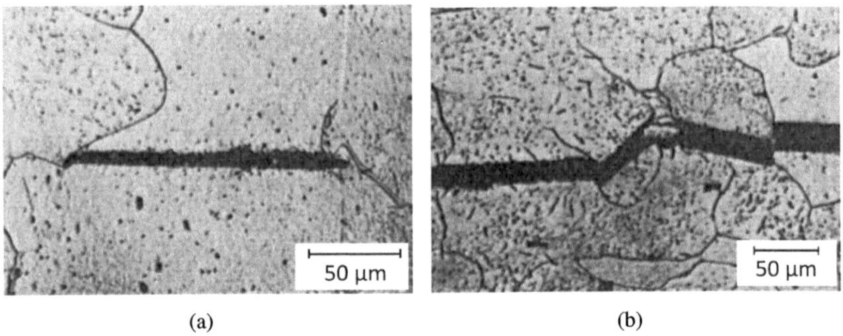

Figure 11.5 (a) Nucleation of a cleavage crack on a carbide particle in a low-carbon steel, followed by propagation (courtesy of Knott). (b) Transgranular propagation of a crack in a low-carbon steel (courtesy of Knott).

A reduction in the ferrite grain size depresses the transition temperature T_c, even though the yield strength increases. This makes grain size reduction a valued mechanism of simultaneously improving strength and toughness. It has been shown by Petch that T_c is linearly related to $(\overline{L})^{-1/2}$ [14], a relationship that can be derived from a model involving the formation of crack nuclei at dislocation pile-ups at grain boundaries. A small grain size limits the number of dislocations that can participate in a pile-up. Bearing in mind that the shear stress at the head of such a pile-up is $n\tau$, where n is the number of dislocations, it follows that a reduced $\overline{L}_\alpha$ is associated with smaller stress concentrations at grain boundaries for a given value of applied stress. This in turn reduces the tendency to nucleate cracks.

11.4 Criteria for the ductile-brittle transition

The starting point of all theories on brittle fracture is the work of Griffith [15], who considered the condition needed for the propagation of a pre-existing crack of length $2c$ in a brittle solid. When the applied stress σ is large enough, the crack propagates releasing strain energy. This energy U_e in the case of remotely loaded thin plates in plane stress conditions is:

$$U_\mathrm{e} = \frac{\pi c^2 \sigma^2}{E} \text{ per unit plate thickness,} \tag{11.3}$$

where E is Young's modulus.[1] The term is negative because this energy is released. However, as the crack creates two new surfaces, each with energy $= 2c\gamma_\mathrm{s}$, there is a positive surface energy term U_s:

$$U_\mathrm{s} = 4c\gamma_\mathrm{s}, \quad \text{where } \gamma_\mathrm{s} = \text{surface energy per unit area.}$$

Griffith showed that the crack would propagate if the increase in surface energy U_s is less than the decrease in elastic energy U_e. The equilibrium position is defined as that in which the change in energy with crack length is zero:

$$\frac{\mathrm{d}U}{\mathrm{d}c} = \frac{\mathrm{d}(U_\mathrm{e} + U_\mathrm{s})}{\mathrm{d}c} = 0. \tag{11.4}$$

This is the elastic energy release rate, usually referred to as G:

$$\therefore \left(-\frac{2\pi c \sigma^2}{E} + 4\gamma_\mathrm{s} \right) = 0,$$

and

$$\sigma_\mathrm{f} = \sqrt{\frac{2\gamma_\mathrm{s} E}{\pi c}}, \tag{11.5}$$

where σ_f is the fracture stress, above which there is a net energy release as the crack propagates. Orowan pointed out that in crystalline solids plastic deformation occurs both during crack nucleation and then at the root of the crack during propagation. This

[1] For a linear elastic solid without a crack, U_e would be half this value, given simply by the area under the elastic stress versus elastic strain curve. However, the stress around an elliptical crack is more complex and an exact solution due to Griffith [15] gives double the value expected from a simple application of Hooke's law.

plasticity at the crack-root blunts the crack and, meaning that more energy is needed to continue crack propagation than indicated by the Griffith analysis. Thus, the Griffith equation is modified to include a plastic work term γ_p:

$$\sigma_f = \sqrt{\frac{E(2\gamma_s + \gamma_p)}{\pi c}}. \tag{11.6}$$

It has been found that $\gamma_p \gg \gamma$, so the condition for crack spreading in a crystalline solid such as iron is:

$$\sigma_f = \sqrt{\frac{E\gamma_p}{\pi c}}. \tag{11.7}$$

The local stress field at the crack tip is usually characterised by a parameter K, the stress intensity factor, which reaches a critical value K_c, when propagation takes place. This critical value is given by:

$$K_c = \sigma_f\sqrt{\pi c}. \tag{11.8}$$

In plane stress conditions:

$$K_c = \sqrt{EG_c},$$

where G_c is the critical release rate of strain energy.

Under plane strain conditions, the critical value of the strain energy release rate is $G_{IC} = \gamma_p$, so it follows that:

$$\sigma_f = \sqrt{\frac{EG_{IC}}{\pi(1-\nu^2)c}}, \tag{11.9}$$

and ν is the Poisson's ratio.

The critical value of stress intensity, K_{IC}, is then related to G_{IC}:

$$K_{IC} = \sqrt{\frac{EG_{IC}}{\pi(1-\nu^2)}}. \tag{11.10}$$

The fracture toughness of a steel often is expressed as a K_{IC} value obtained from tests on notched specimens which are pre-cracked by fatigue and then loaded to fracture in bending or tension. It is a material property that can be used to ensure safe design.

This tensile stress σ_f has been determined for a wide variety of mild steels. It varies approximately linearly with $(\overline{L})^{-1/2}$, Fig. 11.6. There is considerable scatter in the data, probably because the samples in reality contain a distribution of defects. This is conclusive evidence for the role of finer grain sizes in increasing the resistance to crack propagation.

The critical local fracture stress σ_f has been related to two types of structure, as follows:

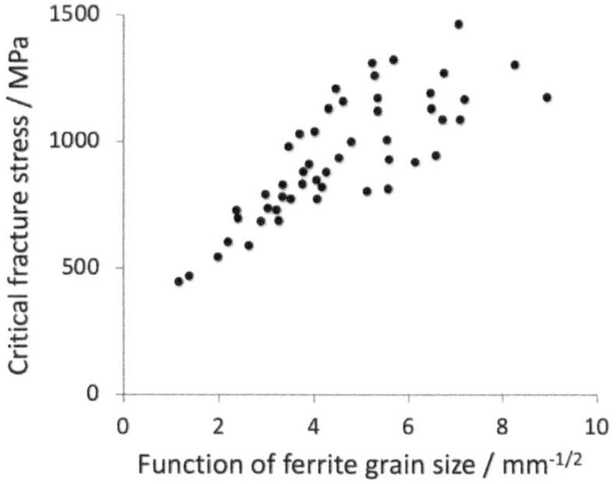

Figure 11.6 Dependence of local fracture stress σ_f on the grain size of mild steel. Data compiled from many sources (courtesy of Knott).

1. For ferritic steels with spheroidal carbide particles:

$$\sigma_f = \sqrt{\frac{\pi E \gamma_p}{2c_0}}, \tag{11.11}$$

where c_0 is carbide diameter.

2. For bainitic and martensitic steels with packets of laths:

$$\sigma_f = \sqrt{\frac{4E\gamma_p}{(1-\nu^2)d_p}}, \tag{11.12}$$

where d_p is packet width and v is Poisson's ratio.

11.5 Practical aspects of brittle fracture

At the onset of fracture, elastic energy stored in the stressed steel is only partly used for the creation of new surfaces and the associated plastic deformation and the remainder provides kinetic energy to the system. Using a Griffith-type model, the speed of crack advance $\dot{c}$ can be shown to be [16]:

$$\dot{c} = \dot{c}_{max}\left(1 - \frac{c_0}{c}\right), \quad \text{where} \quad \dot{c}_{max} = \sqrt{\frac{2\pi}{C_{11}}\frac{E}{\rho}}, \tag{11.13}$$

where in a sheet of unit thickness, $2c_0$ is the critical crack size, c is the crack size at a given instant, ρ is density, C_{11} is a constant and $\dot{c}_{max}$ is the maximum crack growth velocity. The velocity increases with increasing crack size and reaches a limiting value $\dot{c}_{max}$ at large values of c. In practice, $\dot{c}_{max}$ is between 0.4 and 0.5 of the speed of sound in the metal, so brittle fracture occurs with catastrophic rapidity, as many disasters testify.

The phenomenon of brittle fracture became particularly prevalent with the introduction of welding as the major steel fabrication technique. Previously, brittle cracks often stopped at the joints of riveted plates, but the steel structures resulting from welding provided continuous paths for crack propagation. Added to this, incorrect welding procedures can give rise to stress concentrations and also to the formation of weld-zone cracks, which may initiate brittle fracture. Brittle failures of steels have occurred since the second half of the nineteenth century, when steel was used widely for structural work. In recent times, however, failures have been rare because of the extensive use of low-carbon steels and the deep understanding of the welding process and its implementation. A serious brittle failure is illustrated in Fig. 11.7, a constant reminder that human error, inadequate design and lack of scientific control can be disastrous.

Figure 11.7 Brittle fracture of a thick-walled steel pressure vessel (The Welding Institute).

Bearing in mind the temperature dependence of the failure behaviour and the widening use of steels at low temperatures, e.g. in Arctic pipelines, for storage of liquefied gases, rocket fuel-tanks (Chapter 17), etc., it is increasingly necessary to have steels with very low transition temperatures and high fracture toughness. While there are many variables to consider in achieving this end, including the detailed steel-making practice, the composition including trace elements and the fabrication processes involved, the most important is probably grain size refinement. The development of high-strength low-alloy steels or micro-alloyed steels (Chapter 9), in the manufacture of which controlled rolling plays a vital part, has led to the production of structural steels with grain sizes often less than 10 μm combined with yield strengths between 400 and 600 MPa and low transition temperatures. In these steels, to which small concentrations (<0.1 wt%) of niobium, vanadium or titanium are added, the carbon level is usually less than 0.15 wt% and often below 0.10 wt%, so that the carbide phase occupies a small volume fraction. In any case, cementite, which forms relatively coarse particles or lamellae in pearlite, is partly replaced by much finer dispersions of alloy carbides, NbC, etc.

Addition of certain other alloying elements to steel, notably manganese and nickel, results in a lowering of the transition temperature. For example, alloy steels with 9 wt% nickel and less than 0.1 wt% carbon have a sufficiently low transition temperature to be used for large containers of liquefied gases, where the temperature can be as low as 77 K. Below this temperature, austenitic steels have to be used. Of the elements unavoidably present in steels, phosphorus raises the transition temperature and thus must be kept to as low a concentration as possible. Sulphur has a very low solubility in ferrite – it usually is present as manganese sulphide, which has little effect on the transition temperature, but plays an important role in ductile fracture. Oxygen is an embrittling element, even when present in very low concentrations. However, it is removed easily by deoxidation practice involving elements such as manganese, silicon and aluminium.

Finally, the fabrication process is often of crucial importance. In welding it is essential to have a steel with a low carbon equivalent, i.e., a factor incorporating the effects on hardenability of the common alloying elements. A simple empirical expression for carbon equivalent is given in Eq. (13.3). A steel with an equivalent of less than 0.45 should be weldable with modern techniques. The main hazard in welding is the formation of martensite in the heat-affected zone (HAZ), near the weld, which can lead to microcracks. This can be avoided not only by control of hardenability, but also by preheating the weld area to lead to slower cooling after welding or by post-heat treatment of the weld region. However, in some high-strength steels, slower cooling may result in the formation of upper bainite in the HAZ, which encourages cleavage fracture. More detailed discussion is reserved for Chapter 13.

11.6 Hydrogen embrittlement

It has been known for more than a century that minute concentrations of hydrogen embrittle steel. Johnson's paper [17] reached the following conclusions, which remain valid to this day:

1. The hydrogen that embrittles steel is atomic, not molecular.
2. It is *diffusible* hydrogen that embrittles. This is why incredibly small concentrations ($<$ 1 parts per million) of hydrogen in steel have a large effect on toughness. The atoms of hydrogen are attracted towards stress fields of the type associated with a crack tip. Thus, they diffuse there and concentrate, changing the fracture mechanism to the detriment of the steel. Hence the need for diffusible hydrogen for embrittlement. Since a supply of hydrogen must be delivered to the crack tip during the time scale of the fracture process, deformation at large strain rates exhibit less embrittlement than when the loading is applied slowly.
3. It follows that the phenomenon is reversible. If the hydrogen is given the opportunity to diffuse out of the steel, the metal recovers its properties.
4. Stronger steel is more susceptible to embrittlement than softer versions.

There are several explanations for the deleterious effect of hydrogen on the mechanical properties of steel. Hydrogen may encourage the decohesion of the body-centred cubic iron lattice [18,19], although atomistic calculations show that this cannot be more than a minute effect at the concentrations involved [20]. A mechanism that seems counter-intuitive to brittle failure relies on an enhancement of plasticity by hydrogen [21]. When concentrated at a crack tip, the hydrogen is postulated to enhance localised crack tip plasticity by dislocation emission so the fracture surface looks macroscopically brittle but is ductile on a fine scale. Others have argued that hydrogen causes a large increase in the generation of vacancies during plastic deformation, thus hindering ductile deformation [22]; however, positron annihilation experiments on iron and its alloys have not revealed this increase in the number density of vacancies [23].

All of these proposed mechanisms are essentially qualitative and hence lack the ability to predict properties. An alternative view, due to Song and Curtin [24], also involves the attraction and accumulation of hydrogen at the crack tip stress-field, but the consequence this time is the *suppression* of dislocation activity, confirmed using molecular dynamics simulations. The mechanism of dislocation suppression revealed by the simulations is that local regions at the crack tip adopt alternative crystal structures, Fig. 11.8. As a result, the crack tip is not blunted by plasticity and cleavage initiates.

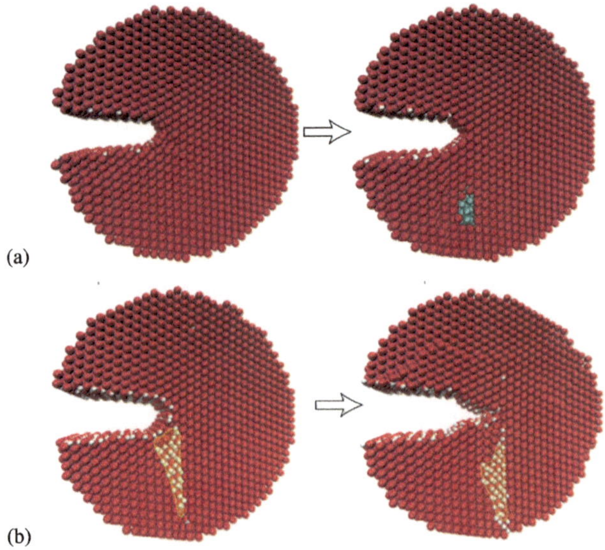

Figure 11.8 Simulated atomic configurations during mode I loading of a crack tip. Hydrogen atoms are coloured white, iron atoms red, except those at dislocation cores which are cyan. (a) At a low hydrogen concentration, the atoms simply segregate to the crack surfaces and do not contribute to embrittlement, so the crack is blunted by dislocation emission. (b) At greater hydrogen concentration, the hydrogen accumulates in a wedge-shaped region, which undergoes a phase transformation that blocks dislocation emission, resulting in a brittle cleavage. Reprinted by permission from [24], Springer Nature.

The continued growth of the initial cleavage event requires a corresponding continuing supply of hydrogen to the crack tip. The crack growth rate ($\dot{c}$) therefore depends on a number of thermodynamic and kinetic parameters:

$$\dot{c} \propto \frac{D_{\rm H}\overline{V}_{\rm p}}{T} K_{\rm I}(\overline{c}_{\rm H})^{5/4}, \tag{11.14}$$

where $D_{\rm H}$ is the hydrogen diffusion coefficient, $\overline{V}_{\rm p}$ is the partial volume of an hydrogen interstitial in iron and determines the magnitude of the interaction of hydrogen with the pressure field at the tip of the crack. This pressure is related to the stress intensity $K_{\rm I}$, which defines the loading of the crack, and $\overline{c}_{\rm H}$ is related to the concentration of hydrogen in the material. The relationship defines nicely the dependence of the crack growth rate on parameters related to concentration, diffusivity and loading during brittle fracture due to hydrogen. It becomes possible using these concepts to predict quantitatively the onset of hydrogen embrittlement in a large variety of steels without fitting parameters.[2]

Hydrogen embrittlement is not in general sensitive to composition *per se*, but to the microstructure and strength of the steel, the problem being most pronounced in strong alloy steels. Fig. 11.9 shows that in steels that have almost identical ductility in the absence of hydrogen behave quite differently when hydrogen is introduced. The decrease in ductility is greatest for the strongest steels for a constant hydrogen content. A stronger steel presents greater barriers to dislocation emission at the crack tip, and hence is more susceptible to embrittlement. On the other hand, macroscopic ductility, as measured in a tensile test in the absence of dissolved hydrogen, depends to a large extent on the ability of the material to work harden and avoid plastic instabilities.

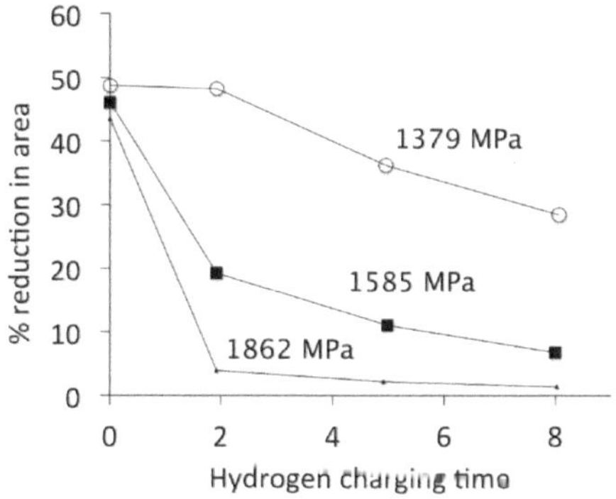

Figure 11.9 Effect of hydrogen charging time (a reflection of the quantity of hydrogen introduced into the steel) on the fracture strength in tension. Data from Frohmberg et al. [26].

There are many ways in which hydrogen can enter into steel, both during the manufacturing process and during service. It is encountered frequently after welding (Fig. 11.10), where it can be introduced by use of damp welding electrodes, leading to cracking which is referred to variously as *underbead* cracking, *cold* cracking and

[2] Other ideas on hydrogen embrittlement have been reviewed by Nagumo [25]. For example, the internal pressure model, where hydrogen accumulates at a suitable defect in the form of molecular hydrogen and the resulting pressure initiates cleavage.

delayed cracking. These phenomena can be avoided by the use of welding electrodes that are oven-dried prior to use, so that embedded moisture that turns into hydrogen in the heat of the welding arc can be minimised.

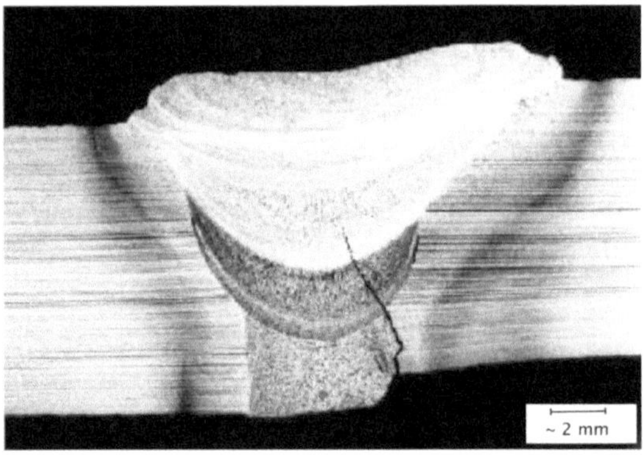

Figure 11.10 Cleavage crack due to hydrogen embrittlement, beginning in the heat-affected zone and then extending into the weld. Reproduced from [27], with permission of P. Woolin of TWI.

11.6.1 Prevention of hydrogen embrittlement

It is diffusible hydrogen that causes embrittlement. Anything that traps the hydrogen inside the steel, or prevents its ingress into the steel, will be effective in ameliorating the embrittlement [28]. Fig. 11.11 summarises the large range of methods available to do this.

A trap provides a favourable environment for hydrogen to reside in, for example, at the coherency strain fields surrounding small carbide particles, at weak interfaces between non-metallic inclusions and ferrite, or by dissolving in particles where the hydrogen has a much greater solubility than the matrix. The potency of trapping is defined as the reduction in free energy when a hydrogen atom is transferred from the matrix to the trap site. An irreversible trap has a large binding energy and hence immobilises the hydrogen over the life time of the component and its specific service conditions. Examples of the latter include vanadium and titanium carbides. A reversible trap is one that is able to easily release the hydrogen under similar circumstances, if the concentration in the matrix drops for whatever reason, and hence is not a particularly effective means for avoiding embrittlement. Nevertheless, both kinds of traps interfere with the transport of hydrogen and dramatically reduce the effective diffusivity of hydrogen through the steel [29]. This is useful in limiting the ingress of hydrogen into the steel, for example, due to corrosion reactions at the steel surface.

In steels, the diffusivity of hydrogen in austenite is orders of magnitude smaller than in ferrite (Fig. 11.12), although its solubility is much greater in austenite. Retained austenite in the microstructure can therefore act as a strong trap for hydrogen. And if the austenite is present in the form of percolating films between the ferritic phase, it will greatly retard the passage of hydrogen through the microstructure [31,32], while

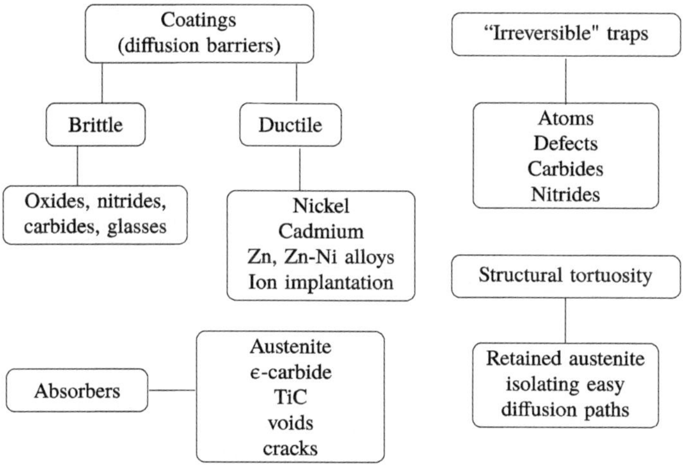

Figure 11.11 Mechanisms available for the modification of steel and its surface to better resist hydrogen by either mitigating its infusion into the steel or by rendering the hydrogen atoms immobile with the steel so that they are prevented from accumulating at stress concentrations [28]. "Irreversible" implies that the rate of de-trapping is small when compared against that of hydrogen infusion and the service life. It is noteworthy that all of the common substitutional-solutes (Co, V, N, W, Si, Mn, Ti, Cr, Al, Mo) added to ferrite reduce the measured diffusion coefficient of hydrogen [30].

at the same time acting as a sink that absorbs hydrogen into itself rather than via strain fields, as is the case for carbides and nitrides.

Many kinds of coatings exist, including alumina, titanium carbide, silicon nitride, titanium oxide, "black" oxide, etc. [28], that have been demonstrated to reduce either the outgassing of hydrogen in vacuum systems, or as diffusion barriers to the ingress of hydrogen. However, the choice of coatings available decreases when they have to perform multiple functions, for example, to resist abrasion and impact. Coatings may also contain defects that locally expose the steel in which case, a sacrificial layer is helpful; Zn-Ni coatings act in this manner. The thickness and integrity of the coating vary with the manufacturing process and has to be compatible with the service conditions of the protected component. The quality of the coating can depend on the chemical composition and structure of the substrate. Ion implantation can be used to favourably alter the surface of the steel.

11.7 Attack by high-pressure hydrogen

The entry of atomic hydrogen into iron is a lot easier during electrochemical reactions than by the dissociation of molecular hydrogen in the gaseous phase, in contact with the iron at the surface. It has been estimated that the cathodic charging at $1\,\mathrm{A\,cm^{-2}}$ would generate atomic hydrogen at a rate equivalent to the exposure of the steel to a H_2 gas pressure of 230 GPa (!) [37]. In reality, the surface of the iron may be covered

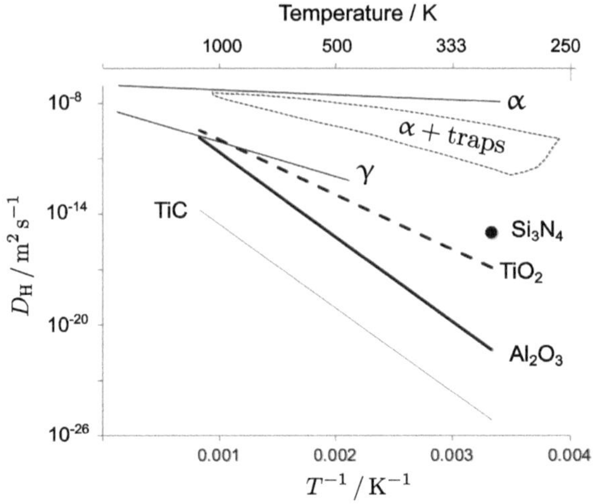

Figure 11.12 Hydrogen diffusion data for TiC [33], TiO_2 parallel to the c-axis [34], Al_2O_3 [35] and Si_3N_4 [36], compared against corresponding diffusivities in steels from Fig. 1.12. γ and α refer to austenite and ferrite, respectively. The data for the α- and γ-iron, in particular, can be sensitive to the chemical composition and microstructure. After Bhadeshia [28].

by adsorbates, oxides, etc., and its crystallographic character is likely to matter. These factors would retard the necessary dissociation of hydrogen.

The classical treatment of the equilibrium between diatomic hydrogen and the metal is Sieverts' law, which relates the solubility to the square root of pressure. This is because in the reaction $H_2 \leftrightharpoons 2[H]$ has the equilibrium constant given by $c^2_{[H]}/p_{H_2}$ (dissolved hydrogen is represented by [H]). Sieverts' law works well for all the pressures of interest in industry, with greater hydrogen solubility predicted when the gas pressure exceeds about $\gtrapprox 1$ GPa; this is illustrated in Fig. 11.13, which is based on calculations that account for the molar volume of hydrogen and the concentration dependence due to interactions between hydrogen atoms [38].

There are applications of steel in the petrochemical industries involving high-pressure hydrogen at elevated temperatures. For example, in the transmission of gas containing partial pressures of hydrogen ($\lessapprox 20$ MPa) at temperatures of the order of 300–500°C. Such steels usually are designed for service at elevated temperatures and therefore they contain a dispersion of carbides. It is known that hydrogen has an affinity for carbon in steels. Its combination with carbon to form methane within the steel causes a loss of structural integrity that can be dangerous [39,40]. The first observation of difficulties of this kind was associated with the Haber–Bosch ammonia production process, where the penetration of hydrogen into cast steel "caused blisters and other dangerous imperfections" [41,42].

The dissociation of $H_2 \rightarrow [2H]$ allows the atomic hydrogen to enter the steel to react with the carbon that is in solution within the ferrite, and when that is depleted sufficiently, methane is produced by the reaction

$$Fe_3C + 2[H] \rightarrow CH_4 + 3Fe.$$

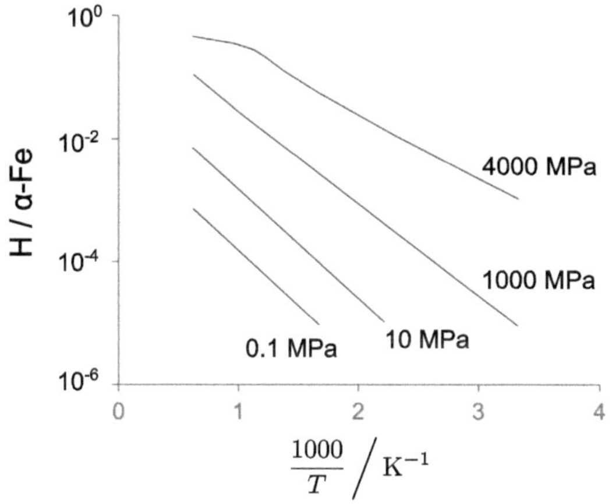

Figure 11.13 Solubility of hydrogen in α-iron, as a function of temperature and pressure. The solubility is expressed as the number of hydrogen atoms per iron atom. Selected data from Sugimoto and Fukai [38].

This decarburisation is accompanied by the formation of methane bubbles within the steel. As the carbon that is in solid solution with the ferrite becomes depleted by reaction with [H], carbides with which the ferrite is in equilibrium dissolve, thus feeding the reaction. The decarburisation as the carbides dissolve, compromises the strength of the alloy when combined with bubble formation. Under load, this can cause the premature fissure of the steel. Unlike normal hydrogen embrittlement, this degradation is not reversible. Some of the features of this chemical-reaction based hydrogen damage are illustrated in Fig. 11.14, taken from the failure investigation following the explosion of the 20 cm diameter pipe [43].

The methane reaction depends on the activity of carbon in the ferrite that is in equilibrium with a carbide phase (Fig. 11.15). More stable carbides reduce that activity and therefore the rate of methane production. In fact, chromium in solution also significantly reduces the diffusivity of carbon in ferrite and in austenite [44], which should in turn retard the decarburisation kinetics. Carbides such as M_2C, M_7C_3, $M_{23}C_6$ are far more thermodynamically stable than cementite.

11.8 Intergranular embrittlement

There are many cases where embrittlement involves the parting of grains at their boundaries, leading to *intragranular failure*, Fig. 11.16a. It is intriguing that in steels, the failure can occur along the *prior* austenite grain boundaries, i.e., between austenite grains that no longer exist. This is because displacive transformations (Widmanstätten ferrite, bainite, acicular ferrite, martensite) involve the coordinated motion of atoms, which cannot be sustained across the austenite grain boundaries during transformation, Fig. 11.16b. Hence, in a fully transformed microstructure, a vestige of the austenite grain boundary remains when the transformation is displacive. In contrast,

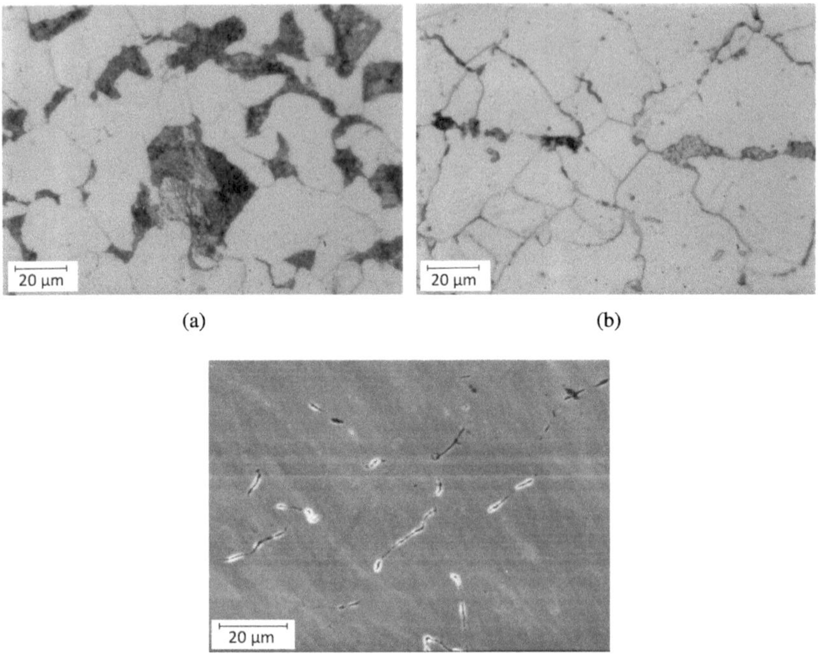

(a) (b)

(c)

Figure 11.14 Degradation of pipe used to convey hydrogen at an excess pressure of 16.3 MPa at 370°C to a reactor that converts viscous oil into more fluid petroleum. The steel has the approximate composition Fe-0.3C-1.2Mn-0.4Cr-0.4Ni-0.15Mo-0.4Cu wt%. (a) Normal microstructure consisting of ferrite and pearlite. (b) Decarburised inner region of the pipe with greatest exposure to hydrogen. (c) Fissures at the grain boundaries, caused by hydrogen attack. Images reproduced from Kim [43] with the permission of Elsevier.

diffusional transformations products are not limited by the austenite grain boundaries, so no trace is left of those boundaries in the final microstructure [47].

This behaviour is encountered in quenched steels, on tempering (*temper embrittlement*), after heating to high austenitising-temperatures (*overheating* and *burning*), and in the *rock candy fracture* in cast steels. These forms of embrittlement are exhibited at or around room temperature. There are, however, other phenomena involving failure along grain boundaries which are essentially high temperature events, e.g., hot shortness during the hot working of steels and high temperature creep failure. It is clear that no one mechanism will explain the various types of embrittlement, but the processes leading to intergranular grain fracture all involve reduced cohesion along the grain boundaries and a strengthening of the grain interiors. This can arise in different ways, but the most relevant appear to be:

1. Segregation of solute atoms preferentially to grain boundaries.
2. Distribution of second-phase particles at grain boundaries.

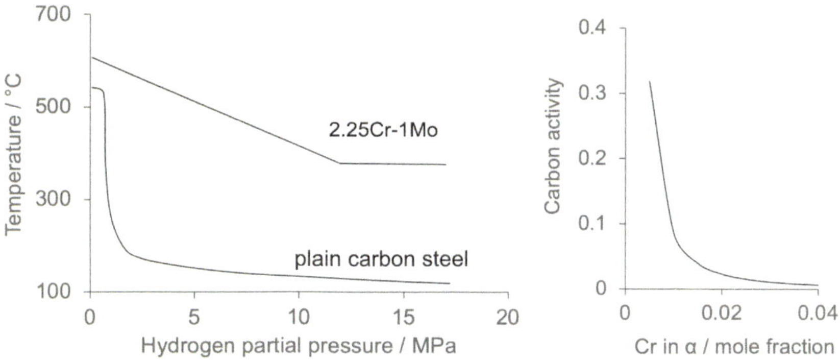

(a) (b)

Figure 11.15 (a) These empirical Nelson curves can, in principle, be used to determine the safe-operating domain (region below the curve) with respect to hydrogen attack for a particular steel. Adapted from [45], where much more detail is available. (b) The influence of the Cr concentration in the ferrite, in equilibrium with $M_{23}C_6$, on the activity of carbon in the ferrite. The calculations are for 525°C, where half the metal atoms in the carbide are iron [46].

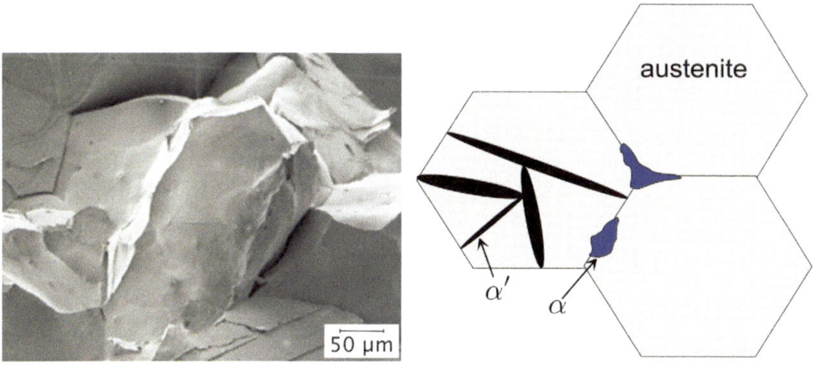

(a) (b)

Figure 11.16 (a) Intergranular embrittlement of ferrite grain boundaries in an Fe-0.17P wt% alloy (courtesy of Jeong In Kim). (b) Displacive transformation products such as martensite (α') do not cross austenite grain boundaries, whereas allotriomorphic ferrite is not limited in this way and hence consumes portions of the γ boundaries across which it grows.

3. Penetration of low melting-temperature liquid metal at grain boundaries (liquid metal embrittlement).

These phenomena reduce the work of fracture, i.e., the $(2\gamma + \gamma_p)$ term in Eq. (11.6) either by lowering the grain boundary energy by segregation or by reducing the plastic work term γ_p by having particles, which more easily provide crack nuclei.

11.8.1 Temper embrittlement

Many alloy steels, when tempered in the range 500–650°C following quenching to form martensite, become progressively embrittled with fracture along the prior-austenite grain boundaries. A similar phenomenon can occur when the steels are continuously cooled through the critical temperature range. The ductile-brittle transition temperature is then raised, the transgranular fracture mode being replaced by an intergranular mode below T_c (Fig. 11.17).

Figure 11.17 Temper embrittlement due to phosphorus, of a low-alloy steel that was heat treated at 550°C for 5 h and then cooled to ambient temperature. The steel has the composition Fe-0.26C-2.11Si-2.27Mn-1.59Cr-0.017P wt% with a microstructure consisting of $\alpha_b + \alpha' + \gamma_r$. Failure is at the prior austenite grain boundaries (image courtesy of Liou-Chun Chang).

This phenomenon is now known to be associated with the segregation of certain elements to the grain boundaries, which reduce the intergranular cohesion of iron. Elements that do not fit well in the substitutional or interstitial sites within the lattice tend to segregate to grain boundaries where there is a greater free volume. This includes solutes such as C, Si, Ge, Sn (periodic table Group IVB), N, P, As, Sb, Bi (Group VB), O, S, Se and Te (Group VIB). With the exception of carbon and silicon, most of these are present as impurity elements in common steels. The segregation of many of these atoms to the boundaries has been demonstrated by Auger electron spectroscopy on specimens fractured intergranularly within the vacuum system of the apparatus. This technique has allowed the precise determination of the concentrations of segregating species at the boundaries, usually expressed in terms of fractions of a monolayer of atoms. These fractions vary between about 0.3 and 1.0 for steels containing the above elements, usually in bulk concentrations well below ≈ 0.1 wt%.

Impurity segregation to an interface is controlled by associated reduction in free energy and the ability to diffuse to the boundary sink. Such *ordering* of impurities is not favoured by high temperatures because the entropy contribution to the free energy $-T\Delta S$ opposes ordering and scales with temperature. Therefore, any segregant tends to evaporate from the boundaries. At low temperatures, the diffusivity is limited, so segregation takes longer. This is why there is a temperature range 500–650°C where impurity segregation to boundaries in steels becomes most prominent, leading to embrittlement.

A useful concept is that if a solute in iron has a greater reduction in energy (Δg_s) when transferred from solid solution to a free surface, than when it is similarly transferred to a grain boundary (Δg_b), then it embrittles the boundary because it becomes favourable to separate the grains [48]. Considering that these two energies are negative for any solute that tends to segregate spontaneously, the potency to embrittle scales with the positive $\Delta g_b - \Delta g_s$. For example, Table 11.1 shows that phosphorus has a strong tendency to embrittle grain boundaries, and first-principles calculations confirm this expectation for special low-Σ boundaries [49]. Σ refers to the fraction of lattice points common to two crystals that share a common origin and are allowed to interpenetrate and fill all space (e.g., [50]). Thus, a low-Σ boundary would, in general, correspond to a low energy boundary. First-principles calculations are often limited to low-Σ boundaries because computational limits mean that they can deal only with a few hundreds of atoms, whereas a general grain boundary would require much greater "supercells".

Some aspects of the mechanism by which impurities embrittle boundaries are known. Krasko and Olson first demonstrated, using first-principles calculations, that upon segregation, atoms including P and S, tend to enhance the bonding of atoms along the plane of the boundary to the detriment of bonding across the boundary plane [51].

Table 11.1 Selected segregation free energies for solutes in iron, from a compilation by Anderson et al. [48].

Segregant	$-\Delta g_b$/kJ mol^{-1}	$-\Delta g_s$/kJ mol^{-1}	$\Delta g_b - \Delta g_s$
Carbon	50-75	73-85	23-10
Tin	30-35	61-87	31-52
Antimony	8-25	83-130	75-105
Sulphur	50-58	165-190	115-132
Phosphorus	32-41	76-80	44-39
Hydrogen	65-68	71-109	6-41

Whether or not embrittlement actually occurs depends on a number of factors other than just $\Delta g_b - \Delta g_s$ [52]. Plastic deformation by slip and separation at grain boundaries are competing mechanisms. Anything which makes slip more difficult generally makes grain boundary failure more likely [53], depending on the ability of the material within the grains to accommodate strain. Factors that should enhance grain boundary failure can be listed as follows:

a) A strong matrix effectively focuses deformation on the boundaries, which then become the weak links;

b) Low testing temperatures because the matrix then becomes stronger, thus reducing the mobility of dislocations;

c) Large grain size [54] since a fixed impurity content is then distributed over a smaller boundary area per unit volume, resulting in a greater concentration of the segregant at the boundaries;

d) A large fraction of high-energy boundaries where impurity segregation is favoured [55,56]. Low-energy boundaries have greater coherency and hence are better able to resist separation, but also have less free volume for impurities to segregate [55]. In a low-carbon Fe-0.006C-1.7Cr-3.5Ni-0.055P wt% steel, the segregation of phosphorus to low-misorientation or Σ3 coincidence site lattice boundaries was found to be minimal. Even boundaries parallel to closely-packed atomic planes resisted the segregation because of the locally small free volume there. Observations like these have defined the subject *grain boundary engineering*, where polycrystalline aggregates are textured to obtain low free-volume interfaces [57].

With the individual elements, the tendency to embrittle appears to increase both with group and period number, i.e., S, Se and Te are the most surface active elements in iron, in ascending order. However, it is doubtful whether they contribute greatly to temper embrittlement, since they combine strongly with elements such as Mn and Cr, which effectively reduce their solubility in iron to very low levels. While the elements in groups IVB and VB are less surface active, they play a greater role in embrittlement because they interact with certain metallic elements, e.g., Ni and Mn, which are common solutes in steels. These interactions lead to the co-segregation of deliberately added solutes and impurity elements to the grain boundaries, and to resultant lowering of cohesion by the impurity element. Analysis of the composition of grain boundaries by Auger spectroscopy has confirmed strong interactions between Ni-Sb, Ni-P, Ni-Sn and Mn-Sb. Fig. 11.18a shows the grain boundary concentrations for three of these interactions in Ni-Cr steels, while the relative effects of Sb, Sn and P on the transition temperature of Ni-Cr steels are shown in Fig. 11.18b.

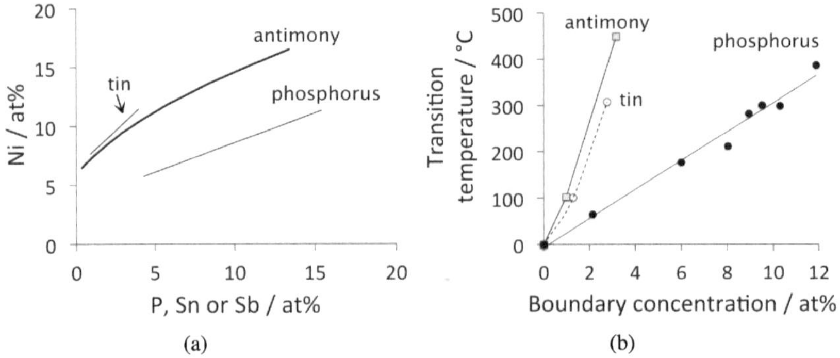

Figure 11.18 (a) Interrelation between concentrations of Sn, Sb and P and of Ni at grain boundaries in Ni-Cr steels of constant hardness and grain size. (b) Effect of grain boundary concentrations of P, Sb and Sn on the ductility of Ni-Cr steels of constant hardness and grain size. After McMahon [58].

Therefore, the driving force for co-segregation to boundaries is a stronger interaction between the alloying element and the impurity element than between either of these and iron. If the interaction is too strong, segregation does not take place. Instead, a scavenging effect is obtained, as exemplified by Ti-P and Mo-P interactions in Ni-Cr steels. In this connection it is well known that molybdenum additions to Ni-Cr steels can eliminate temper embrittlement. A third inter-alloy effect is that one alloying element, e.g., Cr, promotes segregation of Ni and P and of Ni and Sb.

There can be site competition when more than one solute has a tendency to segregate to boundaries. Fig. 11.19a shows how carbon segregation reduces the concentration of phosphorus at the boundaries, an effect beneficial to toughness because phosphorus is more prone to in embrittle boundaries. Niobium and titanium lead to dramatic reductions in the amount of phosphorus at grain boundaries (Fig. 11.19b).

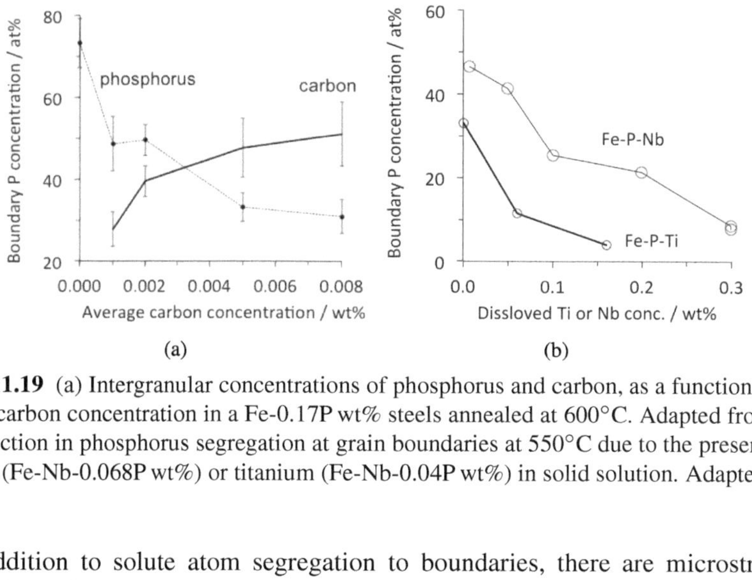

Figure 11.19 (a) Intergranular concentrations of phosphorus and carbon, as a function of the average carbon concentration in a Fe-0.17P wt% steels annealed at 600°C. Adapted from [59]. (b) Reduction in phosphorus segregation at grain boundaries at 550°C due to the presence of niobium (Fe-Nb-0.068P wt%) or titanium (Fe-Nb-0.04P wt%) in solid solution. Adapted from [60].

In addition to solute atom segregation to boundaries, there are microstructural factors that influence the intensity of temper embrittlement. In most alloy steels in which this phenomenon is encountered, the grain boundaries often are decorated with carbides, which can initiate cracks that then propagate along the impurity-enriched boundaries. Indeed, the concentration of impurities in the vicinity of the carbides may be enhanced due to partitioning from the carbide. Interfacial separation has been observed in iron containing coarse grain boundary iron carbide, the interfaces of which contained Sb, As, Sn or P.

The propagation of the grain boundary crack depends on the cohesive strength of the boundary and the deformability of the grain interior. The latter becomes more difficult at low temperatures, as the yield stress of a steel rises sharply with decreasing temperature.

The optimum temperature range for temper embrittlement is between 500°C and 575°C. However, in some steels, embrittlement occurs in the range 250–400°C [61–63]. This phenomenon is called 350°C embrittlement, and occurs at too low a temperature to attribute it to the diffusion of metalloids such as Sb to the austenite grain boundaries. It seems more likely that it could arise from smaller and more mobile atoms, e.g. P, which are rejected during grain boundary growth of iron carbide, which takes place in this temperature range. However, the morphology of the grain boundary Fe_3C, if predominantly sheet-like, could be a prime cause of low ductility in this temperature range.

Stress corrosion cracking involves failure by cracking in the presence of both a stress and of a corrosive medium. It can occur in either a transgranular or an intergranular grain mode. The latter mode appears to be prominent in temper-embrittled steels. For example, a temper-embrittled Cr-Mo steel cracks along the grain boundaries when stressed in a boiling NaOH solution. Use of a heat treatment to remove the temper embrittlement also removes the sensitivity to stress corrosion.

11.9 Irradiation embrittlement

Defects are created in the orderly arrangement of atoms within a steel that is irradiated by high-energy neutrons (>MeV) which originate from the fission of the nuclei of isotopes of uranium or plutonium [64]. This leads to the progressive embrittlement of the steel.

An iron atom has ≈ 56 times the mass of a neutron, but a fast neutron can displace it because it takes only $\approx$25 eV of energy transfer to dislodge the atom from its lattice position. A recoiling atom, following its collision with the neutron, may be forced into an interstitial position and leave a vacancy behind to form a vacancy-interstitial pair; there may even be collision cascades involving many atoms. The interstitial atoms can condense into dislocation loops that consist of a plane of interstitial atoms, the edge of which is the dislocation with Burgers vector pointing normal to the loop. In body-centred cubic iron, these loops have $\mathbf{b} = \langle 100 \rangle_\alpha$, but $\mathbf{b} = \frac{1}{2}\langle 111 \rangle_\alpha$ also occur. The loops are illustrated in Fig. 11.20a, resulting from the irradiation of pure iron with 0.2 dpa of fast neutrons – the microstructure of a typical reactor pressure vessel would be more complicated because it inherently has a large density of interfaces and dislocations even before irradiation, Fig. 11.20b.

Transmutations involving nickel, iron and boron can, for example, cause the formation of helium, which, in association with the excess vacancies, forms bubbles, that in turn cause the material to swell and embrittle. Helium has a negligible solubility in α-iron. Reactions between fast neutrons in the MeV energy range and the iron nucleus lead to the ejection of an α-particle (i.e., a He nucleus), which on dissipating its kinetic energy converts to a helium atom by combining with a couple of electrons from its vicinity. The so-called Fe56(n,α)Cr53 reaction can be written as:

$$\mathrm{n}_0^1 + \mathrm{Fe}_{26}^{56} \rightarrow \mathrm{Fe}_{26}^{57} + \underbrace{\gamma}_{\text{gamma ray}} \rightarrow \mathrm{Cr}_{24}^{53} + \mathrm{He}_2^4$$

where the subscript represents the number of protons and the superscript the total number of protons and neutrons. The Cr_{24}^{53} is a natural isotope of chromium, about 10% in abundance. Reactions like these can be used to estimate the helium production

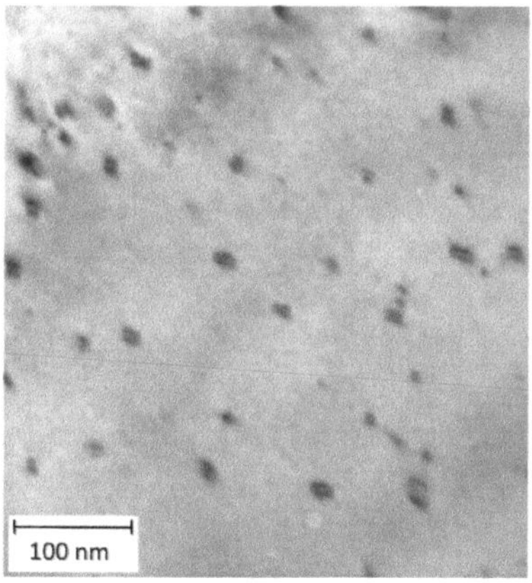

(a)

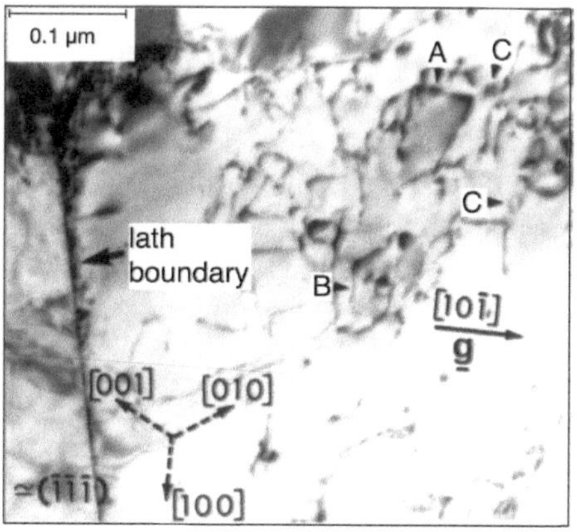

(b)

Figure 11.20 (a) Interstitial dislocation loops observed in pure iron subjected to 0.2 dpa of fast neutrons. After Meslin et al. [65] with permission of Elsevier. (b) 12Cr martensitic steel irradiated to 46 dpa at 465°C. The loops labelled A and B are $\mathbf{b} = \langle 100 \rangle_\alpha$ type, whereas C represents $\mathbf{b} = \frac{1}{2}\langle 111 \rangle_\alpha$. Image reproduced from Little [64] with permission of Taylor & Francis Ltd, http://www.tandfonline.com.

rate as a function of the displacements per atom. Some examples for the production rate per annum that might be expected in the first wall of a fusion reactor are estimated as follows, assuming a neutron wall-loading of 1 MW m^{-2} [66]:

Material	dpa annum^{-1}	He, ppm annum^{-1}
Austenitic steel 316	11.7	160
Ferritic steel ≈0.15C-9Cr-2Mo-1Nb-0.3Mo wt%	11.7	122

Manganese isotopes are also produced during the neutron activation of iron, consistent with the experiments of Fermi [67] where neutrons were first used to induce radioactivity. Fermi found some manganese in a sample that was only iron before irradiation (γ represents gamma-ray):

$$n_0^1 + Fe_{26}^{54} \rightarrow Mn_{25}^{54} + \text{proton} + \gamma$$
$$n_0^1 + Fe_{26}^{56} \rightarrow Mn_{25}^{56} + \text{proton} + \gamma.$$

The creation of long-lived radionuclides from certain alloying elements present in the steel can be problematic when it comes to the end of reactor life, because radioactive waste would need to be managed carefully over long periods of time. Since the neutron fluxes in future fusion reactors are likely to be much larger than those in fission reactors, there has been a concerted effort to substitute Mo with tungsten and Nb with tantalum or vanadium. This is to produce the so-called reduced activation steels [68, p. 358], where the radioactivity on removal from service decays much more rapidly.

The damage caused by neutron bombardment often is characterised in terms of the number of times an atom of iron is displaced from its lattice site, over the period of exposure to the neutron flux. Each atom in a radiation-facing location in a civil nuclear plant may be displaced from its equilibrium position some 20–30 times during the life of the reactor (20–30 displacements per atom). There will be recombinations of point defects and other phenomena that will reduce the number density of defects, so the actual damage rate will be lower than expected on the basis of dpa alone. The flux of point defects during irradiation can cause solute-atom clustering as illustrated in Fig. 11.21, sometimes manifesting as extremely small precipitates. In that steel, the Si, Mn, Ni and Cu became non-uniformly distributed [69]. It is difficult to assume that the clustering is a consequence of the precipitation of the equilibrium phase $Ni_{16}Si_7Mn_6$, because the very small size of the clusters would alter the equilibrium with the matrix (p. 78). Chromium remained homogeneously distributed, attributed to its high solubility in iron.

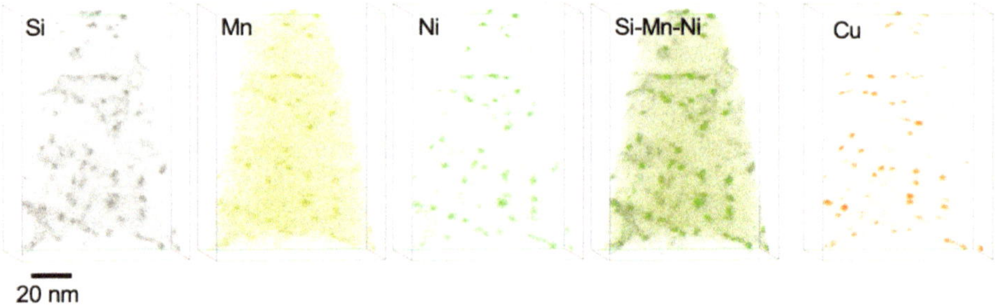

Figure 11.21 Atom-probe tomograms showing the distribution of solute atoms in Fe-0.1C-0.28Si-0.45Mn-8.37Cr-0.21Ni-0,17Cu wt%, following exposure to 3 dpa of fast neutrons at 500°C. There is the clustering of solute atoms, together with their segregation to defects. Image adapted from one supplied by Professor M. J. Swenson, from the work published in [69].

Reactor pressure vessels for civil nuclear power plants are very large objects (Fig. 11.22a), weighing more than 300 tonnes and have multiple roles, one of which is to prevent the release of radiation and associated material in the event of a reactor-core accident. It is therefore vital that the properties of the vessel remain capable throughout the life of the reactor – it is far too expensive and impractical to replace such a vessel and dispose of the resulting waste. Nevertheless, radiation damage hardens and embrittles the steel (Fig. 11.22b). Detailed, quantitative procedures have therefore been developed in order to assess and inspect the structural integrity of reactor pressure vessels on a continual basis [70]. In some cases, an annealing heat treatment at temperatures 450-490°C for a week has been found to be effective in recovering the mechanical properties of embrittled pressure vessel steel [71].

11.10 Ductile or fibrous fracture

11.10.1 General

The higher temperature side of the ductile/brittle transition is associated with a much tougher mode of failure, which absorbs much more energy in the impact test. Failure

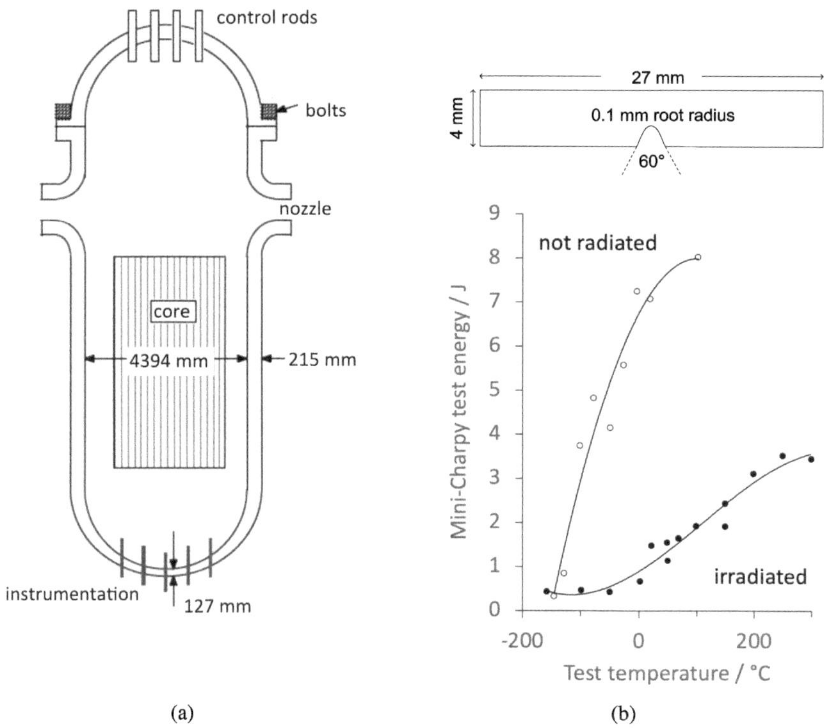

Figure 11.22 (a) Schematic diagram of Sizewell-B reactor pressure vessel, made from forged steel rings that are then welded together. There is considerable complexity within the pressure vessel, not illustrated here. Reproduced from Knott [70] with the permission of the publisher Taylor & Francis Ltd, http://www.tandfonline.com. (b) Comparison of the impact toughness of steel before and after irradiation at 286°C with a fast neutron fluence of 1.11×10^{20} neutrons cm^{-2}. The inset shows the dimensions of the square-sectioned mini-Charpy samples used to generate these data on Fe-0.16C-0.33Si-1.27Mn-1.97Ni-0.63Mo-0.06Cu wt% alloy. Selected data from Kolluri et al. [72].

mode, while is often referred to as ductile fracture, could also be described as rupture, a slow separation process that is transgranular but not markedly crystallographic. Scanning electron micrographs of the ductile fracture surface (Fig. 11.23), in striking contrast to those from the smooth, faceted cleavage surface, reveal a heavily dimpled surface, each depression being associated with a hard particle, either a carbide or non-metallic inclusion.

It is now established that most frequently, ductile failure is initiated by the nucleation of voids at second-phase particles. In steels, these particles are either carbides, sulphide or silicate inclusions. The voids form either by cracking in the particles or by decohesion at the particle/matrix interfaces. It is clear, then, that the volume fractions, distribution and morphology of both carbides and inclusions are important in determining ductile behaviour, not only in the simple tensile test, but in complex working operations. Therefore, significant variables, which determine the ductility of steels,

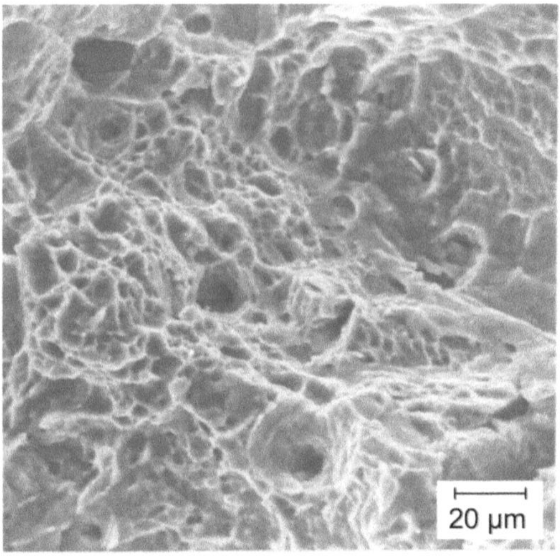

Figure 11.23 Ductile fracture surface of a low-alloy steel showing a combination of large and small voids that link up to lead eventually to the parting of the sample. In commercial steels, the voids usually are nucleated at non-metallic inclusions.

are to be found in the steel-making process, where the nature and distribution of inclusions is partly determined, and in subsequent solidification and working processes. Likewise, the carbide distribution also depends on composition and on steel-making practice, and particularly on the final heat treatment involving the transformation from austenite, which largely determines the carbide size, shape and distribution. Ductile voids can in principle nucleate in pure metals, even in the absence of particles – Orowan [73] showed that a local opening occurs at the intersection of slip planes, especially if slip is localised rather than homogeneous.

In most engineering materials, the formation of voids begins early in a tensile test, as a result of high stresses imposed by dislocation arrays on individual hard particles. Depending on the strength of the particle/matrix bond, the voids occur at varying strains, but for inclusions in steels the bonding is usually weak, so voids are observed at small plastic strains. These elongate under the influence of the tensile stress and eventually, link up with adjacent voids, forming *micronecks*. The mechanism of linking involves strain localisation in the region between adjacent voids, which causes the necks to progressively part (Fig. 11.24), leading to the ultimate ductile-fracture surfaces with a highly dimpled appearance. The second-phase particles (MnS) can be observed in Fig. 11.24.

Many strong steels exhibit poor work-hardening capacity, as shown by relatively flat stress-strain curves in tension. As a result, at high strains the plastic flow localises in shear bands, where intense deformation leads to decohesion, a type of shear fracture. While the detailed mechanism of this process is not clear, it involves the localised interaction of high dislocation densities with carbide particles.

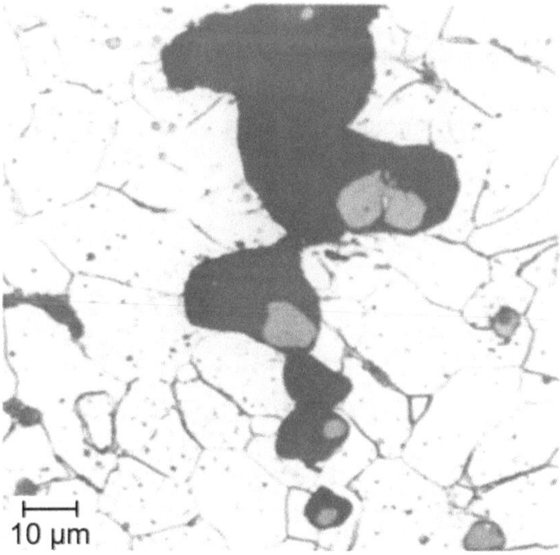

Figure 11.24 Growth of a ductile crack in a free-cutting mild steel containing sulphides. After Smith and Knott [74], reproduced with permission of Elsevier.

11.10.2 Role of inclusions in ductility

It is now generally recognised that the deformability of inclusions is a crucial factor, which plays a major role, not only in service where risk of fracture exists, but also during hot and cold working operations such as rolling, forging, machining. Kiessling has divided the inclusions found in steels into five categories relating to their deformation behaviour:

a) Al_2O_3 and calcium aluminates: these arise during deoxidation of molten steels. They are brittle solids, which are in practical terms undeformable at all temperatures.

b) Spinel-type oxides AO-B_2O_3, where 'A' and 'B' represent different metal atoms: these are undeformable in the range room temperature to 1200°C, but may be deformed above this temperature.

c) Silicates of calcium, manganese, iron and aluminium in various proportions: these inclusions are brittle at room temperature, but increasingly deformable at higher temperatures. The formability increases with decreasing melting point of the silicate, e.g., from aluminium silicate to iron and manganese silicates.

d) FeO, MnO and (FeMn)O: these are plastic at room temperature, but appear gradually to become less plastic above 400°C.

e) Manganese sulphide MnS: this common inclusion type is deformable. There are three main types of MnS inclusion dependent on their mode of formation, which markedly influences their morphology:

- Type I: Globular, formed only when oxygen is present in the melt, e.g., in rimming steels (Fig. 11.25a).

- Type II: Interdendritic eutectic form, familiar in killed (deoxidised) steels (Fig. 11.25b). The eutectic reaction involves the decomposition of sulphur-rich liquid into a mixture of austenite and (Mn,Fe)S.
- Type III: Random angular particles, found in fully deoxidised steels (Fig. 11.25c).

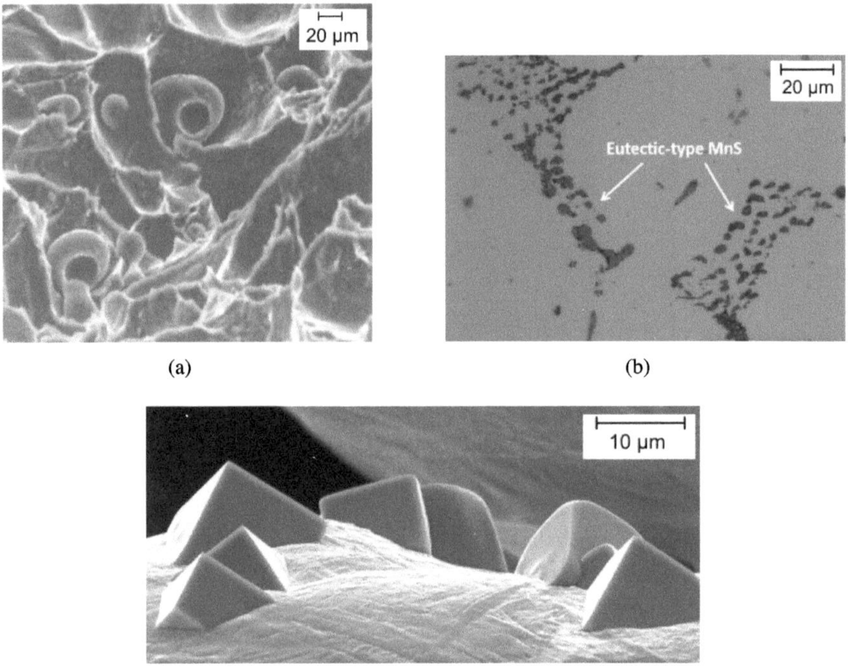

(a) (b) (c)

Figure 11.25 Manganese sulphide inclusions in steels: (a) type 1. After Baker [75], reproduced with the permission from SNCNC. (b) Eutectic type II sulphide. After Lin et al. [76] with permission of Elsevier. (c) Angular type III particles in a steel containing 0.69Mn wt%, 0.008S wt% deoxidised with zirconium. After Aydin et al. [77], Springer Nature.

It is now known that ductile failure can be associated with any of the types of inclusions listed above, from the brittle alumina type to the much more ductile sulphide inclusions. However, coarse inclusions are more damaging because they induce larger stress concentrations, which cause localised rupture and microcrack formation. Some quantitative work has now been done on model systems, e.g., iron-alumina, where the progressive effect on ductility of increasing volume fraction of alumina is readily shown. The reduction in yield stress also observed arises from stress concentrations around the inclusions and is already evident at relatively low volume fractions.

The presence of particles in the size range 1-35 μm broadens substantially the temperature range of the ductile/brittle transition in impact tests and also lowers the energy absorbed during ductile failure, the *shelf energy*. A fine dispersion of non-brittle type inclusions can delay cleavage fracture by localised relaxation of stresses with a concomitant increase in yield stress.

During cyclic stressing, inclusion size and shape influence the nucleation of fatigue cracks; angular particles are worse than spheroids. It has been found in some steels, e.g., ball bearing steels, that fatigue cracks originate only at brittle oxide inclusions, and not at manganese sulphide particles or oxides coated with manganese sulphide. Under such circumstances, the stresses that develop at particle interfaces with the steel matrix as a result of differences in thermal expansion appear to play an important part. The greatest stresses arise in calcium aluminates, alumina and spinel inclusions, which have substantially lower thermal expansion coefficients than steel. These inclusions have the most deleterious effects on fatigue life.

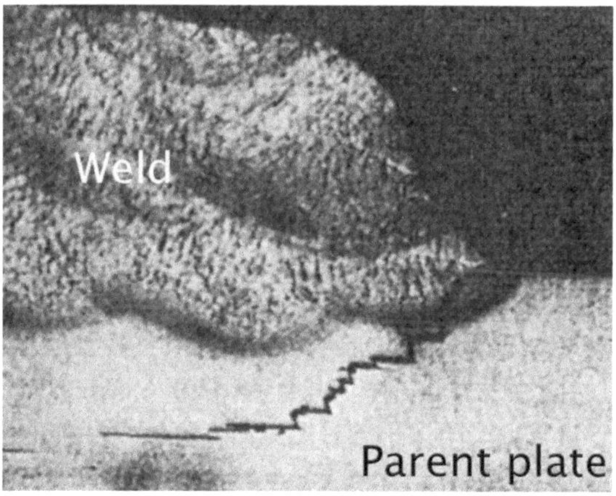

Figure 11.26 Lamellar tearing near a weld (The Welding Institute).

The behaviour of ductile inclusions such as MnS during fabrication processes involving deformation has a marked effect on the ductility of the final product. Types I and III manganese sulphide become deformed to ellipsoidal shapes, while type II colonies rotate during rolling into the rolling plane, giving rise to very much reduced toughness and ductility in the transverse direction. This type of sulphide precipitate is the most harmful, so efforts are made to eliminate it by the addition of strong sulphide forming elements such as Ti, Zr and Ca. The lack of ductility is undoubtedly encouraged by the formation at the inclusion interfaces of voids, because the MnS contracts more than the iron matrix on cooling, and the interfacial bond is probably insufficiently strong to suppress void formation. The variation in ductility with direction in rolled steels can be extreme given the anisotropic shape of the strings of sulphide inclusions. This in turn can adversely affect ductility during many working operations.

Cracking can also occur during welding of steel sheet with low transverse ductility. This takes place particularly in the parent plate under butt welds, the cracks following the line of the sulphide inclusion stringers. The phenomenon is referred to as *lamellar tearing* (Fig. 11.26).

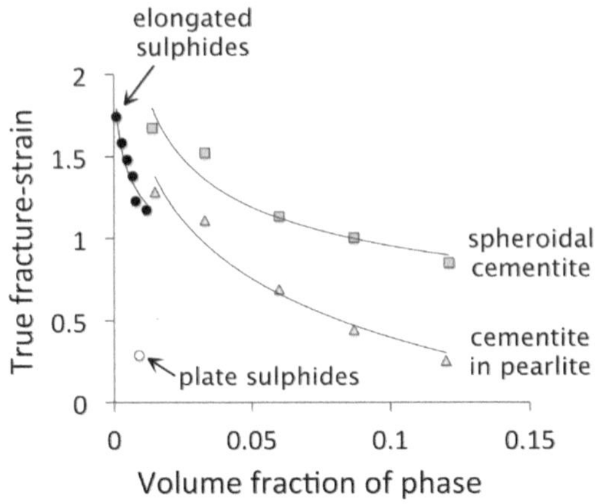

Figure 11.27 Effect of second-phase particles on the ductility of steel. Adapted from Gladman et al. [78].

11.10.3 Role of carbides in ductility

The ductility of steel is influenced by the carbide distribution, which can vary from spheroidal particles to lamellar pearlitic cementite. Comparing spheroidal cementite with sulphides of similar morphology, the carbide particles are stronger and do not crack or exhibit decohesion at small strains, with the result that a spheroidized steel can withstand substantial deformation before voids are nucleated and so exhibits good ductility. The strain needed for void nucleation decreases with increasing volume fraction of carbide and so can be linked to the carbon content of the steel.

Pearlitic cementite does not crack at small strains, but the critical strain for void nucleation is smaller than for spheroidised carbides. In the case of pearlite, that once a single lamella cracks, the crack is transmitted over much of a pearlite colony, leading to well-defined cracks in the pearlite regions. The result is that the normal ductile dimpled fractures are obtained with fractured pearlite at the base of the dimples.

The effects of second phases on the ductility of steel are summarised in Fig. 11.27, where the sulphides are shown to have a more pronounced effect than either carbide distribution. This arises because, in the case of the sulphide inclusions, voids nucleate at an early stage of the deformation process. The secondary effect of the particle shape both for carbides and sulphides is also indicated.

11.10.4 Overheating, burning

Many alloy steels when held in the range 1200–1400°C and subsequently heat treated by quenching and tempering, fail intergranularly along the original austenitic boundaries. There is strong evidence to suggest that this phenomenon is associated with the segregation of sulphur at the austenite grain boundaries at the high temperature, and indeed the phenomenon does not occur when the sulphur content of a steel is less than 0.002 wt%. Sulphur has been shown to be one of the most surface active elements in iron. Work by Goux and colleagues on pure iron–sulphur alloys has shown that an increase in sulphur content from 5 to 25 ppm raises the ductile/brittle transition temperature by over 200°C [79]. Further, Auger spectroscopy on the intergranular grain fracture surfaces has given direct evidence of sulphur segregation. However, this embrittling effect of sulphur as a result of equilibrium segregation is seen only in pure iron and not in steels where there are other impurity elements, and also where interaction of sulphur occurs with alloying elements, notably manganese and chromium.

The presence of manganese substantially lowers the solubility of sulphur in both γ- and α-iron, with the result that when sulphur segregates to high-temperature austenite boundaries, manganese sulphide is either formed there or during subsequent cooling. In either case, the manganese sulphide particles lying on the austenite boundaries are revealed by electron microscopy of the intergranular grain fracture surfaces, where they are associated with small dimples. Typically, the MnS particles are about 0.5 μm, while the dimples are approximately 2-5 μm in diameter. Thus, the grain boundary fracture process is nucleated by the sulphide particles. The mode of fracture is then determined by the size distribution of the particles, which in turn is controlled by the rate of cooling from the austenite temperature, assuming that MnS forms during cooling. With slow cooling rates, the intergranular grain fracture is replaced by cleavage or transgranular fibrous fracture as the grain boundary sulphide distribution is too coarse. Oil quenching from the austenitising temperature does not eliminate the phenomenon, which is accentuated on tempering in the range 600–650°C. This arises from the redistribution of carbides which will strengthen the grain interiors, and by precipitation at the grain boundaries which may further reduce grain boundary ductility.

When very high austenitising temperatures are used (1400–1450°C) extensive MnS precipitate is formed, often in impressive dendritic forms (Fig. 11.28). In extreme cases, partial formation of liquid phase occurs (liquidation), which greatly accentuates the intergranular grain embrittlement during subsequent heat treatment. In the absence of manganese, e.g., in wrought iron, liquid films of the iron–iron sulphide eutectic cause embrittlement during hot working processes down to 1000°C (*hot shortness*). The fact that in normal steels burning occurs only at very high temperatures should not be allowed to detract from its significance. The phenomenon may well intrude in high-temperature working processes such as forging if temperature control is not exact. It can be significant in cast steels, which by definition pass through the burning and overheating temperature range. In many cases, intergranular grain fracture is encountered in cast alloy steels where the as-cast grain structure is clearly involved. Examination of the fractures reveals extensive grain boundary sheets of manganese sulphide, often only 0.2–0.5 μm thick, but covering large areas. Marked embrittlement can occur in the as-cast state or after subsequent heat treatment in the range 500–600°C, and is often referred to as-cast brittleness or rock candy fracture. Precipitation of aluminium nitride may also play an important role in this type of fracture.

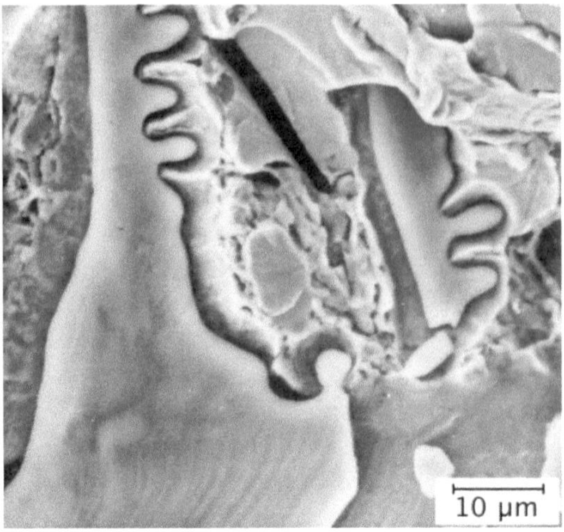

Figure 11.28 Dendrites of sulphide, which melt at a lower temperature than the steel, leading to hot shortness (courtesy of Martin Möser).

11.10.5 Liquid metal embrittlement

Liquid metal embrittlement refers to a phenomenon in which a material that is otherwise well-behaved, suffers from failure when in contact with another which is liquid at a temperature below the melting point of the material.

A liquid (l) is said to wet a grain boundary when the grain boundary energy per unit area (say $\sigma_{\gamma\gamma}$) exceeds that of the two new interfaces that replace the boundary, i.e., $\sigma_{\gamma\gamma} > 2\sigma_{l\gamma}$. Such wetting would lead to embrittlement even without the application of an external stress, because there would then be a reduction in free energy when the liquid penetrates the boundary. However, wetting of this kind is rare and other mechanisms need to be considered, although all mechanisms imply the supply of the liquid or associated vapour to the crack tip [80]. This requirement sets the limits on the temperature range over which the embrittlement can be induced [81]. A second requirement is that the material must be stressed sufficiently to produce plastic deformation, whether by stress concentration or by general yielding [82].

Austenitic steels have long been known to be susceptible to liquid metal embrittlement by zinc [83]. The invention of TWIP steels (Chapter 10) has once again brought the phenomenon to the fore, because the galvanised versions suffer from this phenomenon when resistance spot welded. The process of embrittlement in the case of tensile tests conducted on TWIP steel in contact with liquid zinc appears to be sequential. There is first the penetration of liquid zinc along the austenite grain boundaries (Fig. 11.29) while the sample is stressed, leading to cracking, followed by the further penetration along intact boundaries and more cracking [84].

The type of liquid induced embrittlement described above is but one of a class of liquid metal phenomena, including corrosive attack that is localised at microstructural features in the vicinity of the surface of the steel. The resulting crack-like features can

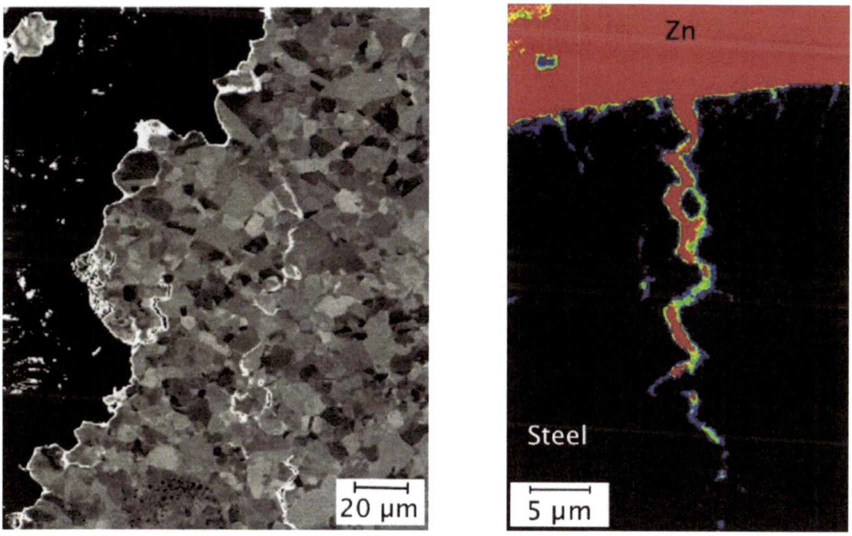

(a) (b)

Figure 11.29 (a) Intergranular fracture of TWIP steel due to liquid-metal embrittlement by zinc (after Beal et al. [85]). (b) Microanalysis image showing the penetration of zinc into a spot-welded TWIP steel, showing a meandering of the crack along the austenite grain boundaries (after Ashiri et al. [86]). Images reproduced with the permission of Elsevier.

then reduce the mechanical performance of the alloy. There may also exist chemical reactions between the liquid metal and the steel that increase brittle behaviour.

11.11 Summary

By understanding what happens when impurities compromise the structural integrity of steels, tremendous improvements have been made in the safe use of steels. Given that steels are manufactured on a vast scale, it is not economical to rid them, during steel making, of elements such as the metalloids to concentrations that are small enough for them to be innocuous. Hydrogen is unique in that, even if due care is exercised during steel production, it is able to penetrate the steel through surface reactions such as those typical in corrosion, or during fabrication and heat treatment.

It is remarkable nevertheless, that methods have been found to control the consequences of these impurities. Phosphorus-induced intergranular embrittlement can be mitigated by alloying the steel with molybdenum. Nascent hydrogen that has entered the steel can be trapped at deliberately engineered sites within the steel. Since hydrogen embrittlement requires the atom to be mobile in the lattice, the traps protect against its pernicious effects. The reaction between hydrogen and carbon in steel can be controlled by implementing more stable carbides.

One aspect worth emphasising is that experiments are no longer required to establish hydrogen embrittlement – the introduction of nascent hydrogen into ferritic steel causes embrittlement, that much is beyond doubt. It would be more productive to focus on methods for the prevention or mitigation of the phenomenon.

These, and many other embrittlement and fracture phenomena are well understood, but sometimes overlooked. Modern steels represent the most advanced materials, which determine the quality of everyday life – they should be applied on the basis of knowledge.

References

[1] A.H. Cottrell, Strengths of grain boundaries in pure metals, in: J.A. Charles, G.C. Smith (Eds.), Advances in Physical Metallurgy, The Institute of Metals, London, U.K., 1990, pp. 181–187.
[2] H. Matsui, S. Moriya, S. Takaki, H. Kimura, Mechanical properties of high purity iron at low temperatures, Transactions of JIM 19 (1978) 163–170.
[3] P. Błoński, A. Kieja, Structural, electronic, and magnetic properties of bcc iron surfaces, Surface Science 601 (2007) 123–133.
[4] C.H. Ersland, I.R. Vatne, C. Thaulow, Atomistic modeling of penny-shaped and through-thickness cracks in bcc iron, Modelling and Simulation in Materials Science and Engineering 20 (2012) 075004.
[5] R. Ayer, R.R. Mueller, T. Neeraj, Electron backscattered diffraction study of cleavage fracture in pure iron, Materials Science & Engineering A 417 (2006) 243–248.
[6] P. Yan, O.E. Güngör, P. Thibaux, M. Liebeherr, H.K.D.H. Bhadeshia, Tackling the toughness of steel pipes produced by high frequency induction welding and heat-treatment, Materials Science & Engineering A 528 (2011) 8492–8499.
[7] H. Inagaki, K. Kurihara, I. Kozasu, Influence of crystallographic texture on the strength and toughness of the controlled rolled high tensile strength steel, Transactions of the Iron and Steel Institute of Japan 17 (1977) 75–81.
[8] M.S. Joo, D.-W. Suh, J.H. Bae, N. Sanchez Moriño, R. Petrov, L.A.I. Kestens, H.K.D.H. Bhadeshia, Experiments to separate the effect of texture on aniostropy of pipeline steel, Materials Science & Engineering A 556 (2012) 601–606.
[9] M.S. Joo, D.-W. Suh, J.-H. Bae, H.K.D.H. Bhadeshia, Toughness anisotropy in X70 and X80 linepipe steels, Materials Science and Technology 30 (2014) 439–440.
[10] T.L. Altshuler, J.W. Christian, Mechanical properties of pure iron tested in compression over the temperature range 2-293 K, Proceedings of the Royal Society A 261 (1967) 253–287.
[11] P.B. Hirsch, Fifth Int. Cong. Crystallography, Cambridge, 1960, Oral report, referred to in numerous publications.
[12] H. Pous-Romero, Quantitative basis for the structure and properties of a critically-important pressure vessel steel, Ph.D. thesis, University of Cambridge, Cambridge, U.K., 2014.
[13] D. Hull, Nucleation of cracks by the intersection of twins in α-iron, Philosophical Magazine 3 (1958) 1468–1469.
[14] N.J. Petch, The influence of grain boundary carbide and grain size on the cleavage strength and impact transition temperature of steel, Acta Metallurgica 34 (1986) 1387–1393.
[15] A.A. Griffith, The phenomena of rupture and flow in solids, Philosophical Transactions of the Royal Society A 221 (1920) 163–197.
[16] J.P. Berry, Some kinetic considerations of the Griffith criterion for fracture I: equations of motion at constant force, Journal of the Mechanics and Physics of Solids 8 (1960) 194–206.

[17] W.H. Johnson, On some remarkable changes produced in iron and steel by the action of hydrogen and acids, Proceedings of the Royal Society of London 23 (1875) 168–179.

[18] A.R. Troiano, Role of H and other interstitials in the mechanical behaviour of metals, Transactions of the ASM 52 (1960) 54–80.

[19] R.A. Oriani, A mechanistic theory of hydrogen embrittlement of steels, Berichte der Bunsengesellschaft für physikalische Chemie 76 (1972) 848–857.

[20] E.J. Song, H.K.D.H. Bhadeshia, D.-W. Suh, Effect of hydrogen on the surface energy of ferrite and austenite, Corrosion Science 77 (2013) 379–384.

[21] C.D. Beachem, A new model for hydrogen-assisted cracking (hydrogen "embrittlement"), Metallurgical Transactions 3 (1972) 437–451.

[22] M. Nagumo, Hydrogen related failure of steels - a new aspect, Materials Science and Technology 20 (2004) 940–950.

[23] K. Sugita, M. Mizuno, H. Araki, Y. Shirai, T. Omura, K. Tomatsu, Y. Sakiyama, Review of positron lifetime studies of lattice defects formed during tensile deformation in a hydrogen environment, ISIJ International 61 (2021) 1056–1063.

[24] E.J. Song, D.W. Suh, H.K.D.H. Bhadeshia, Theory for hydrogen desorption in ferritic steel, Computational Materials Science 79 (2013) 36–44.

[25] M. Nagumo, Fundamentals of Hydrogen Embrittlement, Springer, Singapore, 2016.

[26] R.P. Frohmberg, W.J. Barnett, A.R. Troiano, Delayed failure and hydrogen embrittlement in steel, Technical Report 54-320, Wright-Patterson Air Force Base, Ohio, U.S.A., 1954.

[27] P.H.M. Hart, Hydrogen cracking - its causes, costs and future occurrence, www.twi-global.com/technical-knowledge/published-papers/hydrogen-cracking-its-causes-costs-and-future-occurrence-march-1999/, 1999.

[28] H.K.D.H. Bhadeshia, Prevention of hydrogen embrittlement in steels, ISIJ International 56 (2016) 24–36.

[29] L.S. Darken, R.P. Smith, Behaviour of hydrogen in steel during and after immersion in acid, Corrosion 5 (1949) 1–16.

[30] T. Omura, H. Sawada, K. Kobayashi, Y. Arai, Effects of alloying elements on hydrogen diffusion in iron, ISIJ International 61 (2021) 1287–1293.

[31] A. Turnbull, R.B. Hutchings, Analysis of hydrogen diffusion in a two-phase alloy, Materials Science & Engineering A 177 (1994) 161–171.

[32] L.C.D. Fielding, E.J. Song, D.K. Han, H.K.D.H. Bhadeshia, D.W. Suh, Hydrogen diffusion and the percolation of austenite in nanostructured bainitic steel, Proceedings of the Royal Society of London A 470 (2014) 20140108.

[33] Y. Hatano, T. Nozaki, H. Homma, M. Matsuyama, Diffusion behaviour of hydrogen in TiC, in: Annual Report of Hydrogen Isotope Research Center, vol. 26, Toyama University, Japan, 2006, pp. 31–43.

[34] J.B. Bates, J.C. Wang, R.A. Perkins, Mechanisms of hydrogen diffusion in TiO_2, Physical Review B 19 (1979) 4130–4139.

[35] A.B. Belonoshko, A. Rosengren, Q. Dong, G. Hultquist, C. Leygraf, First-principles study of hydrogen diffusion in α-Al_2O3 and liquid alumina, Physical Review B 69 (2004) 024302.

[36] G.C. Yu, S.K. Yen, Hydrogen diffusion coefficient of silicon nitride thin films, Applied Surface Science 201 (2002) 204–207.

[37] A.G. Quarrell, Hydrogen in steel, in: Symposium on Production, Properties and Applications of Alloy and Special Steels, National Metallurgical Laboratory, Jamshedpur, India, 1956, pp. 256–264.

[38] H. Sugimoto, Y. Fukai, Solubility of hydrogen in metals under high hydrogen pressures: thermodynamical calculations, Acta Materialia 40 (1992) 2327–2336.

[39] H.K. Ihrig, Attack of hydrogen-nitrogen mixtures on steels, Industrial and Engineering Chemistry 41 (1949) 2516–2521.
[40] L.C. Weiner, Kinetics and mechanism of hydrogen attack of steel, Corrosion 17 (1961) 137t–143t.
[41] E.K. Scott, Nitrates and ammonia from atmospheric nitrogen. Lectures III, Journal of the Royal Society of Arts 71 (1923) 900–917.
[42] F. Habashi, History of corrosion research, CIM Bulletin 96 (2003) 88–94.
[43] T.G. Kim, Failure of piping by hydrogen attack, Engineering Failure Analysis 9 (2002) 571–578.
[44] K.E. Blazek, J.R. Cost, Carbon diffusivity in iron-chromium alloys, Transactions of the Japan Institute of Metals 17 (1976) 630–636.
[45] Anonymous, The technical basis document for API RP 941, Technical Report API TR 941-A, American Petroleum Institute, Washington, D.C., USA, 2019.
[46] M.A.M. Alshahrani, S.W. Ooi, M.H. Colliander, G.M.A.M. El-Fallah, H.K.D.H. Bhadeshia, High-temperature hydrogen attack on 2.25Cr-1Mo steel: the roles of residual carbon, initial microstructure and carbide stability, Metallurgical & Materials Transactions A 58 (2022) 4221–4232.
[47] H.K.D.H. Bhadeshia, Models for the elementary mechanical properties of steel welds, in: H. Cerjak, H.K.D.H. Bhadeshia (Eds.), Mathematical Modelling of Weld Phenomena 3, vol. 3, Institute of Materials, London, 1997, pp. 235–255.
[48] P.M. Anderson, J.-S. Wang, J.R. Rice, Thermodynamic and mechanical models of interfacial embrittlement, in: G.B. Olson, M. Azrin, E.S. Wright (Eds.), Innovations in Ultrahigh-Strength Steel Technology, US Army Materials Technology Laboratory, Massachusetts, USA, 1987, pp. 619–650.
[49] R. Wu, A.J. Freeman, G.B. Olson, First principles determination of the effects of phosphorus and boron on iron grain boundary cohesion, Science 265 (1994) 376–380.
[50] H.K.D.H. Bhadeshia, Geometry of Crystals, 2nd edition, Institute of Materials, 2001.
[51] G.L. Krasko, G.B. Olson, Effect of boron, carbon, phosphorus and sulphur on intergranular cohesion in iron, Solid State Communications 76 (1990) 247–251.
[52] H.K.D.H. Bhadeshia, D.W. Suh, Is low phosphorus content in steel a product requirement?, Ironmaking & Steelmaking 42 (2015) 259–267.
[53] T. Dümmer, J.C. Lasalvia, G. Ravichandran, M.A. Meyers, Effect of strain rate on plastic flow and failure in polycrystalline tungsten, Acta Materialia 46 (1998) 6267–6290.
[54] A.V. Nikolaeva, Y.A. Nikolaev, Y.R. Kevorkyan, Grain-boundary segregation of phosphorus in low-alloy steel, Atomic Energy 91 (2001) 534–542.
[55] T. Ogura, T. Watanabe, S. Karashima, T. Masumoto, Dependence of phosphorus segregation on grain boundary crystallography in an Fe-Ni-Cr alloy, Acta Metallurgica 35 (1987) 1807–1814.
[56] S. Kobayashi, S. Tsurekawa, T. Watanabe, G. Palumbo, Grain boundary engineering for control of sulfur segregation-induced embrittlement in ultrafine-grained nickel, Scripta Materialia 62 (2010) 294–297.
[57] T. Watanabe, A new era of grain boundary design and control for high temperature materials, Key Engineering Materials 171–174 (2000) 237–244.
[58] C.J. McMahon Jr, Intergranular fracture in steels, Materials Science and Engineering 25 (1976) 233–239.
[59] H. Erhart, H.J. Grabke, Equilibrium segregation of phosphorus at grain boundaries of Fe-P, Fe-C-P, Fe-Cr-P, and Fe-Cr-C-P alloys embrittlement, Metal Science 15 (1981) 401–408.
[60] H.J. Grabke, K. Hennesen, R. Moller, W. Wei, Effects of manganese on the grain boundary segregation, bulk and grain boundary diffusivity of P in ferrite, Scripta Metallurgica 21 (1987) 1329–1334.

[61] C.L. Briant, S.K. Banerji, Tempered martensite embrittlement in phosphorus doped steels, Metallurgical Transactions A 10 (1979) 1729–1737.
[62] J.P. Materkowski, G. Krauss, Tempered martensite embrittlement in AISI 4340 steel, Metallurgical Transactions A 10 (1979) 1643–1651.
[63] H.K.D.H. Bhadeshia, D.V. Edmonds, Tempered martensite embrittlement: role of retained austenite and cementite, Metal Science 13 (1979) 325–334.
[64] E.A. Little, Development of radiation resistant materials for advanced nuclear power plant, Materials Science and Technology 22 (2004) 491–518.
[65] E. Meslin, M. Lambrecht, M. Hernández-Mayoral, F. Bergner, L. Malerba, P. Pareige, B. Radiguet, A. Barbu, D. Gómez-Briceño, A. Ulbricht, et al., Characterization of neutron-irradiated ferritic model alloys and a RPV steel from combined APT, SANS, TEM and PAS analyses, Journal of Nuclear Materials 406 (2010) 73–83.
[66] H. Ullmaier (author and editor), in: Atomic Defects in Metals, Springer Materials, Heidelberg, Germany, 1991, pp. 385–386, chap. 3.2.
[67] E. Fermi, Radioactivity induced by neutron bombardment, Nature 133 (1934) 757.
[68] H.K.D.H. Bhadeshia, Bainite in Steels: Theory and Practice, 3rd ed., Maney Publishing, Leeds, U.K., 2015.
[69] S.B. Adisa, R. Blair, M.J. Swenson, Comparison of microstructure evolution in Fe^{2+} or neutron-irradiated T91 at 500 °C, Materialia 12 (2020) 100770.
[70] J.F. Knott, Structural integrity of nuclear reactor pressure vessels, Philosophical Magazine 93 (2013) 3835–3862.
[71] M.G. Vassilaros, M.E. Mayfield, K.R. Wichman, Annealing of nuclear reactor pressure vessels, Nuclear Engineering and Design 181 (1998) 61–69.
[72] M. Kolluri, P. ten Pierick, T. Bakker, B. Straathof, A. Magielsen, Z. Szaraz, E. D'Agata, C. Ohms, O. Martin, Influence of Ni-Mn contents on the embrittlement of PWR RPV model steels irradiated to high fluences relevant for LTO beyond 60 years, Journal of Nuclear Materials 553 (2021) 153036.
[73] E. Orowan, Fracture and strength of solids, Reports on Progress in Physics 12 (1949) 185–232.
[74] J. Knott, Micro-mechanisms of fracture and the fracture toughness of engineering alloys, in: Advances in Research on the Strength and Fracture of Materials, Pergamon Press, Oxford, U.K., 1978, pp. 61–92.
[75] T. Baker, Use of scanning electron microscopy in studying sulphide morphology on fracture surfaces, Metallographic and Microstructural Analysis 5 (2016) 69–79.
[76] S.-g. Lin, H.-h. Yang, Y.-h. Su, K.-l. Chang, C.-h. Yang, S.-k. Lin, CALPHAD-assisted morphology control of manganese sulfide inclusions in free-cutting steels, Journal of Alloys and Compounds 779 (2019) 844–855.
[77] E. Aydın, C. Arısoy, M. Sesen, The effect of manganese sulfide inclusions and zirconium additions on the mechanical properties of heavy section cast steel, International Journal of Metalcasting 12 (2018) 383–395.
[78] T. Gladman, B. Holmes, I.D. McIvor, Effects of second-phase particles on strength, toughness and ductility, in: Iron and Steel Institute Conference on Effect of Second-Phase Particles on the Mechanical Properties of Steels, Iron and Steel Institute, London, U.K., 1971, pp. 68–78.
[79] P. Jolly, C. Goux, Influence of certain impurities on intercrystalline embrittlement of iron, Mémoires Scientifiques de la Revue de Métallurgie 66 (1969) 605–617.
[80] M.H. Kamdar, Liquid metal embrittlement, in: C.L. Briant, S.K. Banerji (Eds.), Treatise on Materials Science and Engineering: Embrittlement of Engineering Alloys, vol. 25, Academic Press, New York, USA, 1983, pp. 362–455, chap. 9.

[81] C.F. Old, Liquid metal embrittlement, Journal of Nuclear Materials 92 (1980) 2–15.
[82] M.G. Nicholas, C.F. Old, Liquid metal embrittlement, Journal of Materials Science 14 (1979) 1–18.
[83] R.M. Bruscato, Liquid metal embrittlement of austenitic stainless steel when welded to galvanized steel, Welding Journal, Research Supplement 71 (1992) 455s–459s.
[84] C. Beal, X. Kleber, D. Fabregue, M. Bouzekri, Liquid zinc embrittlement of twinning-induced plasticity steel, Scripta Materialia 66 (2012) 1030–1033.
[85] C. Beal, X. Kleber, D. Fabregue, M. Bouzekri, Embrittlement of a zinc coated high manganese TWIP steel, Materials Science & Engineering A 543 (2012) 76–83.
[86] R. Sshiri, M. Shamanian, H.R. Salimijazi, M.A. Haque, J.H. Bae, C.W. Ji, K.G. Chin, Y.D. Park, Liquid metal embrittlement-free welds of Zn-coated twinning induced plasticity steels, Scripta Materialia 114 (2015) 41–47.

Stainless steel

12

Abstract

A stainless steel is, in its raw form, resistant to stains, discoloration or loss of mass due to rusting. This is because it contains sufficient chromium to form a passive film of oxide on the surface, which isolates the substrate from the reactive environment. The film is able to reform in seconds in the event of damage. With an appropriate combination of alloying elements, stainless steels can be fully austenitic, a mixture of ferrite and austenite, fully ferritic or martensitic. They may or may not be precipitation hardened. The palette of alloys available permits considerable creativity in the application of stainless steel, from the facade of modern buildings, to artistic creations and in critical applications such as nuclear reactors. Not to mention in building space vehicles destined ultimately for Mars.

12.1 Introduction

Some elements extend the γ-loop in the iron-carbon equilibrium diagram (Chapter 4), e.g., nickel and manganese. When sufficient alloying element is added, it is possible to preserve the face-centred cubic austenite at room temperature and pressure, either in a stable or metastable condition. Chromium added alone to a plain carbon steel tends to close the γ-loop and favour the formation of ferrite. This is why nickel is also added whenever the steel is intended to be austenitic, to retard the kinetics of the $\gamma \rightarrow \alpha$ transformation.

The presence of chromium greatly improves the corrosion resistance of the steel by forming a very thin (0.5-4 nm), stable, regenerative and passive chromium-rich oxide film on the surface, in the presence of oxidising agents. The term *passive* means that the reaction of the metal with its surroundings, for example with atmospheric oxygen, forms a protective surface layer that is uniform and stifles further reaction [1]. If the film is damaged, the chromium in the steel reacts again to reform the protective layer, a reaction that takes mere seconds. The actual composition of the oxide film is better represented as $Cr_{2-x}Fe_xO_3$, where x is nearly zero for chromium-rich alloys [2]. The oxide film often is able to adapt further to the environment in which the steel is placed [3], for example, Fig. 12.1. This is why chromium-nickel stainless steels are now the most used materials in a wide range of corrosive environments, both at room and elevated temperatures, in engineering as well as artistic constructions (Fig. 12.2).

During the early stages of corrosion, the iron is oxidised preferentially so the underlying region becomes richer in chromium. The diffusion of chromium through that early-oxide layer is also much slower than of iron, thus adding to the effect. Therefore, it is chromium-rich oxide that eventually predominates at the steel/oxide interface. The composition of this inner layer becomes more enriched in chromium as the average Cr-concentration in the steel increases, but about ≈ 13 wt% of average Cr-concentration is necessary for the steel to become stainless. It is interesting that surface treatments that locally reduce the grain size result in better corrosion resistance because it becomes easier for the chromium to diffuse towards the surface and then combine with oxygen to form the protective oxide [4].

Steels. https://doi.org/10.1016/B978-0-44-318491-8.00018-6

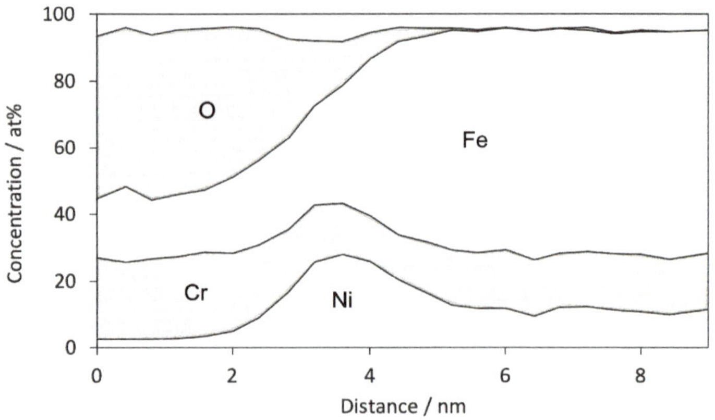

Figure 12.1 Chromium-rich oxide covering a 316 stainless steel that has been exposed to pressurised-water nuclear reactor conditions. Illustrates the variation of chemical composition at the oxide-metal interface – not all species are plotted for the sake of clarity. After Lozano-Perez et al. [5], reproduced and adapted with permission of Elsevier.

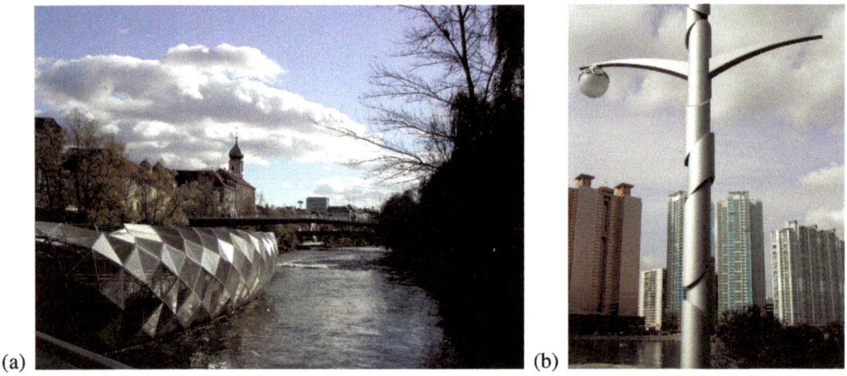

Figure 12.2 Corrosion-resistant artistic creations using stainless steel. (a) A fresh-water island on the river Mur, in Graz, Austria. (b) A street lamp on the sea-port Busan, Republic of Korea.

Added to the remarkable corrosion-resistance that chromium confers to steel, austenitic stainless-steels are readily fabricated and do not undergo a ductile/brittle transition which can cause problems in ferritic steels. This has ensured that they have become a most important group of construction steels, often in very demanding environments. Nevertheless, there also are some important ferritic stainless steels which will be discussed in this chapter. In terms of strength and ductility, stainless steels of all varieties follow the same general trend as the low-alloy steels, Fig. 12.3. Therefore, their elementary mechanical properties are not the key reason for selecting stainless steels, which are 3-4 times more expensive than mild steels. But they do reign supreme in terms of their corrosion-resistance and consequently aesthetics.

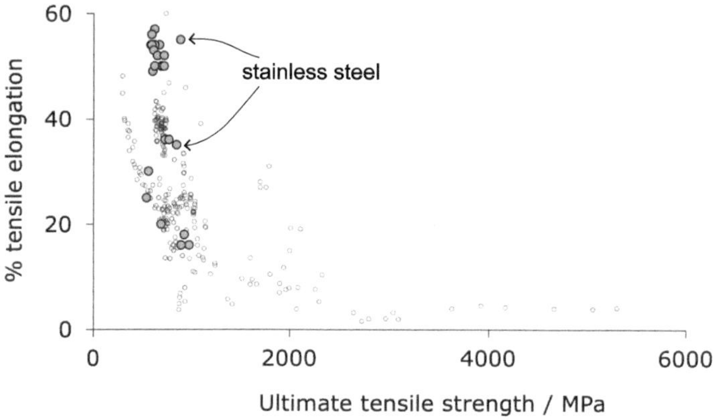

Figure 12.3 The typical combinations of ultimate tensile strength and elongation of commercially available stainless steels, highlighted in the same diagram as for the low-alloy steels described in Fig. 2.1. Similar data for some extraordinary properties achieved in cold-worked stainless steel for cryogenic application are presented in Chapter 17.

12.2 Not always corrosion resistant

While it is self-evident that stainless steels do not ordinarily rust, there are circumstances where they do react and therefore must be applied with care, as illustrated by two examples below.

The electrolysis of water to generate hydrogen can be achieved in many ways, but an efficient method with rapid response involves proton exchange membranes. The operating cell contains porous transport layers that allow water to reach the electrodes and gas to be removed. The environment on the anode side of the cell contains a high concentration of oxygen. Titanium can resist this, but its manufacture in the porous form required and with the platinum coating that prevents it from passivating, is very expensive. Stainless steel mesh, which is much cheaper, releases Fe and Ni in the environment of the cell, thus poisoning the catalyst-coated membrane within. However, applying a Nb/Ti spray-coating on the anode-facing side (Fig. 12.4), which still leaves the mesh porous, has resolved the problem with a dramatic reduction in cost, especially necessary if the process is to be scaled up to the high volume production of hydrogen.

Stainless steel (316L) implants within the human body are not sufficiently corrosion-resistant [7]. This in part is because bodily fluids contain chloride ions that can cause the localised breakdown of the passive film that normally protects the steel. A lack of oxygen may affect the ability of the passive film to maintain itself.

The release of corrosion products into the surrounding tissue induces the formation of fibrous layers some 2 mm in thickness on the surface of the implant, which compromises the connection between the steel and tissue. Coating the steel with gold, for example, is risky because a break in the coated layer produces a small anodic region surrounded by a much larger cathodic area. The corrosion current density is therefore intensified at the break.

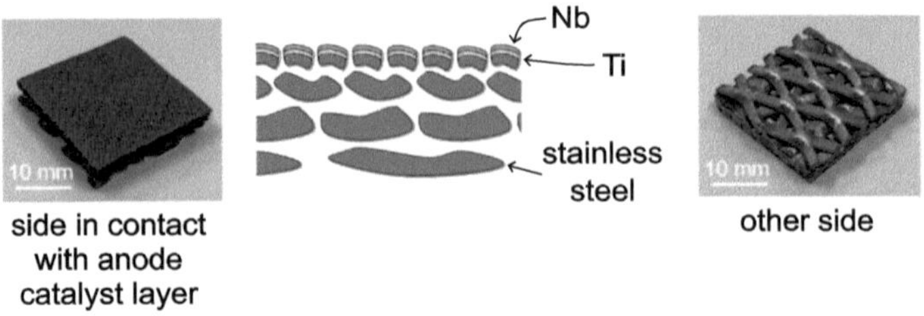

Figure 12.4 Stainless steel mesh, coated on the side facing the anode catalyst, with titanium and niobium, while still allowing permeability through the mesh. After Stiber et al. [6], modified for simplicity under the CC BY 3.0 license https://creativecommons.org/licenses/by/3.0/.

Stainless steels are nevertheless frequently used for non-permanent implants. They are much cheaper than titanium alternatives, which have better corrosion resistance in bodily fluids. The steel has appropriate mechanical properties, and is relatively easy to fabricate. When repairing skeletal fractures, it is in any case, desirable to remove the implant at an appropriate stage of healing. It becomes more difficult to access the implant once it is thoroughly integrated within the bodily tissues, for example, bone overgrowth (Fig. 12.5) makes it more difficult to remove an implant.

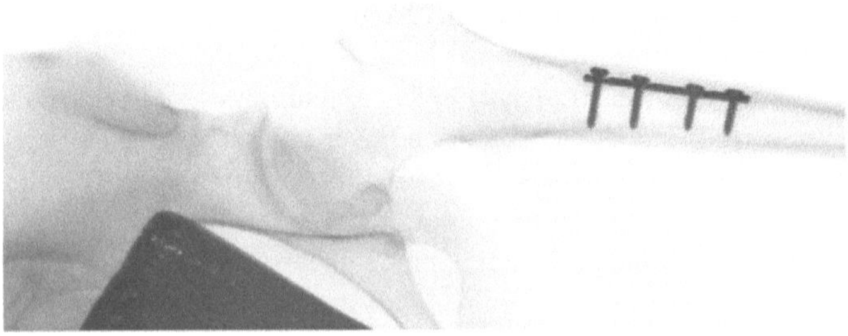

Figure 12.5 Radiograph showing the hip of a patient with an implant, showing bone growing over the retention plate, which would make it difficult to remove the plate at a later stage. Image from Montgomery et al. [8], with the permission of Walters Kluwer Heath. The contrast here has been inverted for clarity.

12.3 The iron–chromium–nickel system

The binary iron-chromium equilibrium diagram (Fig. 12.6a) shows that chromium restricts the austenite phase-field into a γ-loop, such that the alloy is fully ferritic over the whole temperature range below melting, when the chromium content exceeds 13 wt%; the regime where $\alpha+\gamma$ coexist is also narrow, between 12 and 13 wt% Cr. If the precipitation of the intermetallic compound FeCr-σ can be avoided, a miscibility gap develops at the lower temperatures within which a solid solution tends to phase-

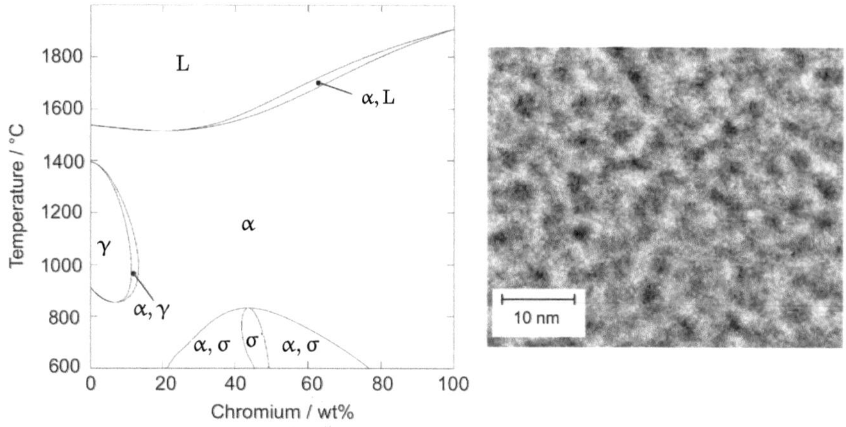

(a) (b)

Figure 12.6 (a) Calculated Fe-Cr equilibrium phase diagram. Chromium has extensive solubility in ferrite. L stands for liquid and the iron-chromium intermetallic compound is sigma phase (tetragonal, space group $P4_2/mnm$), labelled σ. The formation of σ requires considerable diffusion, so it can ordinarily be suppressed by fast-cooling the ferrite. Diagram courtesy of Mathew Peet. (b) Chromium composition fluctuations recorded using electron energy loss spectroscopy in a Fe-36Cr wt% alloy aged at 500°C for 100 h. Image reproduced from Westraadt et al. [9] with the permission of Elsevier.

separate into a Cr-poor and Cr-rich α, i.e., by spinodal decomposition (Fig. 12.6b). The hardness increment associated with this leads to embrittlement, but the process is slow, though it is accelerated when the alloy is subjected to fast-neutron irradiation.

Even though the extent of the γ and $\gamma + \alpha$ phase fields becomes greater with the addition of 0.05 wt% carbon, an alloy containing 19 wt% chromium alloy will contain austenite over a range of temperatures (Fig. 12.7a). The second effect of carbon is to introduce carbides to the structure, as indicated in Figs. 12.7 and 12.8:

$$\theta \equiv M_3C\ (Pnma), \quad \theta_1 \equiv M_{23}C_6\ (Fm\overline{3}m), \quad \theta_2 \equiv M_7C_3\ (Pmcn),$$

where M represents a mixture of metal atoms and the crystallographic space groups are indicated in brackets. Considering equilibria at 973 K as a function of the chromium concentration, the carbon-rich M_7C_3 is at first the stable carbide in ferrite at low w_{Cr}, but with continued increase in Cr concentration, it is replaced by $M_{23}C_6$, which contains less carbon, permitting the carbide to achieve a greater mass fraction, Fig. 12.7b.

Between 0.08 and 0.22C wt%, partial transformation is possible, leading to $\alpha+\gamma$ structures, while above 0.4C wt% the steel can be made fully austenitic (Fig. 12.8) if cooled sufficiently rapidly from the γ-loop region. In austenitic steels, $M_{23}C_6$ is the most significant carbide formed and it can have a substantial detrimental influence on corrosion resistance. This is because there may exist a chromium-depleted zone in the vicinity of the precipitated carbide (Fig. 12.8b), which becomes anodic with respect to the surrounding regions that are richer in chromium, leading to pitting corrosion.

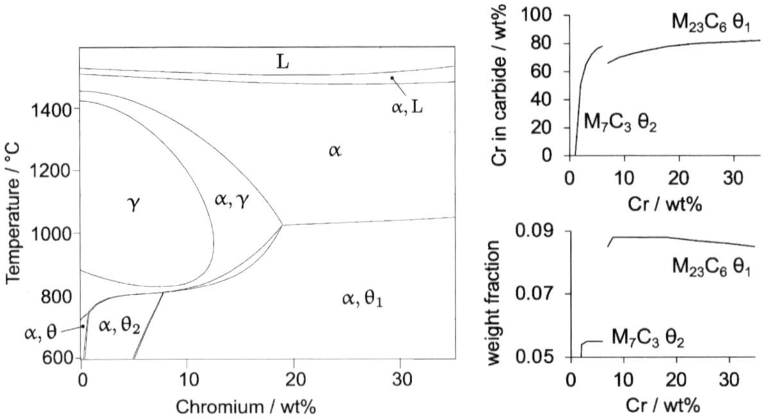

(a) (b)

Figure 12.7 (a) Calculated Fe-Cr-0.05C wt% equilibrium phase diagram. θ, θ_1 and θ_2, stand for cementite, $M_{23}C_6$ and M_7C_3, respectively. Because this is now a section of a *ternary* phase diagram, melting occurs over a range of temperatures even at 0 Cr wt% and the tie-lines connecting the equilibrium compositions of phases do not necessarily lie in the plane of the section. Diagram courtesy of Mathew Peet. (b) Calculated variation in carbide fraction and chromium content as a function of the total Cr for 973 K.

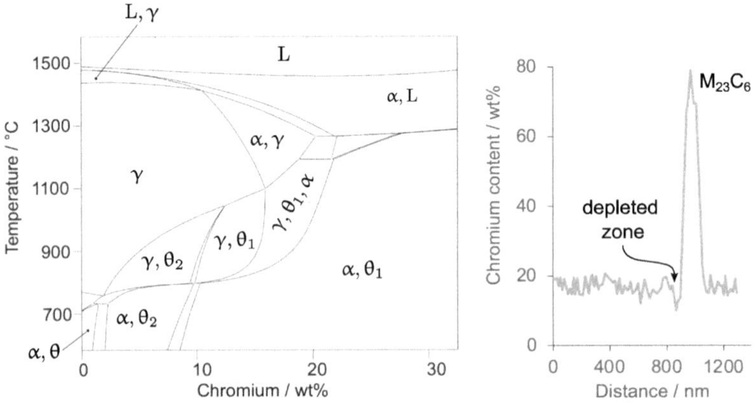

(a) (b)

Figure 12.8 (a) Calculated section through Fe-Cr-0.4C wt% ternary equilibrium phase diagram. Carbides become more prevalent at higher temperatures, coexisting sometimes with austenite as well. Diagram courtesy of Mathew Peet. (b) A chromium-depleted zone on one side of a $M_{23}C_6$ precipitate at an austenite grain boundary in an 18-8 stainless steel. The zone is only on the side where growth is rapid. Selected data from Kaneko et al. [10].

If nickel is added to a low-carbon iron-18Cr wt% alloy, the γ-phase field is expanded until at about 8Ni wt%, the γ-phase persists to room temperature (Fig. 12.9), leading to the familiar group of the so-called 18-8 (18Cr-8Ni) austenitic stainless steels. This particular composition arises because a minimum nickel content is required to retain γ at room temperature. With higher Cr contents, more nickel is needed to stabilise the austenite. Thus, in corrosion-resistant alloys containing 25 wt% Cr, about 15 wt% nickel is needed to retain the austenite under ambient conditions. Lack of complete retention is indicated by the formation of martensite. A stable austenite can be defined as one in which the M_S is lower than room temperature. The 18Cr-8Ni steel, in fact, has an M_S just below room temperature, and on cryogenic cooling, e.g., in liquid nitrogen, about 20% of the austenite transforms into α'-martensite (Fig. 12.10a), with some ε-martensite forming as well; the quantity of martensite is further increased if the steel is plastically deformed [11]. The α' is ferromagnetic, which makes it unsuitable for applications where the magnetic permeability (measure of magnetisation in applied magnetic field) must be very small. For this reason, a more highly alloyed "316LN" is selected for the construction of the CERN proton synchrotron [12]. The composition of the 316LN contains Fe-0.015C-1.4Mn-17.3Cr-12.7Ni-2.5Mo-0.11N wt%, making its austenite stable to deformation at room temperature; the volume fraction of martensite generated by 60% true strain is just 0.002 [13].

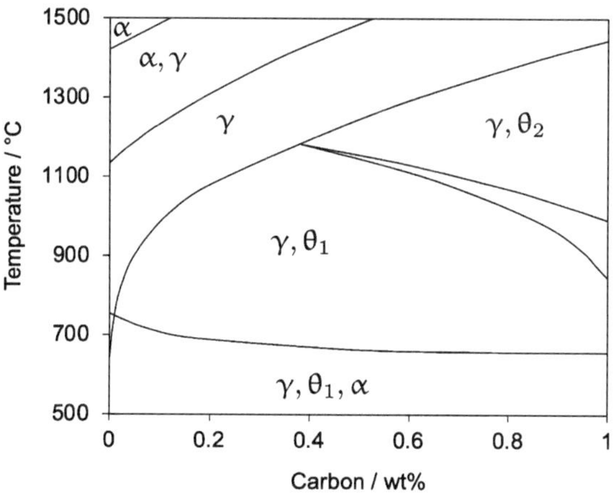

Figure 12.9 Calculated effect of carbon on a section of the quaternary equilibrium phase diagram for 18Cr-8Ni steel as a function of its carbon concentration. θ_1 and θ_2, stand for cementite, $M_{23}C_6$ and M_7C_3 respectively. Diagram courtesy of Mathew Peet.

Fig. 12.10b also shows that the carbide $M_{23}C_6$ exists below about 900°C. However, it goes into solution when the steel is heated to 1100–1150°C and since it requires substitutional atom diffusion in order to form, quenching from the high temperatures results in a precipitate-free austenite at ambient temperature. However, on reheating in the range 550–750°C, $M_{23}C_6$ is re-precipitated preferentially at the grain boundaries.

Manganese expands the γ-loop and can, therefore, be used instead of nickel. However, it is not as strong a γ-former, so greater concentrations are required. In the absence of chromium, around 12 wt% Mn is required to stabilise even higher carbon (1-1.2 wt%) austenite, achieved in Hadfield's steel, which approximates to this

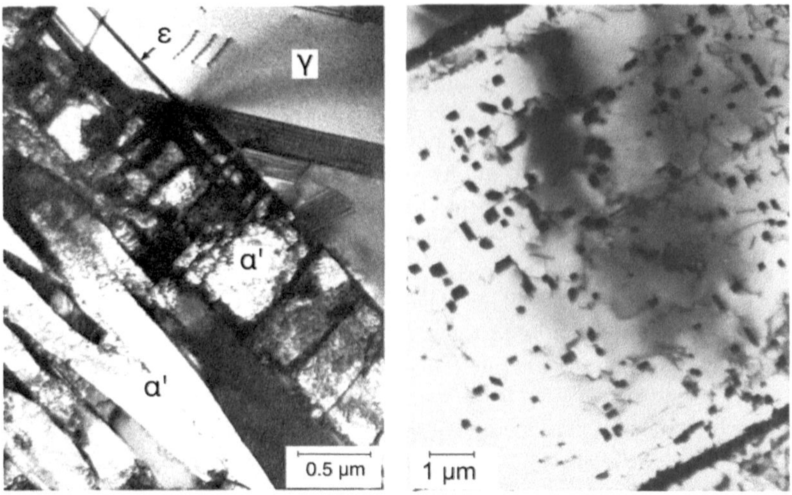

(a) (b)

Figure 12.10 (a) Both α' and ε-martensite in 18/8 stainless steel cooled to −196°C. After Guy et al. [11], reproduced with the permission of Taylor & Francis Ltd, http://www.tandfonline.com. (b) A "304" austenitic stainless steel, which often is referred to as 18/8, and contains a larger quantity of austenite stabilising solutes but only 0.07C wt%, aged at 700°C, showing profuse $M_{23}C_6$ precipitation. After Boeuf et al. [14], reproduced with the permission of Elsevier.

composition. Typically, Cr-Mn steels require 12-15 wt% Cr and 12-15 wt% Mn to remain austenitic at room temperature if the carbon content is low. A modern variety of austenitic steel that are not stainless but utilise large manganese concentrations ($\approx$ 20 wt%), are the TWIP variety where a substantial fraction of the deformation is achieved by mechanical twinning, as described on p. 300.

12.3.1 Nitrogen

Like carbon, nitrogen is a very strong austenite former. Both elements, being interstitial solutes in austenite, are the most effective solid solution strengtheners available. However, their strengthening of austenite is far smaller than that of ferrite. This is because the octahedral interstices are regular, so the interstitial atoms cause only a uniform expansion, in this respect, acting rather like substitutional solutes.

Nitrogen is useful in this respect as it has a lesser tendency to cause intergranular corrosion due to the precipitation of chromium-rich compounds, largely because of its greater solubility in austenite than carbon. Concentrations of nitrogen up to 0.25 wt% are used, which can nearly double the proof strength of a Cr-Ni austenitic steel. Up to 0.7 wt% of nitrogen can be introduced in austenitic steel when melting and processing at ambient pressure. There are high-pressure melting and powder metallurgical techniques whereby greater concentrations of nitrogen can be introduced [15], but obviously, with increased cost.

The alloy 316LN described on p. 351 has < 0.15 wt% of nitrogen; it has wide applications: for containing hydrogen and deuterium, even at elevated temperatures in the range 200–550°C, tubing, conventional power plant, food and beverage processing, and components for cryogenics. Another particularly useful stainless steel containing an even higher nitrogen concentration, has the commercial designation *Cronidur 30*. With the composition Fe-0.33C-15.7Cr-0.25Ni-1Mo-0.36N wt%, the alloy generally is used in the through-hardened condition, with outstanding corrosion resistance and a fine, relatively homogeneous microstructure, particularly in the secondary hardened condition [16–18]. Applications include bearings in cryogenic rocket-turbopumps, helicopters and screw gearings (Fig. 12.11). It can also be used in the induction-hardened condition, where the surface reaches a hardness of 60 HRC, while the core retains a toughness of 70 MPa $m^{1/2}$.

The reason why nitrogen additions improve the corrosion resistance of martensitic stainless steel are not fully understood [19]. It appears that it aids passivation by the formation of CrN at the surface during corrosion, with the chromium component being trivalent Cr^3+ as in the oxide. This is speculated to make it easier for the oxide passivation-film to form. It could also be the case that nitrogen somehow promotes the selective removal of iron.

Figure 12.11 So-called ball screw-gearings made using Cronidur 30, which drive an aeroplane's flap tracks. The alloy has superior corrosion-resistance under salt-spray conditions, due in part to the nitrogen content. After Stein and Hucklenbroich [20], reproduced with the permission of Taylor & Francis Ltd, http://www.tandfonline.com.

12.4 Schaeffler diagram

One of the most convenient ways of representing the effect of various solutes on the basic structure of chromium-nickel stainless steels is the Schaeffler diagram, often used in assessing the likely structure of welding alloys. It plots the compositional limits at room temperature of austenite, ferrite (α, δ) and martensite, in terms of nickel and chromium equivalents (Fig. 12.12). At its simplest level, the diagram shows the domains of existence of the three phases for iron-chromium-nickel alloys. However, the diagram becomes of much wider application when the equivalents of chromium and of nickel are used for the other alloying elements:

$$\begin{aligned}\text{Cr equivalent, castings} &= w_{\text{Cr}} + 2w_{\text{Si}} + 1.5w_{\text{Mo}} + 5w_{\text{V}}\\ \text{Ni equivalent, castings} &= w_{\text{Ni}} + w_{\text{Co}} + 0.5w_{\text{Mn}} + 30w_{\text{C}}\end{aligned} \tag{12.1}$$

$$\begin{aligned}\text{Cr equivalent, welds} &= w_{\text{Cr}} + 1.5w_{\text{Si}} + w_{\text{Mo}} + w_{\text{V}} + 3.5w_{\text{Ti}}\\ \text{Ni equivalent, welds} &= w_{\text{Ni}} + 0.5w_{\text{Mn}} + 30w_{\text{C}} + 30w_{\text{N}}\end{aligned} \tag{12.2}$$

The large influence of C and N relative to that of the metallic elements should be particularly noted. The diagram is useful in determining whether a particular steel is likely to be fully austenitic at room temperature. This is relevant to bulk steels, but particularly to weld metal where it is frequently important to predict the structure in order to avoid weld defects and excessive localised corrosive attack.

The differences apparent in Fig. 12.12 between the welded and cast steel forms of the Schaeffler diagram emphasise that thermal and size effects are not accounted for. Thus, the diagram, originally developed for arc welding, does not correctly estimate the structure for laser welds, where the cooling rates can be much greater [21]. The diagrams should therefore be used as a guide rather than a rigorous tool. The diagrams are now used to assess the phases expected during the additive manufacturing of stainless steel items (p. 418), with the additional complication that there can be considerable chemical segregation during the deposition process.

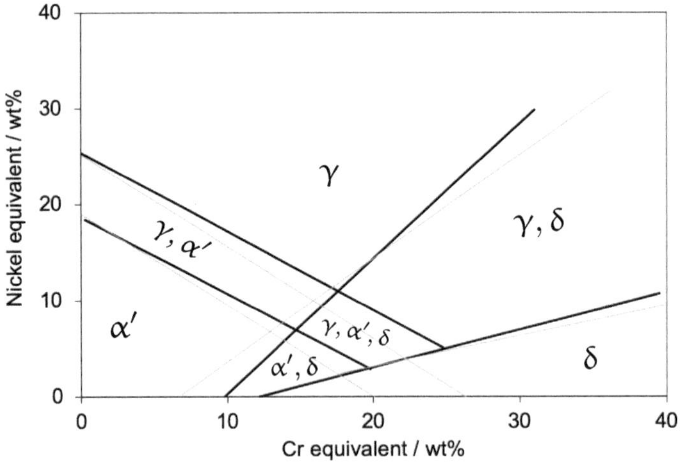

Figure 12.12 Domains of the expected microstructure at room temperature as a function of the chromium and nickel equivalents. The black lines are for castings and the lighter ones for welds. For castings, the equivalent concentrations are given by Eq. (12.1), whereas for welds by Eq. (12.2). Adapted from Schneider [22] and Kakhovskii et al. [23].

12.5 Chromium-rich carbide in Cr-Ni austenitic steels

Simple austenitic steels usually contain between 18 and 30 wt% chromium, 8 to 20 wt% nickel and between 0.03 and 0.1 wt% carbon. The solubility limit of carbon in austenite that is in equilibrium with carbides, is about 0.05 wt% at 800°C, rising to 0.5 wt% at 1100 °C. Therefore, solution treatment between 1050°C and 1150°C puts all the carbon into solution and rapid cooling from this temperature range results in an austenite solid solution at room temperature. However, slow cooling or reheating within the range 550–800°C leads to the rejection of carbon from solution, usually as the chromium-rich carbide, $M_{23}C_6$, even when the average carbon content of the steel is small (<0.05 wt%), Fig. 12.9b.

This carbide nucleates preferentially at the austenitic grain boundaries as faceted particles (Fig. 12.13) or often as complex dendritic arrays. While such precipitation can have an adverse effect on mechanical properties, in particular low-temperature ductility, the most significant result is the depletion of the regions adjacent to the grain boundaries with respect to chromium. This has been revealed directly using high-resolution chemical analysis (e.g., Figs. 12.8b, 12.13b). The surface oxide-film in these regions is thus depleted in chromium and as a result it becomes anodic with respect to its surroundings, so the steel becomes prone to corrosive attack. Consequently, a classic form of intergranular corrosion is experienced, which in severe cases can lead to disintegration of the steel, as illustrated in Fig. 12.13c. This type of corrosion is also experienced in martensitic chromium steels, e.g., 12 wt% Cr steel, in which grain boundary precipitation of $M_{23}C_6$ occurs as well. The corrosion-induced cracking is most potent at boundaries connecting grains that have a large relative crystallographic-misorientation. Boundaries with good atomic fit between the grains have a lower energy per unit area and hence are not readily attacked. This has led to attempts at grain boundary engineering, in which specific crystallographic textures are introduced in order to increase the proportion of low-energy grain boundaries in the steel [24–26]. In the propagation of fracture, it is relevant to consider the connectivity of the most sensitive grain boundaries [27].

$M_{23}C_6$ can precipitate within the austenite grains, particularly at high supersaturations, on dislocations and on solute atom/vacancy clusters. Both the matrix and the carbide have cubic symmetry, and electron diffraction evidence from thin-foil specimens invariably gives the orientation relationship:

$$\{100\}_{M_{23}C_6} \parallel \{100\}_\gamma \qquad \langle 010\rangle_{M_{23}C_6} \parallel \langle 010\rangle_\gamma.$$

The lattice parameter of $M_{23}C_6$ is approximately three times that of austenite, so the electron diffraction patterns are readily identified. The particles usually develop a polyhedral habit, but occasionally in steels deformed at elevated temperatures a more regular cubic morphology is displayed (Fig. 12.14).

As the critical temperature range for chromium carbide nucleation and growth is between 500°C and 850°C, any process that allows the steel to pass slowly through this temperature range will render it sensitive to intergranular corrosion in service. Welding, in particular, provides these conditions in the heat affected zone, leading to localised attack in certain chemical media. It is, therefore, important to have information about the reaction kinetics for the formation of $M_{23}C_6$. Being a typical nucleation

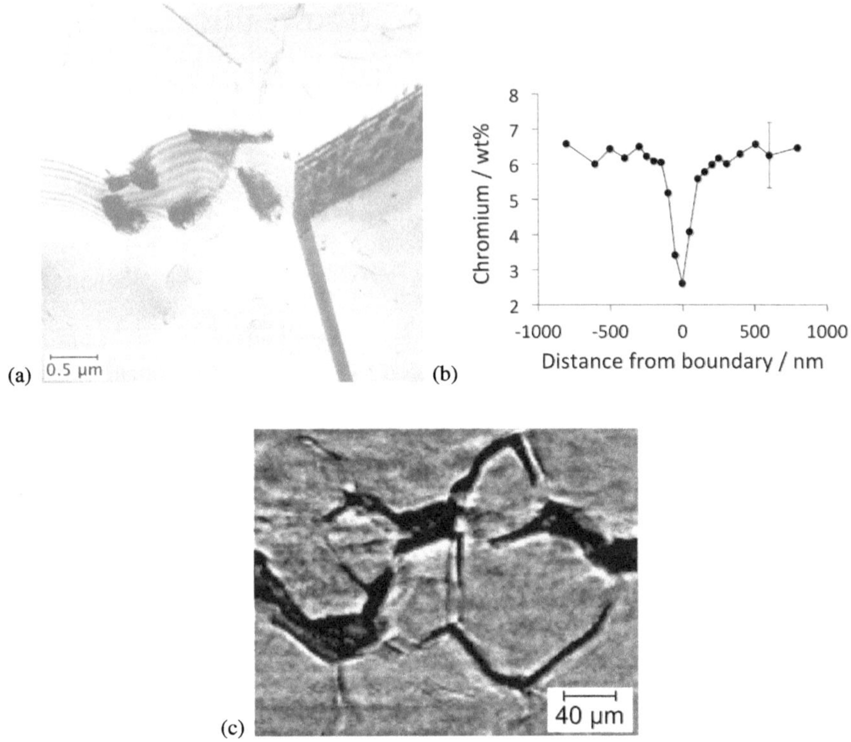

Figure 12.13 Grain boundary precipitation of $M_{23}C_6$ in 316L austenitic stainless steel heat treated at 700°C for 100 h. (a) The precipitation varies along all three of the boundaries shown. (b) A concentration profile across one of the boundaries with precipitates, showing the depletion of chromium in the vicinity of the boundary. Reproduced from Hall and Briant [28] with permission of Springer. (c) Intergranular stress-corrosion cracking in sensitised austenitic stainless steel. Reproduced from Jivkov et al. [27], with permission of Elsevier.

and growth process, the time-temperature-transformation curve is typically C-shaped with the nose at about 750°C (Fig. 12.15). For some steel compositions the minimum time for the formation of $M_{23}C_6$ sufficient to give subsequent intergranular corrosion, i.e. time to achieve *sensitisation*, is as short as 100 s. Several ways of reducing or eliminating the formation of $M_{23}C_6$ are available. The term *stabilisation* is used to describe these processes:

1. Solution treatment: after welding, the steel can be reheated to 950–1100°C to allow $M_{23}C_6$ to redissolve, and further precipitation is then prevented by rapid cooling to avoid the C-shaped curve.
2. Reduction of the carbon content: this can be reduced below 0.03 wt% by modern steel-making methods involving oxygen lancing. For complete immunity from intergranular corrosion in 18-8 steels, a carbon level of 0.015 wt% should not be exceeded.

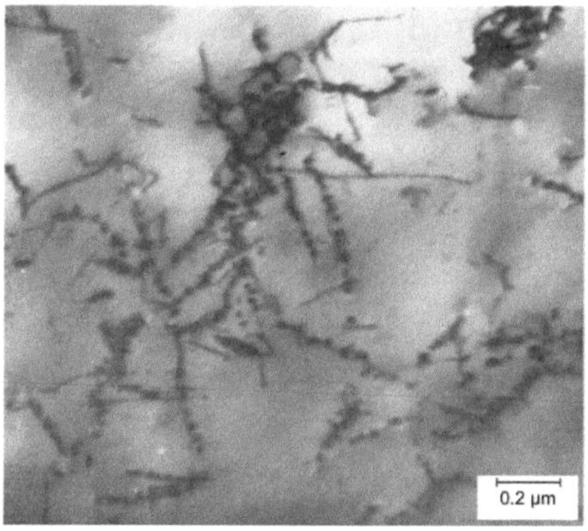

Figure 12.14 Precipitation of $M_{23}C_6$ on dislocations in a 18Cr-12Ni steel after 4 h at 750°C under stress. Image reproduced from Beckitt and Clark [29] with permission of Elsevier.

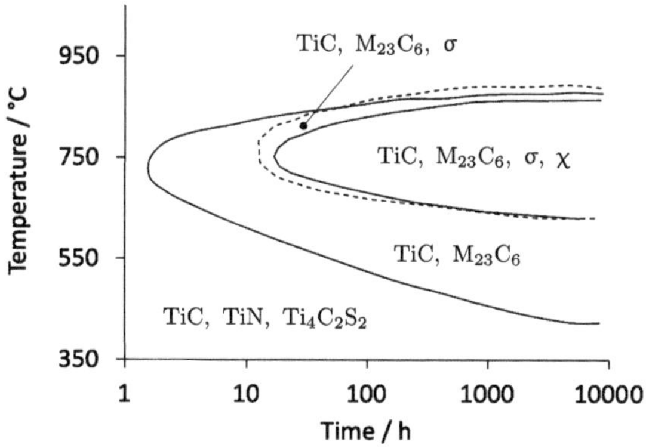

Figure 12.15 Time-temperature precipitation diagram for 316 TM stainless steel, solution-treated and aged as a function of time and temperature. σ-phase is typically an intermetallic compound FeCr, but may contain other solutes such as Mo. The χ-phase is another such compound, but with a different crystal structure (after Grot and Suruiell [30], with permission of Springer Nature).

3. Control of $M_{23}C_6$ reaction kinetics: addition of molybdenum to Cr/Ni stainless steels markedly lengthens the sensitisation time. An increase in nickel content has an adverse effect, while increasing chromium has a beneficial effect.
4. Use of strong carbide-forming elements, niobium and titanium, form carbides, which are much more stable than $M_{23}C_6$, so they preferentially combine with the available carbon and thus lessen the opportunity for $M_{23}C_6$ to nucleate.
5. Introduce a crystallographic texture such that there is a large fraction of low-energy grain boundaries in the steel.

12.6 Precipitation of niobium and titanium carbides

In normal practice, sufficient niobium or titanium is added to the steel to combine with all the carbon, the stoichiometric ratios being:

	Ti:C	Nb:C
Atomic weights	48:12	93:12
Ratios	4:1	8:1.

However, the additions are in excess of these proportions to allow for the fact that, at equilibrium, some Ti or Nb will remain in solid solution in the austenite and some may combine with any nitrogen present. Titanium and niobium carbides are much less soluble in austenite than chromium carbide, so they form at much higher temperatures as relatively stable particles. These should remain relatively inert during commercial heat treatments involving solution temperatures no higher than 1050°C, and thus minimise the possible nucleation of $M_{23}C_6$. However, TiC and NbC have some solubility in austenite at 1050°C and can subsequently precipitate at lower temperatures. During high-temperature processes such as welding, these carbides dissolve to a greater extent in austenite and can then re-precipitate at lower temperatures. Therefore, NbC and TiC do not always form inert dispersions, and are often redistributed by heat treatment. They do, however, have the great advantage of not depleting the matrix of chromium, particularly at sensitive areas such as grain boundaries. The ability to form dispersions of NbC and TiC has a further advantage in that these dispersions can remain very fine at temperatures in the range 500–750°C, and so provide a means of strengthening austenitic steels in this temperature range. The development of creep-resistant austenitic steels owes much to the properties of these carbide dispersions.

The formation of NbC and TiC in austenite is most conveniently studied by subjecting the steel to high-temperature solution treatment (1100–1300°C), followed by rapid cooling to room temperature. On subsequent ageing in the range 650–850°C precipitation takes place. The carbides both have cubic-F lattices with the space group $Fm\overline{3}m$ and lattice parameters within 2-3% of each other, but differing from that of austenite by 20–25%. They both exist over a range of stoichiometry $MC_{0.6}$-$MC_{1.0}$. Precipitation in each case occurs in several different ways.

Grain boundary

Grain boundaries are preferred sites, but because chromium diffuses more rapidly in austenite than does Nb or Ti, $M_{23}C_6$ usually forms first (Fig. 12.16a). This emphasises that NbC or TiC should not be taken into solution if full stabilisation is to be achieved. In Fig. 12.16, TTT curves for $M_{23}C_6$ and (NbTi)C illustrate that, at lower temperatures and shorter times, the chromium carbide forms first, but at longer times it can redissolve and be replaced by (NbTi)C.

Dislocations

NbC and TiC nucleate extensively on dislocations (Fig. 12.14); a significant part of the creep resistance of this group of alloys arises from the alloy carbides that form on dislocations generated by deformation at elevated temperatures. The carbides always

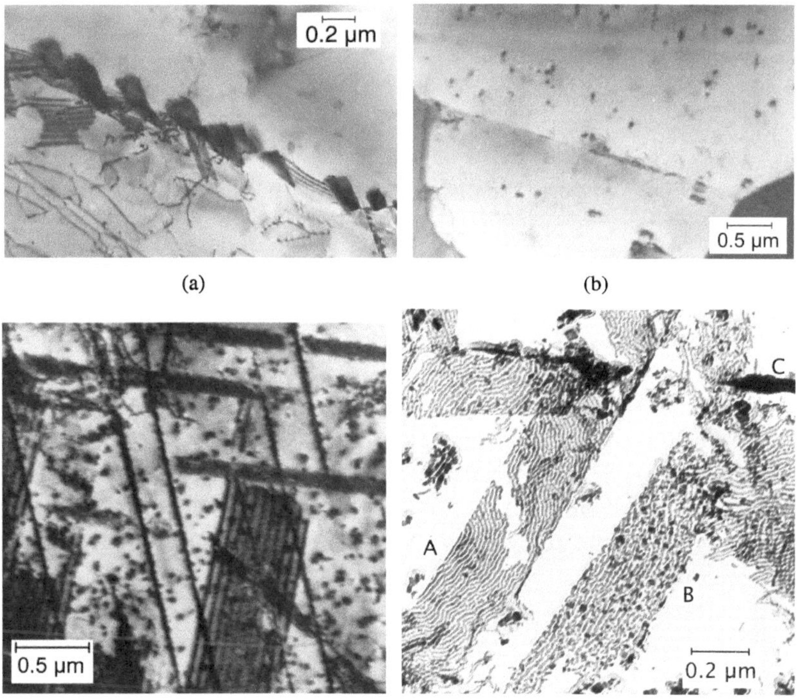

(a) (b)

(c) (d)

Figure 12.16 Different modes of precipitation in austenitic stainless steel. (a) Grain boundary precipitation of $M_{23}C_6$ in a 304 stainless steel containing 0.05 wt% carbon, following ageing at 670°C for 50 h (after Trillo and Murr [31], with permission from Elsevier). (b) Intragranular TiC precipitation in 326 TiM stainless steel following ageing at 750°C for 20 h (after Grot and Suruiell [30], with permission of Springer Nature). (c) TaC precipitates in Fe-18Cr-12Ni-2Ta-0.1C wt% aged at 700°C for 620 h, showing intense precipitation on stacking faults and in the matrix. Image from Froes and Honeycombe [32] with permission of Elsevier. (d) NbN precipitation in association with stacking faults (A,B) and M_6N at 'C' in Fe-18Cr-12Ni-1.25Nb-0.04N aged at 700°C for 500 h (courtesy of Borland).

have a cube-cube orientation relationship with the matrix austenite, as do other MC carbides such as VC, TaC. Since the lattice parameter of austenite is 20–25% less than that of the carbides, a flux of vacancies into the precipitates is needed to reduce internal stresses resulting from growth of the particles. Only a few of these vacancies can be quenched in, so carbide particles will grow most readily in situations where further vacancies are generated, e.g., at dislocation interactions.

Precipitation in association with stacking faults

It often is the case that NbC, TaC and TiC precipitate on $\{111\}_\gamma$ planes as thin discs which exhibit stacking fault contrast in thin foils in the electron microscope (Fig. 12.16). These discs grow very substantially on ageing, e.g., at 700°C. Analysis has shown that the discs are formed by the climb of partial dislocations ($\mathbf{b} = \frac{1}{3}\langle 111\rangle_\gamma$

Frank type), which by climbing generate a continuous source of vacancies. The (NbTi)C precipitate particles nucleate on the partial dislocations and make use of the vacancies in growing, a process which is repeated many times as the partial dislocation escapes from the rows of particles it has nucleated. The final result is an array of discs on $\{111\}_\gamma$ planes, which contain fine dispersions of (NbTi)C. This complex precipitate morphology can occur side by side with normal nucleation on undissociated dislocations.

Matrix precipitation

Random precipitation of (NbTi)C in the matrix, not on dislocations, is occasionally observed, but it is the rarest morphology encountered. The particles still exhibit the cube-cube orientation relationship with the matrix, and are apparently nucleated on solute atom/vacancy clusters. Consequently, they are obtained only after heat treatments, which result in large supersaturations of vacancies in the austenite matrix, i.e., high solution temperatures and rapid quenching (Fig. 12.16b). However, there is some evidence that certain elements, e.g., phosphorus, encourage this type of precipitation by trapping vacancies, the phosphorus atoms being 20% smaller than the other atoms in the austenite solid solution, and so cause localised strain fields.

The carbide morphologies have been presented in decreasing order of occurrence. The evidence suggests that this order is dictated by increasing degree of supersaturation, which is a function of the solution temperature. In practice, high solution temperatures can usually be avoided, except in welding, so grain boundary precipitation and dislocation precipitation are the dominant mechanisms observed.

12.7 Nitrides in austenitic steels

In simple austenitic steels, the role of nitrogen is largely that of a solid-solution strengthening element, although it can replace some of the carbon in $M_{23}C_6$. While higher nitrogen concentrations can be maintained without deleterious precipitation than is the case with carbon, in steels with 0.2-0.3 wt% N, Cr_2N can precipitate at grain boundaries, and also within the grains. Exposure of austenitic steels to air at temperatures greater than 600°C can lead to very high (>1 wt%) nitrogen concentrations under the oxide layer, with coarse Cr_2N matrix precipitation, as well as discontinuous lamellar precipitation at grain boundaries. Such regions often lead to cracks under creep conditions.

In the presence of Nb or Ti, more stable nitrides of these elements are formed, which are much less soluble in austenite than Cr_2N. TiN and NbN, isomorphous with the corresponding carbides, have been identified, and also M_6N, which can eventually replace NbN during ageing (Fig. 12.16d). These phases can precipitate in the range 650–850°C after rapid cooling from high solution temperatures. They may, therefore, occur as a result of welding or in alloys subject to creep conditions at high temperatures. The modes of nucleation of these nitride phases are similar to those of the corresponding carbides, although there are morphological differences.

12.8 Intermetallic precipitation in austenite

Austenitic steels, as a class, possess relatively modest mechanical properties, which are largely outweighed by their excellent corrosion resistance in many media. However, it is desirable to develop stronger alloys, particularly for use at elevated temperatures where deformation by creep needs to be minimised. Carbide dispersions offer one solution, but the volume fraction of precipitate is limited by solubility considerations and there are problems associated with high-temperature ductility and the stability of the dispersions.

The highly alloyed matrices of many austenitic alloys have allowed the development of intermetallic phases as suitable dispersions to achieve elevated temperature strength. The most important of these phases is the γ' primitive-cubic phase $Ni_3(AlTi)$, first found in nickel-base alloys, with an fcc matrix analogous to austenite, containing titanium and aluminium, which can replace each other in the precipitate. The γ' precipitate is obtained in stable austenitic steels, e.g., 20Cr25Ni with an (Al + Ti) content of 1–5 wt% by quenching from a solution temperature of 1100–1250°C, and ageing in the range 700–800°C. The dispersion developed in this way has two important advantages. Firstly, the precipitate particles have the cubic crystal structure well-matched to that of the matrix, with which they have a cube-cube orientation relationship. Moreover, the lattice parameters are similar, so that the interfaces between precipitate and matrix are coherent, and therefore, of low energy. Eq. (9.2) shows that the coarsening rate is related directly to the interfacial energy. Secondly, this type of reaction allows a large volume fraction (0.3-0.5) of precipitate particles to be achieved, the particles being strong, but not brittle when embedded in the austenite.

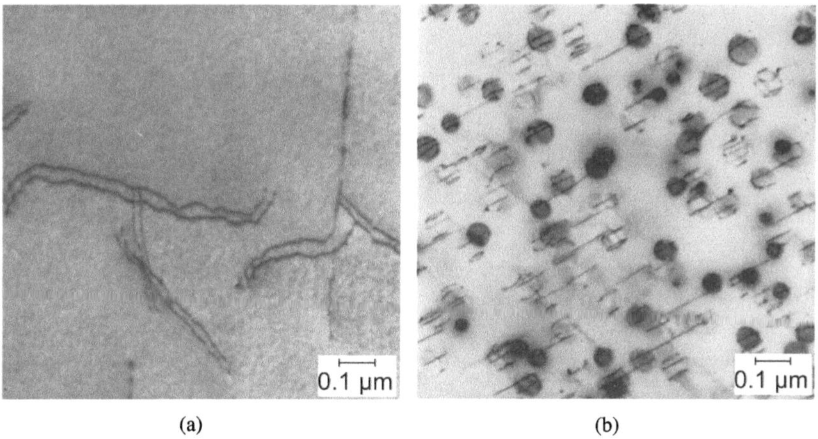

Figure 12.17 Fe-24Ni-17Cr-2Ti wt%. (a) Superdislocations penetrating fine γ' particles, in a sample aged at 675°C for 20 h. (b) Dislocation loops around coarsened γ' precipitates in a sample aged at 775°C for 75 h. After Singhal and Martin [33], reproduced with permission of Elsevier.

The γ' precipitate normally observed in austenite is spherical when the precipitate is very fine (Fig. 12.17a), and indeed there is evidence for the formation of pre-precipitation spherical zones. However, on prolonged ageing at 750°C, the γ' particles gradually adopt a more complex morphology as they lose coherency with the austenitic matrix (Fig. 12.17b). By varying the ratio of Ti to Al in γ', the coarsening characteristics can be substantially modified. Addition of Al to $\gamma' Ni_3Ti$ decreases the lattice parameter from about 3.590 Å for 25Ni-15Cr wt%, resulting in greater stability of the precipitate. However, complete replacement of titanium lowers the γ' parameter to 3.559 Å, which results in an increase in the lattice mismatch with the matrix. This helps to explain why an (Al + Ti) content of 1-1.5 wt% Al and 3-3.5 wt% Ti was found to be optimum for strong austenitic steels, resistant to coarsening.

The γ' phase is not the equilibrium phase in austenitic steels containing Al and Ti. It is replaced eventually by a coarsely dispersed hexagonal phase η (Ni_3Ti) in titanium-containing steels. In steels with a high Al/Ti ratio, the equilibrium intermetallic phase β NiAl is primitive cubic with space group $Pm\bar{3}m$ and a motif of a nickel atom at the cube corners and an aluminium atom at the body-centre (or vice-versa). Both these phases coarsen excessively, so are undesirable constituents of the microstructure in austenitic creep-resistant alloys.

While a number of other intermetallic phases have been observed in austenitic steels, mention will be made only of sigma phase (σ), as it usually has a detrimental influence on mechanical properties at room temperature. The phase, which has a tetragonal lattice with space group $P4_2/mnm$ and lattice parameters 0.8799 and 0.4544 nm [34] dependent on composition, occurs in the binary Fe-Cr system over a wide composition range between 25 and 60 wt% Cr. In CrNi austenitic steel, σ formation is encouraged when the Cr content exceeds 17 wt%, but is discouraged by increasing the nickel content. The phase forms at austenite grain boundaries and requires, for full development, long-term ageing (up to 1500 h) at 750°C. However, in some circumstances, σ has been detected in 25Cr-20Ni steels after 70 h at this temperature. The presence of ferrite in the austenite greatly accelerates the formation of σ, which has been shown to nucleate at the γ/α boundaries (Fig. 12.18). The ferrite is richer in chromium and tends to be preferentially absorbed during the growth of σ phase. Elements such as Mo and Ti can further accelerate σ formation, e.g., in an 18Cr-8Ni-3Mo-1Ti wt% steel, the σ-phase can be formed after only 30 min at 870°C.

12.9 Austenitic steels in practical applications

The commonest austenitic steel is the so-called 18-8 variety, containing around 18 wt% Cr and 8 wt% Ni. It has the lowest nickel content consistent with the achievement of a fully austenitic structure. However, in some circumstances, e.g., after deformation, or if the carbon content is very low, it may transform partially into martensite at room temperature. Several of the most familiar austenitic steel specifications are given in Table 12.1.

Greater stability towards the formation of martensite is achieved by increasing the nickel content, as illustrated in the 301 to 310 types of steel in the Table 12.1. The 18-8 stainless steel owes its wide application to its excellent general resistance to corrosive environments. However, this is substantially improved by increasing the nickel con-

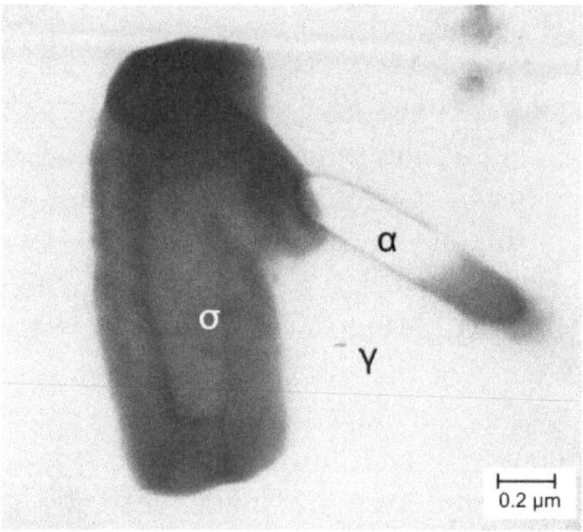

Figure 12.18 Nucleation of sigma phase (σ) at an α/γ interface in a Fe-24Ni-25Cr-0.5Ti wt% stainless steel aged from the austenitic condition at 750°C for 1500 h. After Singhal and Martin [35], reproduced with permission of Elsevier.

Table 12.1 Chemical compositions (wt%) of some austenitic stainless steels.

Element	**Numerical designation of steel type**						
	301	**302**	**304**	**310**	**316**	**321**	**347**
C	0.15 max	0.08 max	0.08 max	0.25 max	0.08 max	0.08 max	0.08 max
N	0.03	0.03	0.03	0.03	0.03	0.03	0.03
Cr	16-18	17-19	18-20	24-26	16-18	17-19	17-19
Ni	6-8	8-10	8-12	19-22	10-14	9-12	9-13
Mo					2-4		
Ti						5 × C	
Nb							10 × C
Mn	1.5	1.5	1.5	1.5	1.5	1.5	1.5

tent, and increasing the chromium gives greater resistance to intergranular corrosion. Austenitic steels are prone to *stress corrosion cracking*, particularly in the presence of chloride ions, where a few p.p.m. can sometimes prove disastrous. This is a type of failure which occurs in some corrosive environments under small stresses, either deliberately applied or as a result of residual stresses in fabricated material. In austenitic steels, it occurs as transgranular cracks which are most easily developed in hot chloride solutions. Stress corrosion cracking is substantially reduced in high nickel austenitic alloys.

Type 316 steel contains 2–4 wt% molybdenum, which gives a substantial improvement in general corrosion resistance, particularly in resistance to *pitting corrosion*, which can be defined as local penetrations of the corrosion-resistant films and which occurs typically in chloride solutions. Recently, some resistant grades with as much as 6.5 wt% Mo have been developed, but the chromium must be changed to 20 wt%

and the nickel to 24 wt% to maintain an austenitic structure. Alloys like these are sometimes known as the *superaustenitic* stainless steels.

Corrosion along the grain boundaries can be a serious problem, particularly when the heat input due to welding allows the precipitation of $M_{23}C_6$ at grain boundaries, resulting in the depletion of the chromium concentration in the vicinity. This type of intergranular corrosion is sometimes referred to as *weld-decay* [36]. To combat this effect, some grades of austenitic steel, e.g., 304 and 316, are made with carbon contents of less than 0.03 wt% and designated 304 L and 316 L. Alternatively, niobium or titanium is added in excess of the stoichiometric amount to combine with carbon so that the chromium-rich $M_{23}C_6$ cannot form, as in types 321 and 347, which are referred to as *stabilised* stainless steels.

The austenitic steels so far referred to are not especially strong or ductile materials, as illustrated on Fig. 12.3. Typically their 0.2% proof strength is about 250 MPa and the tensile strength between 500 and 600 MPa, showing that these steels have substantial capacity for work hardening. It is the other properties of stainless steels, particularly corrosion resistance, that make them so useful. The steels can be made much stronger for special applications, Section 17.4.

12.10 Oxidation resistant stainless steel

The Cr/Ni austenitic steels are resistant to high-temperature oxidation because of the protective surface film, but the usual grades have low strengths at elevated temperatures. Those steels stabilised with Ti and Nb, types 321 and 347, can be heat treated to produce a fine dispersion of TiC or NbC which interacts with dislocations generated during creep. One of the most commonly used alloys is 25Cr-20Ni with additions of titanium of niobium, which possesses good creep strength at temperatures as high as 700°C.

To achieve the best high-temperature creep properties, it is necessary first to raise the room-temperature strength to higher levels. This can be done by precipitation hardening heat treatments on steels of suitable composition to allow the precipitation of intermetallic phases, in particular Ni_3(Al Ti). In Table 12.2, the room-temperature strength of two alloys in this category (A286 and Unitemp 212) after ageing at 700–750°C is compared with that of the simpler standard austenitic alloys, e.g., 304. It can be seen that the strength is more than doubled by the precipitation reaction.

Table 12.2 shows an experimental stainless alloy intended for applications in the temperature range 600–900°C where a protective alumina layer forms to isolate the steel against further oxidation, especially in the presence of water vapour [37]. The alloy is strengthened primarily by Ni_3Al γ' precipitates, but because of the chemical composition rich in niobium, Fe_2Nb Laves phase also forms in addition to some NiAl. The main reason for adding niobium at such a large concentration is to improve the stability of the alumina scale. However, the primary strengthening mechanism at elevated temperatures is through the fine γ' particles (Fig. 12.19), which provide for exceptional creep resistance suitable for high steam temperatures in power plant. One of the problems associated with the application of austenitic stainless steels in such circumstances is the high thermal expansion coefficient and low thermal conductivity,

Table 12.2 Strengthening of austenitic steels at room temperature. Alloy IN744 is not austenitic, but belongs to the class of duplex steels discussed in Section 12.11.

Element	Designation of steel					
	304(N)	347(N)	A286	212	IN744	Experimental
C	0.06	0.08	0.05	0.08	0.05	
N	0.20	0.20				
Cr	18.0	18.0	15.0	13.5	26.0	14
Ni	10.0	11.0	26.0	26.0	6.5	32
Mo			1.2	1.75		
Ti			2.0	3.0	0.3	0.98
Nb		$10 \times C$				2.88
Al			0.15	0.15		4.00
V			0.30			
0.2% σ_Y / MPa	340	415	700	920	570	
σ_{UTS} / MPa	695	710	1000	1300	740	
% Elongation	46	39	25.0[a]	23.0[b]	24	

(N) represents high nitrogen.
[a] Aged at 750°C.
[b] Aged at 700°C.

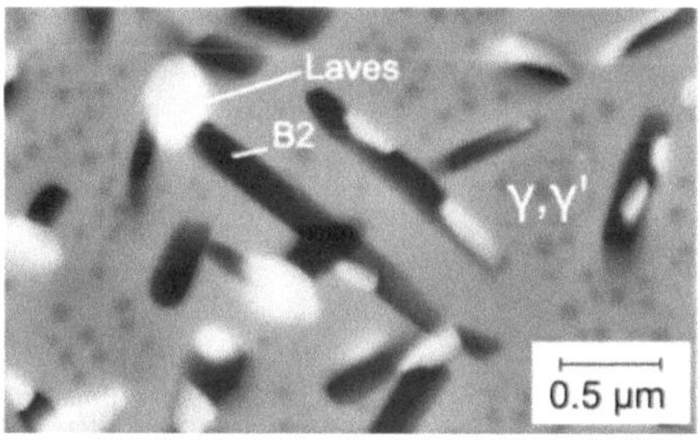

Figure 12.19 The microstructure of the austenitic stainless alloy listed as 'experimental' in Table 12.2 following ageing at 750°C for 2000 h. The Laves phase is Fe_2Nb and 'B2' is NiAl. After Yamamoto et al. [37], reproduced with the permission of Elsevier.

which can combine to give poor thermal fatigue resistance unless the components concerned are not subjected to significant thermal cycling and are used in thin sections.

12.11 Duplex and ferritic stainless steels

In Section 12.3, the importance of controlling the γ-loop in achieving stable austenitic steels was emphasised. Between the austenite and δ-ferrite phase fields there is a restricted $(\alpha+\gamma)$ region, which can be used to obtain two-phase or duplex structures in stainless steels (Fig. 12.20). The structures are produced by engineering a balance between the α-forming elements (Mo, Ti, Nb, Si, Al) and the γ-forming elements (Ni, Mn, C and N). To achieve a duplex structure, it normally is necessary to increase the chromium content to above 20 wt%. However, the exact proportions of α and γ are determined by the heat treatment. It is clear from consideration of the γ-loop section of the equilibrium diagram, that holding in the range 1000–1300°C causes the ferrite content to vary over wide limits. The usual treatment is carried out between 1050°C and 1150°C, when the ferrite content is not sensitive to the subsequent cooling rate.

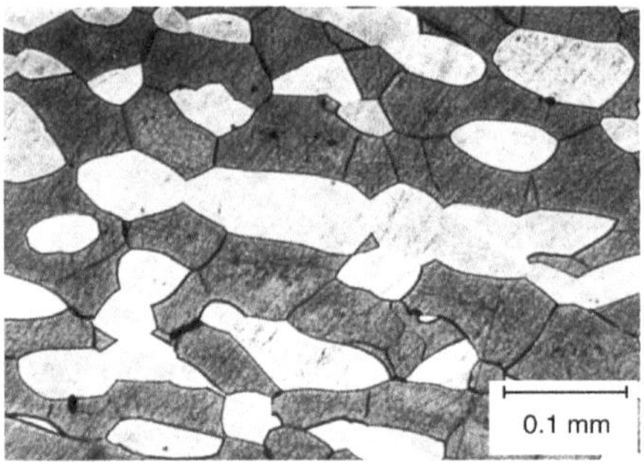

Figure 12.20 Duplex stainless steel with the lighter regions corresponding to austenite and with δ-ferrite in the remaining microstructure. The two phases are present in about equal quantities. Taken from Fe-22Cr-6Ni-2Mn-0.5Si-3Mo-0.14N-0.03C wt% steel, heat treated to 1200°C and water quenched (courtesy of S. Sharafi).

The duplex steels are stronger than the simple austenitic steels, partly as a result of the two-phase structure and also because this leads normally to a refinement of the grain size. Indeed, by suitable thermomechanical treatment between 900°C and 1000°C, it is possible to obtain very fine microduplex structures which can exhibit super-plasticity, i.e., very high ductilities at high temperatures, for strain rates less than a critical value. A typical composition, IN744, is shown in Table 12.2 along with the mechanical properties at room temperature.

A further advantage is that duplex stainless steels are resistant to *solidification cracking*, particularly that associated with welding. While the presence of δ-ferrite may have an adverse effect on corrosion resistance in some circumstances, it does improve the resistance of the steel to transgranular stress corrosion cracking as the ferrite phase is immune to this type of failure.

The super-duplex stainless steels have even better corrosion resistance than the duplex stainless steels. They are therefore used in sea-water applications, one example of which is illustrated in Fig. 12.21. They are particularly superior in their resistance to localised pitting corrosion, because of their larger concentrations of chromium, molybdenum and nitrogen. To maintain the balanced ferrite/austenite microstructure, it is necessary to boost the concentration of austenite stabilising elements such as nickel. Super-duplex stainless steels therefore typically contain 27Cr-7Ni-4Mo-0.3N wt%.

There is another important group of stainless steels, which are essentially ferritic in structure. They contain between 10 and 30 wt% chromium and, by dispensing with the austenite stabilising element nickel, possess considerable economic advantage. These steels, particularly at the higher chromium levels, have excellent corrosion resistance in many environments and are completely free from stress corrosion. Typical compositions are shown in Table 12.3. For some applications, such as the exhaust system

Figure 12.21 A large sea-water pump fabricated using super-duplex stainless steel plates. Adapted from Frances and Byrne [38] under the CC BY license https://creativecommons.org/licenses/by/4.0/.

Table 12.3 Typical compositions (wt%) of some ferritic stainless steels.

Element	Type 409L	Type 430	Type 446	Type 18/2
C	<0.03	0.06	0.08	0.02
Cr	11	17.0	25.0	18.0
Mo				2.0
Mn	1.0	1.0	1.5	
Si	1.0	1.0	1.0	
Ti	$6 \times w_C$			
N				0.25

of a car, the chromium concentration can be limited (e.g., type 409L, Table 12.3), because this provides sufficient corrosion and oxidation resistance while at the same time making the alloy formable; ferrite also has a much smaller thermal expansion coefficient than austenite, so that fatigue induced by temperature variation can be mitigated, Fig. 12.22.

(a) (b)

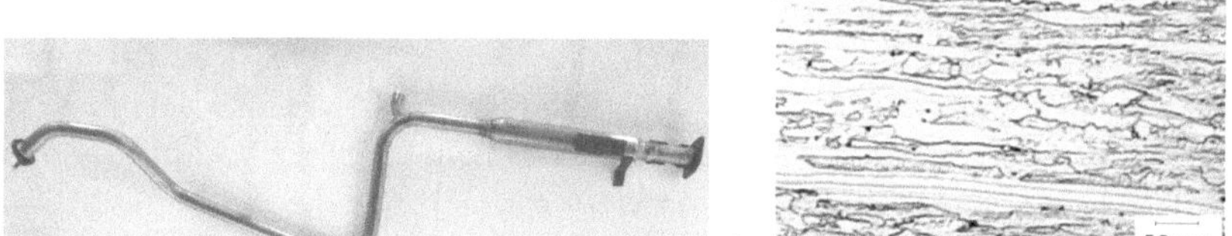

Figure 12.22 (a) An exhaust pipe for an automobile, made using type 409L ferritic stainless steel. (b) The ferritic microstructure in the hot-rolled condition. Micrograph from Ahn et al. [39], reproduced with permission of Elsevier.

These steels do have some limitations, particularly those with larger chromium concentrations, where there can be a marked tendency to embrittlement. This arises partly from the interstitial elements carbon and nitrogen, e.g. a 25 wt% Cr steel will normally become brittle at room temperature if the carbon concentration exceeds 0.03 wt%. An additional factor is that the absence of a phase change makes it more difficult to refine the ferrite grain size, which can become coarse after high-temperature treatment such as welding (Fig. 12.23). This raises still further the ductile/brittle transition temperature, already high as a result of the presence of interstitial elements. Fortunately, modern steel-making methods such as argon-oxygen refining can bring the interstitial contents below 0.03 wt%, while electron beam vacuum melting can do better still.

(a) (b)
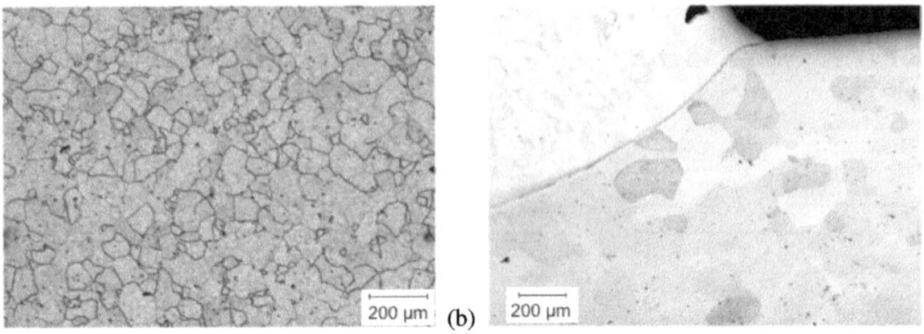

Figure 12.23 Welding-induced grain growth in a ferritic stainless steel of composition Fe-0.015C-17.6Cr-0.2Ni-1.9Mo-0.16Nb wt%. (a) The fine grain structure of the plate. (b) Very coarse ferrite grains in the heat-affected zone of the plate following welding. After Silva et al. [40], with permission of Elsevier.

The ferritic stainless steels are somewhat stronger than austenitic stainless steels, the yield stresses are in the range 300–400 MPa, but they work harden less, so the tensile strengths are similar, ranging from 500 to 600 MPa. However, ferritic stainless steels, in general, are not as readily deep drawn as austenitic alloys because of the overall lower ductility. They are suitable for other deformation processes, such as spinning and cold forging.

Welding causes problems due to excessive grain growth in the heat-affected zone, but recently new low-interstitial alloys containing titanium or niobium have been shown to be readily weldable. The higher chromium ferritic alloys have excellent corrosion resistance, particularly if 1-2 wt% molybdenum is present.

There are two phenomena, which may adversely affect the behaviour of ferritic stainless steels. Firstly, chromium-rich ferrites when heated between 400°C and 500°C develop a type of embrittlement, typically known as the 475°C embrittlement. The most likely cause is the precipitation of a fine coherent chromium-rich phase (bcc α') arising from the miscibility gap in the Fe-Cr system (Fig. 12.6), probably by a spinodal type of decomposition involving $\alpha \rightarrow \alpha_{Fe} + \alpha_{Cr}$, where the subscripts denote the dominant solute in the separated regions. This phenomenon becomes more pronounced with increasing chromium content. Fig. 12.24 illustrates the phenomenon for an exceptionally rich Fe-38.8Cr wt% alloy that is aged at 475°C. The decrease in toughness is dramatic and in a much shorter time than can be explained simply on the basis of hardness changes due to the decomposition of the solid solution. It is likely that the temperature sensitivity of the flow stress increases during ageing, leading to an elevation of the impact transition temperature. Prolonged ageing leads to the formation of σ-phase, the precipitation rate of which increases with the chromium concentration. As in austenite, the presence of sigma-phase can lead to marked embrittlement.

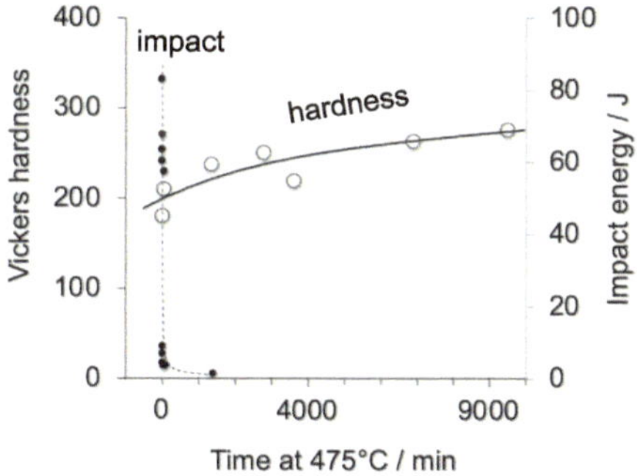

Figure 12.24 Hardness and impact toughness of a Fe-38.8Cr wt% alloy quenched and then aged at 475°C for the period indicated. Selected data from Cortie and Pollak [41].

12.12 Mechanically alloyed stainless steels

Mechanical alloying is a process in which mixtures of fine powders consisting of elemental metals or master alloys are changed into solid solutions, apparently without any melting, Fig. 12.25. The powders are forced to collide with each other and with much larger, hardened steel balls whilst contained in a ball mill. The collisions are energetic, involve large contact pressures and lead eventually to the formation of an intimate solid solution [42,43]. Refractory oxides, commonly yttrium oxide, can be distributed into the mechanically alloyed powder in order to obtain dispersion strengthening. The mechanically alloyed powder is finally extruded to form full density bulk samples in rod, sheet or other useful shapes.

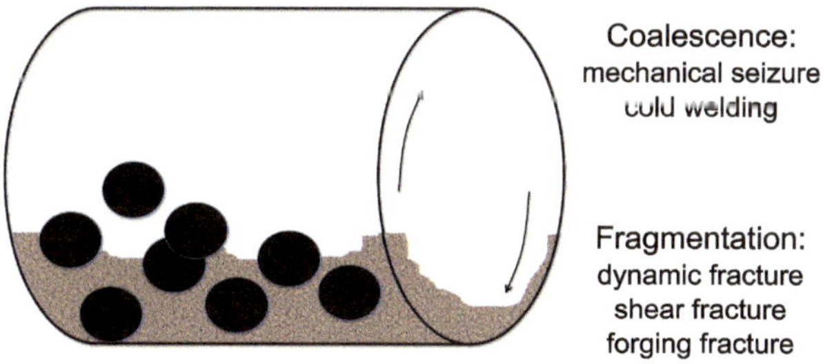

Figure 12.25 Schematic illustration of the mechanical alloying process. The powder, consisting of elemental iron, other elemental alloying additions, and yttria powder, is placed in a ball mill containing cast-iron balls. The rotation of the mill causes the powders to undergo repeated welding and fragmentation, leading to mechanical mixing.

After consolidation by extrusion, the alloys are usually hard and possess an extremely small grain size, typically a fraction of a micrometre (Fig. 12.26a). With one exception (Chapter 15), such a small grain size is impossible to achieve by any other process for bulk samples. The material usually is extremely hard in the as-consolidated state, and has to be softened before further fabrication. The grain boundary area locked into the material gives it a large stored energy, which, under suitable heat-treatment conditions, triggers recrystallisation into a much coarser grain structure. If annealing is carried out by passing the sample through a hot zone (zone annealing), then the recrystallisation front is localised within the high-temperature region. The front then advances at the same rate as the sample, thereby leading to *directional recrystallisation,* in which the microstructure consists of huge columnar grains parallel to the zone annealing direction (Fig. 12.26b). It resembles sometimes the microstructure obtained during directional solidification. The columnar grains are highly anisotropic, usually restricted only by the size of the sample, and can reach lengths in excess of a metre. In some instances, even isothermal annealing can lead to the development of columnar recrystallised grains. The yttria particles introduced during mechanical alloying are not isotropically dispersed, but tend to align parallel to the extrusion direction [44]. Consequently, anisotropic grain boundary pinning leads to the columnar grain growth during recrystallisation. The grain structure can be manipulated by appropriate processing, as illustrated in Fig. 12.27.

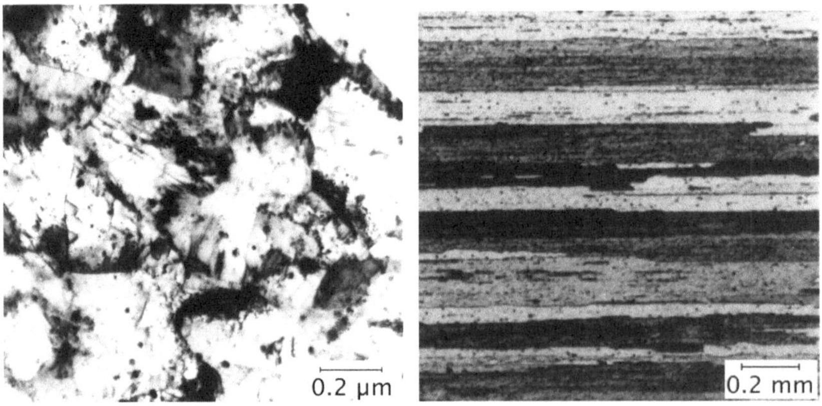

(a) (b)

Figure 12.26 (a) Transmission electron micrograph of a mechanically alloyed steel prior to recrystallisation, showing the fine-grained microstructure. (b) Light micrograph illustrating the elongated recrystallised grain structure.

There are two kinds of mechanically alloyed ferritic stainless steels available commercially, with many other varieties under development for the fusion reactor program (Table 12.4), and even for small, modular nuclear-reactors [45]. The alloy with the highest aluminium and chromium concentrations naturally has a better oxidation resistance. Its oxide content permits its use as a creep-resistant material to temperatures in excess of 1000°C, whereas normal ferritic steels are not used when the service temperature exceeds about 600°C.

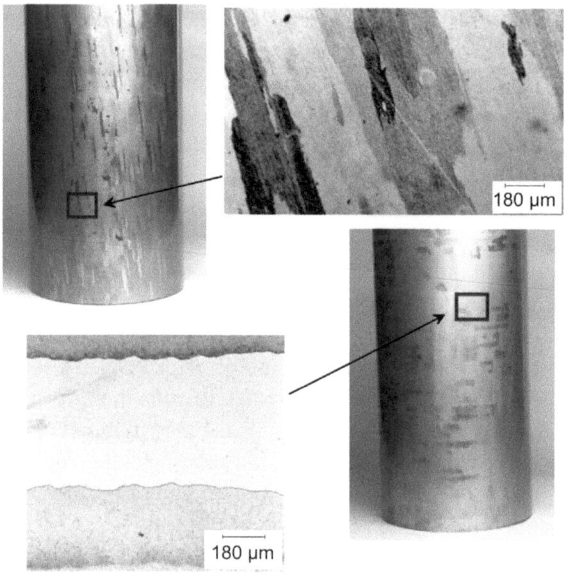

Figure 12.27 Torsion extruded tubes of mechanically alloyed yttria dispersed stainless steel. Heat treatment can then lead to the twisted grain structure along the tube axis, or a circumferential structure, depending on the extrusion conditions. Courtesy of C. Garcia Capdevila.

The ferritic state helps make the steels less susceptible to radiation-induced swelling. The lower chromium alloys (also without aluminium) are therefore designed for nuclear reactor applications. The significant titanium concentration, in the absence of carbon, leads to the precipitation of a bcc-FeCrTiMo intermetallic compound (χ-phase) during ageing at around 800°C. This can further boost the creep strength.

Table 12.4 Chemical compositions of mechanically alloyed oxide dispersion strengthened stainless steels (wt%).

C	Cr	Al	Ti	Mo	Y_2O_3
0.01	20	4.5	0.5	-	0.5
0.01	14	-	1.0	0.3	0.27

12.13 Transformation of metastable austenite

Some austenitic steels are often close to transformation, in that the M_S temperature may be just below room temperature. This is certainly true for low-carbon 18Cr8Ni austenitic steel, which can undergo a martensitic transformation by cooling in liquid nitrogen or by less severe refrigeration (Fig. 12.10a). The application of plastic deformation at room temperature can induce martensite in metastable austenitic steels, a transformation of particular significance when working operations are contemplated. The increase in M_S by cold work is specified by an M_S^σ temperature below which transformation to martensite occurs when the steel is stressed (Fig. 5.23). In general, the higher the alloying element content the lower the M_S and M_S^{σ} temperatures. A distinction is sometimes made between stress induced and plastic-strain induced martensite. However, in all practical scenarios, the strain is achieved by applying a stress, with the stress playing the dominant role [46,47].

The martensite formed in Cr-Ni austenitic steels either by refrigeration or by plastic deformation is similar to that obtained in related steels possessing an M_S above room temperature.

Manganese can be substituted for nickel in austenitic steels, but the Cr-Mn solid solution then has a much lower stacking fault energy. This means that the fcc solid solution is energetically closer to an alternative close-packed hexagonal structure, and that the dislocations tend to dissociate to form broader stacking faults than is the case with Cr-Ni austenites. Under these circumstances, the martensite which forms first is hexagonal in structure (ε-martensite), with a habit plane $\{0001\}_\varepsilon$ parallel to the stacking fault plane $\{111\}_\gamma$ (Fig. 12.28). This phase has been shown to nucleate on stacking faults in which the volume decrease associated with the fcc→hcp change is directed normal to the fault plane [48,49], with the following orientation relationship with austenite:

$$\{0001\}_\varepsilon \parallel \{111\}_\gamma,$$
$$\langle 11\bar{2}0\rangle_\varepsilon \parallel \langle 1\bar{1}0\rangle_\gamma.$$

This type of martensite forms as parallel-sided plates easily confused with annealing twins, common in fcc matrices with low stacking fault energies. α' martensite eventually forms, nucleating at the interface between ε and the austenitic matrix.

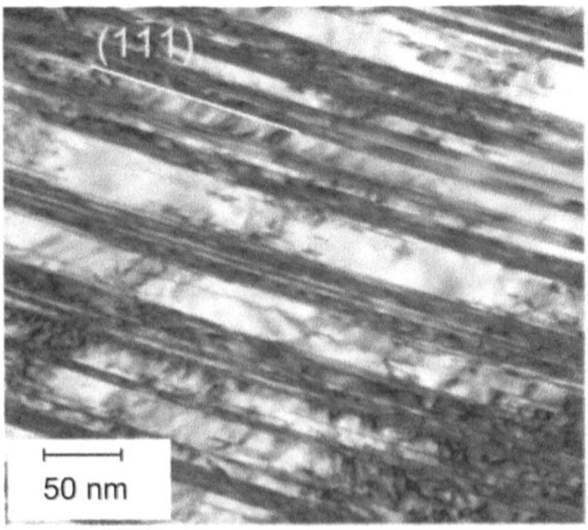

Figure 12.28 The bands present in a deformed austenitic stainless steel consist of a mixture of ε-martensite and mechanical twins, as verified using electron diffraction. After Ullrich et al. [50], reproduced with permission of Elsevier.

Manganese alone can stabilise austenite at room temperature, provided there is sufficient carbon in solid solution. The best example of this type of alloy is the Hadfield manganese steel with 12 Mn-1.2C wt% which exists in the austenitic condition at room temperature and even after extensive deformation does not form martensite. However, if the carbon content is lowered to 0.8 wt%, then M_S^{σ} is above room temperature and transformation is possible in the absence of deformation at 77 K. Both ε and α′ martensites have been detected in manganese steels. Alloys of the Hadfield-type have long been used in wear-resistant applications, e.g., grinding balls, railway points, excavating shovels, and it has often been assumed that partial transformation to martensite was responsible for the excellent wear resistance and toughness required. However, it is likely that the substantial work-hardening characteristics of the austenitic matrix are more significant in this case.

In general, fcc metals exhibit higher work-hardening rates than bcc metals because of the more stable dislocation interactions possible in the fcc structure. This results in the broad distinction between the higher work hardening of austenitic steels and the lower rate of ferritic steels, particularly well exemplified by a comparison of ferritic stainless steels with austenitic stainless steels. Within the austenitic category, however, there are two factors, which influence the extent of work hardening:

a) The stacking fault energy of the matrix, determined by the composition;
b) The stability of the matrix.

The faults in austenite are usually *intrinsic*, i.e., where the stacking sequence of $\{111\}_\gamma$ planes changed locally from

$$ABCABCABC \rightarrow AB\underbrace{CACA}_{\text{hcp}}BC,$$

so in a three-layer thick ε-hcp region there is effect created. Therefore, the stacking fault energy should be related strongly to the difference in the free energies of the γ and ε phases. This interpretation allows the use of the first-principles methods [51] and the thermodynamics of solutions [52] to calculate the effect of solutes in multicomponent steels, on the stacking fault energy, while accounting, for example, for magnetic terms. It also becomes clear that the fault energy is not in general a linear function of solute concentration, nor the sign of the variation is independent of the concentrations of the other solutes. Both carbon and nitrogen can raise or lower the stacking fault energy depending on concentration [53]; the energy goes through a minimum for carbon and a maximum for nitrogen as a function of concentration.

The chromium nickel austenitic steels have stacking faults energies in the range 5–60 mJ m^{-2} (Fig. 12.29), and it would be expected that the highest nickel alloys would show the lowest work hardening, as nickel is one of the elements that raises the stacking fault energy of austenite. Chromium does tend to lower the stacking fault energy of austenite. This can be deduced from the greater tendency for annealing twins to occur in austenites rich in these elements. Plastic deformation of such solid solutions not only produces stable dislocation interactions, but after heavy deformation many fine deformation twins are introduced into the microstructure. Both of these factors contribute to the high flow stresses observed in the deformed alloys. By severe cold working, e.g., up to 80% reduction in wire drawing, the relatively modest yield strengths of ordinary austenitic steels can be raised to over 1200 MPa, although the ductility ambient is then compromised; the ductility can nevertheless be substantial at cryogenic temperatures (Chapter 17).

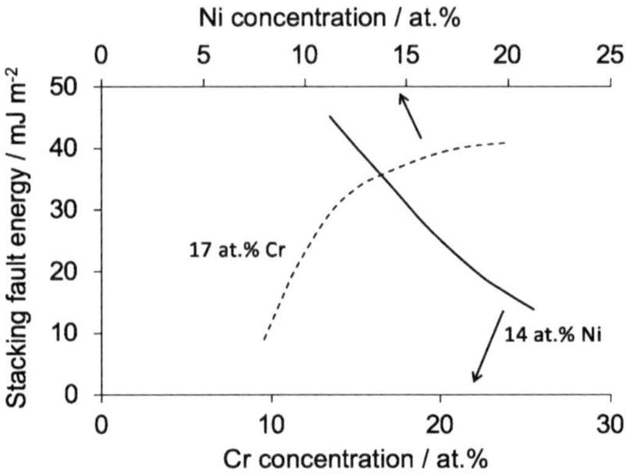

Figure 12.29 Calculated effects of chromium and nickel on the stacking fault energy of austenite at 300 K, selected data from Vitos et al. [51], who also demonstrated the good agreement of the calculations with experimental data.

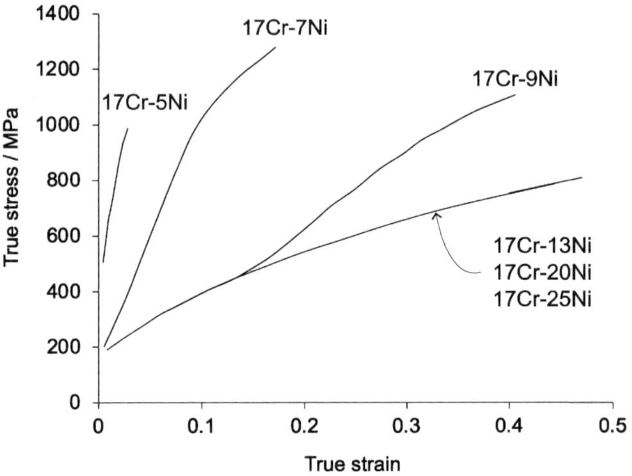

Figure 12.30 Effect of decreasing nickel content on the stress-strain curves of stainless steels (data from Pickering [54]).

The greatest effect on work-hardening rates is undoubtedly the transformation to martensite, as illustrated by the true stress-strain curves of several austenitic steels of decreasing nickel content, i.e., decreasing stability of austenite (Fig. 12.30). By this means yield stresses of well over 1500 MPa can be achieved, e.g., a metastable austenite containing 17Cr-4Ni-3Mn-0.1C wt% after almost complete transformation following 40% deformation at room temperature has a 0.2% proof stress of 1700 MPa. The increase in strength is accompanied by a substantial decrease in ductility, so such steels should not be used for deep drawing, an application where stability of the austenite is essential. In contrast, in stretch forming applications, unstable austenitic steels can be used because the transformation, by raising the work-hardening rate, also increases the extent of uniform straining, as distinct from localised straining, that the steel undergo prior to failure.

The advantages obtainable from the easily fabricated austenitic steels led naturally to the development of *controlled transformation* stainless steels, where the required high strength level was obtained after fabrication, either by use of refrigeration to take the steel below its M_S temperature, or by low-temperature heat treatment to destabilise the austenite. Clearly the M_S–M_F range has to be adjusted by alloying so that the M_S is just below room temperature. The M_F is normally about 120°C lower, so that refrigeration in the range −75°C to −120°C should result in almost complete transformation to martensite. Alternatively, heat treatment of the austenite can be carried out at 700°C to allow precipitation of $M_{23}C_6$ mainly at the grain boundaries. This reduces the carbon content of the matrix and raises the M_S so that, on subsequent cooling to room temperature, the austenite transforms to martensite. This precipitation reaction can be accelerated by designing the steel to include a small volume fraction of δ-ferrite. The δ/γ interfaces then provide very effective nucleating sites for $M_{23}C_6$.

Further heat treatment is then necessary to give improved ductility and a high proof stress; this is achieved by tempering in the range 400–450°C. Typical compositions of these steels and the properties which can be obtained by alternative heat treatments are given in Table 12.5. This category of steels places large demands on metallurgical control, the treatments are complex and the cost high. Consequently, they tend only to be used in critical applications, such as the highly-stressed skins of supersonic aircraft and rocket casings.

Another group of steels has been developed to exploit the properties obtained when the martensite reaction occurs during low-temperature plastic deformation. The TRIP steels were introduced in Chapter 10, with the formation of martensite leading to increases in work-hardening rate and hence of uniform ductility prior to necking. Essentially the principle is the same as that employed in controlled transformation steels, but plastic deformation is used to form martensite and the approach is broader as far as the thermomechanical treatment is concerned.

In one process, the composition of the steel is balanced to produce an M_S^σ temperature above room temperature. The steel is then heavily deformed (~80%) above the M_S^σ temperature, usually in the range 250–550°C, which results in austenite that remains stable at room temperature. Subsequent tensile testing at room temperature gives high strength levels combined with extensive ductility as a direct result of the martensitic transformation which takes place during the test. For example, a steel containing 0.3C-2Mn-2Si-9Cr-8.5Ni-4Mo wt% after 80% reduction at 475°C gives a proof and ultimate tensile strength of 1430 and 1500 MPa, respectively, and an elongation of 50%.

Higher strength levels (proof stress ~2000 MPa) with ductilities between 20% and 25% can be obtained by adding strong carbide-forming elements such as vanadium and titanium, and by causing the M_S^σ temperature to be below room temperature. As in the earlier treatment, severe thermomechanical treatments in the range 250–550°C are then used to deform the austenite and dispersion strengthen it with fine alloy carbides. The M_S^σ temperature is, as a result, raised to above room temperature so that, on mechanical testing, transformation to martensite takes place, giving excellent com-

Table 12.5 Compositions and properties of controlled transformation steels (after Pickering, [55, p. 206]). 'RT' stands for room temperature.

Composition wt%	**Treatment**	**0.2% σ_Y** MPa	σ_{UTS} MPa	**Elongation** %
0.1C-17Cr-4Ni-3Mn	Solution treated 925°C 40% cold reduced tempered 3 h at 450°C	1670	1700	0.5
	Solution treated 950°C refrigerated at −78°C tempered 1 h at 400°C	1200	1440	19
0.06C-16.5Cr-5Ni-2Mn-1.5Mo-2Co-1Al	Solution treated 1050°C aged 2 h at 700°C→ RT, aged 4 h at 450°C	1270	1430	0
	Solution treated 950°C refrigerated at −78°C tempered 4 h at 450°C	1240	1520	21
0.07C-17.5Cr-3Ni-2Mn-2Mo-2Co-1Cu	Solution treated 1050°C aged 2 h at 700°C→RT, tempered 4 h at 450°C	1110	1250	10
	Solution treated 950°C refrigerated at −78°C tempered 4 h at 450°C	1240	1360	20

binations of strength and ductility, as well as substantial improvements in fracture toughness.

Like the controlled transformation steels, the TRIP steels require extremely good metallurgical control but are relatively expensive to make. They only are used in applications where high demands are made on mechanical, as distinct from environmental properties. They do, however, illustrate how a combination of basic principles can be carefully balanced and controlled to achieve outstanding mechanical properties in alloy steels.

12.14 Summary

Chromium has a unique role in stainless steels. It has a sufficiently large solubility in iron to passivate the surface and to do this repeatedly when the oxide layer is damaged. The resistance to pitting corrosion can be improved by molybdenum. Aluminium-rich stainless steels have been developed to resist high temperature oxidation by forming stable alumina scales. These are the key characteristics of stainless steels, which otherwise have ambient temperature mechanical properties that can be achieved in low-alloy steels that are not stainless. In other words, the class itself is defined by the ability of the stainless steel to isolate itself from its environment.

The phase mixtures present in stainless steels and the mechanisms by which these phases form can be manipulated by varying other solutes, given rise to a large variety of alloys suited for different purposes depending on property combinations or economical factors.

References

[1] D. Dulieu, Stay Bright - A History of Stainless Steels in Britain, Outokumpu Stainless Ltd, U.K., 2013.

[2] R.O. Adams, A review of the stainless steel surface, Journal of Vacuum Science & Technology 1 (1983) 12–18.

[3] C.O.A. Olsson, D. Landolt, Passive films on stainless steels - chemistry, structure and growth, Electrochimica Acta 48 (2003) 1093–1104.

[4] T. Olugbade, J. Lu, Enhanced corrosion properties of nanostructured 316 stainless steel in 0.6 M NaCl solution, Journal of Bio- and Tribo-Corrosion 5 (2019) 1–11.

[5] S. Lozano-Perez, D.W. Saxey, T. Yamada, T. Terachi, Atom-probe tomography characterization of the oxidation of stainless steel, Scripta Materialia 62 (2010) 855–858.

[6] S. Stiber, N. Sata, T. Morawietz, S.A. Ansar, T. Jahnke, J.K. Lee, A. Bazylak, A. Fallisch, A.S. Gago, K.A. Friedrich, A high-performance, durable and low-cost proton exchange membrane electrolyser with stainless steel components, Energy & Environmental Science 15 (2022) 109–122.

[7] I. Gotman, Characteristics of metals used in implants, Journal of Endourology 11 (1997) 383–389.

[8] B.K. Montgomery, G. Gamble, S.T. Kha, G.G. Hecht, J.S. Vorhies, J.F. Lucas, Indications for and risks associated with implant removal after pediatric trauma, Journal of the AAOS Global Research & Reviews 6 (2022) 1–9.

[9] J.E. Westraadt, E.J. Olivier, J.H. Neethling, P. Hedström, J. Odqvist, X. Xu, A. Steuwer, A high-resolution analytical scanning transmission electron microscopy study of the early stages of spinodal decomposition in binary Fe-Cr, Materials Characterization 109 (2015) 216–221.

[10] K. Kaneko, T. Fukunaga, K. Yamada, N. Nakada, M. Kikuchi, Z. Saghi, J.S. Barnard, P.A. Midgley, Formation of $M_{23}C_6$-type precipitates and chromium-depleted zones in austenite stainless steel, Scripta Materialia 65 (2011) 509–512.

[11] K.B. Guy, E.P. Butler, D.R.F. West, Reversion of bcc α' martensite in Fe-Cr-Ni austenitic stainless steel, Metal Science 17 (1983) 167–176.

[12] C. Geyari, Design considerations in the use of stainless steel for vacuum and cryogenic equipment, Vacuum 26 (1976) 287–297.

[13] J. Manjanna, S. Kobayashi, Y. Kamada, S. Takahashi, H. Kikuchi, Martensitic transformation in SUS 316LN austenitic stainless steel at RT, Journal of Materials Science 43 (2008) 2659–2665.

[14] A. Boeuf, R. Caciuffo, S. Crico, R. Rebonato, F. Rustichelli, J.P. Morlevat, Time dependence at 550 and 700°C of $M_{23}C_6$ precipitate composition in AISI 304 stainless steel, Materials Letters 2 (1983) 49–52.

[15] J.W. Simmons, Overview: high-nitrogen alloying of stainless steels, Materials Science & Engineering A 207 (1996) 159–169.

[16] M. Heitkemper, A. Fischer, C. Bohne, A. Pyzalla, Laser surface treatment of the high nitrogen steel X30CrMoN15 1, Materialwissenschaft und Werkstofftechnik 31 (2000) 918–922.

[17] F.-J. Ebert, An overview of performance characteristics, experiences and trends of aerospace engine bearings technologies, Chinese Journal of Aeronautics 20 (2007) 378–384.

[18] H.K.D.H. Bhadeshia, Theory of Transformations in Steels, CRC Press, Taylor and Francis Group, London, U.K., 2021.
[19] P.R. Levey, A. van Bennekom, A mechanistic study of the effects of nitrogen on the corrosion properties of stainless steels, Corrosion Science 51 (1995) 911–921.
[20] G. Stein, I. Hucklenbroich, Manufacturing and applications of high nitrogen steels, Materials and Manufacturing Processes 19 (2004) 7–17.
[21] S.A. David, J.M. Vitek, T.L. Heeble, Effect of rapid solidification on stainless steel weld metal microstructures and its implications on the Schaeffler diagram, Welding Journal, Research Supplement 66 (1987) 298s–300s.
[22] H. Schneider, Investment casting of high hot-strength 12% chrome steel, Foundary Trade Journal 108 (1960) 562–563.
[23] N.I. Kakhovskii, V.N. Lipodaev, G.V. Fadeeva, The arc welding of stable austenitic corrosion-resisting steels and alloys, Avtomaticheskaya Svarka 33 (1980) 55–57.
[24] T. Watanabe, A new era of grain boundary design and control for high temperature materials, Key Engineering Materials 171–174 (2000) 237–244.
[25] T. Watanabe, S. Tsurekawa, S. Kobayashi, X. Zhao, L. Zuo, The state-of-the-art of controlling intergranular fracture and brittleness by grain boundary engineering, Key Engineering Materials 345–346 (2007) 619–624.
[26] T. Watanabe, Grain boundary engineering: historical perspective and future prospects, Journal of Materials Science 46 (2011) 4095–4115.
[27] A.P. Jivkov, N.P.C. Stevens, T.J. Marrow, A two-dimensional mesoscale model for intergranular stress corrosion crack propagation, Acta Materialia 54 (2006) 3493–3501.
[28] E.L. Hall, C.K. Briant, Chromium depletion in the vicinity of carbides in sensitized austenitic stainless steels, Metallurgical Transactions A 15 (1984) 793–811.
[29] F. Beckitt, B. Clark, The shape and mechanism of formation of $M_{23}C_6$ carbide in austenite, Acta Metallurgica 15 (1967) 113–129.
[30] S.A. Grot, J.E. Spruiell, Microstructural stability of titanium-modified type 316 and type 321 stainless steel, Metallurgical Transactions A 6 (1975) 2023–2030.
[31] E.A. Trillo, L.E. Murr, Effects of carbon content, deformation, and interfacial energetics on carbide precipitation and corrosion sensitization in 304 stainless steel, Acta Materialia 47 (1998) 235–245.
[32] F. Froes, R. Honeycombe, Conditions controlling matrix and stacking fault precipitation, Acta Metallurgica 15 (1967) 157–159.
[33] L.K. Singhal, J.W. Martin, The mechanism of tensile yield in an age-hardened steel containing γ' (ordered Ni_3Ti) precipitates, Acta Metallurgica 16 (1968) 947–963.
[34] C.C. Hsieh, W. Wu, Overview of intermetallic sigma (σ) phase precipitation in stainless steels, ISRN Metallurgy 2012 (2012) 732471.
[35] L.K. Singhal, J.W. Martin, The formation of ferrite and sigma-phase in some austenitic stainless steels, Acta Metallurgica 16 (1968) 1441–1451.
[36] T.G. Gooch, D.C. Willingham, Weld decay in AISI 304 stainless steel, British Welding Journal 3 (1971) 366–369.
[37] Y. Yamamoto, G. Muralidharan, M.P. Brady, Development of $L1_2$-ordered Ni_3(Al, Ti)-strengthened alumina-forming austenitic stainless steel alloys, Scripta Materialia 69 (2013) 816–819.
[38] R. Francis, G. Byrne, Duplex stainless steels – alloys for the 21st century, Metals 11 (2021) 836.
[39] B.W. Ahn, D.H. Choi, D.J. Kim, S.B. Jung, Microstructures and properties of friction stir welded 409L stainless steel using a Si_3N_4 tool, Materials Science & Engineering A 532 (2012) 476–479.

[40] C.C. Silva, J.P. Farias, H.C. Miranda, R.F. Guimaraes, J.W. Menezes, M.A. Neto, Microstructural characterization of the HAZ in AISI 444 ferritic stainless steel welds, Materials Characterization 59 (2008) 528–533.
[41] M.B. Cortie, H. Pollak, Embrittlement and aging at 475 °C in an experimental ferritic stainless steel containing 38 wt.% chromium, Materials Science & Engineering A 199 (1995) 153–163.
[42] D.R. Maurice, T.H. Courtney, The physics of mechanical alloying: a first report, Metallurgical Transactions A 21 (1990) 289–303.
[43] T.S. Chou, H.K.D.H. Bhadeshia, G. McColvin, I. Elliott, Atomic structure of mechanically alloyed steels, in: Structural Applications of Mechanical Alloying, ASM International, Materials Park, Ohio, USA, 1993, pp. 77–82.
[44] H.K.D.H. Bhadeshia, Recrystallisation of practical mechanically alloyed iron base and nickel base superalloys, Materials Science & Engineering A A223 (1997) 64–77.
[45] E. Getto, B. Baker, B. Tobie, S. Briggs, K. Hattar, K. Knipling, Effect of friction stir welding and self-ion irradiation on dispersoid evolution in oxide dispersion strengthened steel MA956 up to 25 dpa, Journal of Nuclear Materials 515 (2019) 407–419.
[46] I. Tamura, Deformation induced martensitic transformation and transformation-induced plasticity in steels, Metal Science 16 (1982) 245–254.
[47] S. Chatterjee, H.K.D.H. Bhadeshia, TRIP-assisted steels: stress or strain-affected martensitic transformation, Materials Science and Technology 23 (2007) 1101–1104.
[48] J.W. Brooks, M.H. Loretto, R.E. Smallman, In situ observations of martensite formation in stainless steel, Acta Metallurgica 27 (1979) 1829–1838.
[49] J.W. Brooks, M.H. Loretto, R.E. Smallman, Direct observations of martensite nuclei in stainless steel, Acta Metallurgica 27 (1979) 1839–1847.
[50] C. Ullrich, R. Eckner, L. Krüger, S. Martin, V. Klemm, D. Rafaja, Interplay of microstructure defects in austenitic steel with medium stacking fault energy, Materials Science & Engineering A 649 (2016) 390–399.
[51] L. Vitos, J.-O. Nilsson, B. Johansson, Alloying effects on the stacking fault energy in austenitic stainless steels from first-principles theory, Acta Materialia 54 (2006) 3821–3826.
[52] S. Curtze, V.-T. Kuokkala, A. Oikari, J. Talonen, H. Hänninen, Thermodynamic modeling of the stacking fault energy of austenitic steels, Acta Materialia 59 (2011) 1068–1076.
[53] Y.N. Petrov, Effect of carbon and nitrogen on the stacking fault energy of high-alloyed iron-based austenite, International Journal of Materials Research 94 (2003) 1012–1016.
[54] F.B. Pickering, High strength steels and their uses today, in: W.O. Alexander (Ed.), Metallurgical Achievements, Pergamon Press, Oxford, U.K., 1965, pp. 199–258.
[55] F.B. Pickering, Physical Metallurgy and the Design of Steels, Applied Science Publishers, Essex, U.K., 1978, p. 104.

Weld microstructures

13

Abstract

Bridges, ships and much of the infrastructure on which modern civilisation depends, rely on the ability to connect pieces of well-engineered and shaped steel by welding. A localised heat source introduces liquid metal into the gap between the pieces of steel, which on solidification forms an integral joint. The process is complex with phenomena embracing all states of matter that are ordinarily observable, i.e., plasma, gas, liquid and solid. It requires an understanding of many aspects of metallurgy, heat and fluid flow in order to achieve a reliable joint. There also exist processes that do not involve fusion, but nevertheless fall within the remit of welding. Woe betide material developers that neglect the role of joining in practical scenarios.

13.1 Introduction

Fusion welding is of greatest importance in the fabrication of engineering structures. There are many ways in which the process can be implemented, but most involve the deposition of a small amount of molten steel between the components to be joined or coated, the heat source often being an electric arc (Fig. 13.1a,b). There are processes, such as laser or electron beam welding, where the gap between the components to be joined is so small that just running the heat source along the abutting surfaces leads to localised-melting followed by rapid solidification and hence a union. Friction stir welding and friction welding in general do not require a filler material, nor does autogenous arc-welding, in which the fused regions of the metals being welded coalesce to form a joint. But it is worth emphasising that these processes are not, for reasons of productivity, scale, cost and practicality, substitutes for the much more common filler-metal based methods. This is particularly so in the case of steels, where the strength of the material at elevated temperatures makes friction processes difficult to justify.

In the fusion welding processes, it is the solidification of the liquid following the passage of the heat source that welds the components together. The metallurgy of the welded joint can be categorised into two significant regions, the fusion zone and the heat-affected zone (HAZ), Fig. 13.1c. The fusion zone represents both the deposited metal and the parts of the steel component melted during the process – it has a solidification microstructure, but one that is influenced in a variety of ways by initiation at the unfused surfaces of the joint. The HAZ, on the other hand, represents those regions in the close proximity of the weld, where the heat input during welding changes the microstructure without melting the steel. When a weld requires multiple deposits to fill the gap between the components being joined, there are heat-affected zones between each sequential deposit. This chapter describes the development of microstructure in both zones, beginning with the fused regions. Virtually every aspect of phase transformation in steels is relevant to the subject of welding. There is an opportunity for a whole series of phase transformations to occur successively as the weld cools from the vapour and liquid states. Further changes are possible if the weld then has to be

Steels. https://doi.org/10.1016/B978-0-44-318491-8.00019-8

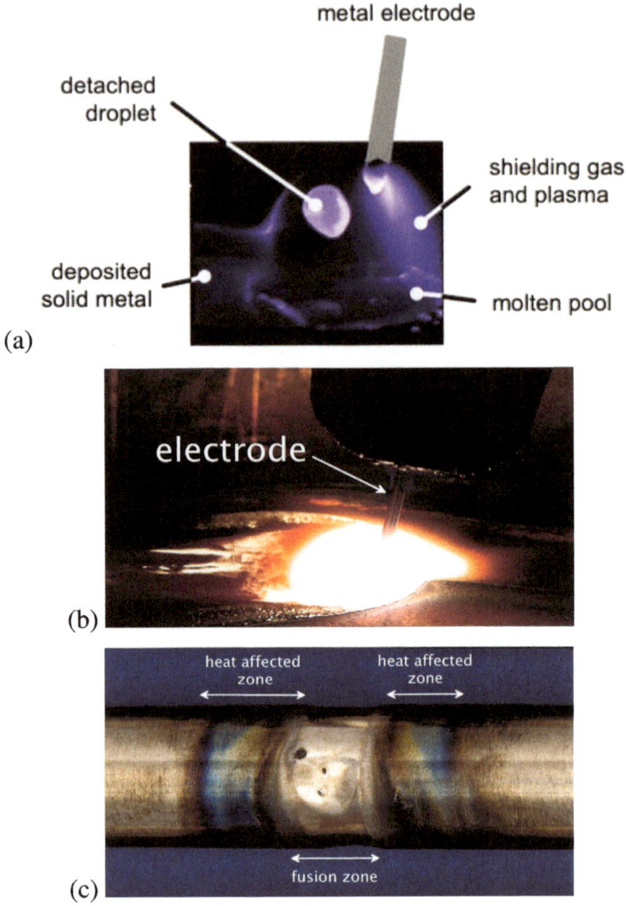

Figure 13.1 A process using a wire electrode that is consumed as the weld is deposited. An electrical arc is struck between a metal electrode and the steel. This melts the consumable electrode with droplets detaching and falling into the weld pool below – some of the steel also melts and contributes to the weld pool while protected by a shield of flowing inert gas (image courtesy of S. Guest, J. Chapuis and P. Mendez). (b) The electrode moves along the welding direction, leaving behind the solidified weld metal. There are many variations on this process [2,3]. Image courtesy of K. Dalaei, ESAB AB. (c) The heat-affected zone highlighted by the oxide interference colours on an arc-welded stainless steel tube. Photograph courtesy of Y. Zhang and S. Condie of TWI Ltd, Cambridge.

heat treated to optimise properties or relieve locked-in stresses caused by the heterogeneous application of heat.

Autogenous arc-welding usually is confined to thin sheets though deep-penetration electron-beams can produce autogenous welds in 10 cm thick steel Fig. 13.2b. The interaction of the electron beam with the steel is such that only a narrow region is melted with the added advantage of limited heat input, which in turn limits the size

(a)

(b)

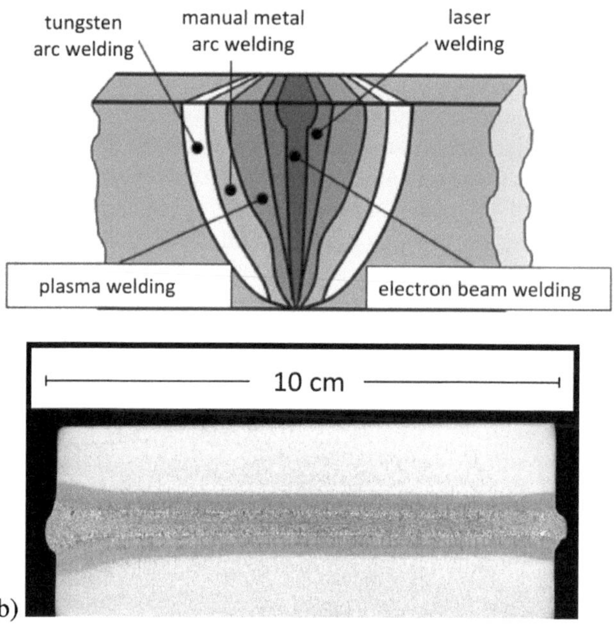

Figure 13.2 (a) Shape of the fusion zones produced by a variety of single-pass welding processes. In the tungsten arc processes, the tungsten electrode is not consumed, but filler metal is fed separately. With the metal arc process, a wire forms both the electrode and feed material. Plasma welding involves a constricted arc. Image from Węglowski et al. [1] with permission of Elsevier. (b) Deep-penetration, autogenous electron-beam weld in steel. The fusion zone and HAZ are narrow compared with the depth of the weld. Micrograph rotated through 90° for presentation. Image courtesy of TWI Ltd, C. Johnston.

of the heat-affected zone. The vapour generated by the electron beam as it hits the steel allows it to penetrate further, the phenomenon referred to as the *keyhole effect*. Electron beam welding must be conducted at reduced atmospheric-pressure in order to avoid beam absorption by air. For single-pass welds, the heat-affected zone width is proportional or even equal, approximately to the thickness of the fusion zone. After all, the width of the fusion zone depends on the energy input per unit length of the weld. This is why, at the top and bottom surfaces of the electron-beam weld the fusion zone and associated HAZs are wider.

13.2 Fusion zone

13.2.1 Weld solidification

Iron and its dilute alloys are ferritic at temperatures just below the melting point, Fig. 1.2. As it cools, the ferrite then transforms into austenite, only to revert back to ferrite on continued cooling. Most steels contain modest concentrations of alloying elements, and hence show similar crystal structure changes as pure iron. This remains the case with some of the richly-alloyed duplex stainless steels, where the formation of ferrite from the liquid, instead of austenite, makes the weld more resistant to solidification-cracking because the δ-ferrite is able to absorb impurities [4].

Weld deposits, therefore begin solidification with the epitaxial growth of columnar δ-ferrite from the hot-grain-structure of the parent plate at the fusion surface, Fig. 13.3a [5–7]. The grains are anisotropic because they grow into the liquid along the principal direction of heat flow, although that direction is not entirely constant given that there is a moving heat source during welding. Crystallography also plays a role because those grains with one of their ⟨100⟩ directions most-parallel to the heat-flow direction grow fastest and hence stifle the growth of less-suitably oriented grains. The width of the columnar grains therefore increases with distance away from the fusion boundary. The crystallographic direction that is favoured during solidification is determined by the anisotropy of the solid/liquid interfacial energy and how sensitive that orientation is to interfacial curvature [8]. In iron alloys, solidification usually is fastest along the ⟨100⟩ direction (e.g., [9]).

For low-alloy steels, the δ-ferrite undergoes a solid-state transformation into austenite as the temperature decreases. The austenite nucleates at the δ-δ grain boundaries and also develops into a columnar austenite grain structure, which resembles that of the original δ-grains (Fig. 13.3b).

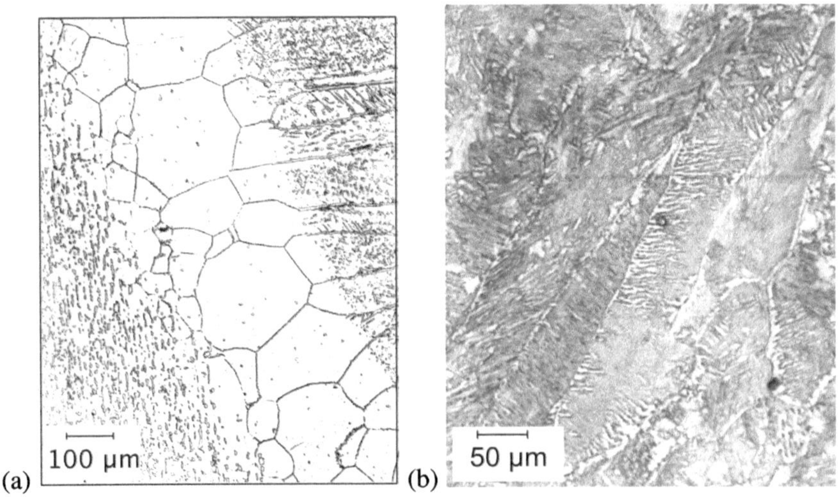

Figure 13.3 (a) Illustration of the epitaxial growth of columnar grains from the fusion boundary of a stainless steel weld (courtesy of Honeycombe and Gooch). (b) Optical micrograph showing the columnar prior-austenite grain structure typical in steel weld deposits (courtesy C. Walsh).

The detailed shape and size of the austenite grains is of importance in the evolution of the final microstructure. The effect of the austenite grain size is two-fold. Firstly, the number density of austenite grain boundary nucleation sites changes inversely with the grain size. Coarse-grained weld deposits therefore have a higher hardenability. The second, and more subtle effect, arises from the columnar shape of the austenite grains, a shape which is like that of a hexagonal prism. The grains are typically about 100 μm wide and about 5000 μm in length [10]. This is quite unlike an equiaxed grain structure and, because of the fewer grain junctions involved, allows the hardenability of a weld to be greater than that of a corresponding wrought alloy.

Solidification does not occur under equilibrium conditions during welding. Solidification-induced chemical segregation, and composition variations due to uncontrolled changes in the welding conditions, make the solidification microstructure inhomogeneous. The amplitude of these variations becomes larger as the alloy concentration increases. The extent of chemical segregation depends on whether solidification occurs as austenite or ferrite, because the ability for solutes to diffuse and therefore to homogenise, is much greater in the case of ferrite. The degree of the partitioning of solute required by solubility differences between the specific solid and liquid phases also is a factor. Therefore, the amplitude of the chemical segregation is much lower when solidification is to ferrite [11].

Mineral fluxes or inert-gas shrouds are employed in order to protect the hot metal against environmental attack during welding. Such protection is not entirely effective, with the result that the oxygen content of welds tends to be much larger than that of wrought steel (Table 13.1), because oxide particles become entrapped in the fusion zone during solidification. As discussed later, these non-metallic particles can serve as heterogeneous nucleation sites and hence are of considerable importance in the evolution of the microstructure and mechanical properties. Table 13.1 reveals some other interesting differences between the plate and weld compositions. The copper concentration of the weld is relatively large because in this case, the welding wire has a copper coating to enable better electrical contact. The silicon concentration in the weld is larger than both the wire and the plate, because the excess silicon is acquired by decomposition of the protective flux. These observations emphasise the complexity of the welding process, in which the chemical composition of the final weld deposit depends on many variables, including the plate, wire and flux compositions.

Table 13.1 A comparison of the chemical composition (wt%) of a submerged arc weld with that of the welded plate and the wire used as the consumable electrode. The welding conditions used were 34 V, 900 A (direct current positive), at a welding speed of 0.005 m s^{-1}, with a calcium silicate flux.

	C	Mn	Si	Cu	Al	N	O
Plate	0.21	1.0	0.2	0.05	0.04	0.01	0.001
Wire	0.14	1.5	0.2	0.31	0.01	0.01	0.001
Weld	0.16	1.1	**0.3**	0.16	0.01	0.01	**0.053**

13.2.2 As-deposited microstructure

The microstructure obtained as the weld cools from the liquid phase to ambient temperature is called the *as-deposited* or *primary* microstructure. Its major components include allotriomorphic ferrite, Widmanstätten ferrite and acicular ferrite (Fig. 13.4). There may also be some martensite, retained austenite or degenerate pearlite. These latter phases occur in small fractions because of the low carbon concentration of the weld metal; they are known by the collective term *microphases*. Bainite, consisting of sheaves of parallel platelets, is not generally found in well-designed welding alloys. Instead, acicular ferrite is induced to nucleate heterogeneously on non-metallic inclusions [12].

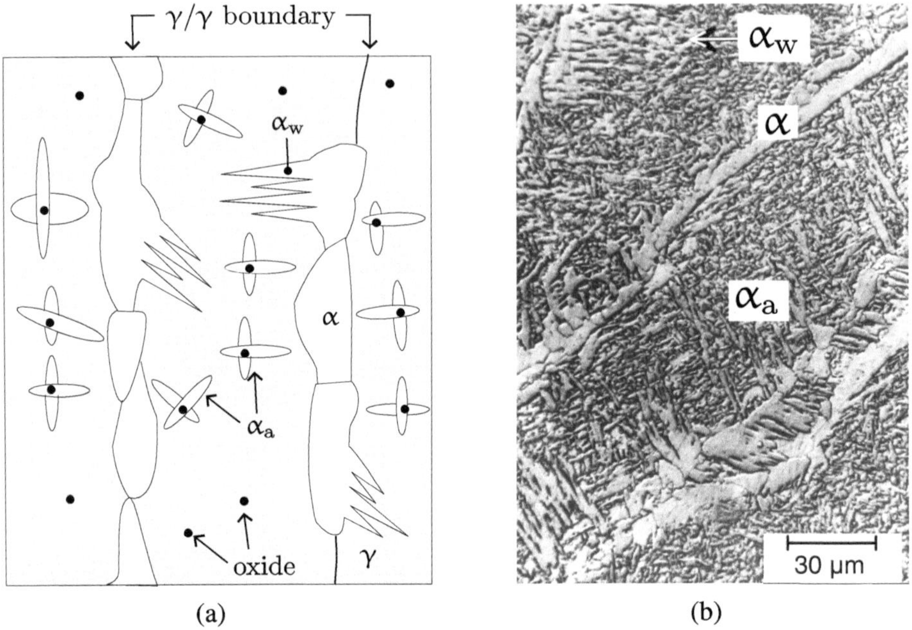

Figure 13.4 (a) Schematic illustration of the essential constituents of the primary microstructure in the columnar austenite grains of a steel weld deposit. (b) Scanning electron micrograph of the primary microstructure of a steel weld (courtesy of G. Rees). The terms α, α_w and α_a refer to allotriomorphic ferrite, Widmanstätten ferrite and acicular ferrite, respectively.

In practice, the gap between the components to be joined often has to, for thick steels, be filled by a sequence of several weld deposits. These multi-run welds have a complicated microstructure (Fig. 13.5). The deposition of each successive layer heat treats the underlying microstructure. Some of the regions of original primary microstructure are reheated to temperatures high enough to cause the reformation of austenite, which during the cooling part of the thermal cycle transforms into a different microstructure. Other regions may simply be tempered by the deposition of subsequent runs. The microstructure of the reheated regions is called the *reheated* or *secondary* microstructure.

13.2.3 Allotriomorphic ferrite

Allotriomorphic ferrite (α) is the first phase to form on cooling the austenite grains below the Ae_3 temperature. It nucleates at the columnar austenite grain boundaries.

Figure 13.5 The macrostructure of a multirun weld, made by sequentially depositing a number of beads in each of the 7 layers (courtesy of Kamellia Dalaei, ESAB AB).

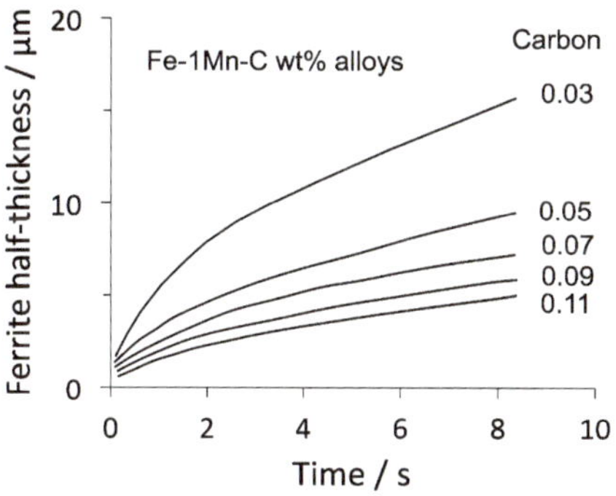

Figure 13.6 An illustration of the parabolic thickening of ferrite during isothermal transformation at 740°C. Each curve represents a Fe-1Mn-C wt% steel with the carbon concentration, as indicated on the diagram.

Because these boundaries are easy diffusion paths, they become decorated with thin, continuous layers of ferrite. The layers then thicken at a rate controlled by the diffusion of carbon in the austenite ahead of the transformation interface. Under isothermal conditions, the ferrite thickness z^* changes parabolically with time t (p. 73):

$$z^* = \alpha_1 t^{1/2}, \tag{13.1}$$

where α_1 is the parabolic rate constant. This is illustrated in Fig. 13.6 for alloys with different carbon concentrations; note that the growth kinetics become sensitive to the carbon concentration as the latter approaches the solubility of carbon in the ferrite, simply because there then is less carbon available to partition into the austenite so diffusion-controlled growth must accelerate. When the average carbon concentration equals the solubility limit in austenite, the growth rate calculated in this way becomes

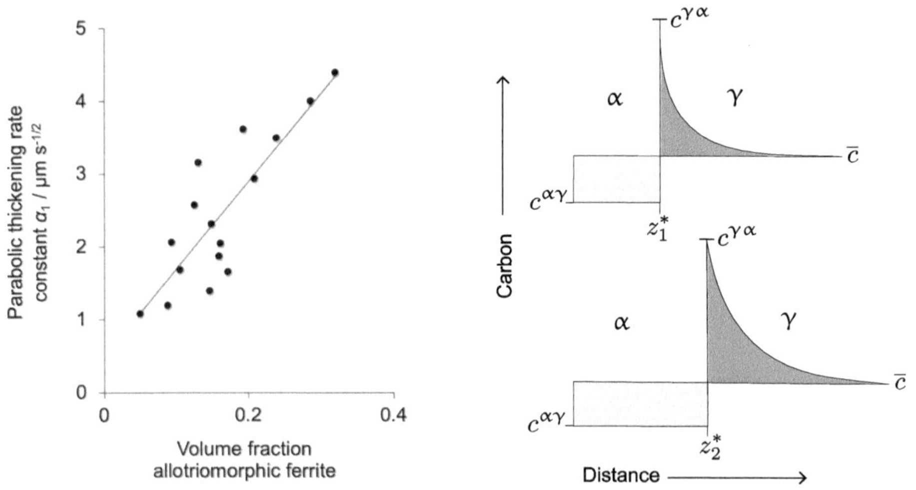

(a) (b)

Figure 13.7 (a) The correlation between the calculated parabolic thickening rate constant (a variable related to the growth rate) and the volume fraction of allotriomorphic ferrite obtained in a series of manual metal arc weld deposits, fabricated using similar welding parameters but with different chemical compositions. The rate constant was calculated for transformation at 700°C. (b) The light-shaded region represents the excess carbon partitioned into the γ as α grows, with an area equal to the dark-shaded region in the austenite. The diffusion distance increases as the ferrite layer thickens ($z_2^* > z_1^*$), making the concentration gradient more shallow, slowing down the diffusion flux away from the interface, and hence the rate of growth.

infinite, but in practice, other factors such as the transfer of the iron atoms across the α/γ interface become rate limiting.

The magnitude of the parabolic rate constant depends on the equilibrium compositions of the austenite and ferrite, and on the diffusivity of carbon in austenite (p. 73). Alloying elements such as manganese, which stabilise austenite, are associated with a smaller value of α_1. Because nucleation is not rate limiting in the case of the majority of low-alloy steel weld metals, the fraction of allotriomorphic ferrite obtained correlates directly with the parabolic rate constant (Fig. 13.7a). Nevertheless, welding does not involve isothermal transformation. The cooling rate of the weld metal once it is in the solid-state is given approximately by [13]:

$$\frac{\mathrm{d}T}{\mathrm{d}t} = \frac{C_{12}}{q\eta}(T - T_\mathrm{i})^{C_{13}}, \tag{13.2a}$$

where q in the case of arc welding is the electrical energy input per unit length of the weld (J m^{-1}), η is the arc energy transfer efficiency (≈ 0.8), T_i is the initial temperature of the steel to be welded and C_i are empirical constants. Most of the development of microstructure occurs over the time interval $\Delta t_{800\text{-}500}$ taken to cool between 800 and 500°C, with

$$\Delta t_{800\text{-}500} = \frac{q\eta}{C_{12}(1 - C_{13})}\left[T - T_{\mathrm{i}}\right]_{500}^{800}. \tag{13.2b}$$

For arc welds, $\Delta t_{800\text{-}500}$ is typically in the range 5–40 s. The parameters are given approximately by:

	C_{12}	C_{13}	η
Manual metal arc	1325	1.60	0.775
Submerged arc	4359	1.51	0.95

We note that, although the discussion here is limited to essential concepts, it is possible to integrate the development of microstructure over a range of temperatures to properly represent welds [14] or steels in general [15].

The fact that during isothermal transformation the thickness of the ferrite varies with the square root of time, means that the rate of growth decreases as the ferrite layer becomes thicker. This is because the distance over which carbon has to diffuse increases with time (Fig. 13.7b). The growth rate for a given alloy goes through a maximum as a function of temperature, because the driving force for transformation increases with undercooling whereas the diffusivity decreases. Consequently, as the weld cools to temperatures less than about 600°C, the diffusional growth of ferrite slows down so much that the layers of allotriomorphic ferrite reach a limiting thickness. The atomic mechanism of the Widmanstätten ferrite transformation does not require the diffusion of substitutional solutes; therefore, its growth is not sluggish at low temperatures. The remaining austenite thus begins to transform into Widmanstätten ferrite (Fig. 13.8).

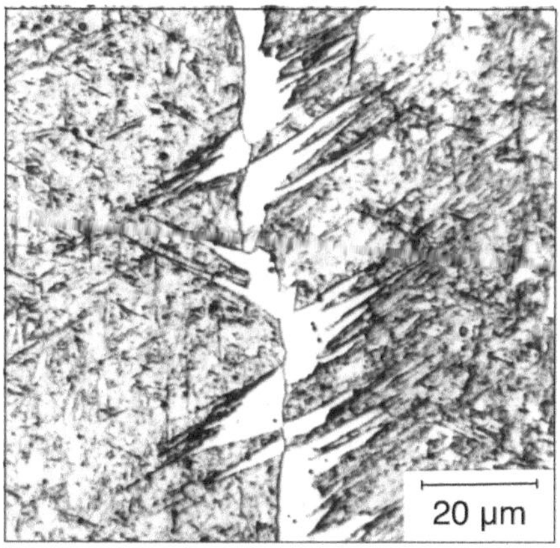

Figure 13.8 Widmanstätten ferrite plates growing from allotriomorphic ferrite in a partially transformed steel weld, which was quenched from the transformation temperature. The matrix is martensitic (courtesy of G. Barritte).

13.2.4 Widmanstätten ferrite and acicular ferrite

Although substitutional solutes and iron atoms do not diffuse during the growth of Widmanstätten ferrite, carbon does partition during transformation. Because of its plate shape, much of the carbon can be accommodated at the sides of the growing plate, so that the plate tip always encounters fresh austenite. This is unlike the case for allotriomorphic ferrite, where the partitioned carbon accumulates ahead of the interface and progressively slows down the rate of growth. Widmanstätten ferrite plates therefore lengthen at a constant rate.

The growth rates are found to be so large for typical weld compositions that the formation of Widmanstätten ferrite usually is completed within a fraction of a second. Hence, for all practical purposes, the transformation can be regarded as being isothermal (Fig. 13.9a).

Unfortunately, the fraction of Widmanstätten ferrite that forms in weld deposits correlates badly with the plate lengthening rate, as illustrated in Fig. 13.9b. [16]. This is because there is an interference between the plates of Widmanstätten ferrite that grow from the austenite grain boundaries, and acicular ferrite plates which nucleate at non-metallic particles dispersed throughout the weld (Fig. 13.10). The formation of Widmanstätten ferrite and acicular ferrite is therefore competitive. Anything that increases the number density of inclusion nucleation sites relative to austenite grain nucleation sites, favours the formation of acicular ferrite at the expense of Widmanstätten ferrite. Hence, the refinement of austenite grain size or a reduction in the oxide content of the weld below a limiting value both lead to a decrease in the acicular ferrite content.

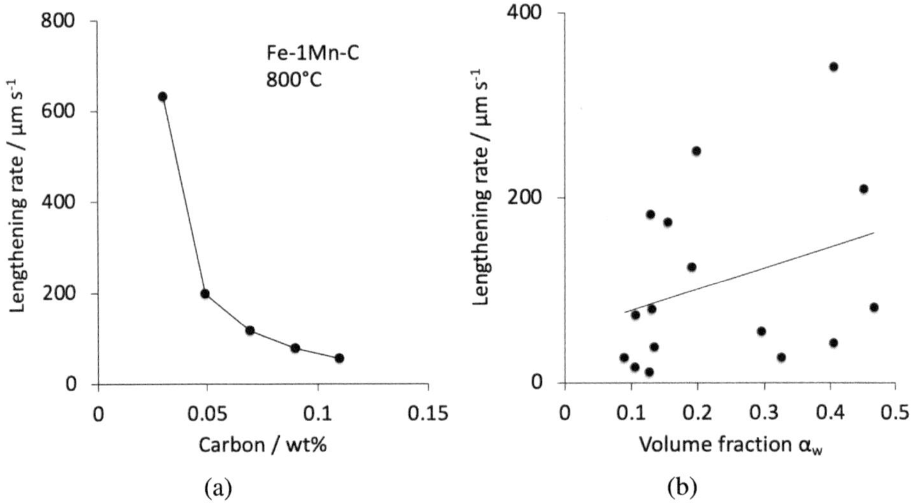

Figure 13.9 (a) The isothermal growth rate of Widmanstätten ferrite in a series of Fe-1Mn-C wt% alloys as a function of carbon concentration. Notice that the growth rates are so large that the plates could grow right across typical austenite grains within a fraction of a second. (b) Poor correlation of the volume fraction of Widmanstätten ferrite against the calculated growth rate.

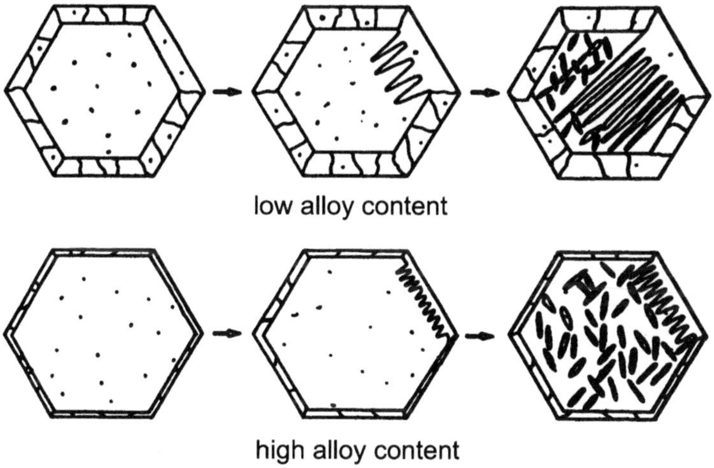

Figure 13.10 Diagrams illustrating the development of microstructure in two weld deposits with different chemical compositions. The hexagons represent cross-sections of columnar austenite grains, the boundaries of which first become decorated with uniform, polycrystalline layers of allotriomorphic ferrite, followed by the formation of Widmanstätten ferrite. Depending on the relative transformation rates of Widmanstätten ferrite and acicular ferrite, the former can grow entirely across the austenite grains or become stifled by the intragranularly nucleated plates of acicular ferrite [14].

By the time the weld deposit cools to about 500°C, most of the austenite has been consumed. The small quantity of remaining austenite (about 5%) is enriched in carbon and either transforms into martensite, or into pearlite, which is degenerate because it does not have the opportunity to establish the lamellar structure associated with the cooperative growth of α and θ. Slower cooling rates favour the formation of pearlite relative to martensite. Some austenite may also be retained at ambient temperatures. Because of their small volume fractions in the overall microstructure, these phases are, in welding terminology, called *microphases*. The microphases are relatively hard and behave in many respects like brittle inclusions. They are, therefore, of importance in determining the toughness and the scatter in the toughness of weld deposits. However, this is not a generic conclusion, the ability of such local zones to be brittle depends also on the following factors [17]:

- **a)** The hardness of the local zone relative to its surroundings. A pearlitic microphase-region would be more tolerable than one consisting of untempered martensite.
- **b)** The size relative to the average microstructural scale, or more precisely, the size of the crack-tip zone sampled by a propagating crack [18].
- **c)** The shape of the zone; thus, films of carbon-rich austenite can be beneficial to toughness whereas large blocks that decompose readily into untempered, high-carbon martensite, are not [19].
- **d)** The number density of the local brittle zones relative to the fracture criteria important in the engineering design. This can also affect the scatter of observed mechanical properties.

13.2.5 Sensitivity to carbon

It is striking that small variations in carbon concentration can have a major influence on the microstructure of welds, especially since the average carbon concentration of a weld is usually kept very low [20]. It is apparent from the previous discussions of the growth rates of allotriomorphic and Widmanstätten ferrite, that the sensitivity of growth kinetics to carbon becomes larger as the concentration of carbon decreases.

These are important observations given that the general trend in the steel industry has been to reduce the carbon concentration, sometimes to levels approaching the maximum solubility of carbon in ferrite in equilibrium with austenite. The rate at which ferrite grows increases sharply as the carbon concentration of the steel approaches this solubility in ferrite. This is because there is then no need for the carbon to diffuse ahead of the γ/α interface, since it can all be accommodated in the ferrite.

Therefore, the effect of carbon is seen to be larger (Figs. 13.6 and 13.9a) when its concentration changes from $0.03 \rightarrow 0.05$ wt%, when compared with the change from $0.09 \rightarrow 0.11$ wt%. Changes in mechanical properties reflect this behaviour, with the strength of low-carbon steels being particularly sensitive to the carbon concentration. This increased sensitivity of the γ/α transformation to carbon at low concentrations, leads to a corresponding decreased sensitivity to substitutional alloying elements. Carbon in effect controls the kinetics of transformation.

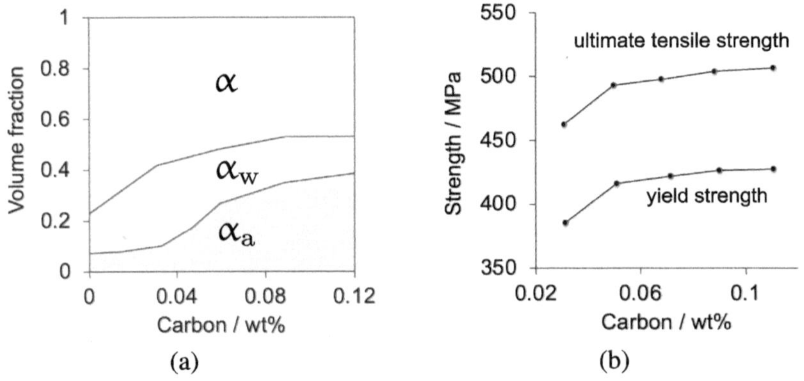

Figure 13.11 Variations in microstructure and mechanical properties as a function of carbon concentration in Fe-1Mn-C wt% steel weld deposit using manual metal arc welding ($1\,\text{kJ}\,\text{mm}^{-1}$).

In welding, the hardenability of the steel often is expressed as a carbon equivalent (CE). The concentration of each solute is scaled by a coefficient which expresses its ability, relative to carbon, to retard the γ/α transformation. Steels with a CE in excess of about 0.4 wt% cannot easily be welded because of their increased tendency to form martensite. There are in fact two popular expressions for the CE, one due to the International Institute for Welding (IIW), and the other attributed to Ito and Besseyo, covering the high and low ranges of carbon, respectively:

$$\text{IIW} > 0.18\,\text{wt\% C},$$
$$\text{CE} = w_\text{C} + \frac{w_\text{Mn} + w_\text{Si}}{6} + \frac{w_\text{Ni} + w_\text{Cu}}{15} + \frac{w_\text{Cr} + w_\text{Mo} + w_\text{V}}{5}\,\text{wt\%}, \tag{13.3}$$
$$\text{Ito-Besseyo} < 0.18\,\text{wt\% C},$$
$$\text{CE} = w_\text{C} + \frac{w_\text{Si}}{30} + \frac{w_\text{Mn} + w_\text{Cu} + w_\text{Cr}}{20} + \frac{w_\text{Ni}}{60} + \frac{w_\text{Mo}}{15} + \frac{w_\text{V}}{10} + 5w_\text{B}\,\text{wt\%}. \tag{13.4}$$

The Ito–Besseyo CE formula has smaller coefficients for the substitutional solutes when compared with the IIW formula. It is believed to be more reliable for low-carbon steels. The IIW formula shows much lower tolerance to substitutional alloying elements than the Ito–Besseyo equation. As already discussed, with low carbon concentrations the kinetics of transformation are more sensitive to carbon than to substitutional solutes. Hence, it is logical that there should be two different empirical expressions for the CE for the low- and high-carbon weldable steels. Fig. 13.11 illustrates that, as expected, both the microstructure and mechanical properties change more rapidly at low carbon concentrations.

13.3 Heat-affected zone

The HAZ is the portion of the material which has not been melted, but where the microstructure and mechanical properties are altered by the heat of welding (p. 383). The width of this zone depends largely on the amount of heat added per unit length of weld. It is much wider for a submerged-arc weld than for a laser weld.

13.3.1 Heat flow

All welding processes involve a source of heat, the prime purpose of which is to cause melting, or in the case of friction stir-welding, to locally plasticise the steel. Subsequent solidification should lead to the formation of an integral joint. Much of the heat manages to diffuse from the fusion zone into the adjacent solid regions. As a consequence, those regions experience a heating and cooling cycle, the severity of which depends on the distance from the fusion boundary (Fig. 13.12). The peak temperature and the heating rate decrease with distance away from the fusion boundary. The cooling rate, on the other hand, is less sensitive to this distance, and can be stated as the time $\Delta t_{800\text{-}500}$ taken to cool over the range 800–500°C. For many weldable steels, this defines the temperature range within which austenite decomposes by solid-state transformation.

The nature of the thermal cycle at any position within the HAZ can be characterised by two parameters, the peak temperature T_P and the time period $\Delta t_{800\text{-}500}$. Both parameters for the heat-affected zone increase with the heat input q:

$$T_\text{P} \propto \frac{q}{r}, \qquad \Delta t_{800\text{-}500} \propto q^n, \tag{13.5}$$

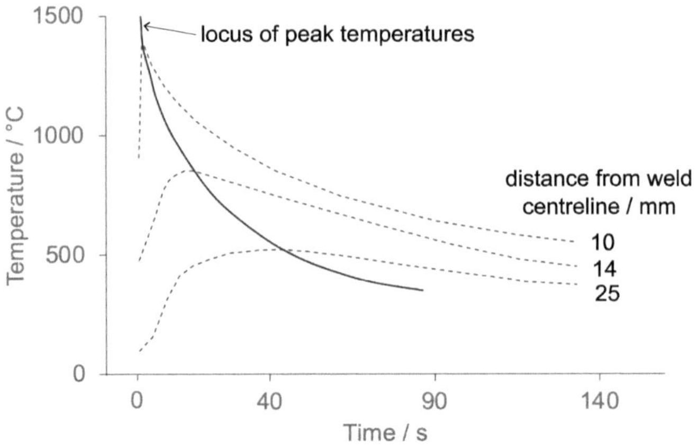

Figure 13.12 Temperature-time curves representing typical thermal cycles experienced in the heat-affected zone of a weld (adapted from data published in the *Welding Handbook* [21]).

where r is the distance from the fusion boundary and n has a value 1 or 2, which depends on whether the component being welded is thick or thin, respectively, compared with the size of the weld bead. The relative thickness determines whether the flow of heat is two- or three-dimensional (Fig. 13.13). The heat input q is per unit length of weld and typically is in the range about 1-5 kJ mm^{-1}.

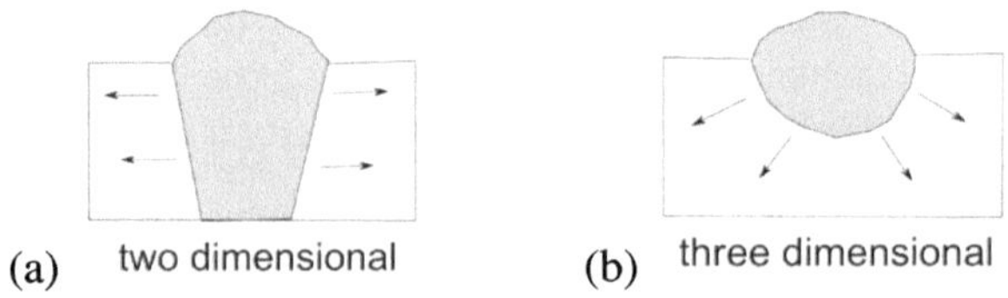

Figure 13.13 Illustration of (a) two-dimensional and (b) three-dimensional heat-flow conditions. Flow into the plane of the diagrams is not illustrated - it is present because the heat source is not stationary.

13.3.2 Microstructural zones

There is a well-defined gradient of microstructure in the heat-affected zone as a function of the distance from the fusion boundary (Fig. 13.14):

1. Those regions immediately adjacent to the fusion boundary are heated to very high temperatures and hence transform completely into austenite. During continuous heating, austenite begins to form at a temperature $Ac_1 \simeq 800°C$, and the samples become fully austenitic at $Ac_3 \simeq 950°C$. These temperatures are different from the corresponding equilibrium temperatures Ae_1 and Ae_3 because they increase with the heating rate. The peak temperatures in the HAZ close to the fusion boundary are well in excess of the Ac_3 temperature of weldable steels (Table 13.2). Con-

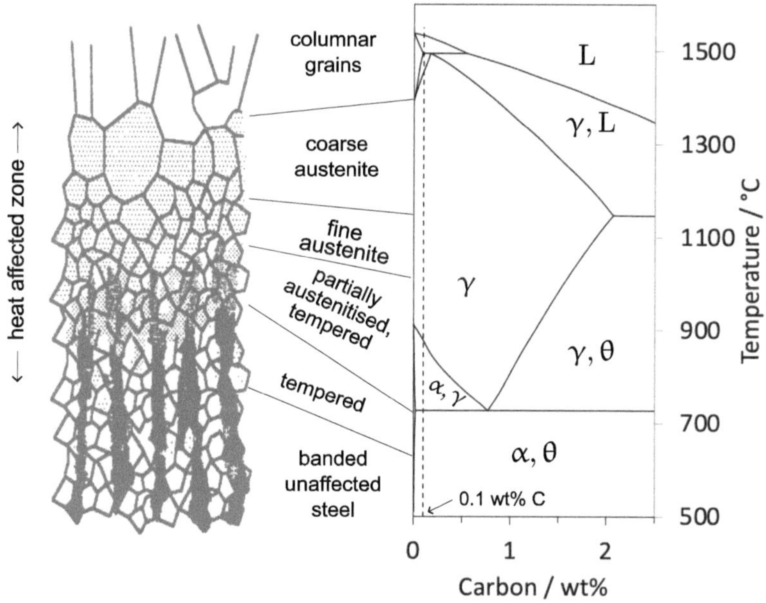

Figure 13.14 Schematic illustration of the microstructural variation to be expected in the heat-affected zone of steel welds. Notice that the microstructural zones do not and should not coincide with the equilibrium transformation temperatures since heating and cooling rates during welding do not approach equilibrium rates.

Table 13.2 Approximate temperature ranges for the variety of microstructural regions within the heat-affected zone of steel welds.

HAZ microstructure	Temperature range
Coarse-grained austenite	1500°C $> T_P >$ 1200°C
Fine-grained austenite	1200°C $> T_P > Ac_3$
Partially austenitised zone	$Ac_3 > T_P > Ac_1$
Tempered regions	$Ac_1 > T_P$

sequently, the austenite that forms is annealed during heating beyond Ac_3, giving rise to a very coarse grain structure. This forms the *coarse-grained austenite zone*.

2. The austenite grain size decreases sharply with distance from the fusion boundary. It is necessary to distinguish this as the *fine-grained zone* because its mechanical properties tend to be superior to those of the coarse zone.
3. As the peak temperature decreases, regions of the HAZ further away from the fusion boundary become only partially austenitic during the heating part of the thermal cycle. The austenite that forms has a rather large carbon concentration, due to the increase in the solubility of carbon in γ with decreasing temperature. The part that does not transform into austenite becomes tempered.

4. When the peak temperature becomes less than the Ac_1 temperature, the only effect of the heat input is to temper the microstructure, the extent of tempering decreasing with distance from the fusion boundary.

The individual microstructures are illustrated in Fig. 13.15 discussed in detail in the sections that follow.

13.3.3 Coarse-grained austenite

The formation of austenite during heating is in many ways different from transformations which occur during cooling below the equilibrium temperature. As discussed in Chapter 3, the formation of ferrite follows a C-shaped curve kinetic behaviour on a time-temperature transformation diagram; the overall transformation rate, therefore, goes through a maximum as a function of the supercooling below the equilibrium temperature. This is because of two opposing effects; the diffusion coefficient decreases as the temperature falls, but the driving force for transformation increases.

During heating, however, both the diffusion coefficient and the driving force increase with temperature. The overall rate of transformation, therefore, increases continuously as the transformation temperature is raised, Fig. 13.16.

For practical purposes, the formation of austenite during heating can be represented by a continuous heating transformation (CHT) diagram, analogous in concept to the continuous cooling transformation (CCT) diagrams so useful in illustrating the formation of ferrite (Fig. 13.17a). The CHT diagram is displaced to longer times when compared with the isothermal transformation diagram for austenite formation. For typical heating rates encountered in the region adjacent to the fusion boundary, the formation of austenite should be completed when the temperature has exceeded the Ac_3 temperature by about 100°C. Since the peak temperature in this zone is much higher than Ac_3, the austenite grains coarsen rapidly as T_P is approached. In steels that are microalloyed, it may necessary for the grain boundary pinning particles (e.g., niobium carbonitrides) to dissolve before substantial grain coarsening occurs. In any case, once the coarsening begins, it proceeds rapidly because the effect of temperature increases exponentially during heating. The austenite grain growth can be expressed conveniently in the form of grain growth diagrams (Fig. 13.17b), which contain contours of equal grain size as a function of the peak temperature and $\Delta t_{800\text{-}500}$.

The importance of the coarse-grained austenite zone is in the mechanical properties that develop as the austenite transforms during the cooling part of the thermal cycle. The coarse grain structure leads to an increase in hardenability, because it becomes easier to avoid intermediate transformation products, so that untempered martensite or other hard phases can form during cooling. The welding process introduces atomic hydrogen into the weld metal, which is able to diffuse rapidly into the heat-affected zone. Hard microstructures are particularly susceptible to embrittlement by hydrogen, the fracture occurring shortly after the weld has cooled to room temperature. This hydrogen-induced phenomenon is called *cold-cracking*. This is why the carbon equivalent of the steel has to be kept low enough to prevent the hardness in the coarse-grained region from becoming unacceptably large.

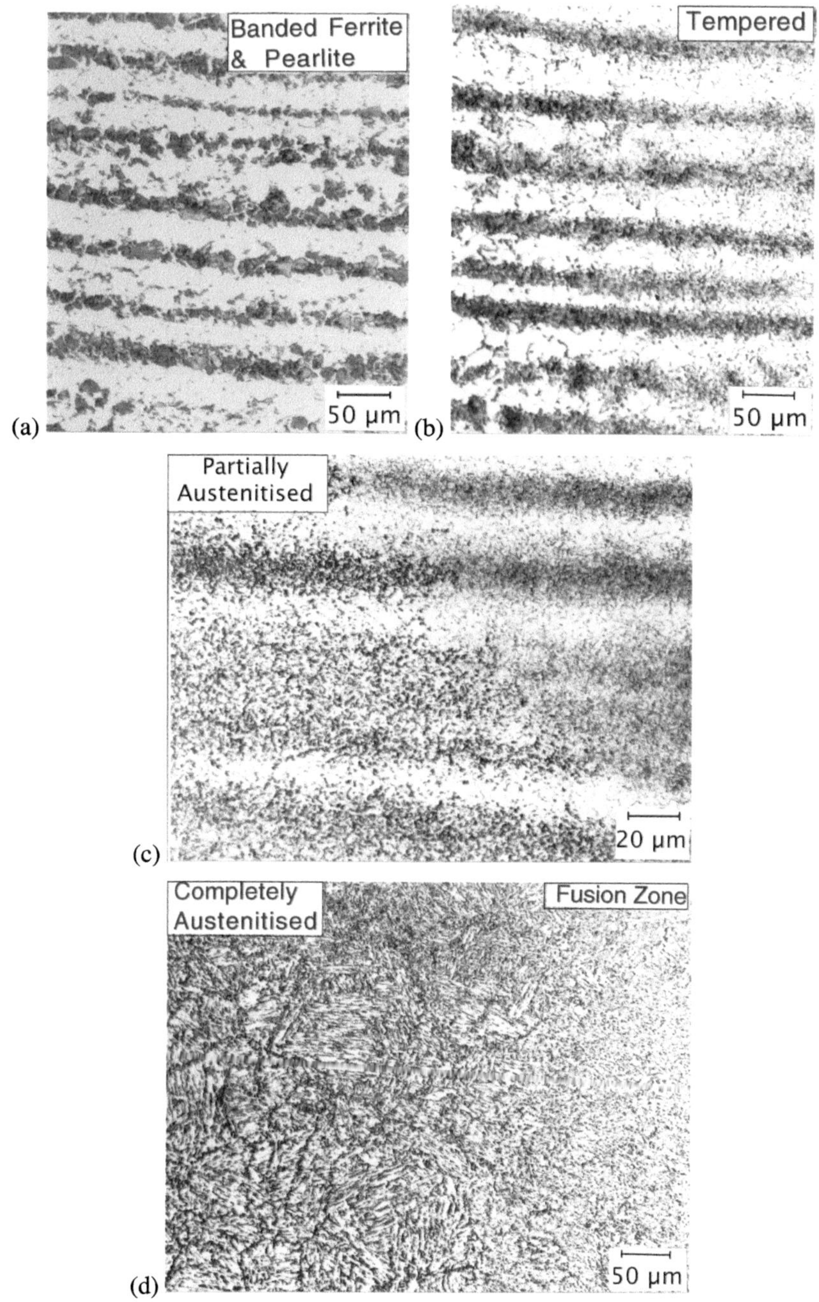

Figure 13.15 The gradient of microstructure in the HAZ of a mild steel plate (courtesy of C. Davis). (a) The plate microstructure far away from the weld, completely unaffected by welding. The bands of ferrite/pearlite are typical of many structural steels that are chemically inhomogeneous. (b) The tempered region. (c) The partially austenitised region. (d) The fully austenitised region, merging into the fusion zone.

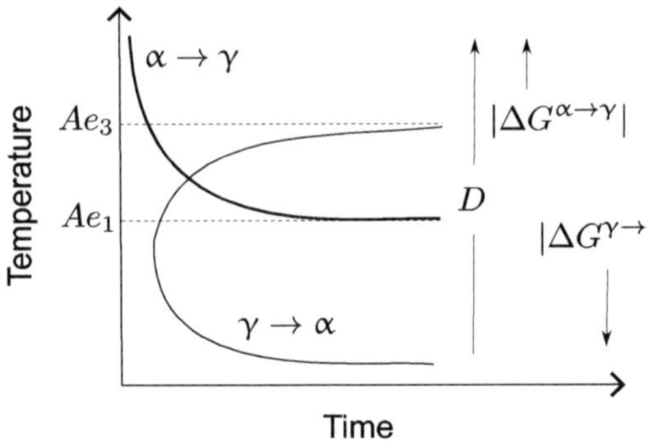

Figure 13.16 A comparison of the TTT curves for the $\gamma \rightarrow \alpha$ transformation, and for the reverse $\alpha \rightarrow \gamma$ transformation. ΔG represents the driving force for transformation and D is the diffusion coefficient.

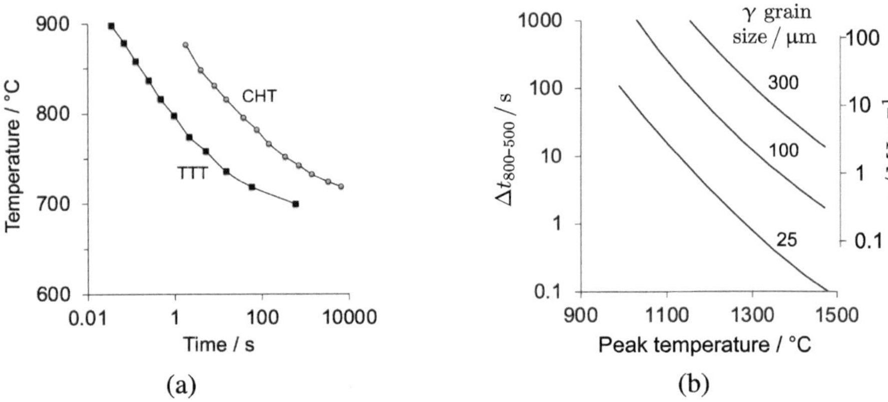

Figure 13.17 (a) The TTT and CHT diagrams for the beginning of austenite growth in a Fe-0.15C-0.5Si-1.5Mn wt% alloy (courtesy of S. Suzuki). (b) Schematic austenite grain growth diagram for the heat-affected zone of a microalloyed steel that is welded. Δt_{8-5} is the cooling time over the temperature range 800–500°C, and the peak temperature reached is plotted on the horizontal axis. The second vertical axis gives the welding heat input. It is assumed that the initial temperature of the steel is 200°C. The lines represent contours of constant austenite grain size (adapted from Ashby and Easterling [22]).

13.3.4 *Fine-grained austenite zone*

This region is typified by austenite grains some 20–40 μm in size. The grain structure and hardenability are, therefore, not very different from those associated with control-rolling operations during the manufacture of the steel. The fine austenite grains thus transform into more desirable ferritic phases with lower hardness values and higher toughness.

13.3.5 Partially austenitic regions and local brittle zones

At a sufficiently large distance from the fusion boundary, the peak temperature is such that the steel cannot transform completely into austenite. The small amount of austenite that does form has a larger carbon concentration. This is because the solubility of carbon in austenite, which is in equilibrium with ferrite, increases as the temperature decreases (Fig. 1.11). The subsequent transformation behaviour of this enriched austenite is then quite different, since it has a higher hardenability.

If the cooling rate is sufficiently large, then the carbon-enriched austenite transforms partially into hard martensite, the remaining austenite being retained to ambient temperature. These minute regions of hard martensite are known as *local brittle zones*. They are located in much softer surroundings consisting of tempered ferrite. Consequently, they do not cause a general reduction in toughness, but lead to an increase in the scatter associated with toughness tests. This is because the test specimen only sometimes samples the local brittle zone, in which case the recorded toughness can be poor. On other occasions, the measured toughness can be much higher, presumably because the test region does not include a local brittle zone. Such scatter in mechanical property data is not only disconcerting, but also makes design difficult because of the existence of a few low values.

When the cooling rate in this region is not high enough to induce martensitic transformation, the carbon-enriched austenite can decompose into a mixture of coarse cementite and ferrite. The cementite particles again constitute local brittle zones and increase the variability in mechanical properties.

13.4 Friction stir welding of steels

Friction stir welding is a solid–state, process [23–25] in which a rotating tool with a shoulder and terminating in a threaded pin, moves along the abutting surfaces of the two plates that are to be joined, Fig. 13.18. Heat generated by friction with the plates, at the shoulder and, to a lesser extent, at the pin surface, softens the material being welded. The flow of this plasticised metal as a result of tool rotation and translation accomplishes the joint. During the process, material is transported from the front of the tool to the trailing edge, where it is forged into a joint.

Since its invention in 1991 [23], friction stir welding has evolved as a technique of choice in the routine joining of aluminium components because the aluminium softens rapidly with heat, making the process easy to implement with steel tools. Its applications for joining difficult metals and metals other than aluminium are growing, albeit at a much slower pace. "Difficult" in this context means metals that do not plasticise easily because they are strong relative to the material used to make the tool. Steel, for example, is strong even when red-hot, but the stirring process requires it to be sufficiently plastic to enable a sound weld to be fabricated. Hence, expensive tools have to be used to survive the high temperatures, between 800–1200°C and severe wear [27], with the possibility that the wear debris becomes incorporated into the weld. Fig. 13.19a illustrates the problem.

The microstructural gradients expected in a friction-stir weld and its proximity are different from conventional welds because the process does not involve melting but

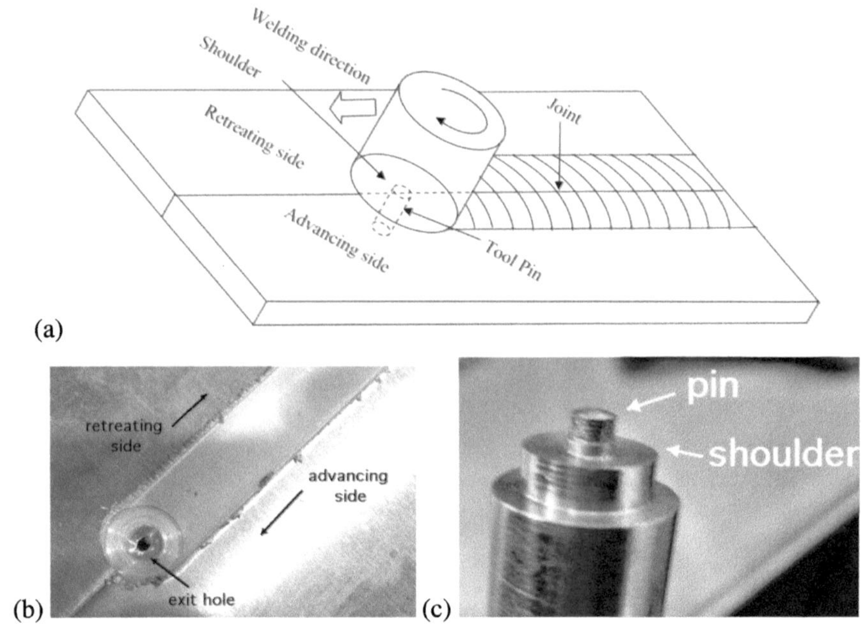

Figure 13.18 (a) The friction stir welding process. The advancing side has the direction of rotation along the welding direction. (b) A friction stir weld between aluminium sheets. (c) A tool with a threaded-pin. After Nandan et al. [26], reproduced with permission of Elsevier.

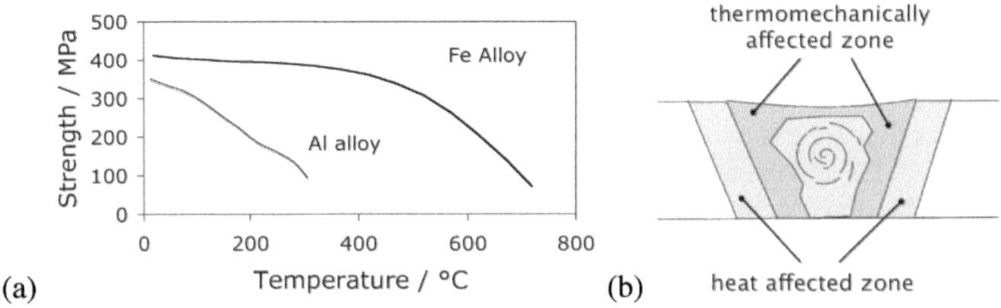

Figure 13.19 (a) Typical temperature dependence of the hot-strength of aluminium alloys and steels [28,29]. (b) Schematic of the microstructural zones expected in friction stir welds.

does include plastic deformation, which is intrinsic to its operation, Fig. 13.19b. The heat-affected zone, however, is as in conventional welds, except that the peak temperatures reached are lower than those associated with fusion welding. Furthermore, the heat-affected zone tends to extend less into the parent plate and the austenite grains created by the heat are relatively fine. The central nugget region containing the flow-pattern, the so-called onion rings, is the most severely deformed region, although it frequently seems to dynamically recrystallise, so that the detailed microstructure may consist of equiaxed grains of austenite that transform on cooling. The layered (onion-ring) structure is a consequence of the way in which a threaded tool deposits material from the front to the back of the weld. It seems that cylindrical sheets of material are extruded during each rotation of the tool, which on a weld cross–section give the characteristic onion-rings.

The thermomechanically-affected zone lies between the HAZ and nugget; it too becomes fully austenitic and, despite the deformation, for most steels, the austenite grain structure is of a recrystallised nature. The top surface of the weld has a different microstructure, a consequence of the shearing induced by the rotating tool-shoulder. Fig. 13.20 shows the variations in microstructure for a low-alloy steel, with banding retained in the partially austenitised region because the austenite forms in the manganese-rich regions of the chemically segregated steel. The thermomechanically affected zone shows a fine microstructure when compared with the stir zone, which reaches higher temperatures. Further details and dependence on alloy composition are discussed in [26].

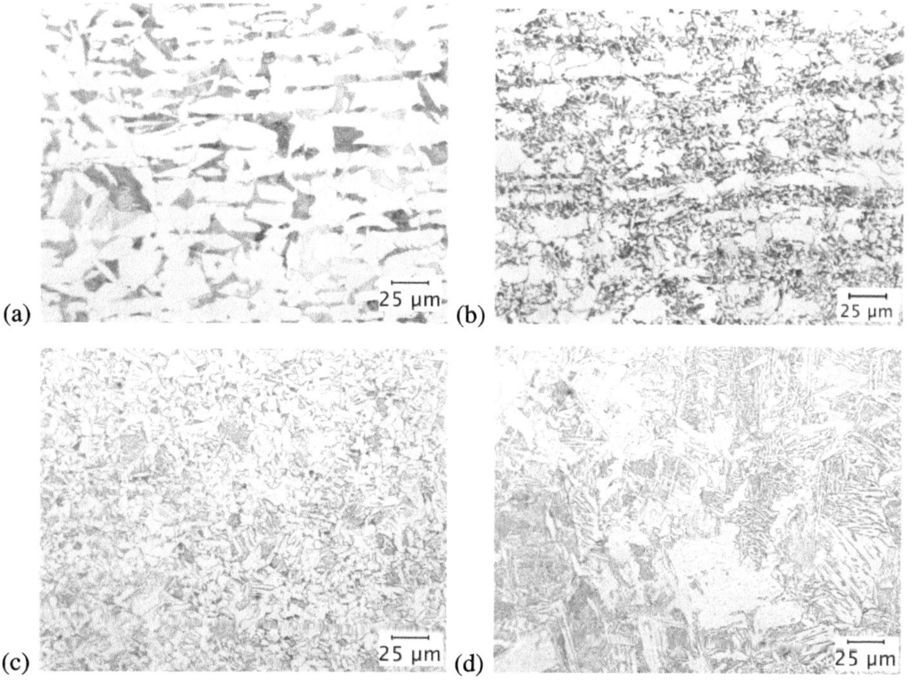

Figure 13.20 Typical microstructure of an FSW weld in a C–Mn steel. (a) Parent steel, with bands of allotriomorphic ferrite and pearlite. (b) Intercritically heated region of HAZ. (c) TMAZ (fine–grained region near stir zone). (d) Stir zone. Micrographs courtesy of Threadgill and Johnson [30].

It is likely that the friction stir welding method will be useful for steels when circumstances arise where conventional methods, which continue to serve so well, cannot be applied. This might include specialised applications [31]. The process is offered to join steel strip-ends before being formed into a coiled tube [32]. The enhanced fatigue properties, reduced susceptibility to galvanic corrosion due to the absence of a filler, and the ability to join difficult-to-weld steels is the basis for the application of friction stir welding in this instance. A variation of the process is applied to steel cutting edges to improve the chipping resistance [33].

For large scale applications, the economic viability of the process has not been demonstrated. An analysis [34] indicates that the cost of the tool (£$\approx$ 2900) would require it to survive a weld length of 120 m to achieve even a 20% reduction in production cost; present tools are not capable of achieving this [27,35]. The friction stir welding of ship steels reduces distortion, but current distortion levels with arc welding are manageable [36]. There are numerous recent reviews on the subject [37–39] with the focus on experiments rather than actual applications.

13.5 Summary

Steels that cannot be welded because they are too strong or if their composition results in the production of brittle phases on the application of heat have their uses, for example, in the manufacture of shafts, bearings, gears and armour. However, the ability to weld vastly increases the possibilities, so it is not surprising that non-weldable steels occupy a hardly noticeable proportion of the total volume of steels manufactured annually.

Wrought steels are produced using thermomechanical processing and careful heat treatment to ensure their properties. However, weld metal usually has to achieve matching or better properties in the cast state. This places limitations on the concentrations and variety of solutes that can be added to weld metal. Impurities such as oxygen are present in most weld metals at greater concentrations than in the steel being joined. Some of the resultant non-metallic particles turn out to be useful in acting as intragranular nucleation sites that stimulate acicular ferrite, which is good for toughness.

Although friction welding, which does not involve melting, has been known for some time, the radically different friction stir welding process has revolutionised the joining of aluminium and its alloys. Whether this process will prove as useful for steels remains to be seen.

References

[1] M.S. Węglowski, S. Błacha, A. Phillips, Electron beam welding-techniques and trends – review, Vacuum 130 (2016) 72–92.

[2] J.F. Lancaster, Metallurgy of Welding, 6th ed., Woodhead Publishing Ltd, Abingdon, U.K., 1999.

[3] S. Kou, Welding Metallurgy, 2nd ed., John Wiley & Sons, Inc., New Jersey, USA, 2020.

[4] E.M. Westin, Hot cracking in duplex stainless steel weldments – a review, Welding in the World 66 (2022) 1483–1499.

[5] G.J. Davies, J.G. Garland, Solidification structures and properties of fusion welds, International Metallurgical Reviews 20 (1975) 83–106.

[6] K.E. Easterling, Solidification microstructure of fusion welds, Materials Science & Engineering A 65 (1984) 191–198.

[7] Ø. Grong, Metallurgical Modelling of Welding, 2nd ed., Maney, London, 1997.

[8] T. Haxhimali, A. Karma, F. Gonzales, M. Rappaz, Orientation selection in dendritic evolution, Nature Materials 5 (2006) 660–664.
[9] M. Rappaz, S. David, J. Vitek, L. Boatner, Analysis of solidification microstructures in Fe-Ni-Cr single-crystal welds, Metallurgical Transactions A 21 (1990) 1767–1782.
[10] B. Gretoft, L.-E. Svensson, H.K.D.H. Bhadeshia, Austenite grain structure of low alloy steel weld deposits, Journal of Materials Science 21 (1986) 3947–3951.
[11] H. Fredriksson, J. Stjerndahl, Solidification of iron-base alloys, Metal Science 16 (1982) 575–586.
[12] D.J. Abson, The role of inclusions in controlling weld metal microstructures in C-Mn steels, Research Report 69/1978/M, The Welding Institute, Abingdon, U.K., 1978.
[13] L.-E. Svensson, B. Gretoft, H.K.D.H. Bhadeshia, Analysis of cooling curves from the fusion zone of weld deposits, Scandanavian Journal of Metallurgy 15 (1986) 97–103.
[14] H.K.D.H. Bhadeshia, L.-E. Svensson, B. Gretoft, Model for the development of microstructure in low alloy steel (Fe-Mn-Si-C) weld deposits, Acta Metallurgica 33 (1985) 1271–1283.
[15] S. Jones, H.K.D.H. Bhadeshia, Competitive formation of inter- and intragranularly nucleated ferrite, Metallurgical & Materials Transactions A 28A (1997) 2005–2103.
[16] H.K.D.H. Bhadeshia, Reliability of weld microstructure and property calculations, Welding Journal, Research Supplement 83 (2004) 237s–243s.
[17] H.K.D.H. Bhadeshia, Local brittle zones and the role of niobium, Materials Science Forum 783–786 (2014) 2129–2135.
[18] J. Neves, A. Loureiro, Fracture toughness of welds – effect of brittle zones and strength mismatch, Journal of Materials Processing Technology 153–154 (2004) 537–543.
[19] H.K.D.H. Bhadeshia, D.V. Edmonds, Bainite in silicon steels: a new composition-property approach I, Metal Science 17 (1983) 411–419.
[20] H.K.D.H. Bhadeshia, L.-E. Svensson, Modelling the evolution of microstructure in steel weld metals, in: H. Cerjak, K.E. Easterling (Eds.), Mathematical Modelling of Weld Phenomena, vol. 1, The Institute of Materials, London, 1993, pp. 109–182.
[21] Anonymous, Welding Handbook, Fundamentals of Welding, vol. 1, 7th edition, American Welding Society, Miami, Florida, 1981.
[22] M.F. Ashby, K.E. Easterling, Diagrams for grain growth in welds, Acta Metallurgica 30 (1982) 1969–1978.
[23] W.M. Thomas, E.D. Nicholas, J.C. Needham, M.G. Murch, P. Temple-Smith, C.J. Dawes, Friction stir butt welding, International Patent Application no. PCT/GB92/02203, 1991.
[24] C.J. Dawes, W.M. Thomas, Friction stir process welds aluminium alloys, Welding Journal 73 (1996) 41–45.
[25] W.M. Thomas, R.E. Dolby, Friction stir welding developments, in: S.A. David, T. DebRoy, J.C. Lippold, H.B. Smartt, J.M. Vitek (Eds.), 6th Int. Trends in Welding Research, ASM International, Materials Park, Ohio, USA, 2003, pp. 203–211.
[26] R. Nandan, T. DebRoy, H.K.D.H. Bhadeshia, Recent advances in friction-stir welding – process, weldment structure and properties, Progress in Materials Science 53 (2008) 980–1023, https://doi.org/10.1016/j.pmatsci.2008.05.001.
[27] A. De, H.K.D.H. Bhadeshia, T. DebRoy, Friction stir welding of mild steel: tool durability and steel microstructure, Materials Science and Technology 30 (2014) 1050–1056.
[28] W.S. Miller, I.G. Plamer, Rapid solidification technology for high-T aluminium alloys, MRS Europe (November 1985) 117–137.
[29] R. Dimitriu, H.K.D.H. Bhadeshia, Hot-strength of creep-resistant ferritic steels and relationship to creep-rupture data, Materials Science and Technology 23 (2007) 1127–1131.
[30] P.L. Threadgill, R. Johnson, Progress in friction stir welding of steels, Technical Report 815/2004, TWI (The Welding Institute), Great Abington, Cambridge, U.K., 2004.

[31] H.K.D.H. Bhadeshia, T. DebRoy, Critical assessment: friction stir welding of steels, Science and Technology of Welding and Joining 14 (2009) 193–196.
[32] J.D. Dubois, Method of manufacturing coil tubing using friction stir welding, Technical Report US20140021244, Global Tubing LLC, 2009.
[33] C.E. Allen, R.A. Flak, S.M. Packer, R.J. Steel, H. Smith, Solid state processing of hand-held knife blades to improve blade performance, Technical Report US8186561, Megastir Technologies, LLC, 2004.
[34] B.G. Amancio, J.F. dos Santos, Solving problems by the use of friction stir, Technical Report W/FT/023.07, European Union Research Fund for Coal and Steel, Brussels, Belgium, 2007.
[35] R. Rai, A. De, H.K.D.H. Bhadeshia, T. DebRoy, Friction stir welding tools, Science and Technology of Welding and Joining 16 (2011) 325–342.
[36] N. McPherson, Private communication to H.K.D.H. Bhadeshia, 2016, Friction stir welding.
[37] M.O. Crăcănel, E. Niţu, D.M. Iordache, Friction stir welding of steel structures — a brief review, Key Engineering Materials 890 (2021) 105–119.
[38] D.G. Mohan, C.S. Wu, A review on friction stir welding of steels, Chinese Journal of Mechanical Engineering 34 (2021) 1–29.
[39] M.M. Ahmed, M.M. El-Sayed Seleman, K. Touileb, I. Albaijan, M.I. Habba, Microstructure, crystallographic texture, and mechanical properties of friction stir welded mild steel for shipbuilding applications, Materials 15 (2022) 2905.

Additive manufacture

14

Abstract

An engineering component is conceived for a function that requires it to satisfy a bank of properties, such as shape, size, weight, aesthetic appeal, structural integrity over the perceived service life, cost, etc. A prototype then has to be made and assessed. If all is well, then a method of manufacture is designed to meet the demand. This entire process can be accelerated by rapid prototyping, in which the component evolves as layers of appropriate geometry are sequentially deposited to create the three-dimensional form. The method has now been renamed *additive manufacturing*, in which the intention is that the printing process represents the final stage of manufacture, not simply prototyping. The evolution of a component by printing permits complex shapes, sometimes with re-entrant angles, to be made more conveniently by altering the shape of each layer that is deposited. The process when applied to metals is necessarily slow and energy intensive. The component may require finishing operations and structural integrity may be affected by the presence of defects introduced during deposition. The repeated passage of a moving heat source means that the temperature in the evolving component is not uniform, leading to residual stresses and distortion. It is important therefore, to focus this promising technology in scenarios where manufacture by conventional methods is not possible.

14.1 Introduction

It usually is necessary to explore the practicality of a component or machine at the design stage, but a successful design is seldom achieved at the first attempt, given the compromises that must be made during planning in order to satisfy conflicting requirements. Prototypes must be created and tested to achieve the best engineering and aesthetic form, but this can be expensive and time-consuming.

The idea of rapid prototyping may have emerged first in software engineering, for code validation by an iterative process that is quick and cheap [1]. In industry, it is the need to rapidly develop products that led to the creation of geometrically complex objects using solid-modelling, that, when coupled to an appropriate robotic-machine, allow the item to be built using layer-by-layer deposition [2].

Rapid prototyping now has morphed into *additive manufacturing* where the objective is not simply to produce a preliminary version of a device, but to manufacture relatively large numbers of components destined for service. This is most difficult to achieve with metals, and steel is no exception. The content of this chapter describes the limited success to date and possible progress towards greater goals. A great deal of development in this subject is directed at process optimisation rather than the materials science because of the difficulties in achieving the intended outcome. It is not the intention here to delve into process parameters, but rather, to focus on metallurgical aspects. A few characteristics of the additive process are as follows:

- It is possible to produce objects that cannot otherwise be made because of re-entrant angles, internal surfaces, designed variations in density etc. Components can be redesigned to take advantage of these capabilities in order to improve performance, reduce the number of parts needed, which may improve reliability, or

Steels. https://doi.org/10.1016/B978-0-44-318491-8.00020-4

to save weight. Fig. 14.1a illustrates a conventional component (puller and punch) used to trim/cut 1 mm thick automotive dual-phase steel sheets. It is made using maraging steel. After the design is optimised for additive manufacture, Fig. 14.1b shows the component serving the same function but at a 70% reduction in weight, again made using maraging steel. The process used laser-based powder bed fusion, with a layer thickness of 30–40 μm [3]. The key therefore, is in recognising how the method can be used for the better engineering of specific items.

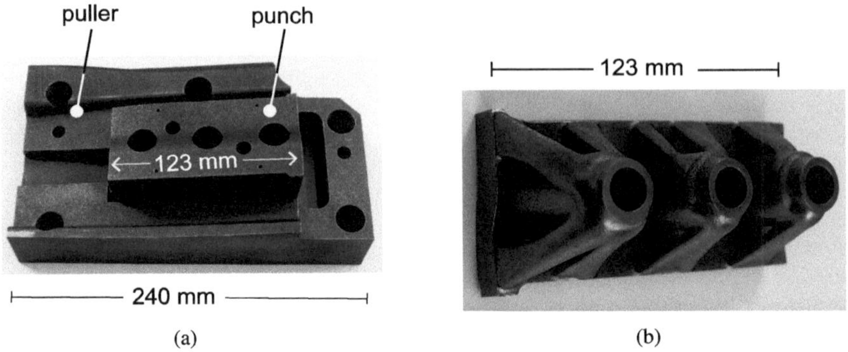

Figure 14.1 (a) Puller and punch assembly. (b) The modified punch, serving the same function as in (a), but with a more complicated shape, because additive manufacturing permitted the manufacture of a stiffer design that is better for the purpose. Images adapted from Asnafi et al. [3,4], under the CC BY 4.0 license, http://creativecommons.org/licenses/by/4.0/. Tools like this are now qualified for use, i.e., at technology readiness level 8.

- The method in principle permits variations in heat input and the grading of chemical composition within the component. The latter aspect is particularly important and already has seen an application in the manufacture of a nozzle.
- The process is energy intensive when compared against mass manufacture and its productivity is necessarily low. But these criticisms may not be relevant if the overall cost of an application is reduced because fewer parts have to be made, or if a new technology emerges due to the method of manufacture.
- It is easy with additive manufacturing to change the design as frequently as necessary. One example is the steel node designed for 3D-printing, illustrated in Fig. 14.2a, for use in the construction industry. By optimising the complex design, the node is under half the normal size because material can be removed in regions where it is not needed, and added where the loading is greater [5]. The emphasis on design rather than making things constrained by the size and shape of available materials has spawned an industry that makes artistic objects not previously possible, for example, a decorative bowl printed in stainless steel and bronze (Fig. 14.2b).

The additive manufacturing method may involve the controlled scanning of a laser over powder that melts or is sintered by this action, the removal of unscanned powder, the placing of another layer of powder for scanning, and repeating this process until the object has evolved into its final form – this is the laser-based powder-bed fusion

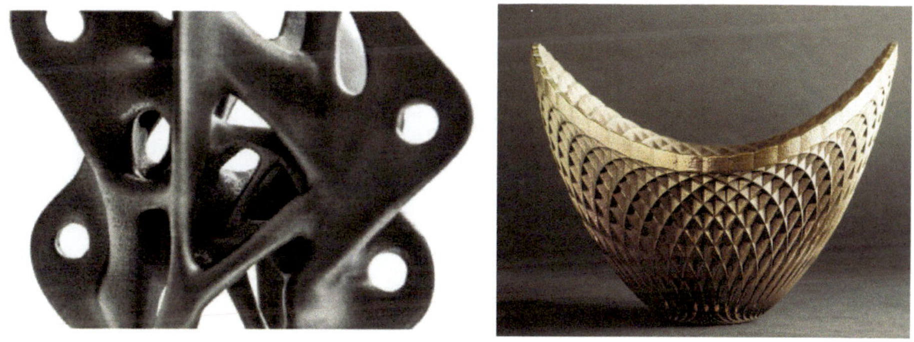

(a) (b)

Figure 14.2 (a) Structural-steel joint about 14 cm high to be used as a lightweight node for joining steel members in a frame. Image reproduced from Paoletti [5] with the permission of Elsevier. (b) Stainless steel and bronze bowl designed by Carl Bass, made using ink-jet binder printing. Image courtesy of Carl Bass.

process. An alternative laser process is illustrated schematically in Fig. 14.3a, in which the metal powder is fed coaxially with the laser as it moves. An example of the size and shape of the powder used to make a duplex stainless-steel object is illustrated in Fig. 14.3b [6]. This laser-based method has a greater build-rate and can be used for larger components.

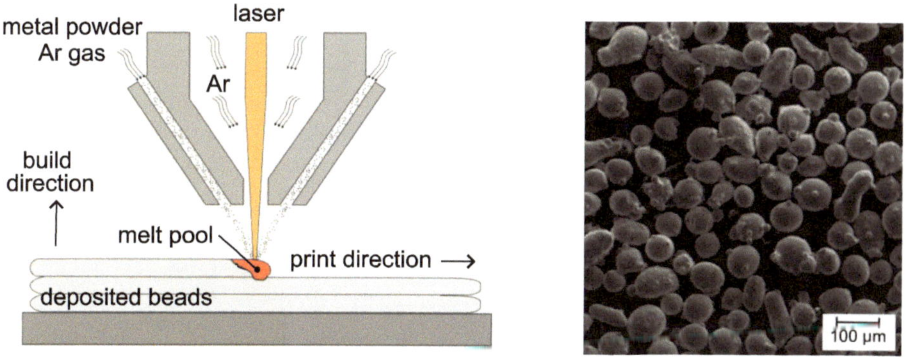

(a) (b)

Figure 14.3 (a) Laser metal-powder directed energy deposition. (b) An illustration of the duplex stainless-steel powder used in a laser additive process. Figures from Edwards et al. [6], with permission of Elsevier.

Another technique utilises an electric arc "such as is ordinarily employed for electric welding, for the formation of ... useful shapes." [7] While the arc is the heat source, a metallic wire can be used as the continuous feedstock. The arc and wire combination are then moved by a robot according to the numerical solid-model implemented within the control system. A conventional arc-welding facility therefore suffices. Such a facility can be used for the production of larger scale objects. The arc welding process itself is illustrated on Fig. 13.1.

There now exist very many processes for additive manufacture, but a comparison between the laser process, in which powder is fed coaxially, and the arc method is useful, though the numbers should be taken as a rough guide, because there are many variables that can be manipulated to alter the values, though not independently:

	laser	arc
Layer height	150-200 μm	1-2 mm
Lateral object-dimensions	350 cm	metres
Surface roughness	0.1-0.3 mm	0.5 mm
Deposition rate (steel)	$2\,\mathrm{kg\,h^{-1}}$	$5\,\mathrm{kg\,h^{-1}}$

It follows that neither of these processes can strictly be regarded as producing components that do not require further preparation, such as surface smoothing. The surface roughness can also determine, for example, the finest cooling channels that can be implemented in additively manufactured components.

The layer height in the case of the arc process depends on the wire feed-speed, the welding speed and electric arc parameters, all of which can be reasonably controlled to give a consistent deposit with little spatter. The height associated with the laser process is determined by both the powder size and the energy input from the laser.

14.2 Macrostructure

There are various scales of heterogeneity in additively manufactured components. Even a single deposited-layer is not uniform, because it consists of a series of parallel strips, each with a width depending on the lateral temperature-distribution of the heat source. Fabricating another layer, which again is a combination of parallel strips, leads to a myriad of fusion and heat-affected zones – the appearance therefore resembles the multipass weld illustrated in Fig. 13.5, if the plates on the sides are ignored.

Fig. 14.4 illustrates a solid object made using the wire-arc process. It is not a component, but rather, an item made to investigate the metallurgy of the multilayer deposit [8–10]. However, it does illustrate many of the features that would be important for a technological application. The object has a rough surface that undulates with the orientation of the deposition direction, reflecting the different layers. Additive manufacturing generally does not create an object that is close to the intended dimensions. This is in contrast to the mass production of routine items, such as rails, which are made by hot-rolling a slab of steel into the required shape, after which there is no further processing other than straightening. The reason straightening is required is because as the rail cools, it does not do so uniformly, leading to residual stresses that cause distortion. The same applies to metal that is deposited. Fig. 14.4a shows that the base plate on which the additively manufactured block was made is significantly curved by distortion along the deposition direction.

This kind of distortion is generic to all additive manufacturing and is a significant problem in achieving the designed shape of the component. In 316L stainless steel, which does not undergo phase transformations in the solid state, the residual stress in a multi-pass deposit due to non-uniform cooling is reported to be 500 MPa, which exceeds its yield strength. In some cases where ductility is limited, the component can crack or even delaminate by fracture between the layers. The magnitudes, sign and spatial variation of such stresses can depend, in principle, on the dimensions of the object being made and the specific method of deposition.

Figure 14.4 Wire arc additively manufactured solid made using wire of composition Fe-0.01C-1.8Mn-0.8Si-18.5Cr-12.2Ni-2.5Mo wt%, a version of 316L austenitic stainless steel with a higher silicon content for deoxidation during welding. (a) Macroscopic image showing the object deposited on a flat plate of 316L stainless steel. Notice how the originally flat plate is now visibly distorted. Some spatter is also noticeable, but in this case it is minimal when compared with the size of the object. Image courtesy of M. Geers [8]. (b) Schematic diagram showing the intended dimensions and idealised shape of the solid deposit and the strip and layer widths.

In ordinary welding, it has long been known that the heterogeneous application of heat and associated localised thermal contraction causes the development of residual stresses, but that this can be negated by expansion due to phase transformation in the solid state [11–13]. The phase transformation should occur at a low temperature, because once it is exhausted, the thermal contraction kicks in again to raise the residual stress. Therefore, martensitic transformation is most useful in this respect. This was exploited in the deposition of a tool made using a steel "H13" ≈Fe-0.4C-5Cr-1.4Mo-1V-1Si wt%, which is used in making hot or cold working dies and molds. It was produced additively in the form of a wedge, using the coaxial laser-powder process with the motivation of reducing machining costs. A 4 mm depth of compressive stress developed on the surface, when the retained austenite transformed into martensite at about 300°C, making it in principle ideal for the die application [14]. It is worth emphasising that this represents an experimental investigation – there are no reports of an actual die made and used in anger.

A comparison of the deposit with the geometry shown in Fig. 14.4b shows that the idealised shape is not obtained in practice, because the conditions at the start and end of each strip deposit are different from those in-between, so the corners and edges become rounded. For a real manufacturing process, the numerical design will require modification to compensate for these kinds of effects so that the final object has the desired shape. In most cases, this is done by trial and error, but continuous monitoring and feedback during manufacture can help.

The methods of additive manufacture can also be used for repairs, but some sort of machining is required to ensure the exact geometry required is obtained in the finished product. Dies made from the H13 maraging tool steel by conventional manufacturing methods, have been repaired successfully using the laser-based additive manufacturing method, Fig. 14.5 [15]. The cost of the die represents some 10-30% of the total price of the forged component. The repaired die was able to produce more forged parts than the original. It should be understood that repair in this manner is not a new concept – weld deposits of various kinds have in the past served this function, for example, on turbine blades or the worn-out teeth of an excavator. Additive enhancement is simply a new name for an established method.

14.3 Powders

The powders used in additive processes are produced by *atomisation*, where a stream of molten metal merges with a jet of a gas or water. This breaks up the stream into droplets that solidify rapidly. The majority of research on the power-based additive manufacturing of steel uses readily available 316L gas-atomised austenitic stainless steel spheroidal-particles in the size range 30–80 μm, Table 14.1. Water-atomised particles tend to be rougher (Fig. 14.6), which reduces their ability to flow. Too fine a particle size also reduces the ease of flow. Good powder-flow is important in ensuring that the manufactured product has a density in excess of 99%, although some porosity is inevitable.

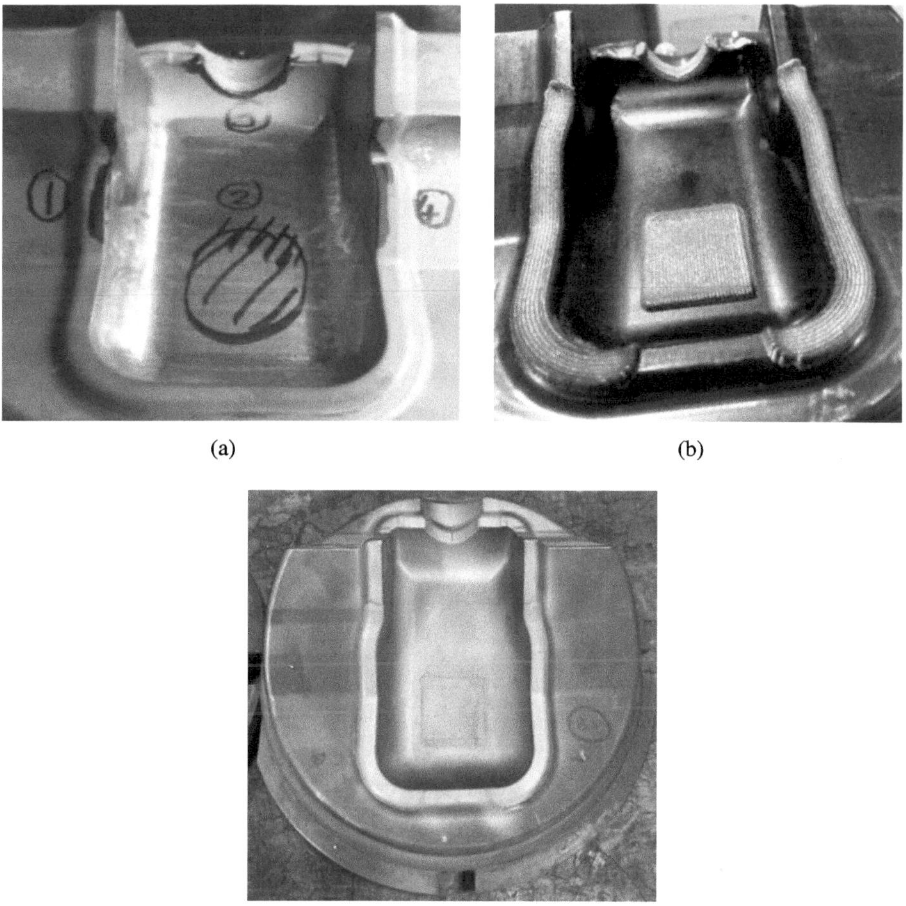

(a) (b)

(c)

Figure 14.5 Hot-forging die, made using the H13 steel in a nitrided condition. (a) Damaged regions marked. (b) After laser metal deposition using a cobalt-based alloy. (c) After machining. Images adapted from Asnafi [15], under the CC BY 4.0 license, http://creativecommons.org/licenses/by/4.0/.

14.4 Thermomechanical processing of additively manufactured steel

It is evident that additive manufacturing can enable convoluted components to be made by progressively depositing thin layers of metal, each with a specific two-dimensional geometry that can change in the subsequent overlays to generate the intended three-dimensional form. It may not in some cases be possible to make such objects by any method other than joining many different components together [19]. The complexity of the objects that can be produced is illustrated in Fig. 14.7. Nevertheless, depending on engineering requirements, the final product may lack structural integrity given

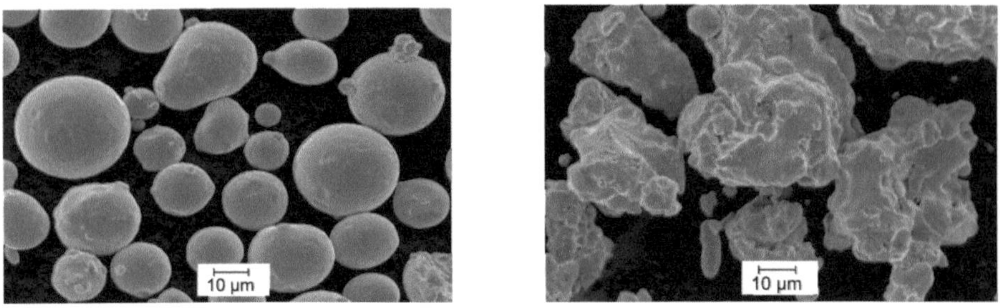

(a) (b)

Figure 14.6 Morphology of 316L austenitic stainless steel powder. (a) Produced by gas atomisation. (b) Produced by water atomisation. Images reproduced from Hoeges et al. [16], with the permission of Elsevier.

Table 14.1 Approximate chemical compositions (wt%) of some of the powders used in the additive manufacture of steels. The powders are likely to contain oxygen concentrations of the order of 0.02 wt%. With the duplex stainless steel, it is interesting that the nitrogen concentration remains intact on deposition [17].

	C	Si	Mn	Cr	Ni	Mo	V	Others	Ref.
L6 martensitic	0.52	0.28	0.96	1.99	0.50	0.11			[18]
316L austenitic	0.02	0.8	0.95	17.3	13.2	2.5			[18]
Duplex stainless	0.03	0.8	1.2	25	7	4		N 0.3, Cu 0.5	[6]
A11 wear resistant	2.45	0.98	0.42	5		1.49	8.28		[18]
H13 tool steel	0.4	1		5		1.4	1		

that it is created in a stepwise process that is less than perfect. A part produced to a specific shape may then be subjected to isostatic pressing or other more dramatic and shape-changing processes such as forging, rolling or extrusion [20].

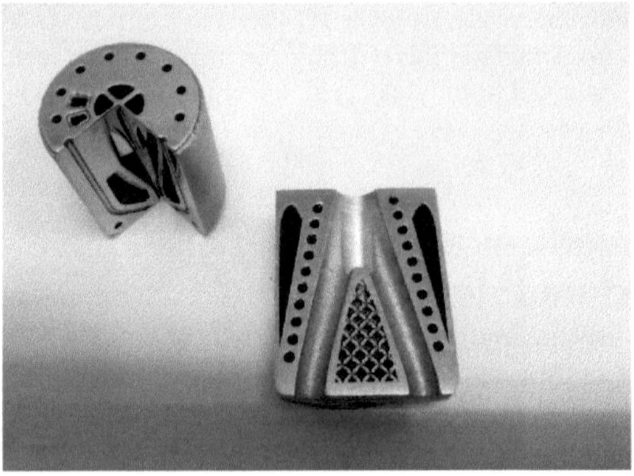

Figure 14.7 Metallic components produced additively at KU Leuven, illustrating the complexity of the internal structure that can be generated using this process.

Additive manufactured objects usually contain porosity, laying defects, surface roughness, delamination, chemical segregation, residual stresses and macroscopic anisotropy induced by the sequential deposition method. All of these lead to significant scatter in measured properties. In some cases, the defects may not matter because the component has sufficient structural integrity to cope with service conditions. However, in general, additional thermomechanical treatment may be required to densify the object or enhance other properties.

Isostatic pressing is often used to reduce porosity and even heal cracks. In principle, this process does not change the shape of the object, which is useful if the additive manufacturing is designed to generate the ultimate shape of the component. The isostatic pressing of steel is usually conducted using 100 MPa pressure at 1150°C for 3 h. The high temperature can help homogenisation to a limited extent, but it may be necessary to implement an additional heat treatment after the isostatic pressing to restore the intended properties. Other aspects of the structure approach equilibrium, for example, the metastable $MnSiO_3$ and amorphous inclusions trapped during the additive manufacture of duplex stainless steel change into the equilibrium MnS [17].

Hot isostatic pressing is not effective in sealing pores that have continuity through the thickness of a thin wall ($\approx$ 1 mm), even when thicker parts become fully densified [18]. After all, the pressure would act also on the exposed surfaces of such pores.

A component additively manufactured, that does not have to have its intended final dimensions can be subjected to deformation to attain the size and shape required. Fig. 14.8 shows a wire-arc additively manufactured cylinder and another one after surface preparation, ready for a rolling process that creates from these objects, a larger diameter ring. This ring-rolling is used frequently in industry, for example, to produce bearing raceways. The steel in the case illustrated has the composition $\approx$Fe-0.08C-0.07Si-0.9Mn wt%, heated to 1100°C prior to deformation [21].

During ordinary ring-rolling, a cylindrical shape is pressed into a short, fat cylinder, which then is profiled and pierced before being subjected to deformation by contact with rotating rolls on either side of the cylinder wall (Fig. 14.8c). This leads to thinning and an increase in the diameter of the seamless ring. Using the additively made preforms illustrated in Fig. 14.8a,b, many of these operations are avoided. After the rolling operation was completed in a few seconds, the component was heat treated at 900°C 30 min to refine the ferrite and pearlite microstructure that is obtained on cooling in air.

It is found that properties of the final product did not depend on the surface finish following deposition, so a smooth surface is not necessary in this case, because of the intensity of the plastic deformation involved during ring rolling. Welding defects featured prominently during tensile testing, but because the welding process itself is well established, it is likely that such problems are easy to overcome.

A significant disadvantage is that the time required to produce each preform by weld deposition was about 4 h, whereas the ring rolling was completed within a minute. This means that the productivity is dramatically reduced, so it is likely that the process would be used either for special applications where the number of rings is limited, for special ring-profiles and when the alloy required is for a niche application. The latter aspect is important because welding wires, especially those that are cored with powder, can easily be made to deposit required chemical compositions. Small quantities of special alloys may be impractical or expensive to make using conventional casting processes.

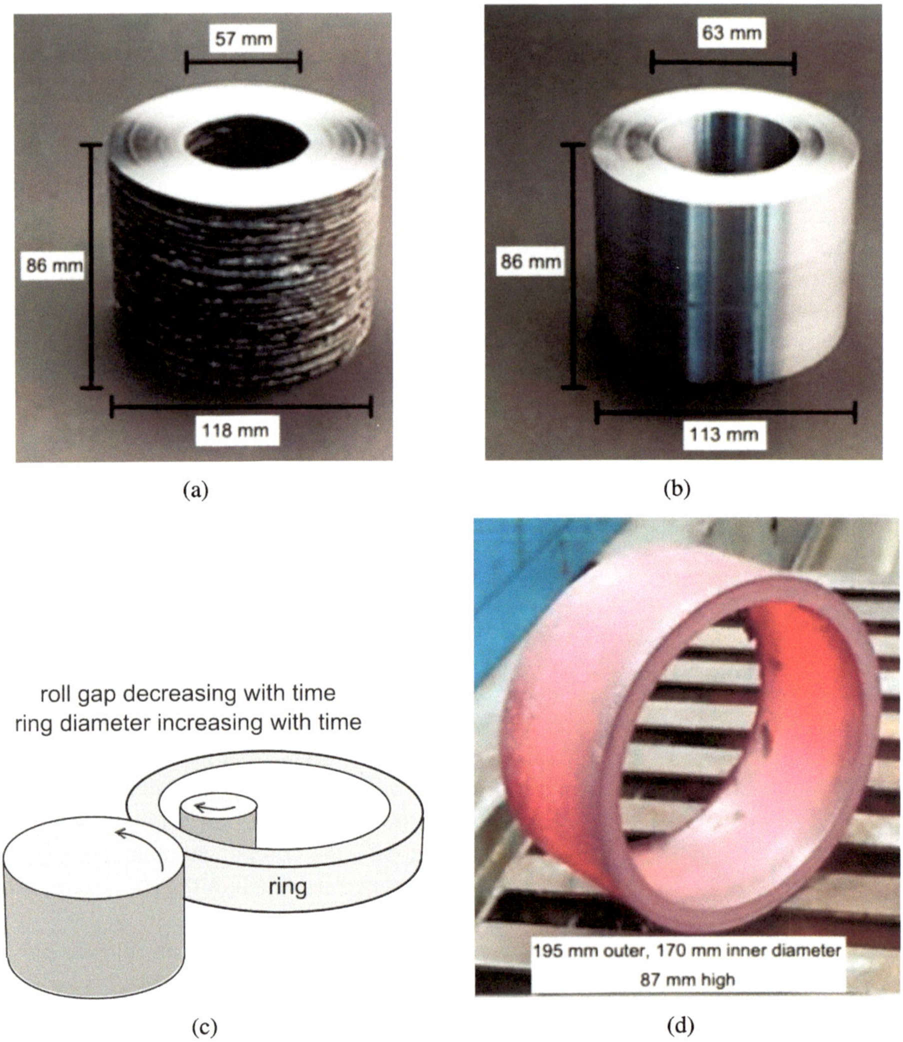

Figure 14.8 (a) Wire arc additively manufactured cylinder, simply cut to length. (b) The cylinder after surface preparation. (c) Simplified diagram of the ring-rolling operation, with the two rollers pushing against each other. (d) After completion of ring rolling. Images adapted from Michl et al. [21] with permission of Elsevier.

14.5 Hybrid processing

It is evident that few additively made components can be used directly without some sort of thermomechanical processing. There is, therefore, a move to integrate additive manufacturing machines with tools that can enhance the process by simultaneous computer-controlled machining or other methods that involve material removal or surface treatment. It then becomes possible to assemble a production line where entire components or parts of an assembled component can be made using the additive method combined with some secondary process, which alleviates any shortcomings.

For example, the integration of the deposition process with numerically controlled milling [22] would make some deposited objects amenable to mass production. In such a process [23], a wire-arc deposited layer is milled before a further layer is deposited in order to provide a clean base for subsequent layers and to maintain a uniform layer-thickness. A final milling gives the accurate profile of the completed part. It has been shown that this reduces the manufacturing time for dies. Sometimes, milling and turning are combined as components of the hybrid process.

The idea of hybrid manufacturing has advanced to the point where integrated additive and "subtractive" machines are commercially available and have been demonstrated to be capable of producing steel jet-engine nozzles, blisks, milling tools, etc., with a high quality of finish, both in terms of shape and roughness [24]. The adjective *subtractive* simply emphasises machining. Fig. 14.9 shows how a laser and a milling tool can be combined in the production process to ensure smooth channels in the final product. As each layer is completed, the computer-controlled milling tool smooths each of the vertical faces to the required precision. This results in a nicely finished three-dimensional article, containing well-defined orifices and channels, that is ready for use when the process is finished.

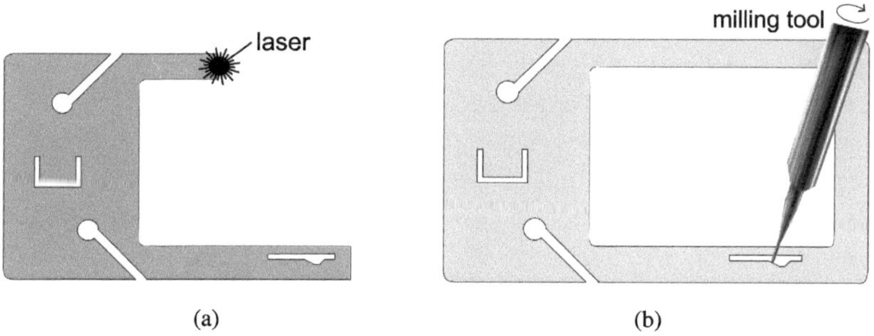

Figure 14.9 (a) Laser causing the fusion of metal in powder layer (not illustrated), to produce a solid layer containing features which when reproduced by depositing further layers, would build in the desired channels. (b) When each layer is completed, a milling tool machines all the vertical faces to the required accuracy before another layer is deposited.

The different ways in which hybrid manufacture has been advantageously implemented for steel components are listed in Table 14.2. In each case, the secondary process describes the treatment implemented after the deposition of each layer.

Table 14.2 A few examples illustrating the combination of processes used to produce better three-dimensional objects that are deposited using additive methods.

Heat source	Secondary process	Materials	Ref.
Laser	Ultrasonic forging	Martensitic steel	[25]
Laser	Cold metal transfer using arc	Austenitic stainless	[26]
Laser	Milling	Martensitic steel	[27]
Laser	Milling, ultrasonic rolling	Mold steel	[28]
Arc	Milling, forming	Stainless steel	[29]
Arc	High-pressure rolling	Mild steel	[30]
Arc	Laser-marking of surface	Stainless steel	[31]
Ink-jet binder	Copper infiltration	Stainless steel	[32]
Ink-jet binder	Iron infiltration	Titanium carbide	[33]

14.6 Microstructure

Solidification in circumstances where the liquid is in contact with a solid-substrate of identical or similar composition, and in the presence of large temperature gradients, leads to the epitaxial growth of columnar grains. The major growth direction tends to be parallel to that of maximum heat flow, but crystallography plays a role, since some orientations are better suited for rapid growth (p. 386). Since the direction of heat flow is not constant because the heat source is not stationary, the columnar grains tend to bend as the heat flux evolves, Fig. 14.10. Furthermore, the fusion surface is not flat in multi-pass deposits and there is some tendency for the columnar grains to grow normal to the solid-liquid interface.

The columnar growth of the same crystals essentially continues across passes since the underlying deposit has selected the grains that are optimally aligned. The large additively-manufactured block illustrated in Fig. 14.4 consisted of 11 layers. Fig. 14.10b shows that many grains traverse multiple layers, making the boundaries between layers difficult to discern in a metallographic section.

Given the preferred growth orientation, it is not surprising that a strong crystallographic texture develops; Fig. 14.10c shows that the $\{100\}_\gamma$ poles are strongly clustered around the build direction (labelled Z), with a broad spread of the orthogonal cube axes along the transverse (X) and deposit (Y) directions. A texture like this would be associated with anisotropy of mechanical properties, but the predicted stress-strain curve of additively manufactured 316L is not especially sensitive to the texture, perhaps a difference of 40 MPa as a function of the test direction once the elastic regime is exceeded [8]. In practice, components are not be loaded to an extent that causes plasticity. Given this, strong crystallographic texture that results in the reduction of the net interface energy in the polycrystalline material is actually beneficial for corrosion resistance [34,35] and in enhancing fatigue resistance [36].

The Schaeffler diagram (Fig. 12.12) is used routinely to assess the combination of phases expected in additively manufactured steels, on the basis of chromium and

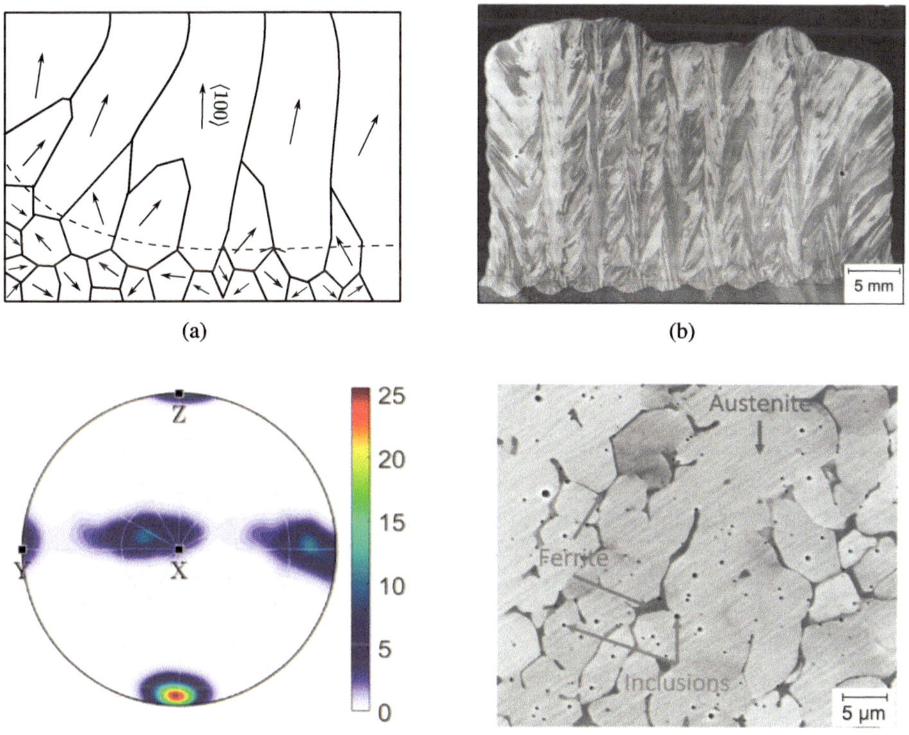

(a) (b) (c) (d)

Figure 14.10 (a) Schematic illustration of the development of a columnar grain structure during the solidification of liquid on a solid substrate. The dashed line represents the fusion surface. (b) Metallography of the cross-section of the 316L austenitic stainless steel object illustrated in Fig. 14.4, with the build direction vertical and the horizontal direction normal to that of deposition. The deposit consists of eleven layers, but many of the grains that predominated as the first layer solidified, continue across the subsequent layers. (c) Stereographic projection showing the $\{100\}_\gamma$ poles relative to the sample axes. (d) The microstructure contains some δ-ferrite and inclusions but is predominantly austenitic. Images reproduced from Belotti et al. [9] under the CC BY 4.0 license, http://creativecommons.org/licenses/by/4.0/.

nickel equivalents. For 316L, the diagram predicts that the solidified structure should contain a small quantity of δ-ferrite, with the remainder being austenite. This is indeed what is obtained, as illustrated in Fig. 14.10d, which shows also some inclusions in the microstructure, entirely to be expected, since the manufacturing process is analogous to welding, where it is not possible to protect the liquid phase entirely from its environment.

Solidification during welding does not occur at equilibrium. It is reasonable therefore to expect that additively manufactured items will not be chemically homogeneous.

14.7 Composites

Manufacturing objects from powder, as in the additive processes, opens up the possibility of making composite materials. The difference with ordinary powder metallurgy, is that the additive method can produce controlled morphologies within the composite.

Iron-TiC metal-matrix composites are used, for example, as blanking-punches that retain their cutting edges over a large number of cycles. Alternatively, they can resist abrasion and other mechanisms of wear. Powder metallurgy is used to produce such composites by hot isostatic-pressing of an appropriate mixture of TiC-Fe.

The composite can also be made using ink-jet printing, which deposits a binder onto a layer of 100 μm thick powder layer, with the build continuing by laying more powder on to the bound layer. The powder is a mixture of titanium and carbon in the ratio appropriate for a particular chemical composition of titanium carbide. The binder is such that it can be cured at about 200°C to produce a *green* component of the correct shape, where the powder is compacted but not yet tightly bound together. The compact is placed on a steel substrate (Fe-0.7C-0.8Mn wt%) and is heated to 1400°C for 4 h in a vacuum furnace in order to generate the titanium carbide. The temperature is then raised to 1550°C for 15 min to allow the steel to melt and infiltrate the green component; this temperature is above the TiC+Fe eutectic temperature of 1450°C. Fig. 14.11 shows the complexity of the component that can be made. TiC has a small contact angle with liquid iron, which therefore is able to wet the carbide particles. The steel is pearlitic in its infiltrated state, so the component can be austenitised and quenched to convert it into martensite, thereby achieving a hardness of about 1600 HV [33]. The process allows the control of the TiC_x stoichiometry, which may vary along the component if the powder mixture is changed during deposition, and the infiltration by iron itself need not be uniform.

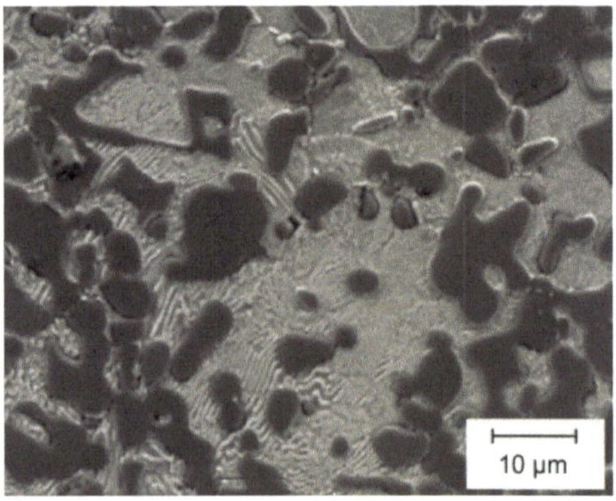

(a) (b)

Figure 14.11 (a) A "green" helical-gear made using 3D-printing of titanium carbide. (b) The infiltrated microstructure of titanium carbide (dark, $TiC_{0.98}$) and iron composite. The iron has absorbed carbon from the carbide and therefore exhibits a pearlitic microstructure. Images from Levy et al. [33] with the permission of Elsevier.

A similar ink-jet binder principle has been used to produce injection moulding tools for polymers in order to integrate cooling channels that follow the profile of the mold core. This facilitates the uniform and more rapid cooling of the polymer during production. The binder is printed on to stainless steel powder, excess unbound powder removed, further such layers deposited to produce the three-dimensional green compact, which is then infiltrated by liquid copper to produce a fully dense object [32].

Carbide-steel composites of many kinds have been investigated using the laser and wire-arc additive processes. Similarly, layers of different iron-alloys can be deposited (e.g., steel, concentrated alloys) to create composites. However, much of this work represents curiosity driven research rather than directed at specific component manufacture. For example, carbon nanotubes have been introduced into 316L stainless steel using laser-based additive deposition [37], resulting in properties that can easily be achieved by other methods. The wear rate of the 316L during dry sliding was reduced, but such steels are not in any case used in wear applications, for which there exist much better alternatives. The process resulted in large amounts of loose nanotubes becoming airborne due to interaction of the powder with the laser heat source – this could represent a safety hazard if inhaled by living creatures [38].

In addition to creating composite materials, the printing process can be used to make components that have two different materials fixed together. The gripping ends of an industrial robot can be heavy, adding to the inertia when the robot-arm executes its functions rapidly. There is, therefore, an advantage in making the grips lighter while maintaining their integrity. Fig. 14.12 shows a composite grip that is 3D-printed in two stages, the light carbon fibre composite and a smaller 17-7PH steel component mechanically attached to produce an overall design that is much lighter than an all-steel component. The printing process used for both parts of the grip is interesting. It involves the extrusion of a mixture of binder and metal through a nozzle, which follows the numerical design to build up layers and produce "green objects" of the correct shape. These are then sintered in the case of the metal or annealed in the case of the composite, to cure the binder. The design allows for shrinkage during sintering. The method leads to smoother objects and there is negligible waste, unlike the process where binder is printed on to a powder layer.

17-7PH is a precipitation hardened stainless steel of approximate chemical composition Fe-0.07C-0.4Si-0.8Mn 17Cr 7Ni-1.25Al wt%. It can be quenched from a fully austenitic condition into martensite, but if some $M_{23}C_6$ precipitates in the austenite, then the steel becomes martensitic on cooling. In this latter condition, it can be precipitation hardened with NiAl particles by annealing at about 500°C.

14.8 Tuned damper

The layer-by-layer evolution of three-dimensional objects can be done on a grand scale. Tall sky-scrapers can sway by elastic deformation with an amplitude that is of the order of a metre at the top, due to the action of wind. This naturally is unacceptable in terms of practicality and can cause nausea. The Taipei 101 sky scraper (509.2 m in height) has a system to ameliorate the problem by damping the oscillations. To achieve

Figure 14.12 One part of a lightweight, composite grip made for an industrial robot. The black part is made from carbon-fibre composite, while the 3D-printed metal part that grips the component is steel. The two materials are mechanically joined together, as illustrated in the lower image. Images with permission of *Markforged.*

this damping, a 660 tonne steel ball that is suspended using steel ropes from the 91st to the 87th floor of the building. The suspended ball at the bottom is connected to eight viscous dampers that absorb the kinetic energy during an excitation, thus greatly lessening the motion of the building. The ball is illustrated in Fig. 14.13, both during its construction stage and as the final assembly.

The ball is made additively, by cutting thick steel plates into circles of the correct dimensions and then joining them together to form the overall-spherical shape by arc welds along the circumference of each plate. This may represent the largest object ever made using additive manufacturing; it certainly represents the largest and heaviest tuned-mass damper in the world. The damper also proved its efficacy during an earthquake of magnitude 6.8 Richter.

14.9 Summary

The additive manufacture of steels is an evolving subject, but it is clear that the process on its own does not serve the function of prototyping or production, without the application of secondary procedures to ensure the required shape and other characteristics. This may be a generic issue for metallic components, because the scale of the surface roughness is significant, and can compromise fine internal features such as cooling channels.

(a)

(b)

Figure 14.13 (a) The steel ball during the construction of the Taipei 101 building. (b) The ball in its finished condition, a tourist attraction in addition to serving as a tuned-mass damper – this particular image courtesy of Jessie Lin of Taipei 101, via the good offices of Professor J. R. Yang.

Additional factors of concern include distortion associated with residual stresses and defect structures inherited from the stepwise build. There may be an anisotropy of mechanical properties, but it has yet to be established that this is sufficiently significant to matter in a real application.

Therefore, while additive manufacturing is not a panacea, there is no doubt that the design freedom it offers will make it a key part of the manufacturing capability of the future.

References

[1] M.R. Poppleton, Formal specification and rapid prototyping: building in quality at the start, Transactions on Information and Communications Technologies 9 (1970) 1–16.

[2] X. Yan, P. Gu, A review of rapid prototyping technologies and systems, Computer Aided Design 28 (1996) 307–318.

[3] N. Asnafi, J. Rajalampi, D. Aspenberg, A. Alveflo, Production tools made by additive manufacturing through laser-based powder bed fusion, BHM Berg- und Hüttenmännische Monatshefte 165 (2020) 125–136.
[4] N. Asnafi, T. Shams, D. Aspenberg, C. Öberg, 3D metal printing from an industrial perspective – product design, production, and business models, BHM Berg- und Hüttenmännische Monatshefte 164 (2019) 91–100.
[5] I. Paoletti, Mass customization with additive manufacturing: new perspectives for multi performative building components in architecture, Procedia Engineering 180 (2017) 1150–1159.
[6] A. Edwards, D. Weisz-Patrault, E. Charkaluk, Analysis and fast modelling of microstructures in duplex stainless steel formed by directed energy deposition additive manufacturing, Additive Manufacturing 61 (2023) 103300.
[7] B. Ralph, Method of making decorative articles, Technical Report 1,533,300, U.S. Patent Office, 1925.
[8] T.F. van Nuland, J. van Dommelen, M.G. Geers, Microstructural modeling of anisotropic plasticity in large scale additively manufactured 316L stainless steel, Mechanics of Materials 153 (2021) 103664.
[9] L.P. Belotti, J. Van Dommelen, M. Geers, C. Goulas, W. Ya, J. Hoefnagels, Microstructural characterisation of thick-walled wire arc additively manufactured stainless steel, Journal of Materials Processing Technology 299 (2022) 117373.
[10] L.P. Belotti, T. van Nuland, M. Geers, J. Hoefnagels, J. van Dommelen, On the anisotropy of thick-walled wire arc additively manufactured stainless steel parts, Materials Science & Engineering A 863 (2023) 144538.
[11] W.K.C. Jones, P.J. Alberry, A model for stress accumulation in steels during welding, Metals Technology 11 (1977) 557–566.
[12] A. Ohta, N. Suzuki, Y. Maeda, K. Hiraoka, T. Nakamura, Superior fatigue crack growth properties in newly developed weld metal, International Journal of Fatigue 21 (1999) S113–S118.
[13] H. Dai, J.A. Francis, H.J. Stone, H.K.D.H. Bhadeshia, P.J. Withers, Characterising phase transformations and their effects on ferritic weld residual stresses with X-rays and neutrons, Metallurgical & Materials Transactions A 39 (2008) 3070–3078.
[14] R. Cottan, J. Wang, Characterization of microstructure and residual stress in a 3D H13 tool steel component produced by additive manufacturing, Journal of Materials Research 29 (2014) 1978–1986.
[15] N. Asnafi, Tool and die making, surface treatment, and repair by laser-based additive processes, BHM Berg- und Hüttenmännische Monatshefte 166 (2021) 225–236.
[16] S. Hoeges, A. Zwiren, C. Schade, Additive manufacturing using water atomized steel powders, Metal Powder Report 72 (2017) 111–117.
[17] A. Iams, J. Keist, L. Giannuzzi, T. Palmer, The evolution of oxygen-based inclusions in an additively manufactured super-duplex stainless steel, Metallurgical & Materials Transactions A 52 (2021) 3401–3412.
[18] S. Riehm, A. Kaletsch, C. Broeckmann, V. Friederici, S. Wieland, F. Petzoldt, Tailor-made net-shape composite components by combining additive manufacturing and hot isostatic pressing, Materials Research Proceedings 10 (2019) 203–209.
[19] T. DebRoy, H.K.D.H. Bhadeshia, Innovations in Everyday Engineering Materials, Springer Nature, Switzerland, 2021.
[20] S.I. Shakil, N.R. Smith, S.P. Yoder, B.E. Ross, D.J. Alvarado, A. Hadadzadeh, M. Haghshenas, Post fabrication thermomechanical processing of additive manufactured metals: a review, Journal of Manufacturing Processes 73 (2022) 757–790.

[21] D. Michl, B. Sydow, M. Babbach, Ring rolling of pre-forms made by wire-arc additive manufacturing, Procedia Manufacturing 47 (2020) 342–348.
[22] M. Merklein, D. Junker, A. Schaub, F. Neubauer, Hybrid additive manufacturing technologies–an analysis regarding potentials and applications, Physics Procedia 83 (2016) 549–559.
[23] A. Sreenathbabu, K.P. Karunakaran, C. Amarnath, Statistical process design for hybrid adaptive layer manufacturing, Rapid Protoyping Journal 11 (2005) 235–248.
[24] E.M. Sefene, Y.M. Hailu, A.A. Tsegaw, Metal hybrid additive manufacturing: state-of-the-art, Progress in Additive Manufacturing 7 (2022) 737–749.
[25] W. Chen, L. Xu, L. Zhao, Y. Han, X. Wang, C. Hu, H. Jing, Application of hybrid additive manufacturing technology for performance improvement of martensitic stainless steel, Additive Manufacturing 51 (2022) 102648.
[26] M. Gong, Y. Meng, S. Zhang, Y. Zhang, X. Zeng, M. Gao, Laser-arc hybrid additive manufacturing of stainless steel with beam oscillation, Additive Manufacturing 33 (2020) 101180.
[27] P. Wüst, A. Edelmann, R. Hellmann, Areal surface roughness optimization of maraging steel parts produced by hybrid additive manufacturing, Materials 13 (2020) 418.
[28] J. Hao, F. Hu, X. Le, H. Liu, H. Yang, J. Han, Microstructure and high-temperature wear behaviour of Inconel 625 multi-layer cladding prepared on H13 mould steel by a hybrid additive manufacturing method, Journal of Materials Processing Technology 291 (2022) 117036.
[29] J.P. Pragana, V.A. Cristino, I.M. Bragança, C.M. Silva, P.A. Martins, Integration of forming operations on hybrid additive manufacturing systems based on fusion welding, International Journal of Precision Engineering and Manufacturing-Green Technology 7 (2020) 595–607.
[30] V. Gornyakov, Y. Sun, J. Ding, S. Williams, Modelling and optimising hybrid process of wire arc additive manufacturing and high-pressure rolling, Materials & Design 223 (2022) 111121.
[31] S. Singh, A. Jinoop, G.T. Kumar, A. Shukla, I. Palani, N. Resnina, C. Paul, K. Prashanth, Laser hybrid wire arc additive manufacturing for fabricating thin sections, Transactions of the Indian National Academy of Engineering 7 (2022) 175–183.
[32] E. Sachs, E. Wylonis, S. Allen, M. Cima, H. Guo, Production of injection molding tooling with conformal cooling channels using the three dimensional printing process, Polymer Engineering and Science 40 (2000) 1232–1247.
[33] A. Levy, A. Miriyev, A. Elliott, S.S. Babu, N. Frage, Additive manufacturing of complex-shaped graded tic/steel composites, Materials & Design 118 (2017) 198–203.
[34] H. Kokawa, Weld decay-resistant austenitic stainless steel by grain boundary engineering, Journal of Materials Science 40 (2005) 927–932.
[35] T. Watanabe, Grain boundary engineering: historical perspective and future prospects, Journal of Materials Science 46 (2011) 4095–4115.
[36] S. Kobayashi, M. Hirata, S. Tsurekawa, T. Watanabe, Grain boundary engineering for control of fatigue crack propagation in austenitic stainless steel, Procedia Engineering 10 (2011) 112–117.
[37] H. Yin, J. Yang, Y. Zhang, L. Crilly, R.L. Jackson, X. Lou, Carbon nanotube (CNT) reinforced 316L stainless steel composites made by laser powder bed fusion: microstructure and wear response, Wear 496 (2022) 204281.
[38] R. Tantra, P. Cumpson, The detection of airborne carbon nanotubes in relation to toxicology and workplace safety, Nanotoxicology 1 (2007) 251–265.

Nanostructured steels

15

Abstract

The prefix *nano* refers to a billionth of some quantity, whereas *structure* represents the way in which objects are organised. Nanostructure, therefore, is an arrangement of objects that are very small indeed. There are additional caveats in our case, because the individual objects are crystals that must fill space so that the resulting agglomerates can be used in engineering structures. In practice, this means that there is a huge density of inter-crystalline interfaces that interrupt the ordered arrangements of atoms, making the steel incredibly strong. However, engineering design requires an appropriate combination of properties, economic viability and the ability to scale-up production. As a consequence, the few applications of nanostructures that have had limited success can all be described within the scope of this chapter.

15.1 Introduction

The idea that a fine structure is always good for ambient temperature properties, and in particular the expectation that strength and toughness should simultaneously improve, became engrained in metallurgy following the unprecedented success of microalloyed, thermomechanically processed steels. We shall see, however, that unexpected difficulties arise when the grain size becomes so small that a substantial proportion of atoms is located at the boundaries. Conventional wisdom then fails, but before discussing the details, it is worth considering what constitutes a nanostructure.

Some guidance is necessary in order to make a rational use of the term *nanostructure* in the present context. For example, the introduction of a large number density of closely spaced precipitates does not correspond to a nanostructure. Such precipitates would be coherent, their strain fields would interact, and, as in the theory of concentrated solid solutions [1], the algebraic mean of their stress fields would tend to zero, so that dislocation lines appear unpinned. The level of strengthening expected would therefore be small. Nickel superalloys are an example, where high-performance versions contain some 70% by volume of coherent γ' precipitates. These contribute to strength, but not to the levels typical of the strongest nanostructured materials.

Fig. 15.1 shows how the grain boundary area per unit volume increases as the grain size is reduced. The relationship is inverse ($S_V = 2/\overline{L}$), so there is a sharp increase as $\overline{L} \lesssim 40\,\text{nm}$. This also means that there is a corresponding dramatic change in properties as the grain size becomes smaller than about 40 nm (or $S_V \gtrsim 0.05\,\text{nm}^{-1}$), as illustrated in Fig. 15.1b. It follows that a grain size less than 40 nm is a good practical definition of when a steel can be said to be nanostructured.

15.2 Why yearn for exceedingly fine grains?

Extremely brittle behaviour occurs when fracture leads only to the creation of new surfaces that are flat on a microscopic scale. Some polycrystalline ceramics fall in

Steels. https://doi.org/10.1016/B978-0-44-318491-8.00021-6

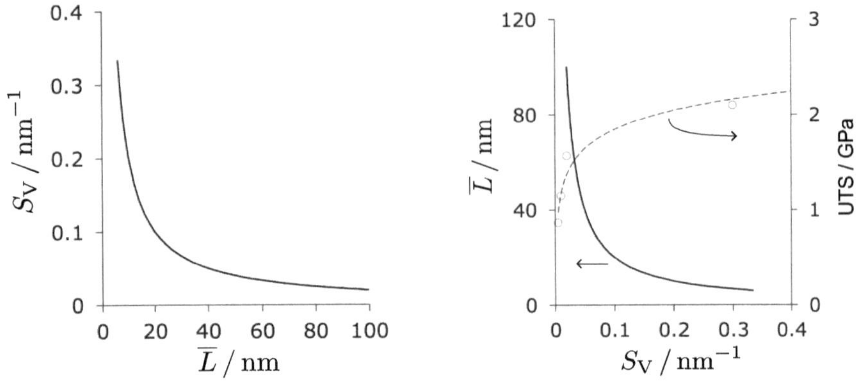

(a) (b)

Figure 15.1 (a) Calculation of the grain boundary surface per unit volume (S_V) as a function of the mean lineal intercept ($\overline{L}$) that defines the size of equiaxed grains. (b) The S_V-$\overline{L}$ data plotted alongside selected experimental data on the ultimate tensile strength as a function of $\overline{L}$ from a compilation by Bhadeshia [2].

this category, but the toughness increases a little as the grain size is reduced. This is because the material can be assumed to contain a crack that is effectively the grain size and the stress concentration at the atomically sharp crack tip scales with the grain size. Brittle cracks occur in planes that are easiest to fracture and are at the same time suitably oriented relative to the applied stress. Therefore, conventional wisdom is that a finer grain size in a structure that is not crystallographically textured, results in the more frequent deflection of the crack tip into cleavage planes that are not optimally oriented with respect to the applied stress [3].

It also is possible that the re-initiation of cleavage in an adjacent grain requires a modicum of plasticity, which would add to the work of fracture per unit area of surface created. This work, in the case of steel that is tested in its brittle regime, is about 14-80 $\mathrm{J\,m^{-2}}$ [4,5], whereas a typical surface energy of the cleavage plane is just $\approx 2.5\,\mathrm{J\,m^{-2}}$ [6], indicating a strong role of microplasticity in the propagation of cleavage cracks.

Most steels contain a substantial quantity of allotriomorphic ferrite and some cementite may be present at the ferrite grain boundaries. This cementite then fractures under stress and the resulting microcrack is the basis for cleavage crack extension in the microstructure as a whole [7]. The cementite thickness is found to scale with the ferrite grain size, as illustrated in Fig. 15.2. Since the stress required to propagate the cleavage crack from the fractured cementite scales with the thickness$^{-1/2}$, the toughness increases as the ferrite grain size is refined [4]. Similar logic must apply to mixtures of bainitic ferrite and carbon-enriched retained austenite – the latter transforms into high carbon, untempered martensite, which is then the source for cleavage fracture. It is well-established that an overall refinement of the mixture ($\alpha_b + \gamma$) leads to an improvement in toughness (Chapter 15).

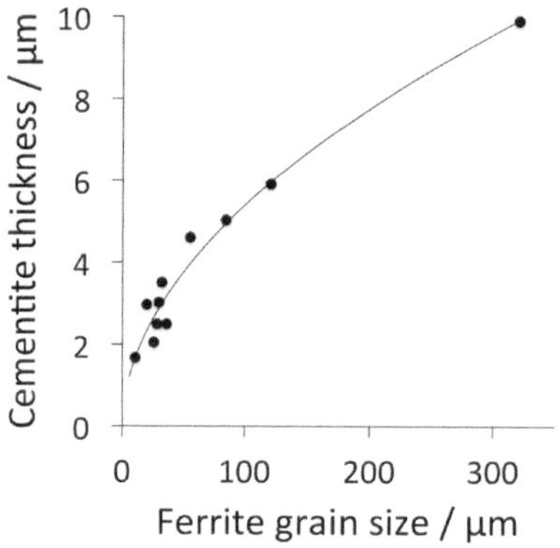

Figure 15.2 The coarsest observed cementite thickness versus the ferrite grain size in a variety of steels. Grain refinement leads also to thinner cementite particles, so the toughness is expected to improve. Data from a compilation by Curry and Knott [4].

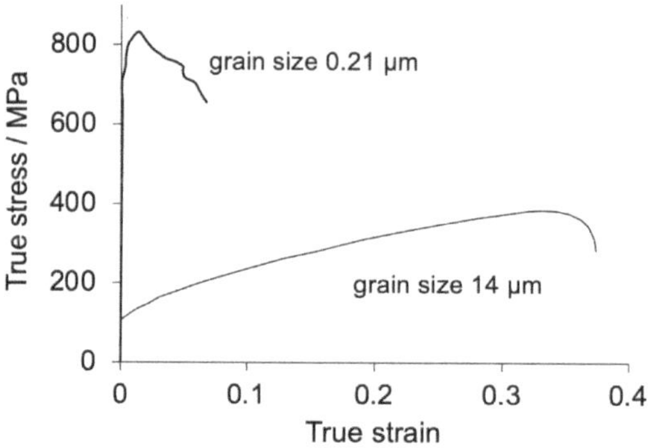

Figure 15.3 Tensile tests on interstitial-free ferritic steel with two different ferrite grain sizes. Selected data from Tsuji et al. [8].

Ductile fracture in tension usually involves the nucleation and growth of voids, which when they coalesce leads to ultimate failure. But tensile elongation contains two components, one of which is the plastic strain that is distributed homogeneously as the sample deforms uniformly. Deformation of course hardens the steel, but at some locality, the rate at which it hardens is not sufficient to compensate for the reduction in area, so the deformation becomes focused, leading to necking. This is the point where the engineering stress reaches a maximum and fracture then becomes inevitable. The majority of elongation should be uniform in a good steel.

Fig. 15.3 shows two tensile tests conducted on fully ferritic steel. The strength obviously increases when the grain size is reduced, but there is no uniform elongation observed because plastic instability sets in as soon as yielding occurs. In contrast, the same steel when it has a coarser grain structure shows steady work hardening and considerable uniform ductility. The mechanism of work hardening is primarily the multiplication of dislocations. The fine-grained sample clearly lacks an ability to work harden because dislocation interactions may not have the space to develop in the same way as in a coarse grained sample, and because the dislocation density may be reduced by the defects sinking into adjacent boundaries. Therefore, a serious shortcoming of ultra-fine grains is the loss of work hardening capacity, which any design procedure must address if nanostructured steels are to be utilised.

15.3 Production of nanostructured steel

15.3.1 Shape preserving deformations

Nanostructured steels often are made in laboratories using severe plastic deformation [9]. The plastic deformation that can be achieved without fracture is frequently in excess of a von Mises strain of $\epsilon_{\mathrm{vM}} = 20$. The von Mises equivalent strain is given by

$$\epsilon_{\mathrm{vM}} = \frac{2}{3}\sqrt{\frac{3}{2}(\epsilon_{xx}^2 + \epsilon_{yy}^2 + \epsilon_{zz}^2) + \frac{3}{4}(\gamma_{xy}^2 + \gamma_{yz}^2 + \gamma_{zx}^2)},$$

where ϵ_{xx} are normal and γ_{xy} are shear strains etc. The very large $\epsilon_{\mathrm{vM}} = 20$ obviously cannot be while the material is in tension, so ingenious processes have been invented to maintain the overall shape without inducing fracture.

A particular shape-preserving process is illustrated in Fig. 15.4a, where the constrained sample material of thickness l is sheared by the conjoint action of torsion and compression. The shear strain per rotation is the displacement $2\pi r$ divided by the thickness l, where r is the radial distance. It follows that the total shear strain for N rotations is given by [10]:

$$\gamma = 2\pi\frac{rN}{l} \qquad \text{with} \qquad \epsilon_{\mathrm{vM}} = \frac{\gamma}{\sqrt{3}}. \tag{15.1}$$

The equation implies that the shear strain is zero at the centre, but this is inconsistent with observations, because there is likely to be some sliding involved between the sample and the torsion support or rotating member [9].

The torsion test-sample is typically less than a centimetre in diameter and a millimetre thick. While such samples are big enough for structural characterisation, the only assessment of mechanical properties is usually with hardness testing. Fig. 15.4b,c shows the *micro*structure obtained in severely sheared samples of Armco iron and a fully pearlitic rail-steel. The emphasis on the adjective "micro" is because the deformation failed in both cases to achieve a structural scale less than 100 nm for the pearlitic steel and 250 nm in the Armco iron. It is not therefore surprising that the Armco iron achieved a hardness of just 350 HV following $\epsilon_{\mathrm{vM}} = 36$. In the case of the pearlite, the deformation led to the fragmentation of the cementite, which was still present after $\epsilon_{\mathrm{vM}} = 16$, with a final hardness of 612 HV.

These properties are a consequence of both the fine structure and the defects such as dislocations, but they are not particularly interesting since these levels of hardness can be achieved routinely through heat treatment of ordinary steels. The technique therefore is limited to basic studies. Although publications (e.g., [9]) refer to the material produced using torsion as 'bulk nanostructured', it is neither bulk nor nanostructured.

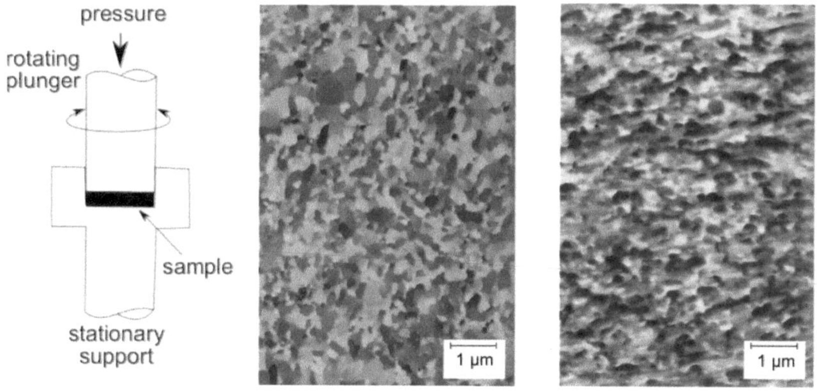

(a) (b) (c)

Figure 15.4 (a) Schematic diagram of compression-torsion severe plastic deformation equipment. The sample experiences severe shear strains and because of the constraint, it retains its essential shape. (b) Commercially pure iron torsion-deformed to $\epsilon_{vM} = 32$. (c) Fully pearlitic steel torsion-deformed to $\epsilon_{vM} = 16$. Adapted images from Wetscher et al. [11], reproduced with the permission of Elsevier.

Equal channel angular processing [12] can be applied to somewhat larger samples, Fig. 15.5. The sample is forced through a die such that it experiences redundant work by being bent through an angle ϕ, but with the bend propagating along the entire length so that the original shape is recovered. The equivalent strain is then given by [13]:

$$\epsilon_{vM} = \frac{2}{\sqrt{3}} N \cot\{\phi/2\} \tag{15.2}$$

The process can be repeated on the same sample in order to accumulate strain. The relatively large samples that can be processed permit a better characterisation of the mechanical properties. Studies on austenitic stainless steel [14] suggest that there is a deterioration in the fatigue properties, even though the yield strength increases due to the refinement of the structure. Similarly, a 0.08C wt% ferritic steel exhibited a refinement of microstructure to about 0.2 μm size, an increase in strength, but a large decrease in elongation (Fig. 15.5c). The strength that is obtained is not remarkable and, again, is routinely accessible in mass produced commercial steels.

Accumulative roll-bonding [8,16] involves plane strain deformation by rolling, of two separate pieces of steel which become cold-bonded as they emerge from the rolls. The final thickness is identical to that of the individual pieces (Fig. 15.6). The von Mises equivalent strain for this case is [17]:

$$\epsilon_{vM} = N \times \frac{2}{\sqrt{3}} \ln\left\{\frac{1}{2}\right\}, \tag{15.3}$$

where N is the number of times the rolled sample is re-introduced into the roll after stacking.

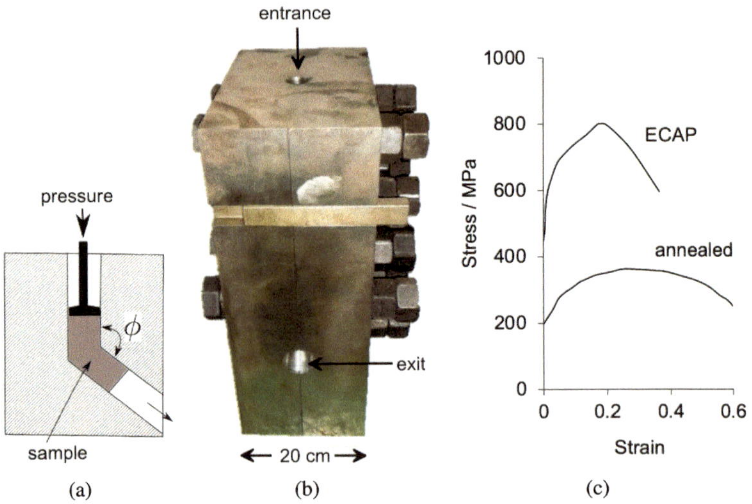

Figure 15.5 (a) Schematic diagram of equal channel angular processing (ECAP). The steel is introduced at the top and is deformed through the angle ϕ but exits the die with the same shape that it started with. The process can be repeated if desired. (b) An actual ECAP die for producing severely deformed cylindrical samples. The die is in two halves, which are bolted together for inspection and maintenance of the cavity. (c) Comparison of the behaviour in tension of a mild steel containing 0.08C wt%, in its annealed and processed conditions (selected data from Fukuda et al. [15]).

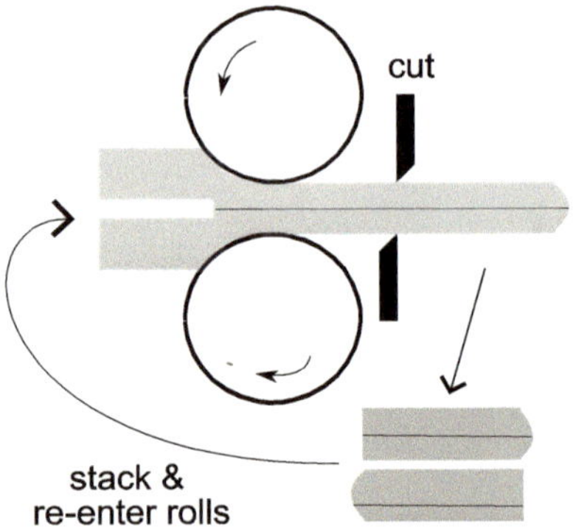

Figure 15.6 Schematic description of accumulative roll bonding. Separate samples of steel are cold-welded by rolling, the emerging steel is the cut, cleaned, stacked and re-rolled a number of times to accumulate plastic strain.

Accumulative roll bonding is an interesting process because, in principle, it can be scaled up and implemented on conventional large rolling equipment that exists already in industry. Obviously, the productivity of conventional single-pass rolling would not be matched, but the cost-benefit analysis depends on whether exceptional properties can be achieved. Table 15.1 lists some of the mechanical properties reported for interstitial-free steel. It is not surprising that the strength increases but ductility decreases, since the material is essentially in a cold-deformed state following accumulative rolling, so much of the ductility that existed in the annealed state is exhausted. It can, of course, be recovered by annealing, but at the expense of strength.

Table 15.1 Some properties of steel made using the accumulative roll-bonding process.

Material	N	UTS / MPa	% elongation	Reference
Interstitial-free steel	0	274	57	[17]
Interstitial-free steel	6	751	6	[17]
Interstitial-free steel	0	255	51	[18]
Interstitial-free steel	8	850	3	[18]
Fe-24Ni-0.3C wt% TRIP steel	0	1017	153	[19]
Fe-24Ni-0.3C wt% TRIP steel	6	1395	39	[19]

15.3.2 Shape altering deformations

One of the most successful engineering products has been steel wire and rope made from pearlitic steels. Attempts have been made to take the process of wire drawing to extremes in the hope of greatly enhancing the strength. It is important to note that because most of these drawn wires are used in making ropes, the usual worry about a loss of ductility associated with strengthening does not apply. This is because the rope is a composite of many strands, where the fracture of an individual strand does not compromise overall safety. Furthermore, individual wires are so thin that plane-strain conditions do not apply, so they may not fail in a brittle manner.

One such product is a wire with a strength in excess of 5 GPa [20]. The chemical composition is Fe-0.2C-0.8Si-1.5Mn wt% and the alloy must be made with sufficient impurity control so that the drawability is not compromised by inclusions, given that the final diameter of the wire is less than 10 μm. The processing begins with 10 mm diameter rods, which are initially martensitic. Intercritical annealing induces the formation of layers of austenite which become enriched in carbon. Quenching to ambient temperature produces a mixture of martensite tempered at the intercritical annealing temperature, untempered high-carbon martensite formed on cooling and some austenite is retained [21].

Following wire drawing to a true strain of 8.85, the initial microstructure of martensite and tempered martensite becomes both chemically and mechanically homogenised by the mechanical work and acquires an ultra-fine dislocation cell structure, as illustrated in Fig. 15.7. It has been shown that much of the strength of the wire comes from this fine structure [22]. Because the cross section of the wire is only 10 μm, it behaves in a ductile manner, failing in tension by necking, in spite of its strength.

Obtaining strength by deformation has the advantage that the material becomes insensitive to size as long as the deformation in homogeneous, which is mostly the case when dealing with wires. Fig. 15.8a shows that the strength of the wire varies much less with diameter than the strength of tiny single-crystals of iron, in which case

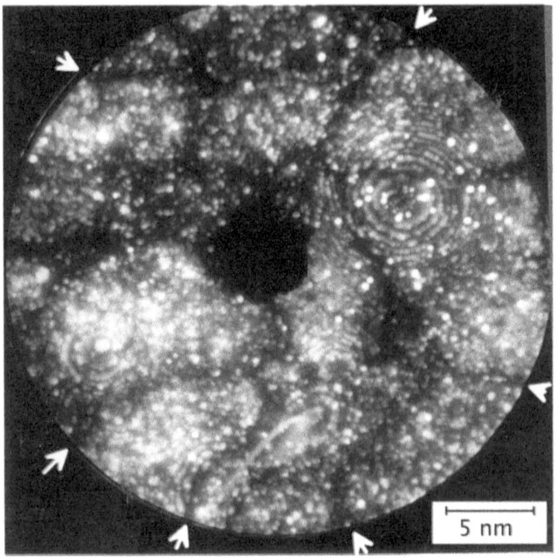

Figure 15.7 Field-ion microscope image of wire in its cold-drawn condition, with the positions of some of the cell boundaries highlighted by the arrows. The black hole approximately in the centre is an artifact of the imaging technique. Each dot represents the position of an atom.

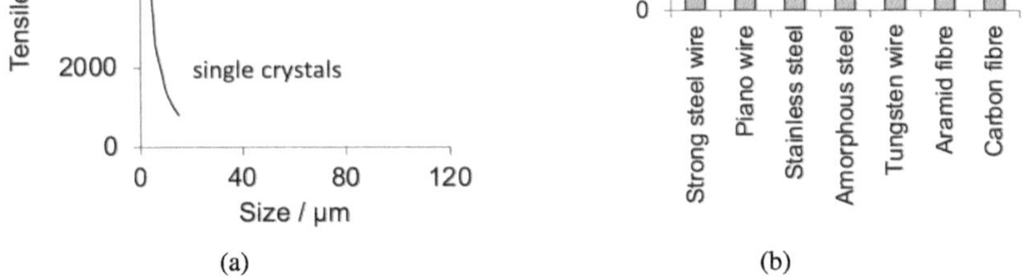

Figure 15.8 (a) A comparison of the size sensitivity of strength, for single crystals of iron (whiskers, Chapter 2) and for wire in its cold-deformed state. (b) A comparison of the strength of the cold-drawn wire against other fibres.

strength relies on perfection through the absence of dislocations. Flawlessness typical in small crystals cannot be maintained as the size increases because of entropy considerations [23]. This, of course, is the reason why carbon nanotubes and graphene are never going to have the strength levels measured on the nanometre dimensions (Section 2.11).

In terms of strength, steel wire competes well with a number of other wires or fibres, Fig. 15.8b, although when it comes to specific strength, which is normalised by the density, the polymer and carbon fibres reign supreme. None of these materials on their own can be large in all dimensions. To produce a nanostructured material that is bulky in all three dimensions, a different approach is required, based on solid-state phase transformation.

15.3.3 Nanostructure without deformation

It often is necessary to make components that are three-dimensional shapes of substantial size and with a combination of properties, including strength, that enables technology. The material and production cost must be reasonable if large quantities of such steel are required. As we shall see, a steel design is possible, the structure of which is intricately subdivided by interfaces and yet is not generated by deformation or rapid cooling, could fulfil these demanding criteria.

A nanostructured material is here defined as one containing an exceptionally high density of strong interfaces, so much so that the strength exceeds 2 GPa – this would require approximately 100 million square metres of interface inside one cubic metre of material [24]. The desire for such materials in the engineering context comes from the yearning for pushing engineering marvels such as the jet engine to extremes of efficiency and performance. Strong nanostructured materials must nevertheless address the following issues if they are to be utilised:

1. It should ideally be possible to make samples, which are large in all dimensions, not simply wires or thin sheets; this is necessary in order to manufacture components with complex shapes, such as bearings, shafts with keyways and armour.
2. The processing should not involve rapid cooling, since the structure would not then be uniform in large sections. Harmful residual stresses and distortion often accompany the rapid cooling of large components.
3. The processing should not involve deformation except when the steel is fully austenitic. Deformation limits the shapes that can be produced and the ductility of the final structure is compromised by cold deformation.
4. The material concerned must be cheap to produce if it is not to be limited to niche applications. A good standard for an affordable material is that its cost must be similar to that of bottled water when considering weight or volume.
5. Unlike formable steels, where crystallographic texture is a key performance parameter, the strong steel needs to be mechanically isotropic for applications where the integrity relies on the whole object rather than a specific feature of that object.

It should be admitted at the outset that strong steels of this kind are unlikely ever to be exploited on a large scale in the context of the 2 billion tonnes of steel used annually, because high strength can only be exploited if smaller cross sections can be tolerated. Reducing the cross section of a component compromises properties such as the engineering stiffness, the susceptibility to buckling, the corrosion life, etc. But for specialist applications such as bearings, where the world wide production amounts to just 60,000 tonnes, and yet, represent components critical in dealing with rotational motion, the value added by specialist steel determines the success of the product.

It turns out that it is possible to accommodate all of the utilisation issues listed above, with the design of a nanostructured steel in which the effective size of each crystal is finer than that of a carbon nanotube [25–28]. The fine crystals are generated by solid-state phase transformation with the following design rules:

1. The material must have a significant work-hardening mechanism to avoid the early onset of plastic instability once yielding occurs;
2. The transformation must be induced at a low homologous temperature[1] in order to avoid coarse microstructures;
3. There should be a mechanism for avoiding an excessive release of heat due to the enthalpy of transformation, since in large components, this heat raises the temperature of the steel and hence negates the need to transform at a low temperature. This effect, where the temperature of the steel rises as the crystal structure changes is known as *recalescence*.
4. The transformation should occur at a slow rate to allow any residual heat of transformation to dissipate, thus maintaining the intended transformation temperature.

Crystals that are fine enough cannot be generated by reconstructive transformation because the atoms are too mobile at the temperature where these reactions occur. And in the case of displacive transformations, martensite can be ruled out for large components because its formation either requires a high cooling rate or sufficient alloying to ensure adequate hardenability. Displacive transformations have the advantage that the energy associated with the shape deformation is stored within the material, thus reducing the enthalpy change during transformation and hence the propensity to recalescence. The bainite transformation is an ideal candidate – after all, large power plant turbines are able to be made with a bainitic microstructure by slow cooling from the austenite phase field. Furthermore, the rate of the reaction can be controlled more effectively than is the case for martensite.

The question, then, is what is the lowest temperature at which bainite can form whilst at the same time avoiding the formation of martensite. The fundamental theory for calculating these transformation temperatures is routinely available [29,30].[2] Fig. 15.9 shows some example calculations for a steel containing 2Si-3Mn wt%, that illustrate how the B_S and M_S temperatures vary as a function of the carbon concentration of the austenite. Surprisingly, there appears to be no lower limit to the temperature at which bainite can be generated, but the time required to initiate the reaction increases dramatically as B_S is reduced, so much so that it would take about a year at 140°C (dashed arrow on Fig. 15.9).

One interesting feature is that the martensite-start and bainite-start temperatures are both suppressed by the addition of carbon (Fig. 15.9). A similar result is obtained for a different steel, Fig. 15.10a. However, if an attempt is made to suppress B_S using substitutional solutes in low-carbon steel, then $B_S \rightarrow M_S$ and at some concentration

[1] A homologous temperature is the absolute temperature divided by the absolute melting temperature of the material concerned.
[2] Algorithms for doing this available on www.phase-trans.msm.cam.ac.uk/map/mapmain.html.

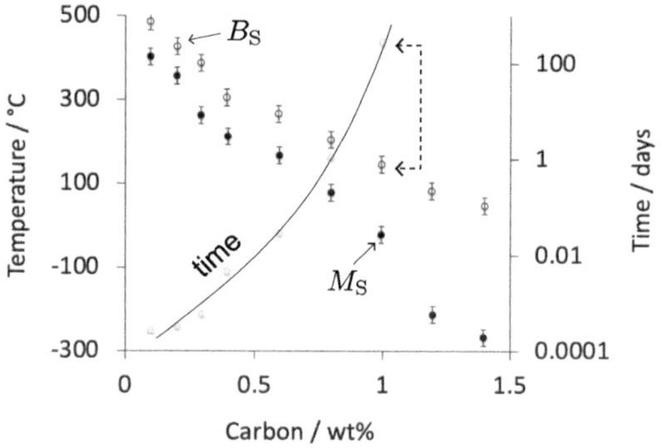

Figure 15.9 Calculated transformation-start temperatures in Fe-2Si-3Mn wt% steel as a function of the carbon concentration, together with the calculated time required to initiate bainite. The double-headed arrow correlates $B_S = 140°C$ with a transformation time of about one year.

only martensitic transformation is possible. In other words, martensite precedes bainite [31]. The reason is that substitutional solutes do not partition during displacive transformations, so bainite and martensite are similarly affected. However, the nucleation of bainite requires the partitioning of carbon, which adds to the available driving force, one of the reasons why bainite forms at higher temperatures than martensite (which is diffusionless in nucleation and growth, p. 210). However, the difference in kinetics vanishes if the carbon concentration is reduced and hence the gap between M_S and B_S decreases to the point where it vanishes [29,32]. The bainite transformation then does not occur at all.

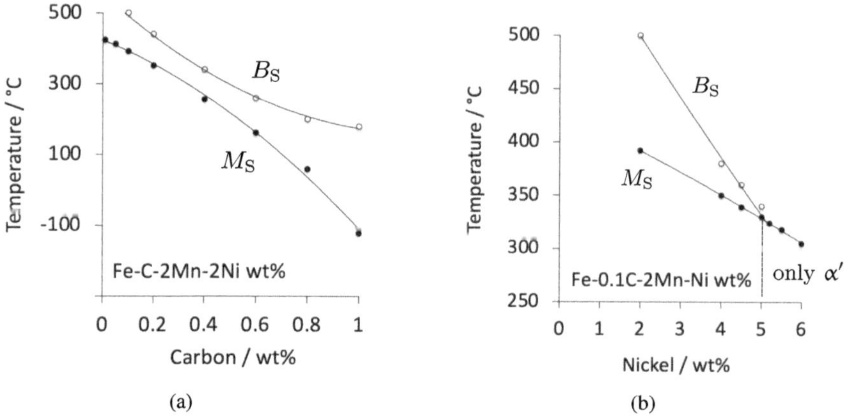

Figure 15.10 Calculated variation in the martensite and bainite-start temperatures as a function of composition. In (b), bainite ceases to form at all when the nickel concentration exceeds 5 wt%.

Apart from their role in the suppression of transformation temperature, substitutional solutes have other roles to play in achieving the totality of requirements for nanostructured steels. Sufficient silicon must be added so that the partitioning of carbon from the supersaturated bainitic ferrite does not lead to cementite precipitation. The carbon then enriches the residual austenite and stabilises it to ambient temperature. A small concentration of molybdenum is added to prevent grain boundary embrittlement by phosphorus [33–38], and chromium to retard the pearlite reaction during cooling to B_S.

Fig. 15.11 shows the structure obtained following isothermal transformation at 200°C, consisting of platelets of bainitic ferrite only 200-400 Å thick, with similarly fine intervening regions of the parent austenite [25,26,28]. This *retained* austenite is important because when it undergoes stress or strain–induced martensitic transformation, it enhances the work–hardening capacity of the material, thereby avoiding the usual problem of fine-grained metals where ductility diminishes as the grain size is reduced.

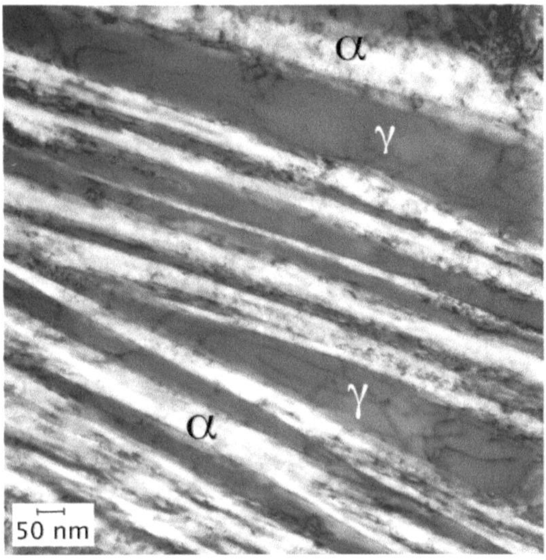

Figure 15.11 A high magnification image of Fe-0.98C-1.46Si-1.89Mn-0.26Mo-1.26Cr-0.09V wt%, transformed at 200°C for 15 days. A corresponding optical micrograph would show the slender plates of ferrite in many different orientations, so the material exhibits isotropic mechanical properties. Stereological measurements show that the true width of the plates is in the range 20-40 nm, which compares with the width of carbon nanotubes.

The bainite obtained by low-temperature transformation is harder than ever achieved, with values in excess of 700 HV. Some strength, ductility and toughness data are illustrated in Fig. 15.12. The simple heat treatment involves the austenitisation of a chunk of steel (at say 950°C), followed by a gentle transfer into an oven at the low temperature (say 200°C) to be held there for ten days or so. There is no rapid cooling – consequential residual stresses are avoided. The size of the sample can be large because the time taken to reach 200°C from the austenitisation temperature is much less than that required to initiate bainite. This is an important commercial

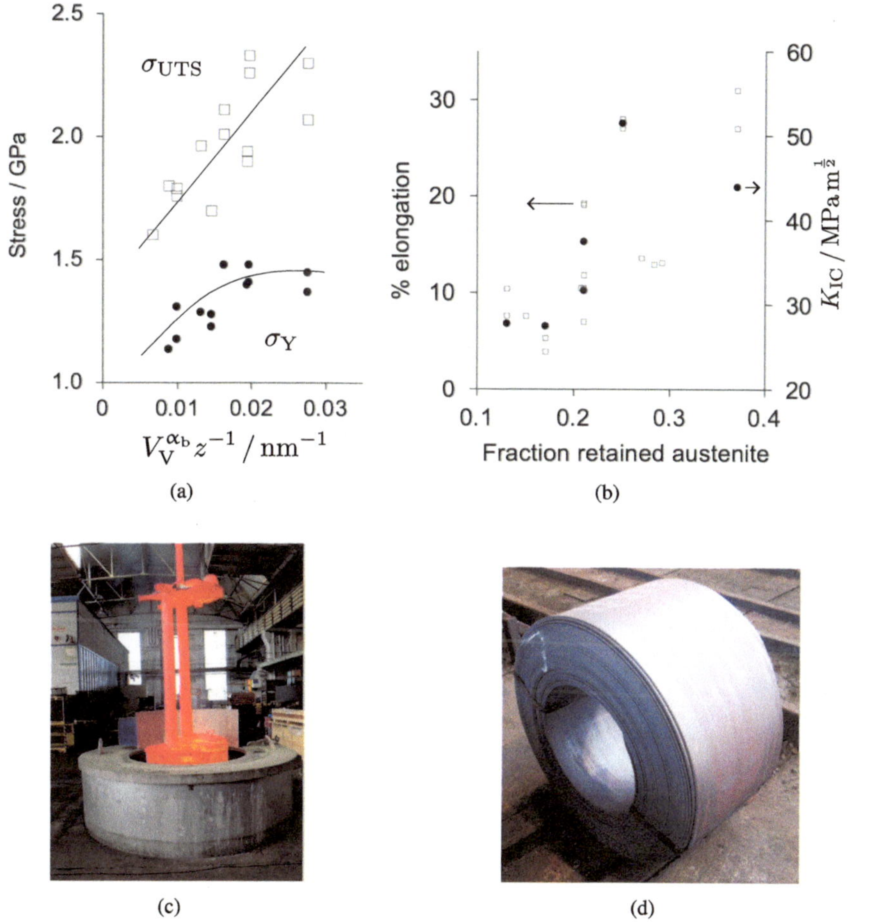

Figure 15.12 Some mechanical properties of a few nanostructured bainitic steels. (a) The ultimate tensile strength and 0.2% proof strength as a function of the volume fraction of bainitic ferrite ($V_V^{\alpha_b}$) divided by the ferrite platelet thickness z. (b) Ductility and toughness K_{IC} represented as crosses. (c,d) Large-scale manufacture of nanostructured bainite.

advantage. And the material can be manufactured in huge quantities with uniform properties across large dimensions (Fig. 15.12c,d).

There are significant limitations to such steels, discussed elsewhere in more detail [30], but they can be summarised as follows:

1. It is not possible to weld the steel on a routine basis, except by using highly specialised techniques which cannot reasonably be applied in practice. This is because the steels of this kind rely on a large carbon concentration resulting in brittle martensite forming in the heat-affected zones of the weld.
2. Although the fracture toughness is good, the Charpy impact toughness is not, for reasons related to the nature of the test [39]. Since the Charpy test forms the basis

of many acceptance criteria, applications of the steel are limited to cases where such tests are not a deciding factor, for example armour.

3. The nanostructured steel is not stable at high temperatures (> 400°C) over long periods because of the tendency for the retained austenite to decompose into a mixture of cementite and ferrite.

Some of these weaknesses apply to all nanostructured metals.

15.4 Detrimental nanostructures in steels

There are particular steels that have to sustain millions of locally applied pulsating loads at contact stress levels that are typically 2 GPa. Bearings fall into this category – when a rolling element moves over a particular location on the raceway (Fig. 15.13a), it induces a momentary system of Hertzian stresses that reach a maximum *below* the contact surface. This stress pulse is felt by a point on the bearing on each occasion that the rolling element traverses, as a combination of cyclic torsion and uniaxial compression superimposed in phase.

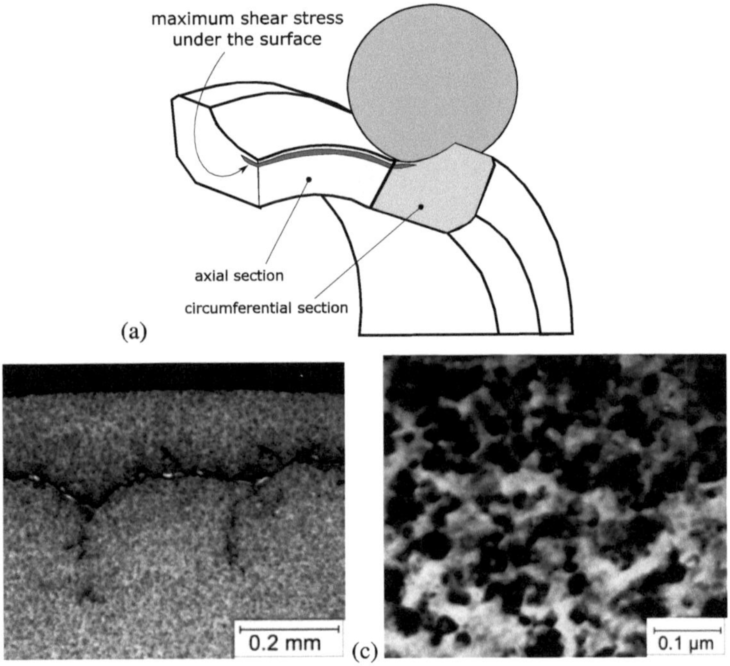

(a) (b) (c)

Figure 15.13 (a) Schematic illustration of a ball traversing a bearing raceway. The contact pressure causes the shear stress to be maximum below the surface, as illustrated in the axial section. (b) A sub-surface crack showing small, disconnected white-etching regions. After Evans et al. [40], with permission of Elsevier. (c) High-resolution image of a white-etching region showing the nanostructure that develops. After Evans et al. [41], with permission of Elsevier.

Bearing steels are relatively brittle because they have to be hard ($\approx$ 650 HV) in order to sustain the high contact stresses. A typical chemical composition contains 1C-1.5Cr wt% and a microstructure of cementite particles in a lightly tempered martensitic matrix [42]. As a result, microcracks develop and the crack faces, when suitably oriented with respect to the stress, beat against each other. This localised mechanical deformation causes the cementite to dissolve and greatly refines the original microstructure. When examined under an optical microscope, these mechanically homogenised regions respond less to chemical attack by etchants compared to the surrounding unaffected material. This is why they are called *white-etching regions*, Fig. 15.13b. Because of the refined structure (Fig. 15.13c) and the fact that the carbon that was originally in the cementite is now in solution, the white-etching regions are extremely hard, typically in excess of 1000 HV. The original crack that initiated the rubbing of crack faces therefore propagates as the white matter develops. When the crack reaches the surface, spalling occurs, leading to bearing failure.

White matter of the kind described here, and varieties in which the deformation involves both rolling and sliding, occurs in many different applications, for example, on rails and during machining.

15.5 Summary

Studies of nanostructured steels have revealed many new phenomena not described here, for example, the persistence of excess carbon within the bainitic ferrite, the nature of the equilibrium between tetragonal ferrite and austenite and the peculiarly small work hardening capacity of single-phase nanostructured metals. Some of these ideas fire the imagination, but the practical consequences of these steels are negligible to date. This is because strength is not the only property of consequence in design, it is usual that a basket of performance parameters must reach specified levels to ensure integrity during service. Many of the processes used are not capable of being scaled; even the nanostructured bainite, which does not have this limitation, has few applications because of the lack of weldability. Thus, from a commercial point of view, the nanostructured steels for which information is openly available, are not particularly relevant.

These conclusions are pessimistic, but the purpose here is to educate so that solutions may or may not emerge in the future, which change the paradigm. It remains important in studies of nanostructured metals to consider whether an appropriate bank of properties can be achieved that will make the material more than just a curiosity. It is necessary to assess whether the production process can be scaled and whether the material itself can be produced in a form that is large in all dimensions.

References

[1] R. Labusch, A statistical theory of solid solution hardening, Physica Status Solidi B 41 (1970) 659–669.

[2] H.K.D.H. Bhadeshia, Nanostructured bainite, Proceedings of the Royal Society of London A 466 (2010) 3–18.

[3] S.M. Wiederhorn, Brittle fracture and toughening mechanisms in ceramics, Annual Reviews in Materials Science 14 (1984) 373–403.
[4] D.A. Curry, J.F. Knott, Effects of microstructure on cleavage fracture stress in steel, Metal Science 12 (1978) 511–514.
[5] L. Rancel, M. Gomez, S.F. Medina, I. Gutierrez, Measurement of bainite packet size and its influence on cleavage fracture in a medium carbon bainitic steel, Materials Science & Engineering A 530 (2011) 21–27.
[6] E.J. Song, H.K.D.H. Bhadeshia, D.-W. Suh, Effect of hydrogen on the surface energy of ferrite and austenite, Corrosion Science 77 (2013) 379–384.
[7] D.A. Curry, Cleavage micromechanisms of crack extension in steels, Metal Science 14 (1980) 319–326.
[8] N. Tsuji, Y. Ito, Y. Saito, Y. Minamino, Strength and ductility of ultrafine grained aluminium and iron produced by ARB and annealing, Scripta Materialia 47 (2002) 893–899.
[9] R.Z. Valiev, R.K. Islamgaliev, I.V. Alexandrov, Bulk nanostructured materials from severe plastic deformation, Progress in Materials Science 45 (2000) 103–189.
[10] R.Z. Valiev, Y.V. Ivanisenko, E.F. Rauch, B. Baudelet, Structure and deformaton behaviour of Armco iron subjected to severe plastic deformation, Acta Materialia 44 (1996) 4705–4712.
[11] F. Wetscher, A. Vorhauer, R. Stock, R. Pippan, Structural refinement of low alloyed steels during severe plastic deformation, Materials Science & Engineering A 387–389 (2004) 809–816.
[12] V.M. Segal, V.I. Reznikov, A.E. Drobyshevski, V.I. Kopylov, Plastic treatment of metals by simple shear, Russian Metallurgy (Metally) 1 (1981) 99–105.
[13] A. Azushima, R. Kop, A. Korhonen, D.Y. Yang, F. Micari, G.D. Lahoti, P. Groche, J. Yanagimoto, N. Tsuji, A. Rosochowski, A. Yanagida, Severe plastic deformation (SPD) processes for metals, CRIP Annals – Manufacturing Technology 57 (2008) 716–735.
[14] M. Greger, V. Vodárek, L.A. Dobrzański, L. Kander, R. Kocich, B. Kuřetogá, The structure of austenitic steel AISI 316 after ECAP and low-cycle fatigue, Journal of Achievements in Materials and Manufacturing Engineering 28 (2008) 151–158.
[15] Y. Fukuda, K. Oh-ishi, Z. Horita, T.G. Langdon, Processing of a low carbon steel by equal-channel angular pressing, Acta Materialia 50 (2002) 1359–1368.
[16] N. Tsuji, Unique mechanical properties of nanostructured metals, Journal of Nanoscience and Nanotechnology 7 (2007) 3765–3770.
[17] Y. Saito, H. Utsunomiya, N. Tsuji, T. Sakai, Novel ultra-high straining process for bulk material development of the accumulative roll-bonding (ARB) process, Acta Materialia 47 (1999) 579–583.
[18] S. Tamimi, M. Ketabchi, N. Parvin, Microstructural evolution and mechanical properties of accumulative roll bonded interstitial free steel, Materials & Design 30 (2009) 2556–2562.
[19] H. Jafarian, Characteristics of nano/ultrafine-grained austenitic trip steel fabricated by accumulative roll bonding and subsequent annealing, Materials Characterization 114 (2016) 88–96.
[20] Anonymous, Kobelco technology review, Technical Report 8, Kobe Steel Ltd, Japan, 1990.
[21] H.K.D.H. Bhadeshia, High-strength steels, in: J.A. Charles, G.W. Greenwood, G.C. Smith (Eds.), Future Developments of Metals and Ceramics, Institute of Materials, London, U.K., 1992, pp. 25–74.
[22] H.K.D.H. Bhadeshia, H. Harada, High-strength (5 GPa) steel wire: an atom-probe study, Applied Surface Science 67 (1993) 326–333.

[23] H.K.D.H. Bhadeshia, Large chunks of very strong steel, Materials Science and Technology 21 (2005) 1293–1302.
[24] H.K.D.H. Bhadeshia, The first bulk nanostructured metal, Science and Technology of Advanced Materials 14 (2013) 014202.
[25] F.G. Caballero, H.K.D.H. Bhadeshia, K.J.A. Mawella, D.G. Jones, P. Brown, Very strong, low-temperature bainite, Materials Science and Technology 18 (2002) 279–284.
[26] C. Garcia-Mateo, F.G. Caballero, H.K.D.H. Bhadeshia, Low-temperature bainite, Journal de Physique. Colloque 112 (2003) 285–288.
[27] F.G. Caballero, H.K.D.H. Bhadeshia, Very strong bainite, Current Opinion in Solid State & Materials Science 8 (2004) 251–257.
[28] C. Garcia-Mateo, F.G. Caballero, H.K.D.H. Bhadeshia, Development of hard bainite, ISIJ International 43 (2003) 1238–1243.
[29] H.K.D.H. Bhadeshia, Rationalisation of shear transformations in steels, Acta Metallurgica 29 (1981) 1117–1130.
[30] H.K.D.H. Bhadeshia, Bainite in Steels: Theory and Practice, 3rd ed., Maney Publishing, Leeds, U.K., 2015.
[31] H.-S. Yang, H.K.D.H. Bhadeshia, Designing low-carbon, low-temperature bainite, Materials Science and Technology 24 (2008) 335–342.
[32] H.K.D.H. Bhadeshia, Hard bainite, in: J.M. Howe, D.E. Laughlin, J.K. Lee, U. Dahmen, W.A. Soffa (Eds.), Solid-Solid Phase Transformations, Pennsylvania, USA, in: TME-AIME, vol. 1, TMS-AIME, Warrendale, Pennsylvania, USA, 2005, pp. 469–484.
[33] C.L. Briant, S.K. Banerji, Intergranular failure in steel: role of grain boundary composition, International Metals Reviews 4 (1978) 164–199.
[34] C.J. McMahon Jr., A.K. Cianelli, H.C. Feng, The influence of Mo on P-induced temper embrittlement in Ni-Cr steel, Metallurgical Transactions A 8 (1977) 1055–1057.
[35] J. Yu, C.J. McMahon Jr., The effects of composition and carbide precipitation on temper embrittlement of 2.25 Cr-1 Mo steel: Part I. Effects of P and Sn, Metallurgical & Materials Transactions A 11 (1980) 277–289.
[36] R. Guillou, M. Guttmann, P. Dumoulin, Role of molybdenum in phosphorus-induced temper embrittlement of 12%Cr martensitic stainless steel, Metal Science 15 (1981) 63–72.
[37] A.J. Rose, F. Mohammed, A.W.F. Smith, P.A. Davies, R.D. Clarke, Superbainite: laboratory concept to commercial product, Materials Science and Technology 30 (2014) 1094–1098.
[38] H.K.D.H. Bhadeshia, D.W. Suh, Is low phosphorus content in steel a product requirement?, Ironmaking & Steelmaking 42 (2015) 259–267.
[39] L.C.D. Fielding, N.G. Jones, J. Walsh, S.V. Boxel, M.S. Blackmur, P.D. Lee, P.J. Withers, H.J. Stone, H. Bhadeshia, Synchrotron analysis of toughness anomalies in nanostructured bainite, Acta Materialia 105 (2016) 52–58.
[40] M.-H. Evans, A.D. Richardson, L. Wang, R.J.K. Wood, W.B. Anderson, Confirming subsurface initiation at non-metallic inclusions as one mechanism for white etching crack (WEC) formation, Tribology International 75 (2014) 87–97.
[41] M.-H. Evans, J.C. Walker, C. Ma, L. Wang, R.J.K. Wood, A FIB/TEM study of butterfly crack formation and white etching area (WEA) microstructural changes under rolling contact fatigue in 100Cr6 bearing steel, Materials Science & Engineering A 570 (2013) 127–134.
[42] H.K.D.H. Bhadeshia, Steels for bearings, Progress in Materials Science 57 (2012) 268–435.

Mathematical modelling

16

Abstract

Mathematical models are not a panacea in the study or design of steels. This is because we simply do not understand many aspects of physical metallurgy, nor can we express all that is known into a quantitative framework. There is no "theory of everything" in any subject. But the fact is that models are useful in reducing the amount of work necessary to create new steels, sometimes quite dramatically, and at other times in pointing experiments into the right direction. But the prize is when a model does not simply reproduce what already is known, but makes unexpected predictions that disrupt the ordinary flow of thinking. We discuss here the essence of the methods now available to cover a wide range of structures and properties.

16.1 Introduction

There is much to be gained by creating theory that is experimentally verifiable and, which does not compromise with the complexity of technology. Such theory can and is used routinely in the design of new steels [1–3]. This is particularly so in the case of alloys where the number of variables involved is sufficiently large to make the problem interesting and the outcomes novel and tangible. However, it is worth emphasising at the outset that mathematical models are rarely sufficiently robust or versatile enough to enable complete solutions. The primary reason for their limitations is our inability to express everything using the language of mathematics. The models that exist are mere tools, rather like microscopes, which help understanding and reduce the task in creating new alloys.

Models can be classified into four main categories [4]:

1. Those which lead to an unexpected outcome that can be verified, for example, the creation of the δ-TRIP steel concept [5] or insight into the barriers to the formation of solid solutions in steel by the mechanical alloying process [6].
2. Those that are created or used in hindsight to explain diverse observations; the quasichemical model that explains why the diffusivity of carbon in austenite is particularly sensitive to concentration is one such case [7].
3. Existing models which are adapted or grouped to innovate materials or processes, one example being the design of blast-resistant naval steels [8,9];
4. Models that are used to express data, reveal patterns, or for implementation in control algorithms; examples that fall in this category have been reviewed [10].

These categories highlight the applications of models, with the emphasis being on quantitative expression. Fig. 16.1 illustrates how the concept of mathematical modelling, as described here, differs from ordinary science, which also yearns for the mathematical formulation of Nature. The ordinary scientific method has the purpose of revealing mechanisms to throw light on the workings of nature, with the possibility of making unexpected discoveries [11]. However, this scheme often involves simplification to a point where complexity, which also is a part of nature is dismissed. The

Steels. https://doi.org/10.1016/B978-0-44-318491-8.00022-8

claim in cosmology that there is a "theory of everything", neglects *entirely* the detail that matters to us in our every day lives, and yet it is exciting because there may be a plausible explanation of the origins of matter from energy that permeated empty space.

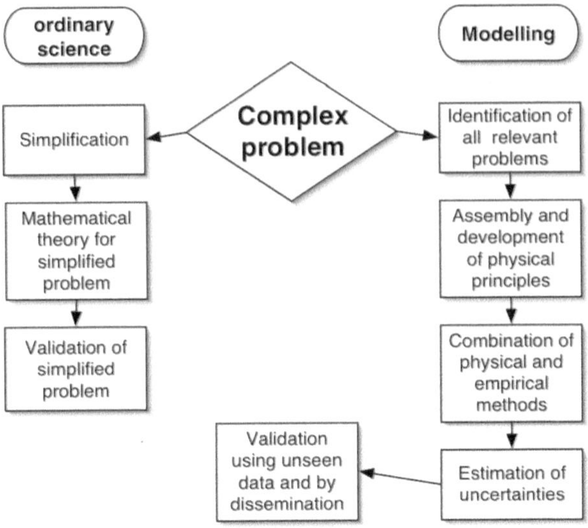

Figure 16.1 Defining qualities of a model compared with the conventional scientific method.

Modelling, in contrast, deals with technology where simplification or dissociation of the problem can make the problem mundane. The method is exciting when it leads to tangible products that can advance technology. We shall discuss the mathematical methods in the context of steel development, all concepts are of generic value.

The process of modelling begins with a need that cannot be met with existing technologies. There is a wide consultation to identify design parameters and tolerances for the problem concerned, Fig. 16.2. For example, in the case of a hydrogen-resistant steel to be developed for undersea applications, two targets might be set, one of which exceeds current technologies and the other much more ambitious and more risky. The thickness of the components produced by forging might range from 100–150 mm so the steel would need to be designed to be insensitive to a wide range of cooling rates, with the latter simulated using finite element methods. The strength, ductility, Charpy toughness to sub-zero temperatures, the crack-tip opening displacement (another standard measure of toughness), the resistance to hydrogen, weldability with and without post-weld heat treatment and structural stability to service up to 250°C over a period of some 25 years, would all need to be specified at the outset.

The modelling tools then are assembled and developed, and if necessary combined with empirical techniques and experience in order to create an overall procedure, taking considerable care to estimate uncertainties. Validation of the model is by testing against unseen data, either publicly available or provided by industry partners, is use-

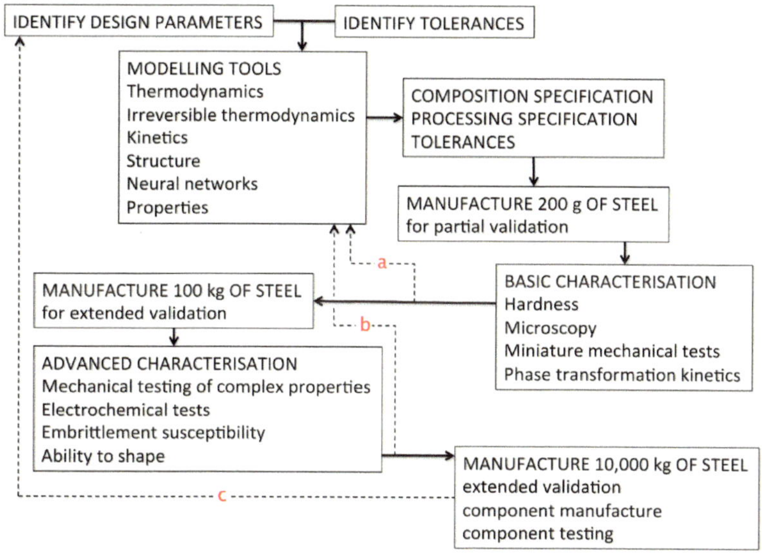

Figure 16.2 Procedure for the efficient design of novel steels. Paths a, b, c are return points in case the ideas generated at the early stages do not work. The costs of the development increase as the samples are scaled to component level testing. The cost of failure also increases in the order of the return paths a < b < c.

ful at this stage to gain confidence in the design procedure. Any principles embedded in the software need to be exposed to a wide range of scenarios to assess limits.

Design in complex problems usually leads to many solutions for the same specification. Therefore, a screening process begins with the manufacture of small samples of steel, typically 200 g each, which then are processed in a limited way. These samples can usually be made in a University environment. A variety of elementary tests can then be carried out to see whether the concepts are proven. If not, then there is a return path to the modelling ("a" in Fig. 16.2), but it may turn out that it simply is not possible to meet the design criteria, in which case there has to be a rethink of the technology. Assuming this is not the case and that the small melts show promise, proper mechanical characterisation requires larger quantities of steel that may also need to be thermomechanically processed, so 100 kg samples can be ordered from specialist suppliers. If all is well, the considerable resources necessary for component-level testing are justified.

In this way, the technological goal is hopefully achieved, and problems may be identified with the models, which in the longer term need to be resolved using the scientific approach. Like ordinary science, the proper use of models leads to insight, but by tackling complexity at the level that is posed, it can reveal issues which are lost during the simplification characteristic of ordinary science. For example, it is unlikely that the existence of a detrimental phase in high–strength steel weld deposits would have been revealed without the creation of a model which uncompromisingly accounts for the full range of variables that influence the Charpy toughness [12–16].

It is important to understand that modelling is not simply an application of a computer program, but rather the combination of a deep understanding of physical principles and the quantitative scientific method.

These issues are illustrated by two examples concerning steels, one an elementary microstructure model, and the other dealing with the mechanical properties of mixed microstructures.

The first example illustrates how phase transformation theory can be used in the optimisation of a microstructure consisting of a mixture of bainite and austenite. The problem is first identified to be associated with the occurrence of large regions of carbon-enriched austenite that are detrimental to toughness. The mechanism of transformation is then utilised to reduce the fraction of this detrimental phase. The major component of this model is the physical metallurgy of the transformation. The model is quantitative, in the sense that it requires the calculation of a phase boundary for a multi-component steel.

The second example helps to understand what is at first sight a strange result, that the strength of a mixed microstructure of martensite and bainite peaks as a function of the volume fraction of martensite. It illustrates how a variety of approximations can be made in order to formulate a model, both by searching the published literature for relationships and data, and by adopting a pragmatic approach.

16.2 Example 1: alloy design

Strong bainitic steels have not always been as successful as quenched and tempered martensitic steels because the coarse cementite particles in bainite are detrimental to toughness (Chapter 6). However, it is now known that the precipitation of cementite during bainitic transformation can be suppressed. This is done by alloying the steel with about 1.5 wt% silicon, which has a low solubility in cementite and therefore retards its growth.

An interesting microstructure results when this silicon-alloyed steel is transformed into upper bainite. The carbon that is rejected into the residual austenite, instead of precipitating as cementite, remains in the austenite and stabilises it down to ambient temperature. The resulting microstructure consists of fine plates of bainitic ferrite separated by carbon-enriched regions of austenite (Fig. 16.3).

The potential advantages of the mixed microstructure of bainitic ferrite and austenite can be listed as follows:

1. Cementite is responsible for initiating fracture in strong steels. Its absence during plastic deformation makes the microstructure more resistant to cleavage failure and void formation.
2. The bainitic ferrite is almost free of carbon, which is known to embrittle ferritic microstructures.
3. The microstructure derives its strength from the fine grain size of the ferrite plates, which are less than a millionth of a metre in thickness. It is the thickness of these

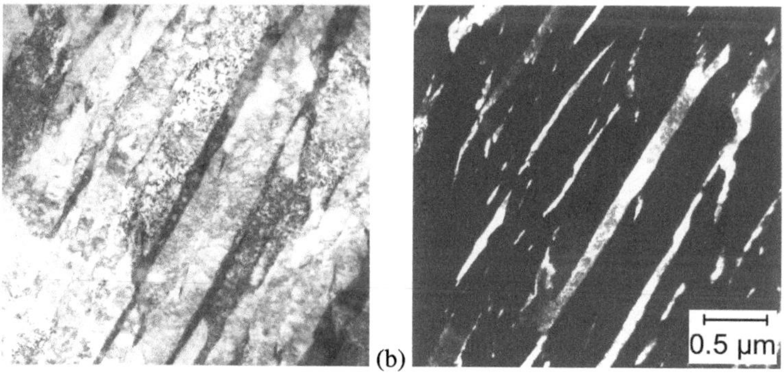

Figure 16.3 Transmission electron micrograph of a mixture of bainitic ferrite and stable austenite. (a) Bright-field image. (b) Retained austenite dark-field image.

plates which determines the mean free slip distance, so the effective grain size is less than a micrometre. This cannot be achieved by any other commercially viable process. It should be borne in mind that grain refinement is the only method available for simultaneously improving the strength and toughness of steels.

4. The ductile films of austenite which are intimately dispersed between the plates of ferrite have a crack blunting effect. They add to toughness by increasing the work of fracture as the austenite is induced to transform into martensite under the influence of the stress field of a propagating crack. This is the TRIP, or transformation-induced plasticity effect (Chapter 12). This effect is also responsible for supplementing work-hardening through the formation of martensite, which helps delay plastic instability during tensile deformation.
5. The diffusion of hydrogen in austenite is slower than in ferrite. The presence of austenite can, therefore, improve the stress corrosion resistance of the microstructure.
6. Steels with the bainitic ferrite and austenite microstructure can be obtained without the use of any expensive alloying additions. Once adequate hardenability is achieved, all that is required is that the silicon concentration should be large enough to suppress cementite.

In spite of these appealing features, the bainitic ferrite/austenite microstructure does not always give the expected good combination of strength and toughness. This is because the relatively large 'blocky' regions of austenite between the sheaves of bainite (Fig. 16.4) readily transform into high-carbon martensite under the influence of stress. This untempered, hard martensite embrittles the steel.

The blocks of austenite clearly are detrimental to toughness, and anything that can be done to reduce their fraction, or increase their stability to martensitic transformation, would be beneficial. Both of these effects are controlled by the T_0' curve of the phase diagram (Chapters 5 and 6). This curve determines the composition of the austenite at the point where the reaction to bainite stops. By displacing the curve to larger carbon concentrations, both the fraction of bainite that can form, and the carbon concentration of the residual austenite can be increased. Modifications to the T_0' curve

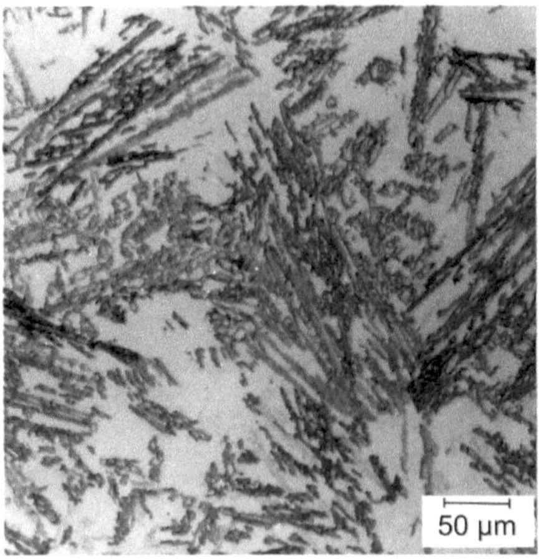

Figure 16.4 Optical micrograph of upper bainite in an Fe-0.43C-3Mn-2.02Si wt% showing the large blocks of untransformed retained austenite between the dark sheaves of bainite.

can be achieved by altering the alloy composition. It therefore is necessary to calculate the effect of substitutional solutes on the T_0' curve.

16.2.1 Calculation of the T_0 curve

Thermodynamic calculations of phase boundaries are now easy to conduct given the popularity of commercially available computer programs and accompanying data (e.g., [17]) and even freely available computer programs [18]. It is instructive, nevertheless to present a simple method for the evaluation of the T_0 temperature for multicomponent steels, where the free energies of austenite and ferrite of the same chemical composition are identical (Chapter 6). The change in free energy as austenite transforms to ferrite is therefore set to zero at the T_0 temperature:

$$\Delta G^{\gamma\rightarrow\alpha}\{T_0\} = 0. \tag{16.1}$$

Zener argued that the free energy difference can be factorised into two components, the magnetic ($\Delta G_{\mathrm{M}}^{\gamma\rightarrow\alpha}$) and non-magnetic ($\Delta G_{\mathrm{NM}}^{\gamma\rightarrow\alpha}$) terms:

$$\Delta G^{\gamma\rightarrow\alpha} = \Delta G_{\mathrm{M}}^{\gamma\rightarrow\alpha} + \Delta G_{\mathrm{NM}}^{\gamma\rightarrow\alpha}. \tag{16.2}$$

The non-magnetic component varies approximately linearly with temperature (Fig. 16.5), but the magnetic component varies non-linearly, becoming nearly zero at low temperatures. However, over a restricted temperature range (in which bainite usually forms), both functions can be represented approximately, as in Table 16.1.

The Zener factorisation of the free energy into magnetic and non-magnetic components helps to account for the effects of alloying elements, via a modification of the temperature at which the free energy is evaluated as follows, bearing in mind that the

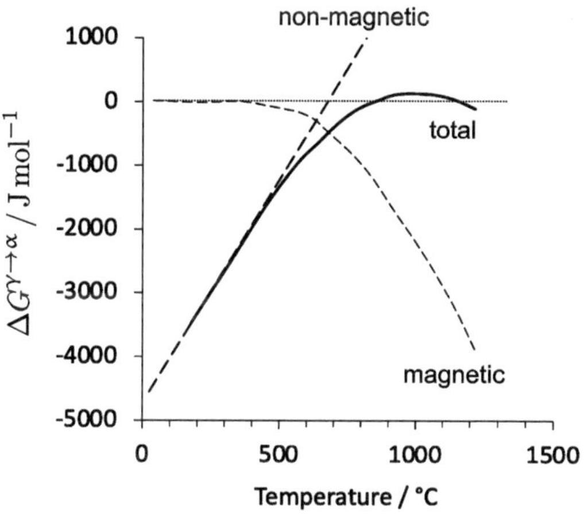

Figure 16.5 Zener's factorisation of the free energy difference between the austenite and ferrite phases into magnetic and non-magnetic components.

Table 16.1 Approximate representations of the free energy components for the $\gamma \to \alpha$ transformation in pure iron.

Function	a	b	Temperature range
$\Delta G_{NM}^{\gamma \to \alpha} = a + bT$ J mol^{-1}	-6660	7	$900 > T > 300$ K
$\Delta G_{M}^{\gamma \to \alpha} = a + bT$ J mol^{-1}	650	-1	$900 > T > 620$ K
$\Delta G_{M}^{\gamma \to \alpha} = a + bT$ J mol^{-1}	0	0	$T < 620$ K

terms in braces are the arguments of functions:

$$\Delta G^{\gamma \to \alpha}\{T\} = \Delta G_{M}^{\gamma \to \alpha}\{T - x\Delta T_{M}\} + \Delta G_{NM}^{\gamma \to \alpha}\{T - x\Delta T_{NM}\}. \qquad (16.3)$$

ΔT_{M} and ΔT_{NM} are temperature changes due to a unit concentration (x) of substitutional solute (Table 16.2). The T_0 temperature is therefore calculated by setting $\Delta G^{\gamma \to \alpha}$ to zero:

$$\Delta G_{M}^{\gamma \to \alpha}\{T_0 - x\Delta T_{M}\} + \Delta G_{NM}^{\gamma \to \alpha}\{T_0 - x\Delta T_{NM}\} = 0. \qquad (16.4)$$

By substituting the expressions listed in Table 16.1, this becomes:

$$a_{NM} + b_{NM}T_0^{Fe} + a_{M} + b_{M}T_0^{Fe} = 0 \text{ for pure iron,}$$

and

$$a_{NM} + b_{NM}(T_0^{FeX} - x\Delta T_{NM}) + a_{M} + b_{M}(T_0^{FeX} - x\Delta T_{M}) = 0 \text{ for an iron alloy.}$$

It follows that the change in the T_0 temperature caused by the addition of a substitutional element is given by the difference between these two equations:

$$\Delta T_0 = \frac{x(b_{\mathrm{NM}}\Delta T_{\mathrm{NM}} + b_{\mathrm{M}}\Delta T_{\mathrm{M}})}{b_{\mathrm{NM}} + b_{\mathrm{M}}}. \qquad (16.5)$$

The effect of several alloying elements can be approximated by assuming additivity:

$$\Delta T_0 = \frac{\Sigma_i x_i(b_{\mathrm{NM}}\Delta T_{\mathrm{NM}_i} + b_{\mathrm{M}}\Delta T_{\mathrm{M}_i})}{b_{\mathrm{NM}} + b_{\mathrm{M}}}. \qquad (16.6)$$

To calculate the shift in the T_0' temperature, we simply set $\Delta G^{\gamma\rightarrow\alpha}$ to the value of the stored energy (say 400 J mol^{-1} for bainite) instead of to zero. The actual T_0' curve for an alloy, rather than just the shift $\Delta T_0'$ relative to pure iron, can be estimated by noting that for an Fe-C alloy, allowing for 400 J mol^{-1} of stored energy:

$$T_0'(\mathrm{K}) \simeq 970 - 80x_{\mathrm{C}}, \qquad (16.7)$$

where x_{C} is the at % of carbon.

We can now proceed to apply this methodology to the design of a tough bainitic ferrite/austenite microstructure.

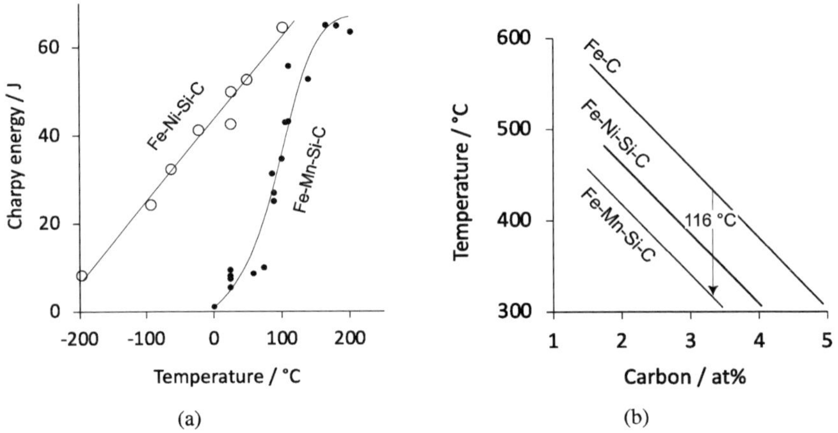

Figure 16.6 (a) Experimentally determined impact transition curves showing how the toughness improves as the amount of blocky austenite is reduced. (b) Calculated T_0' curves for the Fe-C, Fe-Mn-Si-C and Fe-Ni-Si-C steels.

16.2.2 The improvement in toughness

An apparently ideal microstructure consisting of bainitic ferrite and ductile austenite in a Fe-3Mn-2.02Si-0.43C wt% exhibits poor toughness because of the presence of blocky, unstable austenite (Fig. 16.6a). It is necessary to increase the amount of bainitic ferrite in the microstructure and to increase the stability of the austenite. Both of these aims can be achieved by changing the substitutional solute concentration such that the T_0' curve is shifted to higher carbon concentrations (i.e., T_0' is raised at any given carbon concentration).

Table 16.2 Values of ΔT_M and ΔT_{NM} for a variety of substitutional solutes (after Aaronson et al. [19].)

Alloying element	ΔT_M / K per at %	ΔT_{NM} / K per at %
Si	−3	0
Mn	−37.5	−39.5
Ni	−6	−18
Mo	−26	−17
Cr	−19	−18
V	−44	−32
Co	19.5	16
Al	8	15
Cu	4.5	−11.5

Using Eq. (16.6), we see that for the Fe-3Mn-2.02Si-0.43C wt% (2.97 Mn, 3.87Si at%) alloy:

$$\Delta T_0 = \overbrace{\frac{2.97[7 \times (-39.5) + (-1) \times (-37.5)]}{7-1}}^{\text{Mn}} + \overbrace{\frac{3.87[7 \times (0) + (-1) \times (-3)]}{7-1}}^{\text{Si}} = 116\,\text{K},$$

the first term on the right-hand side being the effect of manganese and the second the effect of silicon. Hence, for this alloy Eq. (16.7) can be modified to give:

$$T_0'(\text{K}) \simeq 970 - 80x_C - 116. \tag{16.8}$$

It is shown that manganese has a great influence on lowering the T_0' temperature. An examination of Table 16.2 shows that one possibility is to replace all the manganese with nickel. Thus, for a Fe-4Ni-2Si-0.4C wt% (3.69Ni, 3.85Si at%) alloy, a similar calculation shows that $\Delta T_0 \simeq 72$ K so that:

$$T_0'(\text{K}) \simeq 970 - 80x_C - 72. \tag{16.9}$$

The remarkable improvement in toughness achieved by doing this, without any sacrifice of strength is illustrated in Fig. 16.6, along with the T_0' curves as calculated above.

16.2.3 Precision and limits

The model discussed above has helped in achieving the desired goal of improved toughness, even though the method used is in fact crude. The T_0' curves are not really linear functions of carbon, and the interactions of carbon with the substitutional solutes are not accounted for. Better methods are available [17,20] and could be used when considering a higher degree of optimisation of steel chemistry. The model also does not incorporate kinetics. This could be a disadvantage, because in commercial practice microstructures are generated using complex non-isothermal heat treatments.

It is likely that the exploitation of these concepts would require several iterations to improve precision and adapt the model for complex industrial processing. These examples illustrate the essentials of the modelling technique. Models can be constructed in stages, with significant advances being made at each stage, even though the ultimate problem may not be solved completely. These step-wise successes of the model can be used to justify further development until a point of diminishing returns is reached.

Models can help set the limits to what can or cannot be achieved [21]. Fig. 16.7a is, for carbide-free bainite, a plot of the transformation temperature as a function of the transformation time. The points are experimental data, but the lines, which are calculated using machine learning, define the domain of possible conditions of heat treatment; the two lines define the uncertainty of the calculation.

As an example, it is not possible to complete the bainite reaction at 150°C in 40 min (point marked "X"), it would take much longer. However, it may be useful to have a slow transformation rate when making large components – the material can then reach a homogeneous temperature before transformation starts, encouraging the development of a uniform structure and properties. The point marked "Y" shows that slow transformation is not possible if the steel is designed to transform at a high temperature. Thus, in the interests of uniformity, the alloy composition must be such that the transformation is suppressed to low temperatures. The transformation time for bainite increases as the concentration of austenite-stabilising solutes is enhanced (Fig. 16.7b), as the transformation is suppressed to lower temperatures (Fig. 16.7b).

The strengthening achieved is greater when the structure is generated at low-temperatures/long transformation-times. Fig. 16.7c, in fact, defines the domain of possible combinations of ultimate tensile strength and transformation time. Obviously, it is not possible to achieve 2 GPa strength in a rapidly transforming steel.

In microstructures consisting of just bainitic ferrite and carbon-enriched austenite, tensile failure is known to occur when the austenite content decreases to below a threshold value of about $V_V^\gamma \approx 0.1$ [27]. The initial fraction $V_V^{\gamma\circ}$ of austenite (Fig. 16.7d) will decrease due to deformation-induced martensitic transformation, and the way in which this decomposition occurs as a function of plastic strain can be calculated [28]. It follows that with $V_V^{\gamma\circ}$ and the curve defining how it decreases with plastic strain, the elongation to failure is easy to estimate by assuming that failure occurs when the percolation threshold ($V_V^\gamma \approx 0.1$) is breached. Therefore, steel B, which has a smaller initial fraction of austenite, exhibits a lower elongation ϵ^B than steel A.

The concepts illustrated in Fig. 16.7 can be used before any design procedure begins, to assess whether the properties desired are in fact possible to achieve.

16.3 Example 2: properties of mixed microstructures

A peculiar feature of mixed microstructures of bainite and tempered martensite, is that the strength is found to go through a peak as the volume fraction of marten-

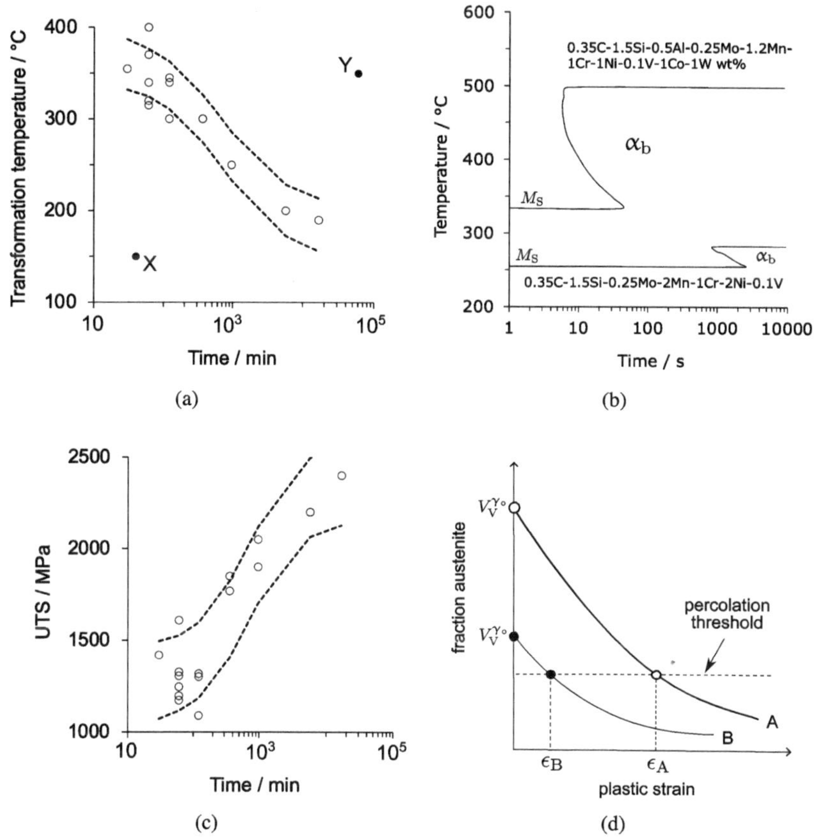

Figure 16.7 Published experimental data on carbide-free bainite (points) [22–26] as a function of transformation time. The curves represent $\pm 1\sigma$ confidence limits obtained by subjecting the data to a neural network analysis. (a) Transformation temperature as a function of isothermal transformation time. (b) Calculated TTT diagrams for two steels. (c) Ultimate tensile strength as a function of isothermal transformation time. (d) The percolation model for the elongation of carbide-free bainite.

site decreases (Fig. 16.8). This is against intuition in that martensite is considered usually to be the strongest microstructure in steels. In this case, the strength should decrease continuously as the fraction of martensite is reduced. However, quantitative modelling, by helping to reveal the mechanisms involved, can explain this anomalous behaviour.

16.3.1 Calculation of the strength of individual phases

It is reasonable to assume that the strength of martensite and bainite can be factorised into a number of intrinsic components, although, as pointed out in Section 2.8, a linear

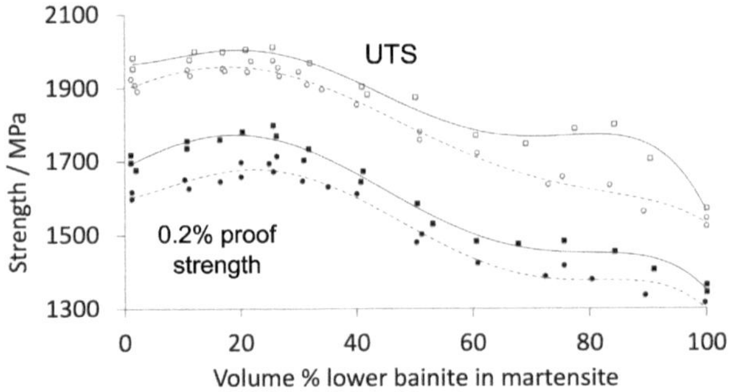

Figure 16.8 Comparison of calculations against experimental data due to Tomita and Okabayashi.

combination of terms may not always be appropriate:

$$\sigma = \sigma_{\mathrm{Fe}} + \sum_i x_i \sigma_{\mathrm{SS}_i} + \sigma_{\mathrm{C}} + K_{\mathrm{L}}(\overline{L})^{-1} + K_{\mathrm{d}}\rho_{\mathrm{d}}^{0.5}, \tag{16.10}$$

where x_i is the concentration of a substitutional solute, which is represented here by a subscript i. The other terms in this equation are as follows:

K_{L}	=	coefficient for strengthening due to lath size, 115 MPa
K_{d}	=	coefficient for dislocation strengthening, 7.34×10^{-6} MPa
σ_{Fe}	=	strength of pure, annealed iron, 219 MPa at 300 K
σ_{SS_i}	=	substitutional solute (i) strengthening
σ_{C}	=	solid solution strengthening due to carbon
ρ_{d}	=	dislocation density, typically 10^{16} m^{-2}
$\overline{L}$	=	measure of the ferrite plate size, typically 0.2 μm.

The individual strengthening contributions are discussed below.

Table 16.3 Strengthening contributions (MPa) of pure iron and other factors as a function of temperature and solid solution strengthening terms for ferrite, for one wt% of solute. The data are for a strain rate of 0.0025 s^{-1}, calculated as in [29].

	200°C	**100°C**	**Room temperature (23°C)**	**−40°C**	**−60°C**
Fe	215	215	219	355	534
Si	78	95	105	70	−44
Mn	37	41	45	8	−57
Ni	19	23	37	−2	−41
Mo	-	-	18	-	-
Cr	7.8	5.9	5.8	7.4	15.5
V	-	-	4.5	-	-
Co	1.0	1.8	4.9	9.1	5.8

16.3.2 Iron and substitutional solutes

Pure body-centred cubic iron in a fully annealed condition makes an intrinsic contribution σ_{Fe} to the overall strength. Substitutional solutes do not partition during the displacive growth of either martensite, bainite or Widmanstätten ferrite, so their concentrations are fixed by the composition of steel as a whole. Solid solution strengthening contributions, σ_{SS} can be estimated as a function of temperature and strain rate from published data. Table 16.3 shows that whereas the strength of pure iron increases as the temperature is reduced, strengthening due to substitutional solutes often goes through a maximum as a function of temperature. Indeed, there is some solution softening at low temperatures because the presence of a foreign atom locally assists a dislocation to overcome the Peierls barrier at low temperatures.

16.3.3 Carbon

Bainitic ferrite ordinarily has only a small amount of carbon dissolved in interstitial solution, often assumed to be less than 0.02 wt%. Martensite, on the other hand, can have concentrations well in excess of the average in the alloy, since the prior formation of bainite enriches the residual austenite according to the following relationship derived from a balance of mass. The total carbon concentration in the alloy is the sum of the concentrations in the austenite (x_γ) and bainitic ferrite (x_{α_b}):

$$\overline{x} = x_\gamma V_{\mathrm{V}}^{\gamma} + x_{\alpha_b} V_{\mathrm{V}}^{\alpha_b}, \tag{16.11}$$

where V_{V}^{γ} and $V_{\mathrm{V}}^{\alpha_b}$ are the volume fractions of austenite and bainitic ferrite, respectively. It follows that:

$$x_\gamma = \frac{\overline{x} - V_{\mathrm{V}}^{\alpha_b} x_{\alpha_b}}{1 - V_{\mathrm{V}}^{\alpha_b}}, \tag{16.12}$$

where x_γ is the concentration in the residual austenite before it transforms into martensite, which is important in determining the hardness of the martensite. Solid-solution theory indicates that the strength increment due to carbon dissolved in ferrite should vary with the square root of the carbon concentration [30]:

$$\sigma_{\mathrm{SS_C}} = 1722.5 \times w_{\mathrm{C}}^{1/2} \text{ MPa}. \tag{16.13}$$

16.3.4 Dislocations

When martensite or bainite form at high temperatures, the shape change due to shear transformation causes plastic deformation, and hence the accumulation of dislocations in both the parent and product phases (Chapter 6). The extent of the plasticity depends on the yield strength, and hence on the temperature. It has therefore been suggested [31] that the dislocation density (ρ_{d}) of both martensite and bainite can be represented empirically as a function of temperature alone, for the temperature range 570-920 K:

$$\log_{10}\{\rho_d\} = 9.2840 + \frac{6880.73}{T} - \frac{1780360}{T^2}, \tag{16.14}$$

where T is the transformation temperature in Kelvin and ρ_d is stated in units of m^{-2}. The strengthening σ_ρ due to dislocations is given by:

$$\sigma_\rho = 0.38\,\mu b(\rho_d)^{0.5} \simeq 7.34 \times 10^{-6}(\rho_d)^{0.5}, \quad \text{MPa} \tag{16.15}$$

where μ is the shear modulus and b is the magnitude of the Burgers vector.

16.3.5 Lath size

Martensite and bainite grow in the form of very fine plates or laths. The resulting grain size strengthening σ_L is defined as:

$$\sigma_L \simeq 115(\overline{L})^{-1} \quad \text{MPa}, \tag{16.16}$$

where $\overline{L}$ (μm) is the mean linear intercept measured on random sections. This is not the classical Hall–Petch relation (Chapter 2), but another relation due to Langford and Cohen, because at the typically sub-micrometre grain sizes, the mechanism of yield is different, involving the initiation of dislocation sources in the grain boundaries.

16.3.6 Martensite composition and transformation temperature

The excess carbon in the bainitic ferrite partitions into the residual austenite, which then transforms to martensite. The carbon concentration of the martensite can therefore be calculated using equation (16.11). The martensite-start temperature of the residual austenite can be estimated using Eq. (5.2):

$$M_S = \overline{M}_S - 539(x_\gamma - \overline{x}) \quad {}^\circ\text{C}, \tag{16.17}$$

where the concentrations are in wt%, the temperatures in centigrade and $\overline{M}_S$ is the martensite-start temperature of austenite with the average composition of the alloy.

The different contributions to the strength of martensite are illustrated in Fig. 16.9. Carbon is a major contributor since it causes a severe, asymmetrical distortion of the martensite crystal structure and hence interacts strongly with the movement of dislocations. The dislocation density itself makes a significant contribution to the overall strength.

16.3.7 Strength of mixed microstructures

The normal way to calculate the strength of a multiphase alloy is to use a rule of mixtures, i.e., to calculate a mean strength from the strength of each component phase

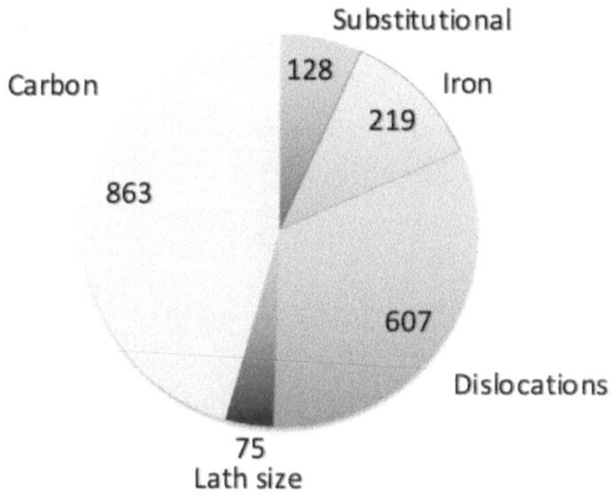

Figure 16.9 Calculated components of the room-temperature strength (MPa) of virgin martensite in Fe-0.4C-0.2Si-0.71Mn-1.9Ni-0.25Mo-0.88Cr wt% alloy. This is a typical strong steel of the type used in the manufacture of gears, gun barrels, etc.

weighted by its volume fraction. However, this is not adequate for the present purposes because of constraint effects. It is well established in fracture mechanics that the yield strength is increased by plastic constraint. This is why a weak brazing alloy can bond much stronger samples as long as the thickness of the braze material is small enough to be constrained throughout by the surrounding stronger matrix. Indeed, the strength of the joint increases as the thickness of the braze layer decreases.

Dispersions of bainite plates form in austenite which subsequently transforms to much stronger martensite. The problem can be modelled assuming that deformation of the bainitic ferrite is constrained by the harder martensite in the same way as the braze material is constrained by the surrounding matrix [32]. The constraint can, therefore, be modelled using experimental data available from brazed joints in high-strength steels. The data in a normalised form are summarised in Fig. 16.10a. The vertical axis is the joint strength normalised with respect to that of the unconstrained braze material; the horizontal axis is the braze thickness normalised relative to a thickness value where the restraint effect vanishes.

To analyse the properties of a mixed microstructure, it can be assumed that the normalised braze thickness is equivalent to the volume fraction of bainite. Under this assumption and the form of the normalised strength versus normalised thickness plot (Fig. 16.10a), the strength of constrained bainite may be represented by the equation:

$$\sigma \simeq \sigma_0[0.65\exp\{-3.3V_V^{\alpha_b}\}+0.98] \leq \sigma_M \quad \text{MPa}, \tag{16.18}$$

where σ and σ_0 represent the strengths of constrained and unconstrained bainite, respectively, and σ_M is the strength of the martensite – that of bainite is always less than or equal to that of martensite.

When the volume fraction $V_V^{\alpha_b}$ of bainite is small, its strength nearly matches that of martensite (Fig. 16.10b), always remaining above that of bainite on its own. The

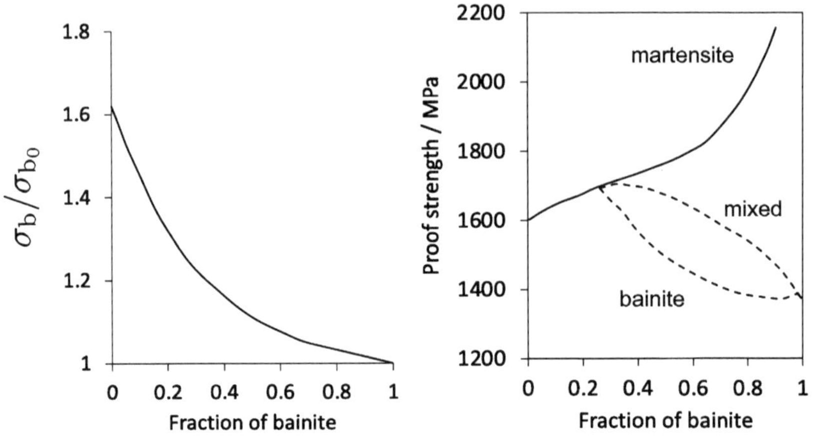

(a) (b)

Figure 16.10 (a) Plot of the normalised strength of a brazed joint versus the normalised thickness of the brazing material, the latter being identified with the fraction of bainite in a martensitic matrix. (b) The strength contributions of bainite and martensite in the mixed microstructure.

strength of martensite continues to increase with the fraction of bainite, as the carbon concentration of the residual austenite from which it grows, increases.

Fig. 16.11 shows how the strength of the mixed microstructure is predicted. Line 'a' on Fig. 16.11 shows that a rule of mixtures cannot account properly for the variations observed. The agreement between calculation and experiment improves (curve b) as allowance is made for the change in the strength of martensite as carbon partitions into the austenite due to the formation of bainite. The consistency between experiment and theory becomes excellent as constraint effects are also included in the calculations (curve c).

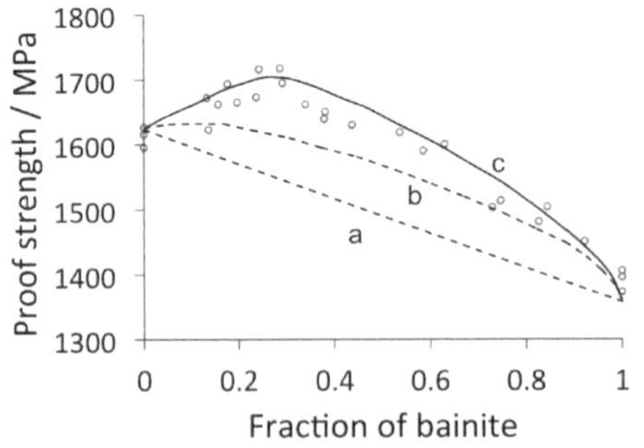

Figure 16.11 Comparison of calculations against experimental data due to Tomita and Okabayashi. The continuous lines represent tests done at 203 K and the dashed line for measurements at 287 K.

16.4 Methods

The two examples described in the preceding sections are necessarily simplified presentations of quite complex models. It is useful to illustrate some of methods that are now common in the mathematical modelling of steels. It is worth emphasising that in general it is a combination of methods that leads to useful solutions, with the optimum approach to a problem being one that is interdisciplinary, with a bank of requirements outlined before the search for solutions. A design problem is always more interesting when there are many solutions to the same requirements. In mathematics, a Pareto front is the set of all solutions in multi-objective optimisation that are considered equally good [33].

16.4.1 Electron theory

A metal is created when atoms are brought so close together that the electrostatic repulsion in transferring a valency electron between the adjacent atoms is offset by the gain due to the delocalisation of electrons. This enables the valency electrons to move within the metal. The delocalised electrons feel a weak electrostatic field from the positively charged cores of atoms because of repulsion by the core electrons. The valence electrons are also screened from each other by positive holes that surround them. All this makes it possible to introduce approximations, which allow a single-electron wave function to be exploited in calculating the energy of an electron gas in a metal.

These electrons in the metallic bonding can move without being scattered by the partly screened potential of the positive ion cores because the latter provide a *periodic* potential whose effect is simply to modulate the free-electron wave function. Difficulties arise only when the electrons satisfy the Bragg condition within the metal. This introduces *band gaps* in the distribution of electron energies. The metallic state can exist only if the valency bands are partly filled.

Using these concepts, the energy of the electron gas can be expressed in terms of the potential due to the ion cores, Coulomb interactions, kinetic energy and exchange and correlation effects. It is then possible to calculate with an input of the electronic charge and the atomic number of the element, properties such as the cohesive energy of crystals, the elastic moduli, magnetic and acoustic properties. The calculations are known as *first principles* or *ab initio* calculations because they do not require experimental inputs other than the charge on the electron and the types of atoms involved. The calculations are limited to small numbers of atoms because they are computer intensive.

Fig. 16.12 shows some calculations of the cohesive energy at 0 K of two allotropic forms of iron, fcc and the hypothetical structure diamond cubic. In each case, the cohesive energy goes through a minimum, which gives the expected density of the allotrope. The calculation of the diamond form of iron shows how it is possible, using electron theory, to estimate the properties of phases which do no exist in reality. Such a form would have a density of only $5\,\mathrm{g\,cm^{-3}}$, but unfortunately, the energy difference relative to the stable forms of iron is simply too large, meaning that it would be improbable for the fcc$\rightarrow$diamond transition to be induced, e.g., by alloying.

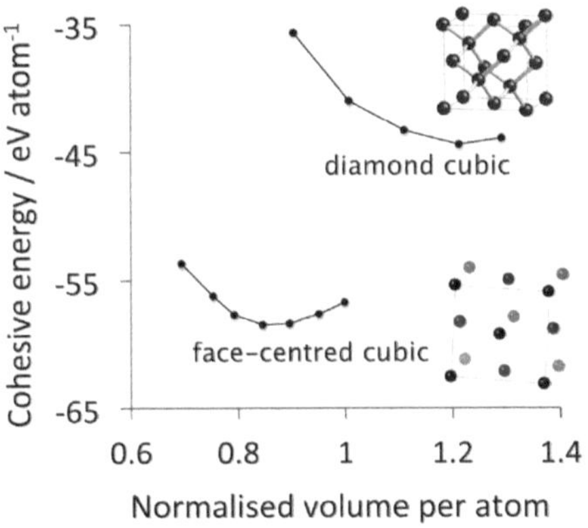

Figure 16.12 The cohesive energy at 0 K and zero pressure, versus the volume per atom divided by the volume of an iron atom, for two crystal structures of iron. Selected data from Paxton et al. [34].

Calculations like these are now applied routinely in the context of steels, particularly when clear experiments are impossible – a few examples are listed here for illustration purposes:

1. A useful concept that can only be verified using electron theory, states that grain boundaries are embrittled if it is more favourable for a solute to segregate to a free surface [35]. Thus, it has been confirmed using first-principles calculations, that phosphorus does in fact have a high tendency to embrittle special boundaries [36]. The calculations prove that the embrittlement is not due to a reduction in Fe-Fe bonding due to the presence of the phosphorus.
2. Fine titanium carbides in automotive sheet-steel coarsen less if molybdenum is added. This is so even though molybdenum is not favoured in the TiC. The coarsening occurs when the hot steel is coiled as it comes out of the rolling mill. First-principles calculations elucidate the mechanism, that even though Mo in TiC is thermodynamically unfavourable, its presence reduces the misfit of the carbide with the parent phase, thus making it easier to nucleate [37]. This is why coarsening is retarded because the coarsening rate scales with the interfacial energy per unit area.
3. One theory for the embrittlement of steel by hydrogen is that it reduces the cohesive energy of iron, making it easier to cleave [38]. Iron cleaves on the $\{100\}_\alpha$ planes. To establish whether this is a significant effect, first-principles estimates of the change in the energy between hydrogen in the dissolved state and when hydrogen is at the $\{100\}_\alpha$ surface were calculated [39]. If there is a reduction in energy on exposing the cleavage plane, then the material can be considered embrittled. As illustrated in Fig. 16.13, the calculations revealed only a small dependence of the surface energy on hydrogen for the classical cleavage plane in ferrite, much smaller than the profound embrittlement observed at minute concentrations of hydrogen.

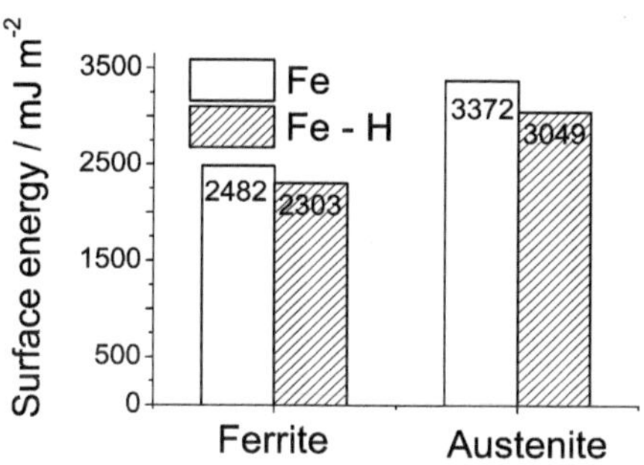

Figure 16.13 Calculated surface energy of ferrite and austenite for pure metal and hydrogen-containing system.

4. Silicon has a negligible solubility in cementite. The design of bainitic steels containing retained austenite relies on this because the addition of silicon suppresses cementite and hence the austenite enriched in carbon, is retained. The suppression occurs when cementite can only grow by a para-equilibrium mechanism, in which case it is forced to absorb silicon. There are no experimental thermodynamic data for silicon in cementite. Therefore, first-principles calculations were used to generate the data for subsequent use in phase diagram calculations for alloy design [40,41].

16.4.2 Phase diagram calculations and thermodynamics

Given experimentally determined thermodynamic data, it is possible to estimate in multicomponent, multiphase alloys, the stable phases, their equilibrium fractions and equilibrium chemical compositions as a function of temperature, pressure, magnetic fields and the detailed composition of the alloy. In other words, all the information plotted on phase diagrams.

The free energy of a phase α is simply the weighted mean of the free energies of its component atoms (μ_i), which for a binary solution containing components "A" and "B" is:

$$G^{\alpha} = (1 - x)\mu_{\mathrm{A}}^{\alpha} + x\mu_{\mathrm{B}}^{\alpha},$$

where x is the mole fraction of B. μ_i^{α} is known as the *chemical potential* of component i in phase α. Although this equation is expressed for a binary solution, it is generally true that equilibrium between any number of phases in contact and containing any number of components is defined by:

$$\mu_i^{\alpha} = \mu_i^{\beta} = \ldots \qquad \text{for} \quad i = 1, 2, 3, \ldots \qquad \text{and} \qquad \text{phase} = \alpha, \beta \ldots$$

The chemical potential must be uniform everywhere at equilibrium.

There are many thermodynamic methods, which express the chemical potential as a function of the mixing of solutes in a phase. Most of these methods either are too simple or so complex that they cannot easily be generalised. Therefore, in the computer calculations, the deviation of the free energy of mixing from that of an ideal solution, i.e., the excess Gibbs free energy, is written as an empirical polynomial equation:

$$\Delta_e G_{AB} = x_A x_B \sum_i L_{AB,i}(x_A - x_B)^i,$$

where L_i are measured interaction coefficients, in this case for a binary solution. In a ternary solution interaction, the term would be of the form:

$$\begin{aligned}\Delta_e G_{ABC} = {} & x_A x_B \sum_i L_{AB,i}(x_A - x_B)^i \\ & + x_B x_C \sum_i L_{BC,i}(x_B - x_C)^i \\ & + x_C x_A \sum_i L_{CA,i}(x_C - x_A)^i.\end{aligned}$$

The advantage of this kind of a polynomial becomes clear, since the relation reduces to the binary problem when one of the components is set to be identical to another, e.g., $B \equiv C$. The method can be extended to deal with any number of components, with the great advantage that few coefficients have to be changed when the data due to one component are improved. It is therefore adopted in many of the phase diagram calculation programs available commercially.

Although thermodynamics usually is associated with the state of equilibrium, the calculation method can also be used to estimate constrained equilibria, e.g., para-equilibrium (Chapter 3) and diffusionless transformation involving solute-trapping (Chapter 5). Fig. 16.14 illustrates calculated isothermal Fe-Cr-C phase diagrams for both the equilibrium and para-equilibrium states – notice the dramatic change when substitutional solutes are not allowed to partition between the phases.

There is another subtle application of thermodynamics in the design of steels, dealing with steady-state processes in which the system is not at equilibrium but an appropriate observer may not perceive change. An example is diffusion across a constant gradient. Neither the flux nor the concentration at any point changes with time, and yet the free energy of the system is decreasing since diffusion occurs to minimise free energy. The rate at which energy is dissipated is the product of the temperature and the rate of entropy production (i.e., $T\sigma_e$):

$$T\sigma_e = JX,$$

where J is a generalised flux of some kind and X a generalised force. In the case of an electrical current, the heat dissipation is the product of the current (J) and the electromotive force (X). Provided that flux-force sets can be expressed in this way, it is found that $J \propto X$ for small deviations from equilibrium. In the case of the electrical current, this leads to Ohm's law, where the current is proportional to the electromotive force.

This concept can be applied to the case where a number of irreversible processes occur simultaneously. In a ternary Fe-Mn-C alloy, the diffusion flux of carbon depends

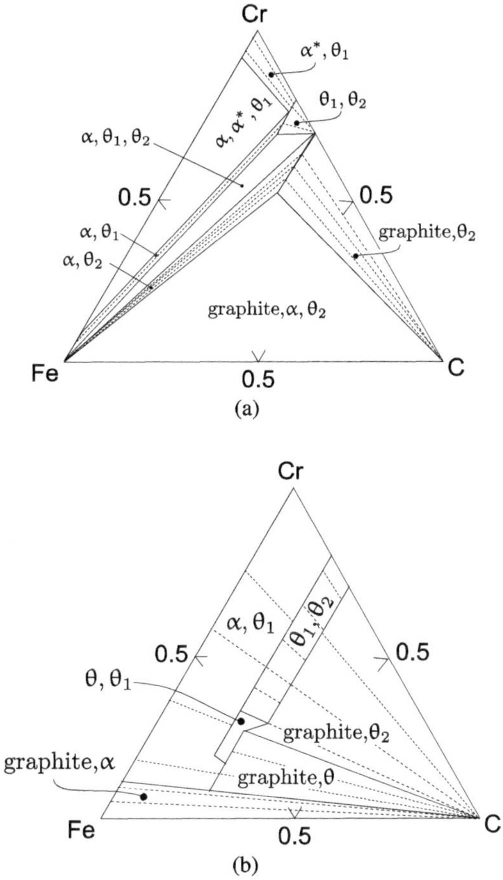

Figure 16.14 Isothermal sections of the Fe-Cr-C system. The body-centred cubic phase is ferrite and "M" stands for a mixture of iron and chromium atoms in a variety of carbide phases (courtesy of J. Robson). θ_1 and θ_2 represent $M_{23}C_6$ and M_7C_6 respectively. (a) Equilibrium. (b) Para-equilibrium.

not only on the gradient of carbon, but also on that of manganese. Thus, a uniform distribution of carbon tends to become inhomogeneous in the presence of a manganese concentration gradient. When there is more than one dissipative process, the total energy dissipation rate is the sum of all the dissipations:

$$T\sigma_e = \sum_i J_i X_i,$$

with

$$J_i = M_{ij} X_j \quad \text{with} \quad i, j = 1, 2, 3 \ldots$$

and it is the cross coefficients M_{ij} $(i \neq j)$ that drive the diffusion of carbon in a gradient of manganese. The theory is applied widely in computer calculations of the kinetics of phase transformations in steels.

16.5 Kinetics

Almost all the solid-state transformations in steels involve nucleation and growth. The theories for these two processes are well established and have been described in Section 3.6.5. The evolution of the volume fraction requires the additional treatment of *impingement* between particles which nucleate at different locations. The change in the real volume of the product phase (α) can be related to the extended volume using Eq. (3.13) that was introduced in Chapter 3 as:

$$\mathrm{d}V^{\alpha} = \left(1 - \frac{V^{\alpha}}{\mathrm{V}}\right)\mathrm{d}V_{\mathrm{e}}^{\alpha},$$

where it is assumed that the microstructure develops at random. The subscript "e" refers to extended volume, V^{α} is the volume of α and V is the total volume.

However, it often is the case that different transformation products occur simultaneously from the parent phase, albeit at different rates. The extended volume idea can be generalised to this scenario [42–45]. Suppose α and β precipitate simultaneously, then the relation between extended and real space becomes a coupled set of two equations:

$$\mathrm{d}V^{\alpha} = \left(1 - \frac{V^{\alpha} + V^{\beta}}{\mathrm{V}}\right)\mathrm{d}V_{\mathrm{e}}^{\alpha} \quad \text{and} \quad \mathrm{d}V^{\beta} = \left(1 - \frac{V^{\alpha} + V^{\beta}}{\mathrm{V}}\right)\mathrm{d}V_{\mathrm{e}}^{\beta}, \quad (16.19)$$

which must generally be solved numerically.

When martensitic transformation occurs under the influence of an externally applied stress, particular crystallographic orientations that comply better with the applied stress are stimulated first, but this does not rule out the formation of other orientations. The martensite-start temperature of the optimally oriented martensite (M_{S_1}) will be greater than of a less favoured orientation (M_{S_2}). Both orientations of martensite will grow below M_{S_2} and the theory described in equation (16.19) can be used to follow the evolution of the volume fractions of each kind of martensite as a function of temperature. Fig. 16.15 illustrates such calculations as an example of the simultaneous transformation model [46].

There has in recent years been much prominence given to the phase field method as an alternative technique for calculating the evolution of microstructure [47]. This begins with the description of the entire microstructure in terms of an *order parameter*. The precipitate and matrix each have a particular value of the order parameter and the interface between them is located by the position where the order parameter changes from its precipitate-value to its matrix-value. The range over which it changes is the *width* of the interface. The set of values of the order parameter over the whole microstructure is the *phase field*.

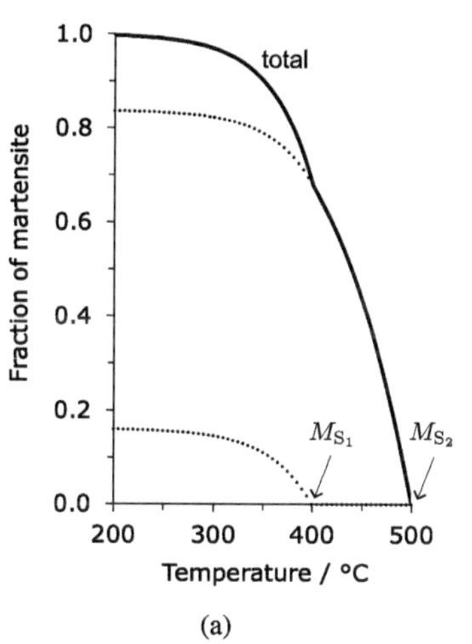

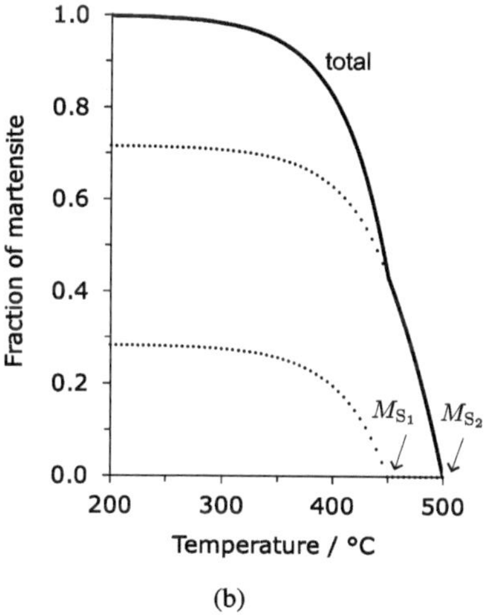

Figure 16.15 Two martensite-start temperatures, (a) 500°C and 400°C, (b) 500°C and 450°C.

The free energy per atom is then written for the whole of the (heterogeneous) phase field as a single functional and the evolution of microstructure with time is assumed to be proportional to the variation of this functional with respect to the order parameter.

The method has been extremely successful in dealing with spinodal reactions and in the modelling of solidification, but its utility with respect to solid-state reactions, such as are important in steels, has yet to be demonstrated. The definition of the width of the interface and associated coefficients, and handling nucleation are two difficulties that require fitting to experimental data. On the other hand, effects such as the overlap of diffusion fields are natural outcomes.

16.5.1 Finite difference method

The finite difference is a discrete analogue of a derivative. Consider one-dimensional diffusion in a concentration gradient along a coordinate z (Fig. 16.16) The concentration profile is divided into slices, each of thickness h. The matter entering a unit area of the face at "a" in a time increment τ is given approximately by $J_a = -D\tau(c_1 - c_0)/h$. That leaving the face at 'b' is $J_b = -D\tau(c_2 - c_1)/h$. If c_1' is the new concentration in slice 1, then the net gain in solute is $(c_1' - c_1)h$, so that:

$$c_1' - c_1 = \frac{D\tau}{h^2}(c_0 - 2c_1 + c_2). \tag{16.20}$$

This allows the concentration at a point to be calculated as a function of that at the two neighbouring points. By successively applying this relation at each slice, and advancing the time τ, the entire concentration profile can be estimated as a function of time.

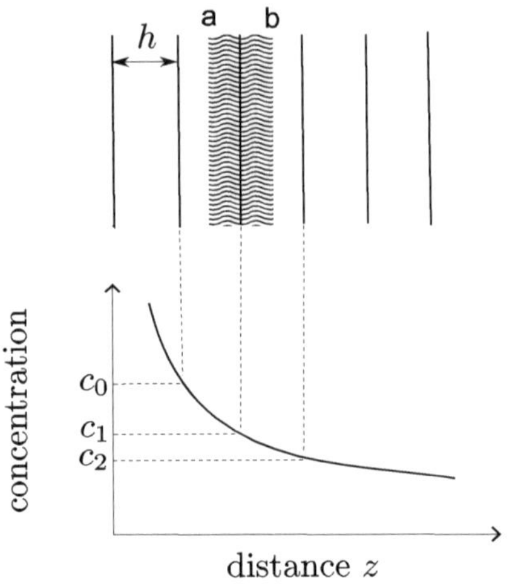

Figure 16.16 Finite difference representation of diffusion.

The approximation is that the concentration gradient within each slice was assumed to be constant. This approximation is better for smaller values of h, but at the expense of computation time. The accuracy can be assessed by changing h and seeing whether it makes a significant difference to the calculated profile.

16.6 Finite element method

In this, continuous functions are replaced by piecewise approximations. The consequence of applying force to a body represented as a set of springs is illustrated here, assuming that the force F in each spring varies linearly with the displacement δ, with the constant of proportionality labelled the stiffness k. The body is at rest at equilibrium, so for the case illustrated in Fig. 16.17a, $F_1 = -F_2$ so that:

$$\begin{bmatrix} F_1 \\ F_2 \end{bmatrix} = \begin{pmatrix} k & -k \\ -k & k \end{pmatrix} \begin{bmatrix} \delta_1 \\ \delta_2 \end{bmatrix},$$

The forces at the nodes of the springs illustrated in Fig. 16.17b are therefore

$$\begin{bmatrix} F_1 \\ F_2 \\ 0 \end{bmatrix} = \begin{pmatrix} k_1 & -k_1 & 0 \\ -k_1 & k_1 & 0 \\ 0 & 0 & 0 \end{pmatrix} \begin{bmatrix} \delta_1 \\ \delta_2 \\ \delta_3 \end{bmatrix}$$

$$\begin{bmatrix} 0 \\ F_2 \\ F_3 \end{bmatrix} = \begin{pmatrix} 0 & 0 & 0 \\ 0 & k_2 & -k_2 \\ 0 & -k_2 & k_2 \end{pmatrix} \begin{bmatrix} \delta_1 \\ \delta_2 \\ \delta_3 \end{bmatrix},$$

$$\begin{bmatrix} F_1 \\ F_2 \\ F_3 \end{bmatrix} = \underbrace{\left(\begin{pmatrix} k_1 & -k_1 & 0 \\ -k_1 & k_1 & 0 \\ 0 & 0 & 0 \end{pmatrix} + \begin{pmatrix} 0 & 0 & 0 \\ 0 & k_2 & -k_2 \\ 0 & -k_2 & k_2 \end{pmatrix} \right)}_{\text{Component stiffnesses}} \begin{bmatrix} \delta_1 \\ \delta_2 \\ \delta_3 \end{bmatrix}$$

$$\equiv \underbrace{\begin{pmatrix} k_1 & -k_1 & 0 \\ -k_1 & k_1 + k_2 & -k_1 \\ 0 & -k_2 & k_2 \end{pmatrix}}_{\text{Overall stiffness}} \begin{bmatrix} \delta_1 \\ \delta_2 \\ \delta_3 \end{bmatrix}.$$

This illustrates how the properties of the simple elements can be combined to yield an overall response function for a more complex body, which is extremely useful when dealing with intricate industrial problems.

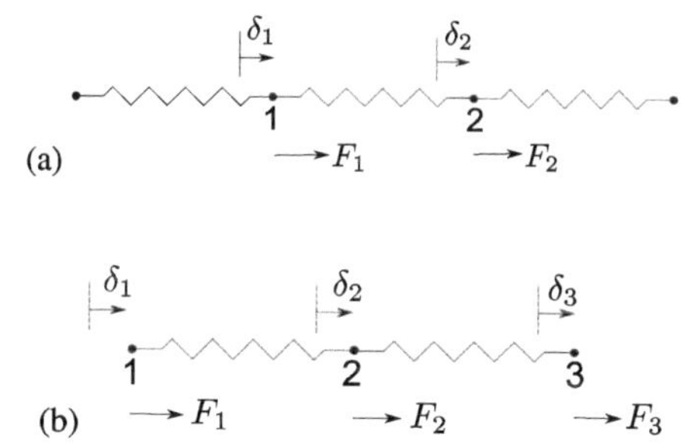

Figure 16.17 Forces on springs.

16.7 Neural networks

The usual approach when dealing with difficult problems is to correlate the results against chosen variables using linear regression analysis. A more powerful method of empirical analysis involves the use of neural networks [48,49], which have had tremendous success in the quantitative treatment of structure-property relationships [10].

In linear regression, fitting data to a specified relationship yields an equation which relates the inputs x_j via weights w_j and a constant θ to obtain an estimate of the output $y = \sum_j w_j x_j + \theta$. Equations like these are used widely in industry, e.g., in the formulation of the *carbon equivalents* (Section 13.2.5). It is well understood that there is risk in using the relationships beyond the range of fitted data, but the risk is not quantified.

With neural networks, the input data x_j are again multiplied by weights, but the sum of all these products forms the argument of a flexible mathematical function, often a hyperbolic tangent. The output y is therefore a non-linear function of x_j. The exact shape of the hyperbolic tangent can be varied by altering the weights (Fig. 16.18a). The weights are changed systematically until a best-fit description of the output is obtained as a function of the inputs; this operation is known as *training* the network.

Further degrees of non-linearity can be introduced by combining several of these hyperbolic tangents (Fig. 16.18b), so the neural network method is able to capture almost arbitrarily non-linear relationships.

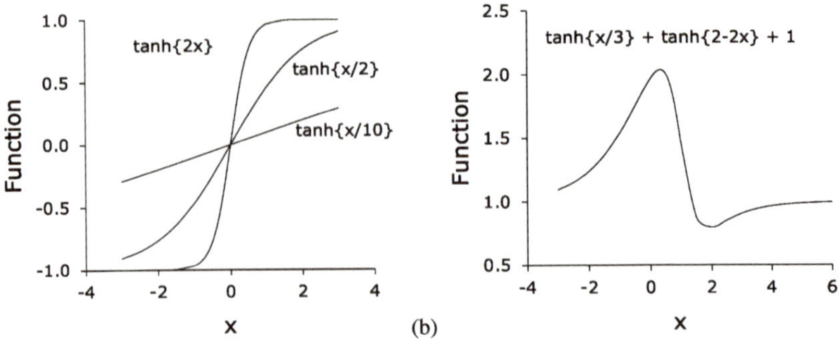

(a) (b)

Figure 16.18 (a) Three different hyperbolic tangent functions; the "strength" of each depends on the weights. The diagram shows how flexible a hyperbolic tangent is. (b) A combination of two hyperbolic tangents to produce a more complex model. Such combinations can be continued indefinitely to produce functions of ever greater complexity.

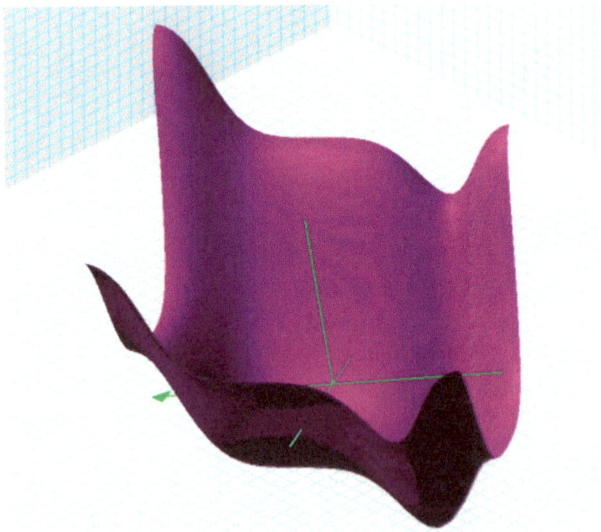

Figure 16.19 Variation in the output z (vertical axis) as a function of two input variables x, y (horizontal axes), the whole surface was generated using just four hyperbolic tangent functions, $z = 0.8[\tanh\{wx - 2\} + \tanh\{x^2 - w\} + \tanh\{wy + 2\} + \tanh\{y^2 - w\} + 1]$.

Fig. 16.19 illustrates the complexity of the surface that can be produced when representing the output (vertical axis) as a function of two inputs using just four hyperbolic tangents.

A potential difficulty with the ability to produce complex, non-linear functions is the possibility of overfitting of data. To avoid this difficulty, the experimental data can be divided into two sets, a *training* dataset and a *test* dataset. The model is produced using only the training data. The test data are then used to check that the model behaves itself when presented with previously unseen data. This is illustrated in Fig. 16.20, which shows three attempts at modelling noisy data for y as a function of x. A linear model (Fig. 16.20a) is too simple and does not capture the real complexity in the data. An over complex function such as that illustrated in Fig. 16.20c accurately models the training data, but generalises poorly. The optimum model is illustrated in Fig. 16.20b. The training and testing errors are shown schematically in Fig. 16.20d. Not surprisingly, the training error tends to decrease continuously as the model complexity increases. It is the minimum of the test error that allows one to choose the model that generalises best to unseen data.

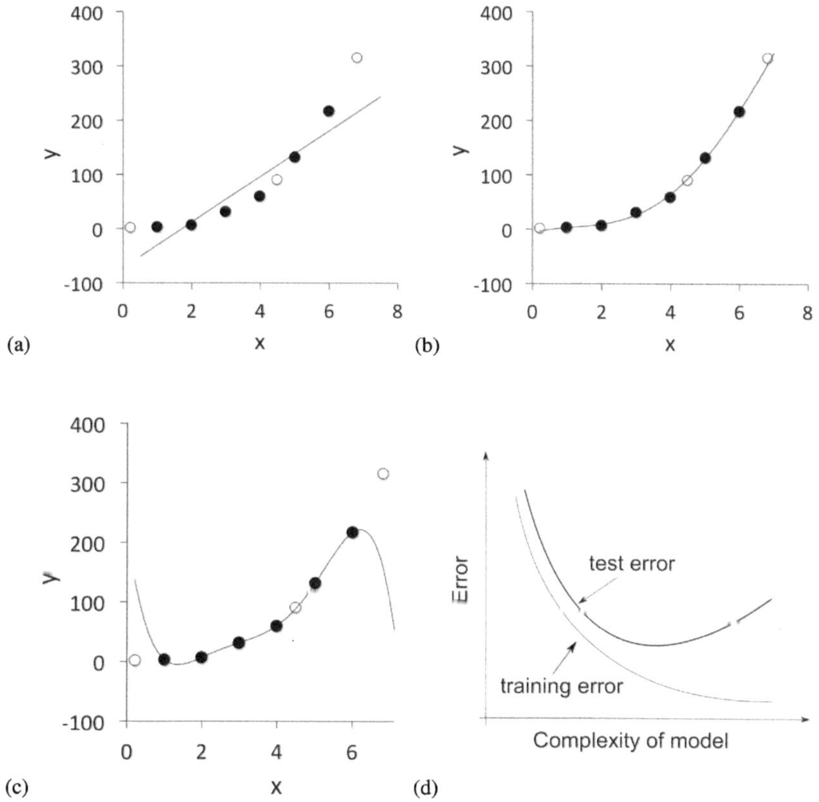

Figure 16.20 Variations in the test and training errors as a function of model complexity, for noisy data. The filled points were used to create the models (i.e., they represent training data), and the circles constitute the test data. (a) A linear function which is too simple. (b) A cubic polynomial with optimum representation of both the training and test data. (c) A sixth-order polynomial which generalises poorly. (d) Schematic illustration of the variation in the test and training errors as a function of the model complexity.

A neural network like this can capture interactions between the inputs because the hidden units are non-linear. Appropriate measures must be taken to avoid overfitting. For complex networks, it is also important to consider the *modelling uncertainty*, i.e., what is the range of models which can adequately represent the known data? A large modelling uncertainty corresponds to the case where these models behave differently when extrapolated. Just a few examples of new metallurgical concepts emerging directly from this kind of machine learning are as follows:

1. The δ-TRIP steel concept discovered by machine learning [5,50,51], where a chemical composition and a set of properties were reached that permitted α- to be replaced by δ-ferrite resulting from the fact that the steel could never be made fully austenitic. This also meant that it could be resistance spot-welded even though the carbon concentration was large [52,53].
2. A novel fire-resistant welding alloy designed using machine learning, with immediate success and implementation [54].
3. A novel, cheap nickel-alloy for large-scale application in power plants, designed using machine learning and thermodynamics [55–57].

16.8 Summary

It is possible now to design an innovative steel and to introduce it into service within five years. This is because there is a realisation in academia that there is more than just the measurement of a few properties that makes a usable material. At the same time, industry appreciates the need to devote vast resources in order to take an elementary idea to a stage where the manufacturing capability and technological readiness levels are sufficient. The process towards success can be summarised in stages that are often labelled as readiness levels, which are paraphrased here:

1. Principles established based on science or creativity.
2. Technology concept developed, needs defined.
3. Basic characterisation.
4. Prototype development and validation.
5. Major investment to prove component-level parameters and scaling.
6. Demonstration in real circumstances.
7. Fully fledged commercial product, which may be developed further with service experience and new knowledge.

If all this is to be achieved in the five-year timeframe, then it usually is the case the mathematical models play a role in testing options. The models are rarely sophisticated enough to deal with complex materials, so experiments and thought processes based on experience and past literature are essential.

References

[1] G.B. Olson, Genomic materials design: the ferrous frontier, Acta Materialia 61 (2013) 771–781.

[2] H.K.D.H. Bhadeshia, Computational design of advanced steels, Scripta Materialia 70 (2014) 12–17.
[3] G.B. Olson, C.J. Kuehmann, Materials genomics: from CALPHAD to flight, Scripta Materialia 70 (2014) 25–30.
[4] H.K.D.H. Bhadeshia, Mathematical models in materials science, Materials Science and Technology 24 (2008) 128–135.
[5] S. Chatterjee, M. Murugananth, H.K.D.H. Bhadeshia, δ-TRIP steel, Materials Science and Technology 23 (2007) 819–827.
[6] A.Y. Badmos, H.K.D.H. Bhadeshia, Evolution of solutions, Metallurgical & Materials Transactions A 28 (2000) 2189–2193.
[7] R.H. Siller, R.B. McLellan, The variation with composition of the diffusivity of carbon in austenite, TMS-AIME 245 (1969) 697–700.
[8] A. Saha, G.B. Olson, Computer-aided design of transformation toughened blast resistant naval hull steels: part I, Journal of Computer-Aided Materials Design 14 (2007) 177–200.
[9] A. Saha, J. Jung, G.B. Olson, Prototype evaluation of transformation toughened blast resistant naval hull steels: part II, Journal of Computer-Aided Materials Design 14 (2007) 201–233.
[10] H.K.D.H. Bhadeshia, Neural networks in materials science, ISIJ International 39 (1999) 966–979.
[11] P.W. Anderson, Local moments and localized states, Nobel Lecture, Bell Telephone Laboratories, Princeton University, 1977.
[12] M. Murugananth, H.K.D.H. Bhadeshia, E. Keehan, H.O. Andrén, L. Karlsson, Strong and tough steel welds, in: H. Cerjak, H.K.D.H. Bhadeshia (Eds.), Mathematical Modelling of Weld Phenomena 6, IOM Communications, London, U.K., 2002, pp. 205–230.
[13] E. Keehan, H.-O. Andrén, L. Karlsson, M. Murugananth, H.K.D.H. Bhadeshia, Microstructural and mechanical effects of nickel and manganese on high strength steel weld metals, in: S.A. David, T. DebRoy (Eds.), Trends in Welding Research, ASM International, Materials Park, Ohio, USA, 2002, pp. 695–700.
[14] E. Keehan, L. Karlsson, H.-O. Andrén, Influence of C, Mn and Ni on strong steel weld metals: part 1, effect of nickel, Science and Technology of Welding and Joining 11 (2006) 1–8.
[15] E. Keehan, L. Karlsson, H.-O. Andrén, H.K.D.H. Bhadeshia, Influence of C, Mn and Ni on strong steel weld metals: part 2, increased impact toughness, Science and Technology of Welding and Joining 11 (2006) 9–18.
[16] E. Keehan, L. Karlsson, H.-O. Andrén, H.K.D.H. Bhadeshia, Influence of C, Mn and Ni on strong steel weld metals: part 3, increased strength, Science and Technology of Welding and Joining 11 (2006) 19–24.
[17] J.O. Andersson, T. Helande, L. Hoglund, P. Shi, B. Sundman, Thermo-calc & DICTRA, computational tools for materials science, Calphad 26 (2002) 273–312.
[18] U. of Cambridge, NPL, Materials Algorithms Project, www.phase-trans.msm.cam.ac.uk/map/mapmain.html, 2013.
[19] H.I. Aaronson, H.A. Domian, G.M. Pound, Thermodynamics of the austenite – proeutectoid ferrite transformation II, Fe-C-X alloys, TMS-AIME 236 (1966) 768–780.
[20] NPL, MTDATA, Software, National Physical Laboratory, Teddington, U.K., 2006.
[21] S. Khare, K.Y. Lee, H.K.D.H. Bhadeshia, Carbide-free bainite: compromise between rate of transformation and properties, Metallurgical & Materials Transactions A 41A (2010) 922–928.
[22] H.K.D.H. Bhadeshia, D.V. Edmonds, Bainite in silicon steels: a new composition-property approach II, Metal Science 17 (1983) 420–425.

[23] L.C. Chang, The rolling/sliding wear performance of high silicon carbide-free bainitic steels, Wear 258 (2005) 730–743.

[24] F.G. Caballero, H.K.D.H. Bhadeshia, K.J.A. Mawella, D.G. Jones, P. Brown, Very strong, low-temperature bainite, Materials Science and Technology 18 (2002) 279–284.

[25] C.G. Mateo, F.G. Caballero, Ultrahigh-strength bainitic steels, ISIJ International 45 (2005) 1736–1740.

[26] V.T.T. Miihkinen, D.V. Edmonds, Tensile deformation of two experimental high-strength bainitic low alloys steels containing silicon, Materials Science and Technology 3 (1987) 432–440.

[27] H.K.D.H. Bhadeshia, Properties of fine-grained steels generated by displacive transformation, Materials Science & Engineering A 481–482 (2008) 36–39.

[28] M. Sherif, C. Garcia-Mateo, T. Sourmail, H.K.D.H. Bhadeshia, Stability of retained austenite in TRIP-assisted steels, Materials Science and Technology 20 (2004) 319–322.

[29] A.A.B. Sugden, H.K.D.H. Bhadeshia, A model for the strength of the as-deposited regions of steel weld metals, Metallurgical Transactions A 19A (1988) 1597–1602.

[30] J.W. Christian, The strength of martensite, in: A. Kelly, R. Nicholson (Eds.), Strengthening Methods in Crystals, Elsevier, Netherlands, 1971, pp. 261–329.

[31] M. Takahashi, H.K.D.H. Bhadeshia, Model for transition from upper to lower bainite, Materials Science and Technology 6 (1990) 592–603.

[32] C.H. Young, H.K.D.H. Bhadeshia, Strength of mixtures of bainite and martensite, Materials Science and Technology 10 (1994) 209–214.

[33] N. Chakraborti, Genetic algorithms in materials design and processing, International Materials Reviews 49 (2004) 246–260.

[34] A.T. Paxton, M. Methfessel, H.M. Polatoglou, Structural energy-volume relations in first-row transition metals, Physical Review B 41 (1990) 8127–8138.

[35] P.M. Anderson, J.-S. Wang, J.R. Rice, Thermodynamic and mechanical models of interfacial embrittlement, in: G.B. Olson, M. Azrin, E.S. Wright (Eds.), Innovations in Ultrahigh-Strength Steel Technology, US Army Materials Technology Laboratory, Massachusetts, USA, 1987, pp. 619–650.

[36] R. Wu, A.J. Freeman, G.B. Olson, First principles determination of the effects of phosphorus and boron on iron grain boundary cohesion, Science 265 (1994) 376–380.

[37] J.H. Jang, C.H. Lee, Y.U. Heo, D.W. Suh, Stability of (Ti, M) C (M=Nb, V, Mo and W) carbide in steels using first-principles calculations, Acta Materialia 60 (2012) 208–217.

[38] A.R. Troiano, Role of H and other interstitials in the mechanical behaviour of metals, Transactions of the ASM 52 (1960) 54–80.

[39] E.J. Song, H.K.D.H. Bhadeshia, D.-W. Suh, Effect of hydrogen on the surface energy of ferrite and austenite, Corrosion Science 77 (2013) 379–384.

[40] G. Miyamoto, J.C. Oh, K. Hono, T. Furuhara, T. Maki, Effect of partitioning of Mn and Si on the growth kinetics of cementite in tempered Fe-0.6 mass% C martensite, Acta Materialia 55 (2007) 5027–5038.

[41] J.H. Jang, I.G. Kim, H.K.D.H. Bhadeshia, Substitutional solution of silicon in cementite: a first-principles study, Computational Materials Science 44 (2009) 1319–1326.

[42] S. Jones, H.K.D.H. Bhadeshia, Competitive formation of inter- and intragranularly nucleated ferrite, Metallurgical & Materials Transactions A 28A (1997) 2005–2103.

[43] S. Jones, H.K.D.H. Bhadeshia, Kinetics of the simultaneous decomposition of austenite into several transformation products, Acta Materialia 45 (1997) 2911–2920.

[44] J.D. Robson, H.K.D.H. Bhadeshia, Modelling precipitation sequences in power plant steels: part I, kinetic theory, Materials Science and Technology 13 (1997) 631–639.

[45] J.D. Robson, H.K.D.H. Bhadeshia, Modelling precipitation sequences in power plant steels, part 2: application of kinetic theory, Materials Science and Technology 28A (1997) 640–644.

[46] H.K.D.H. Bhadeshia, Multiple, simultaneous, martensitic transformations: implications on transformation texture intensities, Materials Science Forum 762 (2013) 9–13.

[47] R.S. Qin, H.K.D.H. Bhadeshia, Phase field method, Materials Science and Technology 26 (2010) 803–811.

[48] D.J.C. MacKay, Practical Bayesian framework of backpropagation networks, Neural Computation 4 (1992) 448–472.

[49] D.J.C. MacKay, Bayesian interpolation, Neural Computation 4 (1992) 415–447.

[50] H.L. Yi, K.Y. Lee, H.K.D.H. Bhadeshia, Stability of ferrite in δ-TRIP steel, Materials Science and Technology 27 (2011) 525–529.

[51] H.L. Yi, K.Y. Lee, H.K.D.H. Bhadeshia, Stabilisation of ferrite in hot-rolled in δ-TRIP steel, Materials Science and Technology 27 (2011) 525–529.

[52] H.L. Yi, K.Y. Lee, J.H. Lim, H.K.D.H. Bhadeshia, Spot weldability of δ-TRIP steel containing 0.4 wt% C, Science and Technology of Welding and Joining 15 (2010) 619–624.

[53] H.L. Yi, Review on δ-transformation-induced plasticity (TRIP) steels with low density: the concept and current progress, Journal of Materials 66 (2014) 1759–1769, https://doi.org/10.1007/s11837-014-1089-6.

[54] K. Ichikawa, Y. Horii, R. Motomatsu, M. Yamaguchi, N. Yurioka, Mechanical properties of weld metal of fire resistant steel by large heat input submerged arc welding, Quarterly Journal of Japan Welding Society 14 (1996) 27–32.

[55] F. Tancret, H.K.D.H. Bhadeshia, D.J.C. MacKay, Design of a creep-resistant nickel-base superalloy for power-plant applications. Part 1 – mechanical properties modelling, Materials Science and Technology 19 (2003) 283–290.

[56] F. Tancret, H.K.D.H. Bhadeshia, Design of a creep-resistant nickel-base superalloy for power-plant applications. Part 2 – phase diagram and segregation, Materials Science and Technology 19 (2003) 291–295.

[57] F. Tancret, T. Sourmail, M.A. Yescas, R.W. Evans, C. McAleese, L. Singh, T. Smeeton, H.K.D.H. Bhadeshia, Design of a creep-resistant nickel-base superalloy for power-plant applications. Part 3 – experimental results, Materials Science and Technology 19 (2003) 296–303.

Visionary use of steels ($CO_2 \downarrow$) 17

Abstract

Imagine a scenario in which steel production can, overnight, be reduced by 22%. By utilising better steel, a building has been constructed with *existing* technologies, to use much less steel without compromising its function. How remarkable would it be for the environment if this could be extrapolated across the whole of the steel-consuming industry? In another scenario, the most ordinary steel has been used to regenerate devastated coral-reefs, with stunning outcomes. Gigantic rockets are now constructed of steel, in such a way that both the booster stages and spaceships can land intact for reuse, giving a better meaning to recycling. Ships collide and such accidents can lead to pollution and loss of life. One solution is to make them double-hulled, but a better approach seems to be to use a superior steel that as a single-hull can resist rupture in realistic scenarios, and at the same time use less steel.

17.1 Introduction

It is blindingly obvious that the carbon dioxide burden needs to be reduced rapidly to avoid changes in sea levels and climate, that primarily endanger the well-being of the poorest in the world. Metal production is a significant contributor. Steel, in particular, is a notable culprit because of the indecent quantities produced today. While there is much talk of making steel using hydrogen or by electrolysis, these ideas are essentially pipe dreams, because there is no possibility of meeting the huge demand using energy from non-polluting sources. And certainly not given the time available to mitigate the CO_2 problem before it becomes irreversible.

One way to make a big dent in the pollution associated with steels, is to reduce the quantity without compromising the function of the engineered product – one such case, based on clever design, has been described on p. 132. The goal can also be achieved by using better steel – and it is shown here that this does not require research, simply the use of existing *better steels*. The feeling that this might be an expensive option is also shown quantitatively not be justified. The idea that recycling can help is taken a stage further, where re-use is emphasised and demonstrated to be worthwhile, because recycling itself has an environmental cost. An example is given of how the imaginative use of the most ordinary of steels has led to a huge environmental boost, and another where marine safety is improved using less steel.

Steel needs to become more expensive so that it is used intelligently, taking full advantage of known metallurgy and engineering. A tax on steel content, whether that represents domestic or imported items, would work.

17.2 Buildings

Half of all the steel produced is used in making buildings and infrastructure. Steels offer properties at an affordable cost. Aesthetics, the ease of assembly, an efficient use

Steels. https://doi.org/10.1016/B978-0-44-318491-8.00023-X

of space, recyclability, fire resistance are just some of the attributes that make them indispensable. Steel simply has no competitors.

Nevertheless, the extraction of iron from its ore accounts for 7% of CO_2 emissions, a circumstance that no longer should be tolerated. The consumption of steel needs to be reduced without compromising the quality of life that we are used to. Legislation would help achieve this – the introduction of mandatory fuel efficiency standards related to transportation, stimulated the development of steels with a bank of properties that enable weight reduction without sacrificing structural integrity or safety. Legislation, unfortunately, takes time and consensus, a luxury no longer available if a warming catastrophe that most affects the weakest in the world is to be avoided.

It is demonstrable that the use of smaller quantities of more expensive steels in the construction industry can reduce both overall cost and pollution. This does not require research, simply common sense and education, as illustrated here for a building in which ordinary alloys are replaced by better steel, so that the total consumed is reduced as is the overall cost.

17.2.1 Improved steel

Construction steels have to be shaped for rigidity, must be weldable and available in quantity. In buildings, they would need to be coated in order to provide resistance to a fire over a period of one hour, depending on the height and intended use of the construction. A typical construction steel has the composition Fe-1Mn-0.3Si-0.22C wt%, supplied in a hot-rolled condition in the form of I-beams with a yield strength of about 290 MPa. The limited carbon-concentration ensures weldability by avoiding the formation of martensite in the heat-affected zone of the welds. Fig. 17.1a shows the microstructure consisting of the expected mixture of ferrite ($\overline{L}_{\alpha} = 38\,\mu\text{m}$) and pearlite.

An alternative steel with the same Mn, Si, C concentrations, but also containing 0.03 wt% Nb as a *microalloying* addition is commercially available. During hot-rolling in the austenitic condition, the niobium precipitates as stable carbides that hinder the growth of austenite grains, so that the transformed microstructure becomes refined with a ferrite grain size of just $\overline{L}_{\alpha} = 22\,\mu\text{m}$. Since the pearlite forms after the ferrite, its distribution is also altered favourably.

A reduced microstructural scale is one of the few methods of enhancing both the strength and toughness of the steel. The steel illustrated in Fig. 17.1b has a yield strength of about 402 MPa and the toughness, as measured by Charpy tests conducted at ambient temperature, increases from an average of 108 J for Fe-1Mn-0.3Si-0.22C wt% to 168 J for the microalloyed variant. This represents a dramatic enhancement of properties, but the cost of the microalloyed steel is greater by 6% relative to that which does not contain niobium – however, the story does not end there.

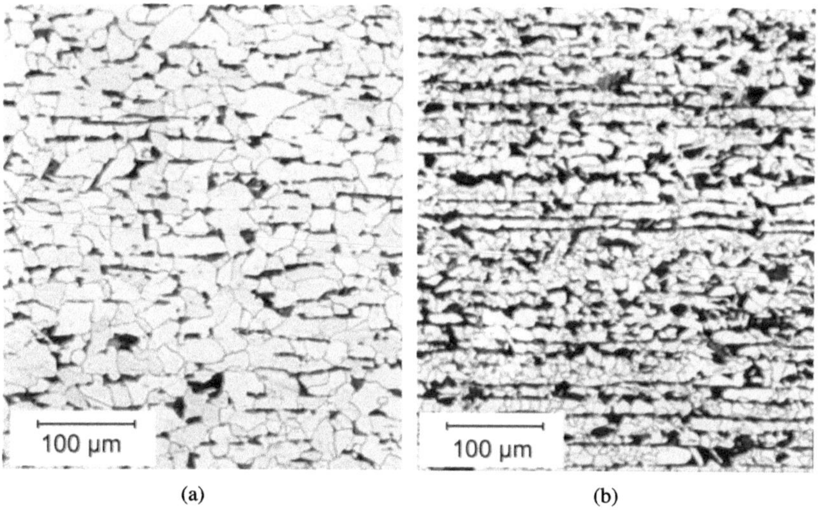

Figure 17.1 (a) Steel Fe-1Mn-0.3Si-0.22C wt% in the hot-rolled condition. The microstructure consists of a mixture of ferrite and pearlite with a ferrite grain size of 38 μm. (b) Microalloyed version of the same steel, but with an addition of 0.03 wt% of niobium, resulting in a refined structure. After Silvestre et al. [1].

17.2.2 Building with microalloyed steel

An industrial building has been fabricated using microalloyed steel, which can be compared against an equivalent construction using ordinary steel. Both had exactly identical applied load, internal building area, volume, span, height, length and width, although obviously, the beams and columns are larger when using the low-strength steel, Fig. 17.2. The total quantity of strong steel used in constructing the building was 284,000 kg, whereas the same building would have required 363,000 kg of ordinary steel. This corresponds to a saving of 130,000 kg of CO_2 emissions and 1.8 GJ of energy [1].

The total cost of the building was reduced 17% by using the more expensive microalloyed steel, Table 17.1, which made it possible to achieve the engineering goals and elegant design using 22% less steel. The stages of construction and the completed building at CBMM in Brazil are illustrated in Fig. 17.3.

Table 17.1 Relative monetary cost of constructing the building using the two kinds of steels.

Steel type	Total weight %	Unit price	Total cost, %
Ordinary	100	100	100
Microalloyed	78	106	83

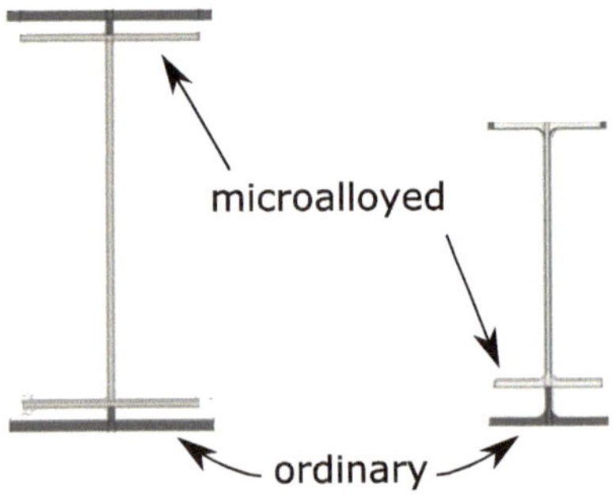

Figure 17.2 The beams on the left are for columns and those on the right for cross-members. The diagrams are approximately to scale. It is obvious that because of the strength, less steel is needed to design the same building in the case of the microalloyed steel. This leads to a saving of some 79 tonnes of steel.

Figure 17.3 Construction of a sinter plant at CBMM in Brazil. (a) Construction site and foundations. (b) Columns and beams. (c) Crane installation. (d) Completed building. Images courtesy of Marcos Stuart.

17.3 Coral reef regeneration

Corals are marine invertebrate animals that propagate either by the growth of detached fragments or sometimes sexually by the combination of sperm and eggs. The basic entity is a *polyp,* which is shaped somewhat like a bulbous cylinder, the top opening being a mouth with a ring of tentacles [2]. The underside of the polyp has structures that deposit pure limestone, which then attracts calcium from seawater to deposit $CaCO_3$ crystals, the basis of coral reefs. These reefs support a large variety and number density of sea-creatures and have a role in buffering coastlines against turbulent seas.

Coral reefs can degrade or die completely for different reasons, but blast fishing with dynamite and coral mining in particular led to the destruction of a large areas of reef in the sea near the shores of Badi island, Indonesia (Fig. 17.4). The blast fishing kills not only sea creatures but breaks the fragile calcium-carbonate coral skeletons that provide food and shelter for marine organisms, so the thriving ecosystem is vandalised. The shock waves leave expanses of unstable coral rubble that is moved around by the currents, so recovery is minimal or entirely absent [3].

Figure 17.4 Coral reef damaged some 30-40 years ago due to fishing with dynamite, and later due to large-scale coral mining for house construction. The illustrated region is located in the sea near Badi Island, Indonesia. Picture from The Ocean Agency, reproduced with permission from Jos van Oostrum.

The regeneration of the damaged reef began in 2013 using frames as in Fig. 17.5, made from reinforcing steel bar that was coated first with a water-based non-toxic solution containing natural tannins, that coverts rust into stable black oxide, which bonds with the steel. The frame was then dipped into resin and coated with sand to provide a rough base for coral recruitment. Live coral fragments collected sympathetically from a shallow reef were then attached to the frame, constantly dowsed in sea water until deployment to the damaged sites [3]. The design of the frame allowed each one to be located on the sea bed with natural anchoring due to the density of steel. Close-packed arrays of 550 hexagonal frames could cover an area of 300 m^2. The total cost of installing 11,000 frames was \$174,000, with occasional maintenance during the first 3 months following installation, requiring an estimated \$3 per frame.

The amazing transformation of the damaged reef into a vibrant colony with thriving sea-life, is illustrated in Fig. 17.6. Coral transplantation after attachment to steel modules sunk in Indonesian waters was associated with a 100% survival rate in a short term study at another location in Indonesian waters [4]. The use of steel structures in coral regeneration has increased dramatically in recent times, although the number of active restoration projects remains rather small [5].

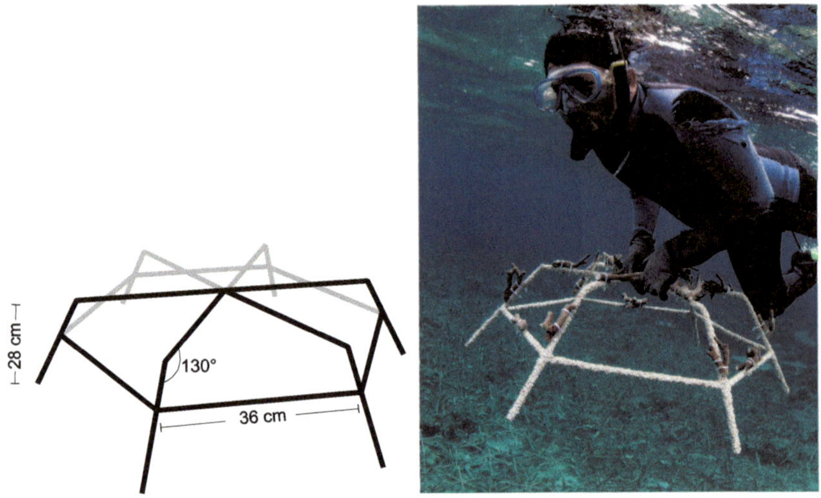

(a) (b)

Figure 17.5 (a) Design of frame, made from bars that are used routinely to reinforce concrete. (b) The steel frame is covered with resin and then sand to make its surface rough for the coral. Bits of coral are then tethered to it, and the frame then placed on to the sea bed. Picture from The Ocean Agency, reproduced with permission from Jos van Oostrum.

Slag, a by-product of steelmaking using the electric-arc furnace, was used to regenerate coral [6]. This slag contains predominantly iron and calcium oxides and some silica. Large *slabs* about 25→35 cm diameter and 15 cm thick with living coral segments glued on were immersed into a region of the Red Sea. Calcium carbonate precipitated on the surfaces so the slabs became covered within three months, forming a suitable substrate for some forms of sea life, but the composition of the slag itself did not change over time. There was some success in nurturing the development of the planted corals, while at the same time, making effective use of industrial waste. Because of the thin coating of precipitated calcium carbonate, the slabs soon fitted in with the normal appearance of the sea bed. The growth of limestone can also be induced using electrified steel rods that cause dissolved minerals in the sea water to deposit in a manner that promotes coral reef [7], but this naturally adds to complexity.

Figure 17.6 The regenerated reef. Small parts of the steel frames still visible on careful examination. Picture from The Ocean Agency, reproduced with permission from Jos van Oostrum.

17.4 Starship

The German V2 rockets that later seeded the USA and USSR space programmes, were about 14 m tall with their fuel and oxygen tanks made using an aluminium-magnesium alloy to reduce weight. In contrast, the first Atlas rocket was about 23 m in height. It relied on stainless-steel tanks so thin that when empty, they had to be pressurised with nitrogen gas to prevent them from collapsing under their own weight by elastic buckling. The tanks were therefore often referred to as *balloon tanks*.

The critical stress, which must not be exceeded to avoid bucking of a cylindrical shell under uniform axial compression varies directly with the thickness of the shell [8]. The effect of internal pressure P is to provide an axial tensile stress that increases the magnitude of the critical buckling-load by $P \times \pi r^2$, where r is the radius of the pressure vessel. The pressure-induced hoop stress, which is twice the value of the axial stress, further inhibits buckling. The pressure can also mitigate the destabilising effect of imperfections in the cylinder [9].

The SpaceX Starship, currently undergoing launch trials, consists of a spacecraft on top of a powerful *Super Heavy* rocket with a combined height of 120 m. It forms a transportation system said to be capable eventually of taking people and cargo into Earth orbit, the Moon and Mars, Fig. 17.7. Both of the parts of Starship are designed to be reusable, with even the Super Heavy booster able to re-enter Earth's atmosphere and land at its launch site. Super Heavy is the most powerful launch vehicle ever developed, some 69 m in height and 9 m in diameter. Its long tanks can hold some 3400 t of propellant, whereas the spacecraft on top holds about 220 t of liquid methane which is separated from 780 t of liquid oxygen by a common dome, Fig. 17.8.

(a) (b)

Figure 17.7 The stainless steel Starship, which consists of a spacecraft (a) which then is mounted on to a stainless steel booster-rocket (b). Images courtesy of SpaceX via Charles Kuehmann.

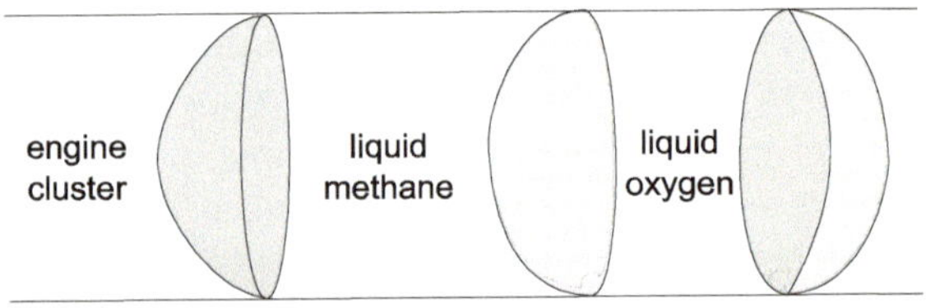

Figure 17.8 Schematic of the main body of the Super Heavy booster, where the propellant (liquid oxygen and liquid methane) is held, here illustrated in a horizontal position. The spacecraft part will also contain some smaller "header" tanks that are not shown; these serve as reserves during landing when the main tanks are partially empty. They can therefore feed the main engines with precision even when the landing manoeuvres involve large changes in rocket orientation.

Stainless steel is used generously in the construction of both the structural parts of the Starship, which at the time of writing is the biggest and most powerful rocket ever built. Therefore, the stainless steel is thicker at ≈ 4 mm than in the Atlas era, where the tanks were less than 1 mm thick and had to be inflated with nitrogen to maintain shape. The elastic-buckling problem is, as a result, greatly reduced in the Starship assembly. The design ensures that it would not buckle under its own weight without having to pressurise it. This is in spite of the fact that there is no internal support structure within the tank, which is a cylinder with a dome at the top, a dome at the bottom and a common-dome in the middle, the latter separating the fuel and oxygen.

Nevertheless, a small positive internal-pressure is implemented to enhance the safety of personnel working in the vicinity of the un-fuelled rocket. Because of the size of the rocket, uneven heating from sunlight can cause negative pressure which promotes buckling. Only when filling the tanks or operating in strong winds are the tanks under significant pressure.

Liquid oxygen boils at 90 K and liquid methane at 112 K, so the cryogenic properties of the steel become really important. To achieve a strength-to-weight ratio that is competitive with aluminium alloys, the stainless steel is used in a cold-worked condition with $\approx 50\%$ rolling reduction prior to fabrication. A compilation of the relevant properties of two stainless steels, designated 301 and 304L (Chapter 12) is presented in Fig. 17.9. Cold deformation makes both steels stronger, with the resistance to plastic deformation increasing as the temperature is reduced below ambient. The strength-to-weight ratio over that temperature range is about 200-250 $\mathrm{MPa\,g^{-1}\,cm^3}$, which compares well with the most advanced Al-Li-Cu-Mg alloys used in the aerospace industries [10].

Fig. 17.9 shows also that 304 stainless steel in the deformed condition has a large ductility even at cryogenic temperatures, whereas the 301 variety does not. This presumably is related to the greater stability of the austenite in 304 because of its higher nickel concentration relative to 301 (Table 12.1), and its smaller overall strength in the cold-deformed condition relative to 301. Since the properties at cryogenic temperatures are most relevant for the Starship, cold-deformed stainless steel is used in its construction. Although 304 stainless steel was used in the early stages of rocket development, SpaceX now has a proprietary variety that has the advantage of the good ductility of 304 at low temperatures and the better cold-working response of the 310 stainless steel.

The cost of stainless steel per kg is roughly 2% percent that of the aluminium-lithium alloys or of carbon-fibre composites. There is about 35% scrap in the fabrication of a fuel tank using carbon-fibre composite and the fabrication itself is complex [19]. The aluminium alloys have to be friction-stir welded, which in the context of long rockets such as the Starship, requires the installation of custom-made machines that have a long lead-time, together with significant inspection after welding. The cold-rolled stainless steel for the Starship is joined together using off-the-shelf robotic laser welders in a simple process, illustrated in Fig. 17.10. Long segments of sheets are bent and welded longitudinally to create cylinders, each with a single vertical weld. These are then welded circumferentially to create the columnar part of the rocket.

Welding naturally creates a heat-affected zone, which is narrow in the case of laser welding, but nevertheless softens the region adjacent to the fusion boundary. However, this zone cold-works preferentially when the tank is pressurised at cryogenic temperatures. The stainless steel has a high cold-working coefficient that permits just a small plastic strain to harden up the heat-affected zone.

The cold-rolled stainless steel used for the Starship is intended also for the Tesla Cybertruck™, designed to exploit the strength of the steel while at the same time present an elegant overall-shape. The shape employs 3 mm thick planar panels that do not require forming operations other than bending (Fig. 17.11). This saves on in-

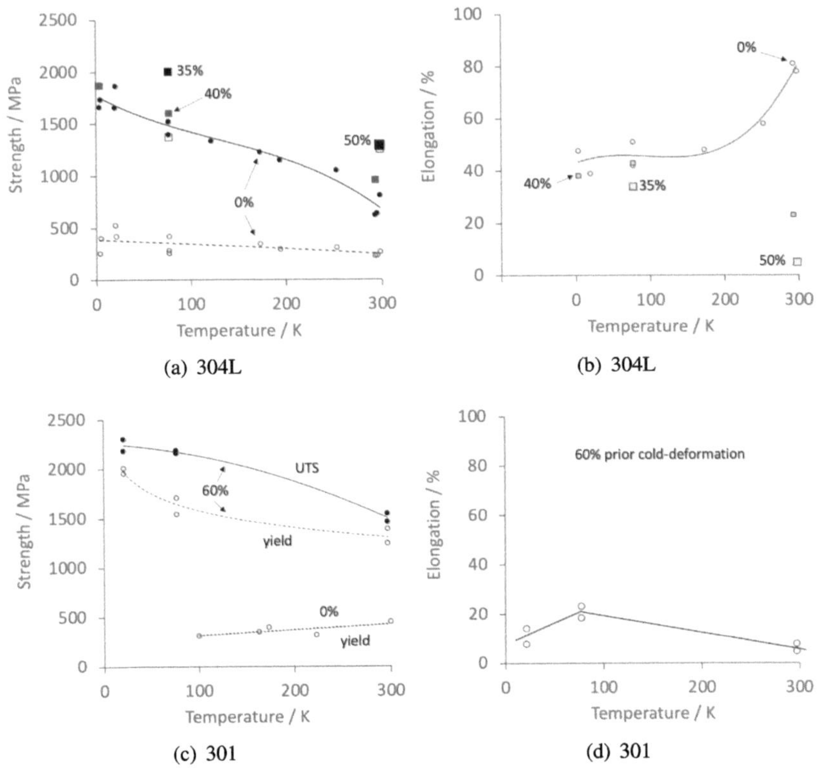

Figure 17.9 Mechanical properties of stainless steels as a function of temperature. The percentages indicate the degree of cold-deformation prior to tensile testing. (a) Yield (open points) and UTS (filled points) of 304L stainless steel. (b) Elongation of 304L. (c) Strength of 301 stainless steel that is 60% cold-deformed prior to tensile testing. (d) Corresponding elongation data for 301 stainless steel. Data compiled from [11–18].

vestment in forming-press dies and machines. Stamping is avoided by using lasers for cutting the hard-steel sheet. The electric truck does not require painting.

Simulations indicate that the angular shape of the body, necessitated by the use of the hard steel, has an aerodynamic drag coefficient between 0.3–0.4, which is less than many of the commercially available 'pickup' trucks [20]. Furthermore, it has a low aerodynamic lift coefficient of 0.11, which augers well for stability and maintenance of contact with the road surface.

17.5 Robust ships

Collisions between ships account for about a half of the accidents and failures that occur in the marine environment, with a prevalence of incidents near coastlines or in narrow sea-channels. In the worst cases, this can lead to a loss of life or pollution when the structural integrity of either of the ships involved is compromised. The number of

Figure 17.10 Circumferential autogenous laser-welds on abutting stainless steel cylinders that eventually form a part of the Starship. Image courtesy of SpaceX via Charles Kuehmann.

Figure 17.11 The Tesla Cybertruck, the body of which is made using strong, cold-rolled stainless steel. Image of prototype from 2020, reproduced in truncated form, under the CC BY 2.0 license courtesy of Steve Juvetson, https://creativecommons.org/licenses/by/2.0/deed.en. Notice the angular construction that does not require forming tools to create the body.

major collisions is not large, typically less than 100 per annum – out of a population of about 103,000 that traverse the oceans. Nevertheless, the consequences of ship collisions can be huge, with unlimited liabilities. The pollution risk is large during collisions with oil tankers and one mitigation is to increase the number of tanks so that the volume of oil that is released accidentally decreases (Fig. 17.12). However, that would typically result in an increase in the amount of steel needed by some 500 tonnes with an accompanying increase the complexity of the piping infrastructure.

A second option is to have two steel layers forming the vulnerable body of the ship, the so-called double-hulled vessels, in the hope that this reduces the chance of both hulls being perforated in the event of a collision. There are, therefore, two watertight layers between the crude oil and the surrounding sea. In Fig. 17.12, large values of Δx add Δy in double-hulled structures must clearly reduce the chance of oil leakage. However, any increase in the width y of the ship may reduce its propulsion efficiency, whereas large values of Δx and Δy at constant x limit the amount of oil that can be stored. Crude oil carriers must, nevertheless, be double-hulled to be permitted transit and access to ports in most coastal waters.

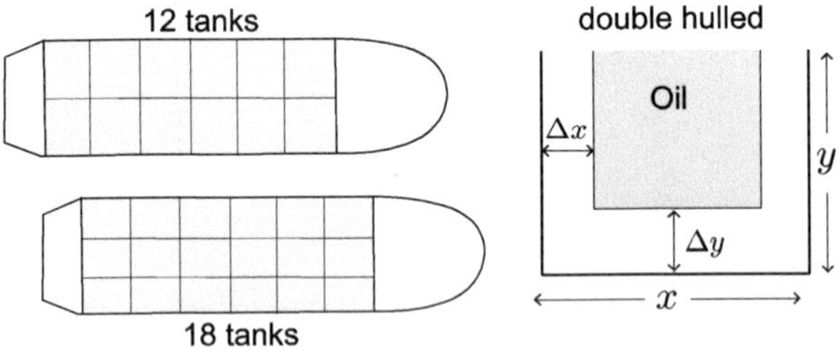

Figure 17.12 Measures to mitigate the release of oil in the event of the fracture of the hull of a maritime tanker.

Another approach is to design a steel with enhanced resistance to the rupture of the hull during a collision so that there is a greater uniform plastic-strain prior to the onset of a necking instability [21]. This also permits more energy to be absorbed by the steel before fracture. These so-called high-ductility steels are made using thermomechanical processing, sometimes with two-stages of accelerated cooling, so that the final microstructure consists of ferrite grains that are about 10 μm is size, with a residue of about 0.2 volume fraction consisting of a mixture of pearlite and bainite, Fig. 17.13 [21]. The manganese sulphide inclusion content is also controlled so that long stringers are absent [22].

Fig. 17.14 shows representative tensile test data from a variety of thermomechanically processed steels used in the construction of ship hulls. The chemical compositions are approximately Fe-0.15C-0.15Si-1.1Mn-0.01Nb wt% [21] or Fe-0.05C-0.05Si-1.6Mn-0.01Nb-0.01Ti,wt% [23], which makes them easy to weld. The conventional steel has a microstructure that is a mixture of ferrite ($\approx$ 150 HV) and pearlite (264 HV). The other two steels have been processed in such a way that the hard ($\approx$ 290 HV) regions consist of a mixture of pearlite and bainite – this improves the elongation by delaying the onset of plastic instability, by a mechanism not dissimilar to that associated with the $\alpha' + \alpha$ dual phase steels for automotive applications (p. 293), although the yield point is not eliminated in the case of the ship steels because of the structure of the hard phase is a mixture of $\alpha_b + \mathrm{P}$.

Many analyses using non-linear elasto-plastic finite element methods [21,22,24–26], together with large-scale experiments [27] have established that the ductile steels perform a lot better than conventional hull-steel in collision scenarios. In particular circumstances, it has been demonstrated that the critical velocity required to rupture a

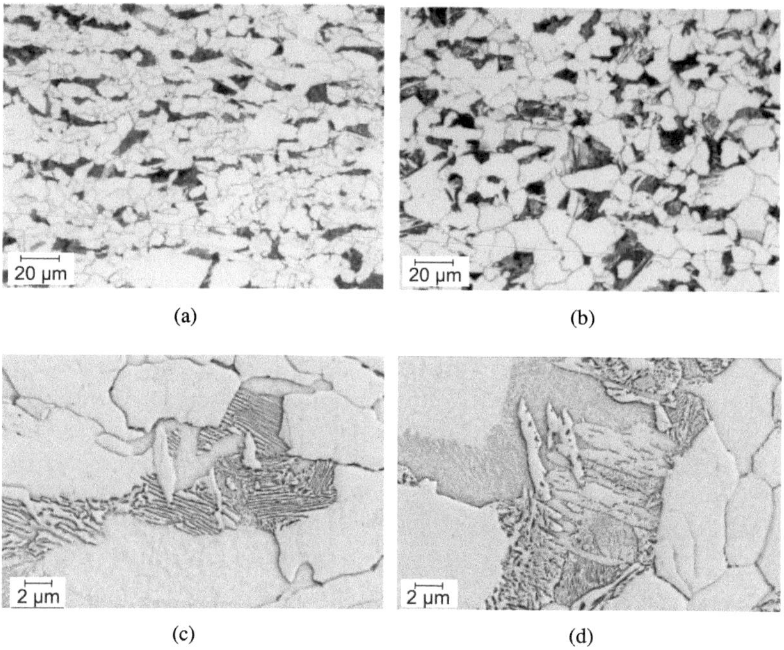

Figure 17.13 Fe-0.15C-0.15Si-1.1Mn-0.01Nb wt%. (a,c) Structure of the steel, obtained by hot-rolling down to 810°C, followed by cooling to obtain the ferrite and pearlite mixture. (b,d) Structure of the same steel, hot-rolled down to 840°C, cooled to transform partially to ferrite, and then cooled rapidly to generate a mixture of pearlite and bainite from the remaining austenite. Images reproduced with permission from Hirota et al. [21].

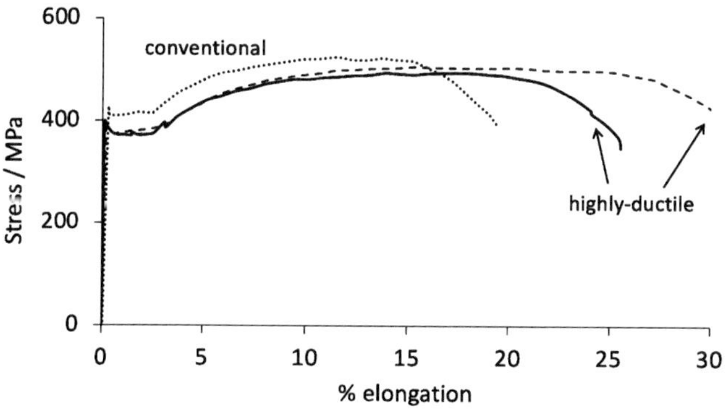

Figure 17.14 Tensile data for three thermomechanically processed steels used in the construction of large ships. Data compiled from [21,22].

double-hulled carrier increases from about 3-6 knots to 12 knots when identical ships collide, one hitting the other at 90° along its mid-length [28]. The success is reflected in the reduction of insurance costs for large ships that are made using ductile steel.

The ductile steel has been used for increased collision protection in the construction, for example, of the *Orange Phoenix* bulk carrier, which is about 300 m long and 50 m wide, capable of carrying some 200,000 tons of cargo. The special-steel plates, which constitute about 15% of the total used to construct the ship, have been used in the cargo hold, fuel oil tank sections and other areas where collision safety is important.

17.6 Summary

The climate crisis is apparently a race that we can win [29], but wise words on their own will have the opposite effect. Steel production at the current indecent-levels cannot be tolerated. Environmentally friendly steel production simply is not possible within the time scales available and for the quantities consumed. Therefore, it is imperative that less steel is consumed. It is possible to use available but better-steel to reduce dramatically the amount necessary to put up a building. The better steel may be more expensive but the overall cost is reduced because less is needed in order to achieve the same engineering goals. And there is no research needed to achieve this reduction, the better steels already are available but not applied. In general, the cost of steel needs to be increased, whether that is by implementing taxes on consumption, or by changing and enforcing construction codes.

Radical changes in design can assist progress towards the efficient use of steel. A well-designed additively manufactured node for structural steel joints has half the size of conventional nodes, not only making it lighter by depositing material where it is most needed, but also by creating an aesthetically pleasing shape, p. 408. Similarly, an industrial punch well-designed using additive manufacturing resulted in the use of 70% less maraging steel, p. 408. For more than a century, the manufacture of cars has involved the forming of steel. Thus, large forming presses made from steel and housed in large buildings are routine in that industry. It also is difficult to form strong steel, which limits the strength that can be exploited. Creating a vehicle that uses only flat sheet to make an innovative, aerodynamically efficient shape can, and has, eliminated the need for presses, because the angular-shape can be engineered by simply bending the sheet at straight character-lines. Therefore, such a vehicle can be made using steel that has a hardness 420–500 HV.

The ability to reuse rockets and spaceships, or to repair using well-established processes (p. 412) is an aspect that needs to be engrained into manufacturing.

It has been demonstrated that using better steel for ship construction has reduced the quantity of steel used and the maritime insurance costs because of the enhanced rupture-resistance of the new material. Steel replacing the aluminium-lithium alloys used in the manufacture of large rockets has a smaller thickness because of its strength; the CO_2 emissions in producing the aluminium-based alloys are substantially greater. And the creative use of very small quantities of steels that take advantage of density can regenerate the marine environment, as demonstrated in the context of coral reefs.

It follows from this discussion that a combination of the application of existing better-steels, legislation, better design, better manufacturing methods and creativity can lead to a large reduction in steel consumption in a manageable time period, without any need for research. Steel should not be treated as a commodity and its monetary cost needs to be increased.

References

[1] L. Silvestre, R. Pimenta, M. Nogueira, L. Queiroz, H. Salles, A. Jordao, R. Ribeiro, M. Pimenta, J. Conceicao, L. Rocha, F. Buratto, E. Fonseca, High strength steel as a solution for the lean design of industrial buildings, Journal of Materials Research and Technology 1 (2012) 35–41.

[2] C. Sheppard, Coral reefs: a very short introduction, Oxford University Press, Oxford, U.K., 2021.

[3] S.L. Williams, C. Sur, N. Janetski, J.A. Hollarsmith, S. Rapi, L. Barron, S.J. Heatwole, A.M. Yusuf, S. Yusuf, J. Jompa, F. Mars, Large-scale coral reef rehabilitation after blast fishing in Indonesia, Restoration Ecology 27 (2019) 447–456.

[4] S.N. Hamzah, S. Nursinar, The success of coral rehabilitation through transplantation using spider modules (case study: Botutonuo marine area, Bone Bolango Regency), Aquaculture, Aquarium, Conservation & Legislation 14 (2021) 3023–3031.

[5] T.B. Razak, L. Boström-Einarsson, C.A.G. Alisa, R.T. Vida, T.A. Lamont, Coral reef restoration in Indonesia: A review of policies and projects, Marine Policy 137 (2022) 104940.

[6] T.A. Mohammed, H. Aa, E.E.-A. Ma, E.-M. Khm, Coral rehabilitation using steel slag as a substrate, International Journal of Environmental Protection 2 (2012) 1–5.

[7] D.D. Pelasula, M.P. Patria, S. Wouthuyzen, A. Syahailatua, Key success factors and problems in coral transplantation: A review, Aquaculture, Aquarium, Conservation & Legislation 16 (2023) 1–24.

[8] L. Hart-Smith, Buckling of thin cylindrical shells under uniform axial compression, International Journal of Mechanical Sciences 12 (1970) 299–313.

[9] Anonymous, Acoustic structures manual, volume II, Technical Report N76-76167, NASA, Alabama, USA, 1975.

[10] R.J. Rioja, J. Liu, The evolution of Al-Li base products for aerospace and space applications, Metallurgical and Materials Transactions A 43 (2012) 3325–3337.

[11] G. Sanderson, D. Llewellyn, Mechanical properties of standard austenitic stainless steels in the temperature range minus 196 to plus 800 C, Journal of the Iron and Steel Institute 207 (1969) 1129–1140.

[12] H. McHenry, The properties of austenitic stainless steel at cryogenic temperatures, Technical Report NBSIR 83-1690, National Bureau of Standards, Colorado, USA, 1983.

[13] T. Ogata, K. Nagai, K. Hiraga, K. Ishikawa, H. Irie, Effect of EB-weld and cold-rolling on low temperature strength and toughness of austenitic stainless steels, in: R.P. Reed, T. Horiuchi (Eds.), Austenitic Steels at Low Temperatures, Plenum Press, London, U.K., 1983, pp. 211–220.

[14] M. Milad, N. Zreiba, F. Elhalouani, C. Baradai, The effect of cold work on structure and properties of AISI 304 stainless steel, Journal of Materials Processing Technology 203 (2008) 80–85.

[15] A. Hedayati, A. Najafizadeh, A. Kermanpur, F. Forouzan, The effect of cold rolling regime on microstructure and mechanical properties of AISI 304L stainless steel, Journal of Materials Processing Technology 210 (2010) 1017–1022.

[16] Y.B. Das, A.N. Forsey, J. Kelleher, S. Kabra, M.E. Fitzpatrick, T.H. Simm, S. Gungor, R.J. Moat, The influence of temperature on deformation-induced martensitic transformation in 301 stainless steel, Materials Science and Technology 34 (17) (2018) 2114–2125.

[17] S. Zhang, Z. Jiang, H. Li, B. Zhang, S. Fan, Z. Li, H. Feng, H. Zhu, Precipitation behavior and phase transformation mechanism of super austenitic stainless steel S32654 during isothermal aging, Materials Science and Engineering A 137 (2018) 244–255.

[18] Q. Lu, J. Zheng, G. Huang, K. Li, H. Ding, Z. Wang, S. Cheng, Enhancing combined cryogenic mechanical properties of metastable austenitic stainless steel by warm forming, Journal of Materials Processing Technology 291 (2021) 117017.

[19] E. Arevalo, Elon Musk shares SpaceX will use a new stainless-steel alloy for starship, https://www.tesmanian.com/blogs/tesmanian-blog/starshipstainlesssteelalloy, 2020.

[20] J.G. Hortelano-Capetillo, J. Vázquez, E. Bañoslopez, J. Alfaro-Ayala, Drag and lift force analysis for the Cybertruck Tesla vehicle, Journal of Technological Engineering 5 (2021) 9–16.

[21] K. Hirota, S. Nakayama, T. Okafuji, K. Nakashima, K. Hase, H. Shiomi, S. Tsuyama, Improved structural redundancy of hull structure in collision by applying steel plates with high uniform elongation, in: The Twenty-second International Offshore and Polar Engineering Conference, OnePetro, 2012, pp. 763–768.

[22] T. Okawa, K. Ichikawa, K. Yanagita, H. Shirahata, A. Inami, K. Ishida, Y. Yamada, T. Inai, Development and adoption of steel plate (NSafe^{tm}-Hull) for shipbuilding with improved collision safety, Technical Report 110, Nippon Steel & Sumitomo Metals, 2015.

[23] K. Nakashima, K. Hase, T. Eto, Development of shipbuilding steel plate with superior low temperature toughness for large heat input welding, Technical Report 20, JFE, Japan, 2015.

[24] Y. Yamada, S. Tozawa, T. Arima, K. Ichikawa, N. Oda, K. Kamita, H. Suga, Effects of highly ductile steel on the crashworthiness of hull structure in oblique collision, in: Proceedings of the ICCGS, vol. 15, University of Ulsan, Korea, 2016, pp. 217–223.

[25] Y. Yamada, H. Ochi, K. Ichikawa, Risk analysis of ship collision considering striking ship velocity and plate ductility, in: Developments in the Collision and Grounding of Ships and Offshore Structures, CRC Press, 2019, pp. 364–374.

[26] K. Ichikawa, Highly-ductile steel plates for shipbuilding securing crashworthiness, Bulletin of the Iron and Steel Institute of Japan 26 (2021) 92–97.

[27] T. Okawa, K. Ichikawa, H. Shimanuki, S. Nakamura, N. Oda, Y. Yamada, Verification of crashworthiness of highly ductile steel by large-scale model tests, in: Developments in the Collision and Grounding of Ships and Offshore Structures, CRC Press, 2019, pp. 34–39.

[28] Y. Yamada, S. Tozawa, T. Arima, K. Ichikawa, K. Kamita, H. Suga, Comparative study on the application of highly ductile steel to the sides structures for improving structural integrity of double hull tankers in ship-ship collision, in: Proceedings of PRADS2016, 2016.

[29] U. Nations, The climate crisis – a race we can win, www.un.org/sites/un2.un.org/files/2020/01/un75_climate_crisis.pdf, 2020.

Acronyms

bcc	Body-centred cubic
bct	Body-centred tetragonal
CE	Carbon equivalent
fcc	Face-centred cubic
P-LE	Partitioning local equilibrium
NP-LE	Negligible partitioning local equilibrium
ILS	Invariant-line strain
IPS	Invariant-plane strain
ECAP	Equal channel angular processing
FATT	Fracture assessed transition temperature
HAZ	Heat-affected zone of welded joints
hcp	Hexagonal close-packed
HREM	High-resolution electron microscopy
HSLA	High-strength low-alloy steels
HTP	High-temperature processed
HV	Vickers hardness
KN	Knoop hardness
KS	Kurdjumov–Sachs
LEFM	Linear-elastic fracture mechanics
NW	Nishiyama–Wasserman
p.p.m.	Parts per million by weight
RE	Rare earth additions
SCR	Stress corrosion cracking resistance
SSAW	Self-shielded arc weld
TRIP	Transformation-induced plasticity
TTT	Time-temperature-transformation
CCT	Continuous cooling transformation
CHT	Continuous heating transformation
ULCB	Ultra-low carbon bainite
UTS	Ultimate tensile strength

Nomenclature

α	Allotriomorphic or idiomorphic ferrite
α'	Martensite
α_1	One-dimensional parabolic thickening rate constant
α_a	Acicular ferrite
α_b	Bainite
α_{lb}	Lower bainite
α_w	Widmanstätten ferrite
α_{ub}	Upper bainite
γ	Shear strain
γ_p	Plastic work done in creating crack surfaces, per unit area
γ_s	Surface energy per unit area
γ	Austenite
γ'	Carbon-enriched austenite
γ_r	Retained austenite
Υ	gamma ray
δ	Body-centred cubic iron that precipitates in liquid
$\dot{\epsilon}$	Strain rate
ϵ_p	Plastic strain
η	Arc energy transfer efficiency in welding
ζ	Uniaxial dilatation normal to the habit plane
θ	Cementite
θ_1	$M_{23}C_6$
θ_2	M_7C_3
κ	Mean % planar misfit between inclusion and ferrite
λ	Interatomic spacing
μ	Shear modulus
μ_i	Chemical potential of element i
ν	Poisson's ratio
ξ	Volume fraction normalised by the equilibrium volume fraction
ρ	Density
ρ_d	Dislocation density
ρ_m	Mobile dislocation density
ρ_L	Mobile dislocation density at the lower yield point
ρ_U	Mobile dislocation density at the upper yield point
σ	Applied stress
σ_0	Friction stress
σ_r	Stress as a function of the distance r ahead of the crack tip
σ_c	Critical stress to trigger a dislocation source
σ_e	Entropy production rate
σ_{Fe}	Strength of pure annealed iron
σ_g	Strengthening due to grain boundaries
σ_{iy}	Instantaneous flow stress at any particular stage of a test
σ_N	Normal stress on the habit plane
σ_p	Work of fracture, per unit area of crack surface
σ_r	Stress corresponding to a unit dislocation velocity
σ_{SS}	Solid solution strengthening due to substitutional solutes

σ_s	Saturation value of σ_{iy} in a fatigue test
σ_{UTS}	Ultimate tensile strength
σ_Y	Yield or 0.2% proof strength
$\sigma_{\theta\alpha}$	Interfacial free energy per unit area for the boundary between θ and α
τ	Incubation time corresponding to a detectable degree of transformation. Alternatively, the shear stress resolved along the shear direction
τ^*	Athermal resistance to dislocation motion
τ_o	Temperature dependent resistance to dislocation motion
a_α	Lattice parameter of phase identified in subscript
Ac_3	Temperature at which a sample becomes fully austenitic during heating
Ae_1	Temperature separating the $\alpha+\gamma$ and α phase fields for a specific alloy
Ae_3	Temperature separating the $\alpha+\gamma$ and γ phase fields for a specific alloy
Ae_4	$\gamma \rightleftharpoons \delta$ transition temperature
Ar_1	Temperature at which austenite completes its transformation to ferrite during cooling
Ar_3	Temperature at which austenite begins to transform to ferrite during cooling
b	Magnitude of Burgers vector of dislocation
B_F	Bainite-finish temperature (not well defined)
B_S	Bainite-start temperature
$\mathbf{B}$	Matrix representing the Bain deformation
$\overline{c}$	Average concentration
c	Concentration, or half the length of crack
$c^{\gamma g}$	Concentration in γ that is in equilibrium with gas (g)
$c_i^{\alpha\theta}$	Concentration of element i in phase α in equilibrium with phase θ
C_i	Constants, with $i = 1, 2, 3\ldots$
X	Generalised force that drives a flux
d	Diameter
d^I	Ideal critical diameter in Grossman test
$\mathbf{d}$	Displacement vector for an invariant-plane strain
$D^{\gamma}_{C,V}$	Diffusion coefficient of solute identified by subscript in phase identified by superscript. The terms V or B refer to volume or boundary diffusion.
E	Young's modulus
f^*	Attempt frequency
f_N	Empirical function in nucleation theory
Δg	Free energy of formation of a single vacancy
Δg_∞	Free energy change per atom
ΔG	General term representing driving force
$\Delta G^{\gamma\rightarrow\alpha}$	Free energy change for transformation without composition change, $G^\alpha - G^\gamma$
ΔG_{chem}	Chemical driving force
ΔG_{mech}	Mechanical driving force
ΔG_m	Molar Gibbs free energy change on transformation; alternatively, the maximum molar Gibbs free energy change accompanying nucleation
ΔG_m	Molar Gibbs free energy
ΔG_{strain}	Coherency strain energy during nucleation
G	Growth rate
G^*	Activation free energy for nucleation, or for interfacial motion
G_{SB}	Stored energy of Widmanstätten ferrite
G_{SW}	Stored energy of bainite
G_s	Strain energy per mole
δ	Grain boundary half-thickness
Δh	Enthalpy of formation of a single vacancy
H^*	Activation enthalpy during plastic deformation
I_V	Nucleation rate per unit volume
J	Diffusion flux, or flux in general
k	Boltzmann constant

k_A	Constant in the Avrami equation
k_g	Constant relating lath size to strength
k_y	Hall-Petch onstant relating grain size to yield strength
K	Bulk modulus of elasticity
K_{IC}	Critical stress intensity in mode 1 loading
K_I	Stress intensity in mode 1 loading
$\overline{L}$	Mean intercept length in stereology, grain size
$\overline{L}_\gamma$	Austenite grain size
L	Average distance moved by dislocations during plastic deformation
L_p	Interparticle distance, mean spacing of neighbouring particles on plane
L_S	Lower bainite start temperature
m	Magnitude of shape deformation
m_i	Mass fraction of element i
M_σ	Temperature below which martensite can be induced by stress
M_F	Temperature at which 95% martensite is obtained
M_S	Martensite-start temperature
M_S^o	Martensite-start temperature for infinitely large austenite grain size
n	Time exponent in the Avrami equation
N_V	Martensite plates per unit volume of sample
N_V^γ	Martensite plates per unit volume of austenite
p_{dis}	Probability of dislocations being absorbed into grain boundary
q	heat input during welding
Q	Activation energy
$\mathbf{Q}$	Matrix representing an inhomogeneous lattice-invariant deformation
$\overline{r}_o$	Mean particle size
$\overline{r}_t$	Mean particle size following isothermal coarsening for a time t
r	Radius of a disc; alternatively, the distance ahead of a crack tip; alternatively the tip radius of a growing plate, particle radius, atomic radius
r^*	Critical size beyond which an embryo can grow with a reduction in free energy
r_c	Critical radius of particle at which its growth rate becomes zero
r_p	Mean particle radius intersected by plane
R	Universal gas constant
$\mathbf{R}$	Matrix for rigid body rotation
Δs	Entropy of formation of a single vacancy.
s	Shear component of the IPS shape deformation
s_A	Apparent shear strain
S	Interlamellar spacing in pearlite, with critical value S_C
S_V	Interfacial area per unit volume
$\mathbf{S}$	Deformation matrix in the crystallographic theory of martensite
$\Delta t_{8\rightarrow 5}$	Time required to cool from 800 to 500°C in the context of welding
t	Time; alternatively, the thickness of a disc
t_i	Time interval for step i in a series of isothermal heat treatments
t_a	Time required to reach a given fraction ξ of isothermal transformation
t_c	Time required for a sub-unit to reach a limiting size
t_d	Time required to decarburise a plate of bainite
t_p	True thickness of a plate
T	Temperature
T_0	Temperature at which γ and α of identical composition have the same free energy
T_0'	As T_0, but accounting for the stored energy of ferrite
T_θ	Temperature at which the T_0 curve and $\gamma/\gamma+\theta$ phase boundary intersect
T_{crit}	Temperature at which half the dislocation core sites are occupied by solute
T_E	Eutectoid temperature
T_P	Peak temperature
U_i	Binding energy between solute and defect

U_O	Strain energy due to interstitial in octahedral hole
U_{Te}	Strain energy due to interstitial in tetrahedral hole
U_e	Elastic energy
U_s	Surface energy
v^*	Activation volume
v_ℓ	Plate lengthening rate
v_a	Volume per atom
v_i, v_0	Interfacial velocity
v_P	Growth rate of pearlite
V	Volume of a sample
V^α	Volume of phase α
V_e^α	Extended volume of phase α
V_τ	Volume per particle
V_m	Molar volume
V_S	Sheaf lengthening rate
V_u	Volume of sub-unit
V_V	Volume fraction of material identified by superscript
w	Thickness of a bainite sub-unit
w_i	Weight percent of element i
$\overline{x}$	Average mole fraction of carbon in an alloy
$x^{\alpha\gamma}$	Mole fraction of carbon in α in equilibrium or paraequilibrium with γ
$x^{\gamma\alpha}$	Mole fraction of carbon in γ in equilibrium or paraequilibrium with α
x_α^α	Mole fraction of phase α
x_{Ae_3}	Carbon concentration given by the Ae_3 curve
$x_{T_0'}$	Carbon concentration given by the T_0' curve
y_C	Ratio of carbon to all other atoms, $x/(1-x)$
z	Coordinate normal to the interface plane
z^*	Position of the interface along coordinate z
z_d	Effective diffusion distance

Index